Mathematical Methods for Engineering and Science

Merle C. Potter • Brian F. Feeny

Mathematical Methods for Engineering and Science

Second Edition

Springer

Merle C. Potter
Department of Mechanical Engineering
Michigan State University
East Lansing, MI, USA

Brian F. Feeny
Department of Mechanical Engineering
Michigan State University
East Lansing, MI, USA

https://doi.org/10.1007/978-3-031-26151-0

This Springer imprint is published by the registered company Springer Nature Switzerland AG
The registered company address is: Gewerbestrasse 11, 6330 Cham, Switzerland
(www.springernature.com/mycopy)

"From MCP: to my wife, Gloria, who lets me cook and always loves what I make!"
"From BFF: to my wife, Kwi, who lets me eat and always cooks what I love!"

Contents

Preface xiii

1 Ordinary Differential Equations 1

1.1 Introduction 1

1.2 Definitions 2

1.3 Differential Equations of First Order 3

1.3.1 Separable Equations, *3*
1.3.2 Exact Equations, *6*
1.3.3 Integrating Factors, *8*

1.4 Physical Applications 9

1.4.1 Simple Electrical Circuits, *9*
1.4.2 The Rate Equation, *11*
1.4.3 Fluid Flow, *13*
1.4.4 Dynamics, *14*

1.5 Linear Differential Equations 19

1.6 Homogeneous Second-Order Linear Equations with Constant Coefficients 21

1.7 Spring–Mass System—Free Motion 25

1.7.1 Undamped Motion, *26*
1.7.2 Damped Motion, *29*
1.7.3 The Electrical Circuit Analog, *34*

1.8 Nonhomogeneous Second-Order Linear Equations with Constant Coefficients 36

1.9 Spring–Mass System—Forced Motion 39

1.9.1 Resonance, *42*
1.9.2 Near Resonance, *43*
1.9.3 Forced Oscillations with Damping, *45*

1.10 Periodic Input Functions—Fourier Series 49

1.10.1 Even and Odd Functions, *53*
1.10.2 Half-Range Expansions, *57*
1.10.3 Forced Oscillations, *59*

1.11 The Cauchy Equation 61
1.12 Variation of Parameters 64
1.13 Miscellaneous Information 66

2 Power-Series Methods **76**

2.1 Power Series 76
2.2 Linear Differential Equations with Variable Coefficients 80
2.3 Legendre's Equation 89
2.4 The Method of Frobenius 93
2.4.1 Distinct Roots Not Differing by an Integer, *94*
2.4.2 Double Roots, *96*
2.4.3 Roots Differing by an Integer, *100*
2.5 Bessel's Equation 103

3 Laplace Transforms **117**

3.1 Introduction 117
3.2 The Laplace Transform 118
3.3 Laplace Transforms of Derivatives and Integrals 128
3.4 Derivatives and Integrals of Laplace Transforms 132
3.5 Laplace Transforms of Periodic Functions 135
3.6 Inverse Transforms—Partial Fractions 139
3.6.1 Unrepeated Linear Factor, *139*
3.6.2 Repeated Linear Factor, *139*
3.6.3 Unrepeated Quadratic Factor, *140*
3.6.4 Repeated Quadratic Factor, *140*
3.7 Solution of Differential Equations 144

4 Matrices and Determinants **159**

4.1 Introduction 159
4.2 Matrices 160
4.3 Addition of Matrices 161
4.4 The Transpose and Some Special Matrices 163
4.5 Matrix Multiplication—Definition 167
4.6 Matrix Multiplication—Additional Properties 169
4.7 Determinants 171

4.8 The Adjoint and the Inverse Matrices 177

4.9 Solution of Simultaneous Linear Algebraic Equations 182

4.9.1 Nonhomogeneous Sets of Linear Algebraic Equations, *182*

4.9.2 Homogeneous Sets of Linear Algebraic Equations, *184*

4.9.3 Solutions to Sets of Linear Equations by MATLAB, *186*

4.10 Least-Squares Fit and the Pseudo Inverse 192

4.11 Eigenvalues and Eigenvectors 201

4.12 Eigenvalue Problems in Engineering 209

4.12.1 Moments of Inertia, *209*

4.12.2 Stress, *213*

4.12.3 Linear Dynamic Systems and Stability, *216*

5 Vector Analysis 228

5.1 Introduction 228

5.2 Vector Algebra 228

5.2.1 Definitions, *228*

5.2.2 Addition and Subtraction, *230*

5.2.3 Components of a Vector, *230*

5.2.4 Multiplication, *232*

5.3 Vector Differentiation 241

5.3.1 Ordinary Differentiation, *241*

5.3.2 Partial Differentiation, *246*

5.4 The Gradient 249

5.5 Cylindrical and Spherical Coordinates 259

5.5.1 Cylindrical Coordinates, *259*

5.5.2 Spherical Coordinates, *263*

5.6 Integral Theorems 267

5.6.1 The Divergence Theorem, *267*

5.6.2 Stokes's Theorem, *270*

6 Partial Differential Equations 282

6.1 Introduction 282

6.2 Wave Motion 284

6.2.1 Vibration of a Stretched, Flexible String, *284*

6.2.2 The Vibrating Membrane, *286*

6.2.3 Longitudinal Vibrations of an Elastic Bar, *288*

6.2.4 Transmission-Line Equations, *289*

6.3 The d'Alembert Solution of the Wave Equation 292

6.4 Separation of Variables 296

6.5 Diffusion 308

6.6 Solution of the Diffusion Equation 311

6.6.1 A Long, Insulated Rod with Ends at Fixed Temperatures, *311*
6.6.2 A Long, Totally Insulated Rod, *315*
6.6.3 Two-Dimensional Heat Conduction in a Long, Rectangular Bar, *319*

6.7 Electric Potential About a Spherical Surface 324

6.8 Heat Transfer in a Cylindrical Body 326

6.9 Gravitational Potential 330

7 Complex Variables 337

7.1 Introduction 337

7.2 Complex Numbers 337

7.3 Elementary Functions 343

7.4 Analytic Functions 349

7.5 Complex Integration 353

7.5.1 Green's Theorem, *353*
7.5.2 Cauchy's Integral Theorem, *356*
7.5.3 Cauchy's Integral Formula, *360*

7.6 Series 365

7.6.1 Taylor Series, *365*
7.6.2 Laurent Series, *366*

7.7 Residues 373

8 Numerical Methods 386

8.1 Introduction 386

8.2 Finite-Difference Operators 388

8.3 The Differential Operator Related to the Difference Operators 392

8.4 Truncation Error 397

8.5 Numerical Integration 400

8.6 Numerical Interpolation 407

8.7 Roots of Equations 409

8.8 **Initial-Value Problems—Ordinary Differential Equations** 412
8.8.1 Taylor's Method, *413*
8.8.2 Euler's Method, *413*
8.8.3 Adams' Method, *414*
8.8.4 Runge–Kutta Methods, *414*
8.8.5 Direct Method, *418*

8.9 **Higher-Order Equations** 421

8.10 **Boundary-Value Problems—Ordinary Differential Equations** 428
8.10.1 Iterative Method, *428*
8.10.2 Superposition, *429*
8.10.3 Simultaneous Equations, *429*

8.11 **Numerical Stability** 432

8.12 **Numerical Solution of Partial Differential Equations** 432
8.12.1 The Diffusion Equation, *433*
8.12.2 The Wave Equation, *434*
8.12.3 Laplace's Equation, *435*

Bibliography 446

Appendix A 447

Table A1 **United States Engineering Units, SI Units, and Their Conversion Factors** 447

Table A2 **Gamma Function** 448

Table A3 **Error Function** 449

Table A4 **Bessel Functions** 450

Appendix B Introduction to MATLAB 454

B.1 **Introduction** 454

B.2 **Real and Complex Numbers** 456

B.3 **Vectors and Matrices** 458

B.4 **Formate and Scientific Notation** 464

B.5 **Programming Loops** 466

B.6 **Plotting** 468

B.7 **String Arrays** 470

B.8 MATLAB Files, Input and Output **471**

B.8.1 Setting the Path, *471*

B.8.2 Script Files, or m-Files, *471*

B.8.3 Input Files and Output File, *473*

B.8.4 Interactive Input and Output, *474*

B.9 Functions **475**

B.10 The Workspace Browser **477**

B.11 Final Remarks **478**

Answers to Selected Problems **479**

Index **493**

Preface

The purpose of this book is to introduce students of engineering and science to several mathematical methods often essential to the successful solutions of practical problems. The methods chosen are those most frequently used in typical physics and engineering applications. The material is not intended to be exhaustive. Each chapter introduces a subject area that can be found in books that treat the subject in greater depth. The reader is encouraged to consult such books should more study be desired in any of the areas introduced.

Perhaps it would be helpful to discuss the motivation that led to the writing of this text? Undergraduate education in the physical sciences has become more advanced and sophisticated since the advent of the space age. During this period, mathematical topics usually reserved for graduate study have become part of the undergraduate program. It is now common to find an applied mathematics course, usually covering one topic, that follows differential equations in the engineering and physical science curricula. Choosing the contents for this additional mathematics course is often difficult. In each of the physical science disciplines, different phenomena are investigated, which result in a variety of mathematical tools. To be sure, several outstanding textbooks exist that present advanced and comprehensive treatments of these methods. However, these texts are usually written at a level too advanced for the undergraduate student, and the material is so exhaustive that it inhibits the effective presentation of the mathematical techniques as a tool for analysis. This book was written to provide for an additional mathematics course after differential equations, to permit several topics to be introduced in one quarter or semester, and to make the material comprehensible to the undergraduate.

Ordinary differential equations, including several physical applications, are reviewed in Chapter 1. Fourier series are also presented in this chapter so that differential equations describing the behavior of systems with periodic input functions can be solved. The solution of ordinary differential equations using power series is the subject of Chapter 2. Subsequent chapters introduce Laplace Transforms, Matrices and Determinants, Vector Analysis, Partial Differential Equations, Numerical Methods and Complex Variables. The material is presented so that any subset of these topics may be included in a three- or four-credit course. The actual instructional pace may vary with the topics chosen and the completeness of coverage. The style of presentation is such that the step-by-step derivations may be followed by the reader with a minimum of assistance from the instructor. Liberal use of examples and homework problems should aid the student in the study of the mathematical methods presented.

There are several additions to this edition of the text. Additional applications of separable, first-order linear, and exact ordinary differential equations are included. Applications of matrix methods have been added, including eigenvalue problems and least-squares fitting of data. Problems have also been added to help students exercise these topics. MATLAB, as a common and useful software tool, has been integrated into the text primarily in matrix computations and numerical solution of differential equations. An Introduction to MATLAB has been added as a section in the Appendix and should help students who lack experience with the package.

Finally, the title has been revised from *Engineering Analysis* to *Mathematical Methods for Engineering and Science*, reflecting that the content includes mathematical methods that are not only applicable in engineering and physics, but may also be useful for advanced studies in a variety of disciplines, such as mathematical and theoretical biology, neuroscience, physical chemistry, data science, and economics.

MERLE C. POTTER
BRIAN F. FEENY

1 Ordinary Differential Equations

Outline

1.1 Introduction
1.2 Definitions
1.3 Differential Equations of First Order
 1.3.1 Separable Equations
 1.3.2 Exact Equations
 1.3.3 Integrating Factors
1.4 Physical Applications
 1.4.1 Simple Electrical Circuits
 1.4.2 The Rate Equation
 1.4.3 Fluid Flow
 1.4.4 Dynamics
1.5 Linear Differential Equations
1.6 Homogeneous Second-Order Linear Equations with Constant Coefficients
1.7 Spring–Mass System—Free Motion
 1.7.1 Undamped Motion
 1.7.2 Damped Motion
 1.7.3 The Electrical Circuit Analog
1.8 Nonhomogeneous Second-Order Linear Equations with Constant Coefficients
1.9 Spring–Mass System—Forced Motion
 1.9.1 Resonance
 1.9.2 Near Resonance
 1.9.3 Forced Oscillations with Damping
1.10 Periodic Input Functions—Fourier Series
 1.10.1 Even and Odd Functions
 1.10.2 Half-Range Expansions
 1.10.3 Forced Oscillations
1.11 The Cauchy Equation
1.12 Variation of Parameters
1.13 Miscellaneous Information

Problems

1.1 INTRODUCTION

Differential equations* play a vital role in the solutions to many problems encountered when modeling physical phenomena. All the disciplines in the physical sciences, with their own unique interests representing a variety of physical situations, require that the student be able to derive the necessary mathematical equation (often a differential equation) and then solve the equation to obtain the desired solution. We shall consider a variety of physical situations that lead to differential equations, using representative problems from several disciplines, and standard methods used to solve the equations will be developed.

*It is assumed that a course in differential equations precedes the course in which this text is used. This chapter is meant to be a quick review of the parts of significance to the science student.

M. C. Potter and B. F. Feeny, *Mathematical Methods for Engineering and Science*,
https://doi.org/10.1007/978-3-031-26151-0_1

An equation involving one or more derivatives of a function is a *differential equation*. The solution of a differential equation is an expression involving the dependent and independent variables, free of derivatives and integrals, which when substituted back into the differential equation results in an identity. The solution is valid in some domain in which it is defined and differentiable; however, it may or may not be stable. If it is unstable, it may lead to a second solution. Questions of stability will not be considered in this book. Often, exact solutions are difficult, if not impossible, to determine, and then approximate solutions are sought or the problem is solved on the computer through the use of numerical methods such as MATLAB. We shall present some exact solutions in this chapter and several numerical methods in Chapter 8. A textbook on differential equations should be sought for completeness.

1.2 DEFINITIONS

An *ordinary differential equation* is one in which only total derivatives appear. A *partial differential equation* is one that involves partial derivatives. If a dependent variable is a function of only one independent variable, such as $f(x)$, an ordinary differential equation would result; however, if the dependent variable depends on more than one independent variable, such as $f(x, y, z)$, a partial differential equation may describe the phenomenon of interest (see Chapter 6).

The dependent variable is usually the *unknown* quantity sought after in a problem, or it leads directly to the desired quantity. For example, the lift on an airfoil is the quantity desired; to determine the lift we would solve a partial differential equation to find the unknown velocity $v(x, y)$, from which we could calculate the pressure and subsequently the desired lift.

The *order* of a differential equation is equal to the order of the highest derivative. An equation is *linear* if it contains only terms of the first degree in the dependent variable and its derivatives; if it contains a term that involves combinations of derivatives or products of the dependent variable, it is *nonlinear.* A differential equation is *homogeneous* if it can be written in a form such that all terms contain the dependent variable or one of its derivatives. The equation

$$x^2\frac{d^2f}{dx^2} + x\frac{df}{dx} + (x^2 - 4)f = 0 \tag{1.2.1}$$

is a linear, homogeneous, ordinary differential equation of second order. The equation

$$\frac{\partial^4 u}{\partial x^4} + 2\frac{\partial^4 u}{\partial x^2 \partial y^2} + \frac{\partial^4 u}{\partial y^4} = 0 \tag{1.2.2}$$

is a linear, homogeneous, partial differential equation of fourth order. The equation

$$\frac{d^2f}{dx^2} + 4f\frac{df}{dx} + 2f = \cos x \tag{1.2.3}$$

is a nonlinear, nonhomogeneous, ordinary differential equation of second order. The *degree* of an equation is the degree of the highest ordered derivative that occurs, if the derivatives can be written in polynomial form; for example, $\sin(d^2f/dx^2)$ has no degree, whereas $(df/dx)^2$ is of degree 2.

Some differential equations have relatively simple solutions. For example, the differential equation

$$\frac{du}{dx} = f(x) \tag{1.2.4}$$

has the solution

$$u(x) = \int f(x)dx + C \tag{1.2.5}$$

where the constant C is a constant of integration. The integration process is actually a process of finding a solution to a differential equation. Integrating n successive times would provide the solution to the nth-order differential equation

$$\frac{d^n u}{dx^n} = f(x) \tag{1.2.6}$$

The differential equations to be considered in this chapter will possess solutions which are obtained with more difficulty than the two above; however, there will be times when simple equations such as that of Eq. 1.2.4 do model phenomena of interest.

A *general solution* of an nth-order differential equation is a solution that contains n arbitrary constants. For example,

$$f(x) = Ax + B(x^3 + 1) \tag{1.2.7}$$

where A and B are arbitrary constants, is the general solution of the second-order equation

$$(2x^3 - 1)\frac{d^2 f}{dx^2} - 6x^2 \frac{df}{dx} + 6xf = 0 \tag{1.2.8}$$

A *specific solution** results when values are assigned to the constants A and B.

In most physical applications it is the specific solution that is of interest. For a first-order equation it is found by satisfying a given *initial condition,* the condition that at some x_0 the solution $f(x_0)$ has some prescribed value f_0. The differential equation together with the initial condition is called an *initial-value problem.* The problem is solved by determining the specific solution.

When a second-order differential equation is being solved, two conditions are required to determine the two arbitrary constants. If the conditions are known at one value of the independent variable, an initial-value problem results; if the conditions are known at two different values of the independent variable, a *boundary-value problem* results.

1.3 DIFFERENTIAL EQUATIONS OF FIRST ORDER

1.3.1 Separable Equations

Many first-order equations can be reduced to

$$g(u)\frac{du}{dx} = h(x) \tag{1.3.1}$$

*This is often called a "particular solution"; however, such terminology is sometimes used in the solution of nonhomogeneous equations, and, to avoid confusion, a "specific solution" results when the arbitrary constants are evaluated.

which can be written as

$$g(u)du = h(x)dx \tag{1.3.2}$$

This nonlinear, first-order equation is *separable* because the variables u and x are separated from each other. By integrating, the general solution of Eq. 1.3.1 is found to be

$$\int g\,du = \int h\,dx + C \tag{1.3.3}$$

where C is the constant of integration. This technique should be used when first attempting to solve a nonlinear differential equation.

Certain equations that are not separable may be made separable by a change of variables. For example, consider the equation

$$\frac{du}{dx} = f\left(\frac{u}{x}\right) \tag{1.3.4}$$

Let the new dependent variable be

$$v = \frac{u}{x} \tag{1.3.5}$$

Then, with $v = v(x)$ and $u = u(x)$, differentiation gives

$$\frac{du}{dx} = x\frac{dv}{dx} + v \tag{1.3.6}$$

Substitute back into Eq. 1.3.4 and there results

$$x\frac{dv}{dx} + v = f(v) \tag{1.3.7}$$

This can be put in the separated form

$$\frac{dv}{f(v) - v} = \frac{dx}{x} \tag{1.3.8}$$

For a particular $f(v)$ this equation may now be solved to find $v(x)$ and, in turn, $u(x)$.

Example 1.1

Find the general solution to the differential equation

$$x\frac{du}{dx} + u^2 = 4$$

Solution

The equation is separable and is written as

$$\frac{du}{4 - u^2} = \frac{dx}{x}$$

To aid in the integration we write

$$\frac{1}{4 - u^2} = \frac{1/4}{2 - u} + \frac{1/4}{2 + u}$$

Our equation becomes

$$\frac{1}{4}\frac{du}{2-u}+\frac{1}{4}\frac{du}{2+u}=\frac{dx}{x}$$

This is integrated to give

$$-\frac{1}{4}\ln(2-u)+\frac{1}{4}\ln(2+u)=\ln x+\frac{1}{4}\ln C$$

where $1/4\ \ln C$ is the constant, included because of the indefinite integration. This is put in the equivalent form,

$$\frac{2+u}{2-u}=x^4 C$$

which can be written as

$$u(x)=\frac{2(Cx^4-1)}{Cx^4+1}$$

If the constant of integration had been chosen as just plain C, an equivalent but more complicated expression would have resulted. We chose $1/4\ \ln C$ to provide a simpler solution.

Example 1.2

Determine the general solution to the differential equation

$$xu\frac{du}{dx}-u^2=x^2$$

Solution

The equation in the given form is not separable and it is nonlinear. However, by dividing by x^2 the equation can be put in the form

$$\frac{u}{x}\frac{du}{dx}-\frac{u^2}{x^2}=1$$

This is in the form of Eq. 1.3.4, since we can write

$$\frac{du}{dx}=\frac{1+(u/x)^2}{u/x}$$

Define a new dependent variable to be $v=u/x$, so that

$$\frac{du}{dx}=x\frac{dv}{dx}+v$$

Substitute back into the given differential equation and obtain

$$v\left(x\frac{dv}{dx}+v\right)-v^2=1$$

This can be put in the separable form

$$v\,dv = \frac{dx}{x}$$

Integration of this equation yields

$$\frac{v^2}{2} = \ln x + C$$

Substitute $v = u/x$ and obtain $u(x)$ to be

$$u(x) = \sqrt{2}x(C + \ln x)^{1/2}$$

This method of substitution leads to a separable equation whenever the variables in all the terms of the differential equation form products to the same power. In this example the power is 2; if it were some other power, for example the equation $x^2u(du/dx) + u^3 = x^3$, the method would have yielded a solution.

1.3.2 Exact Equations

A first-order equation of the form

$$N(x,u)\frac{du}{dx} + M(x,u) = 0 \tag{1.3.9}$$

can be written as

$$M(x,u)dx + N(x,u)du = 0 \tag{1.3.10}$$

This equation is *exact* if the left-hand side is an exact differential, that is, if

$$M(x,u)dx + N(x,u)du = \frac{\partial\phi}{\partial x}dx + \frac{\partial\phi}{\partial u}du = d\phi = 0 \tag{1.3.11}$$

where $\phi = \phi(x, u)$. The solution is

$$\phi = C \tag{1.3.12}$$

where C is an arbitrary constant. From Eq. 1.3.11 we see that

$$M = \frac{\partial\phi}{\partial x}, \qquad N = \frac{\partial\phi}{\partial u} \tag{1.3.13}$$

Differentiate to obtain

$$\frac{\partial M}{\partial u} = \frac{\partial^2\phi}{\partial u\,\partial x}, \qquad \frac{\partial N}{\partial x} = \frac{\partial^2\phi}{\partial x\,\partial u} \tag{1.3.14}$$

Assuming that the order of differentiation can be interchanged, in general, an acceptable assumption, there results

$$\frac{\partial M}{\partial u} = \frac{\partial N}{\partial x} \tag{1.3.15}$$

This is the test used to determine if a first-order differential equation is an exact equation. The function $\phi(x, u)$ is found by solving the two first-order partial differential equations 1.3.13. Integration of these equations leads to

$$\phi(x,u) = \int M(x,u)dx + f(u) = \int N(x,u)du + g(x) \tag{1.3.16}$$

Determination of $f(u)$ (or $g(x)$) is demonstrated in an example.

Example 1.3

Find the specific solution of the differential equation

$$(2 + x^2u)\frac{du}{dx} + xu^2 = 0 \quad \text{if} \quad u(1) = 2$$

Solution

The differential equation is found to be exact by identifying

$$N = 2 + x^2u, \qquad M = xu^2$$

Appropriate differentiation results in

$$\frac{\partial N}{\partial x} = 2xu, \qquad \frac{\partial M}{\partial u} = 2xu$$

$\frac{\partial N}{\partial x} = \frac{\partial M}{\partial u}$ demonstrate exactness.

Thus, the equation is exact. The solution is $\phi = C$, where

$$M = \frac{\partial \phi}{\partial x}, \qquad N = \frac{\partial \phi}{\partial u}$$

To find an expression for $\phi(x, u)$ we first solve

$$xu^2 = \frac{\partial \phi}{\partial x}$$

This is integrated to obtain

$$\phi = \frac{x^2u^2}{2} + f(u)$$

This is then differentiated to find

$$\frac{\partial \phi}{\partial u} = x^2u + \frac{df}{du} = 2 + x^2u$$

Thus,

$$\frac{df}{du} = 2$$

or

$$f = 2u$$

The solution for $\phi(x, u)$ is thus found to be

$$\phi = \frac{x^2u^2}{2} + 2u = C$$

The solution $u(x)$ is determined from the equation above to be

$$u(x) = -\frac{2}{x^2} \pm \frac{1}{x^2}\sqrt{4 + 2x^2C}$$

Using the condition $u(1) = 2$, we have

$$2 = -2 \pm \sqrt{4 + 2C}$$

This requires that $C = 6$. The specific solution is

$$u(x) = -\frac{2}{x^2} + \frac{2}{x^2}\sqrt{1 + 3x^2}$$

Note that the plus sign is retained, so that $u(1) = 2$.

1.3.3 Integrating Factors

Certain equations that are not exact may be made exact by multiplying by a function $F(x, u)$, called an *integrating factor.* For nonlinear equations this factor is found by inspection; however, for linear equations the integrating factor is known. Consider the most general form of a linear first-order differential equation

$$\frac{du}{dx} + f(x)u = g(x) \tag{1.3.17}$$

which can be solved by the use of the integrating factor

$$F(x) = e^{\int f(x)dx} \tag{1.3.18}$$

Multiplying the equation by the integrating factor gives

$$\left[\frac{du}{dx} + fu\right]e^{\int f\,dx} = ge^{\int f\,dx} \tag{1.3.19}$$

which can be rewritten as

$$\frac{d}{dx}[ue^{\int f\,dx}] = ge^{\int f\,dx} \tag{1.3.20}$$

This has the form that the variables separate if we consider the quantity in brackets as a new variable and multiply both sides by dx. We then have, using Eq. 1.3.18,

$$d[uF(x)] = gF(x)dx \tag{1.3.21}$$

This is then integrated to yield

$$uF = \int gF\,dx + C \tag{1.3.22}$$

arriving at the solution

$$u(x) = \frac{1}{F(x)}\left[\int g(x)F(x)dx + C\right] \tag{1.3.23}$$

This is the general solution of all first-order, linear differential equations.

Example 1.4

Solve the linear differential equation

$$x^2 \frac{du}{dx} + 2u = 5x$$

for the general solution.

Solution

The differential equation is first order and linear but is not separable. Thus, use an integrating factor to aid in the solution. Following Eq. 1.3.17, the equation is written in the form

$$\frac{du}{dx} + \frac{2}{x^2} u = \frac{5}{x}$$

The integrating factor is provided by Eq. 1.3.18 to be

$$F(x) = e^{\int (2/x^2) dx} = e^{-2/x}$$

Equation 1.3.22 then provides the solution

$$u(x) = e^{2/x} \left[\int \frac{5}{x} e^{-2/x} dx + C \right]$$

The integral in this expression cannot be integrated, although an integration by parts should be attempted; hence, the solution is left as is.

1.4 PHYSICAL APPLICATIONS

There are abundant physical phenomena that can be modeled with first-order differential equations that fall into one of the classes of the previous section. We shall consider several such phenomena, state or derive the appropriate describing equations, and provide the solutions. Other applications will be included in the Problems.

1.4.1 Simple Electrical Circuits

Consider the circuit in Fig. 1.1, containing a resistance R, inductance L, and capacitance C in series. A known electromotive force $v(t)$ is impressed across the terminals. The differential equation relating the current i to the electromotive force may be found by applying Kirchhoff's second law,* which states that the voltage impressed on a closed loop is equal to the sum of the voltage drops in the rest of the loop. Letting q be the electric charge on the capacitor and recalling that the current i flowing through the capacitor is related to the charge by

$$i = \frac{dq}{dt} \tag{1.4.1}$$

*Kirchhoff's first law states that the current flowing into any point in an electrical circuit must equal the current flowing out from that point.

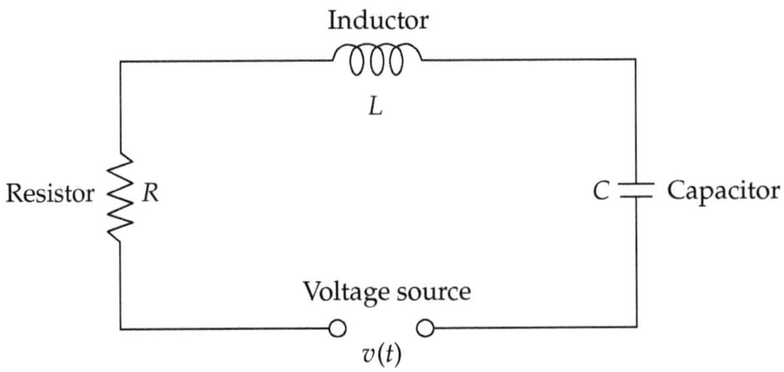

FIGURE 1.1 An *RLC* circuit.

we can write

$$v(t) = L\frac{d^2q}{dt^2} + R\frac{dq}{dt} + \frac{1}{C}q \tag{1.4.2}$$

where the values, q, v, L, R, and C are in physically consistent units—coulombs, volts, henrys, ohms, and farads, respectively. We consider L, R, and C to be constant. In the equation above we have used the following experimental observations:

$$\begin{aligned} \text{voltage drop across a resistor} &= iR \\ \text{voltage drop across a capacitor} &= \frac{q}{C} \\ \text{voltage drop across an inductor} &= L\frac{di}{dt} \end{aligned} \tag{1.4.3}$$

Differentiating Eq. 1.4.2 with respect to time and using Eq. 1.4.1, where i is measured in amperes, we have

$$\frac{dv}{dt} = L\frac{d^2i}{dt^2} + R\frac{di}{dt} + \frac{1}{C}i \tag{1.4.4}$$

If dv/dt is nonzero, Eq. 1.4.4 is a linear, nonhomogeneous, second-order differential equation.

If there is no capacitor in the circuit, Eq. 1.4.4 reduces to

$$\frac{dv}{dt} = L\frac{d^2i}{dt^2} + R\frac{di}{dt} \tag{1.4.5}$$

Integrating, we have (Kirchhoff's second law requires that the constant of integration be zero)

$$L\frac{di}{dt} + Ri = v(t) \tag{1.4.6}$$

The solution to this equation will be provided in the following example.

Example 1.5

Using the integrating factor, solve Eq. 1.4.6 for the case where the electromotive force is given by $v(t) = V \sin \omega t$.

Solution

First, put Eq. 1.4.6 in the standard form

$$\frac{di}{dt} + \frac{R}{L}i = \frac{V}{L}\sin \omega t$$

Using Eq. 1.3.18 we find that the integrating factor is

$$F(t) = e^{(R/L)t}$$

According to Eq. 1.3.22 the solution is

$$i(t) = e^{-(R/L)t}\left[\int \frac{V}{L}\sin \omega t\, e^{(R/L)t} dt + C\right]$$

where C is the constant of integration, not to be confused with the capacitance. Simplification of this equation yields, after integrating by parts,

$$i(t) = V\left[\frac{R\sin \omega t - \omega L\cos \omega t}{R^2 + \omega^2 L^2}\right] + Ce^{-(R/L)t}$$

If the current $i = i_0$ at $t = 0$, we calculate the constant C to be

$$C = i_0 + \frac{V\omega L}{R^2 + \omega^2 L^2}$$

and finally that

$$i(t) = V\left[\frac{R\sin \omega t - \omega L\cos \omega t}{R^2 + \omega^2 L^2}\right] + \left[i_0 + \frac{V\omega L}{R^2 + \omega^2 L^2}\right]e^{-(R/L)t}$$

In this example we simplified the problem by removing the capacitor. We could also consider a similar problem where the capacitor is retained but the inductor is removed; we would then obtain a solution for the voltage. In Section 1.7 we consider the solution of the general second-order equation (1.4.4) where all components are included.

1.4.2 The Rate Equation

A number of phenomena can be modeled by a first-order equation called a *rate equation.* It has the general form

$$\frac{du}{dt} = f(u,t) \tag{1.4.7}$$

indicating that the rate of change of the dependent quantity u may be dependent on both time and u. We shall derive the appropriate rate equation for the concentration of salt in a solution. Other rate equations will be included in the Problems.

Consider a tank with volume V (in cubic meters, m^3), containing a salt solution of concentration $C(t)$. The initial concentration is C_0 (in kilograms per cubic meter, kg/m^3). A brine containing a concentration C_1 is flowing into the tank at the rate q (in cubic meters per second, m^3/s), and an equal flow of the mixture is issuing from the tank. The salt concentration is kept uniform throughout by continual stirring. Let us develop a differential equation that can be solved to give the concentration C as a function of time. The equation is derived by writing a balance equation on the amount (in kilograms) of salt contained in the tank:

$$\text{amount in} - \text{amount out} = \text{amount of increase} \tag{1.4.8}$$

For a small time increment Δt equation 1.4.8, becomes, assuming that the concentration of the solution leaving is equal to the concentration $C(t)$ in the tank,

The amount of salt in a volume V is CV. The volume leaving the tank in time Δt is $q\Delta t$.

$$C_1 q\,\Delta t - Cq\,\Delta t = C(t + \Delta t)V - C(t)V \tag{1.4.9}$$

The volume V of solution is maintained at a constant volume since the outgoing flow rate is equal to the incoming flow rate. The equation above may be rearranged to give

$$q(C_1 - C) = V\frac{C(t + \Delta t) - C(t)}{\Delta t} \tag{1.4.10}$$

Now, if we let the time increment Δt shrink to zero and recognize that

$$\lim_{\Delta t \to 0} \frac{C(t + \Delta t) - C(t)}{\Delta t} = \frac{dC}{dt} \tag{1.4.11}$$

we arrive at the rate equation for the concentration of salt in a solution,

$$\frac{dC}{dt} + \frac{q}{V}C = \frac{qC_1}{V} \tag{1.4.12}$$

The solution will be provided in the following example.

Example 1.6

The initial concentration of salt in a 10-m^3 tank is 0.02 g/m^3. A brine flows into the tank at 2 m^3/s with a concentration of 0.01 g/m^3. Determine the time necessary to reach a concentration of 0.011 g/m^3 in the tank if the outflow equals the inflow.

Solution

Equation 1.4.12 is the equation to be solved. Using $q = 2$, $V = 10$, and $C_1 = 0.01$, we have

$$\frac{dC}{dt} + \frac{2}{10}C = \frac{2 \times 0.01}{10}$$

The integrating factor is

$$F(t) = e^{\int (1/5)dt} = e^{t/5}$$

The solution, referring to Eq. 1.3.23, is then

$$\begin{aligned} C(t) &= e^{-t/5}\left[\int 0.002 e^{t/5} dt + A\right] \\ &= 0.01 + Ae^{-t/5} \end{aligned}$$

where A is the arbitrary constant. Using the initial condition there results

$$0.02 = 0.01 + A$$

so that

$$A = 0.01$$

The solution is then

$$C(t) = 0.01\left[1 + e^{-t/5}\right]$$

Setting $C(t) = 0.011$, we have

$$0.011 = 0.01[1 + e^{-t/5}]$$

Solving for the time, we have

$$0.1 = e^{-t/5}$$

or

$$t = 11.51 \text{ s}$$

1.4.3 Fluid Flow

In the absence of viscous effects it has been observed that a liquid (water, for example) will flow from a hole with a velocity

$$v = \sqrt{2gh} \qquad \text{m/s} \tag{1.4.13}$$

where h is the height of the free surface of the liquid above the hole and g is the local acceleration of gravity (usually assumed to be 9.81 m/s^2). Bernoulli's equation, which may have been presented in a physics course, will yield the result above. Let us develop a differential equation that will relate the height of the free surface and time, thereby allowing us to determine how long it will take to empty a particular reservoir. Assume the hole of diameter d to be in the bottom of a cylindrical tank of diameter D with the initial water height h_0 meters above the hole. The incremental volume ΔV of liquid escaping from the hole during the time increment Δt is

$$\begin{aligned} \Delta V &= vA\,\Delta t \\ &= \sqrt{2gh}\,\frac{\pi d^2}{4}\Delta t \end{aligned} \tag{1.4.14}$$

This small volume change must equal the volume lost in the tank due to the decrease in liquid level Δh. This is expressed as

$$\Delta V = -\frac{\pi D^2}{4}\Delta h \tag{1.4.15}$$

Equating the two expressions above and taking the limit as $\Delta t \to 0$, we have

$$\frac{dh}{dt} = -\sqrt{2gh}\,\frac{d^2}{D^2} \tag{1.4.16}$$

This equation is immediately separable and is put in the form

$$h^{-1/2}dh = -\sqrt{2g}\,\frac{d^2}{D^2}dt \tag{1.4.17}$$

which is integrated to provide the solution, using $h = h_0$ at $t = 0$,

$$h(t) = \left[-\sqrt{\frac{g}{2}}\frac{d^2}{D^2}t + \sqrt{h_0}\right]^2 \tag{1.4.18}$$

The time t_e necessary to drain the tank completely would be (set $h = 0$)

$$t_e = \frac{D^2}{d^2}\sqrt{\frac{2h_0}{g}} \tag{1.4.19}$$

Additional examples of physical phenomena will be included in the Problems.

1.4.4 Dynamics

Forces F are often functions of displacement x or velocity v. Then the application of Newton's second law, $F = ma$, where m is the mass of a particle or body, naturally produces differential equations, since the acceleration $a = dv/dt = d^2x/dt^2$. Depending on the nature of the forces, the resulting differential equations can sometimes be separable, exact, or first-order linear. It may require some manipulation to put the equations in separable or exact form. It is sometimes useful to use the chain rule and express $a = dv/dt = (dv/dx)(dx/dt) = v(dv/dx)$. The following examples provide illustrations.

Example 1.7

A moving train accelerates under constant power, $P_0 = Fv$, such that the force on the train is $F = P_0/v$, where $v \neq 0$. (a) Neglecting resistance forces, find $v(t)$. (b) Given a mass $m = 150000$ kg, and a power of $P_0 = 2400$ kilowatts, find the time to reach a speed of 40 meters per second from rest.

Solution

a) From a free-body diagram of forces, applying Newton's second law that the sum of forces is equal to mass times acceleration, and knowing that acceleration is the derivative of velocity with respect to time, we have

$$\frac{P_0}{v} = ma = m\frac{dv}{dt}$$

The differential equation is separable, such that

$$\frac{P_0}{m}dt = vdv$$

Integrating both sides yields

$$\frac{P_0}{m}t + C = \frac{1}{2}v^2$$

Thus the general solution is

$$v(t) = \sqrt{\frac{2P_0}{m}t + 2C}$$

b) For the above train at constant power, given a mass $m = 150000$ kg, and a power of $P_0 = 2400$ kilowatts, find the time to reach a speed of 40 meters per second from rest.

Using the general solution for $v(t)$, we first find the integration constant. Knowing that the velocity is $v(0) = 0$ at time $t = 0$, we have $C = 0$. Therefore $v^2 = (P_0/m)t$. Solving for t leads to $t = (m/2P_0)v^2$. Thus the time to reach 40 meters per second is

$$t = \frac{(150\,000 \text{ kg})(40 \text{ m/s})^2}{2(2400 \times 10^3 \text{ W})} = 50 \text{ s}$$

where watts (W) were substituted with the equivalent kg m^2/s^3 to simplify the units.

Comment: Notice that the force $F = P_0/v$ was defined for $v \neq 0$, yet the train in this problem starts at $v(0) = 0$. The integration of the separable equation, and indeed the solution $v(t)$, did not break down at $v = 0$. However, at the onset of motion of this train, this ideal model requires infinite force, which would not be achievable in a real train.

Example 1.8

Use the chain rule $dv/dt = (dv/dx)(dx/dt) = v(dv/dx)$ to modify the differential equation in Example 1.7. Then find the distance covered by the train in Example 1.7 to go from a speed of 0 to 40 m/s.

Solution

Using the chain rule, we can modify the differential equation of motion as

$$\frac{P_0}{v} = m\frac{dv}{dt} = m\frac{dv}{dx}\frac{dx}{dt} = mv\frac{dv}{dx}$$

Making use of the chain rule has cast the relationship in terms of v and x, which directly pertain to the information to be obtained in this problem. The differential equation is separable, such that

$$P_0 dx = mv^2 dv$$

Integrating both sides, we have

$$\frac{P_0}{m}x + C = \frac{v^3}{3}$$

or

$$x = \frac{mv^3}{3P_0} - \frac{mC}{P_0}$$

which provides the distance as a function of velocity. At $x = 0$, we have $v = 0$, which implies that the integration constant $C = 0$. The distance covered to achieve the speed of 40 m/s is then

$$x = \frac{mv^3}{3P_0} = \frac{(150\,000\ \text{kg})(40\ \text{m/s})^3}{3(2400 \times 10^3\ \text{W})} = 1333\ \text{meters}$$

Some equations in mechanics can be cast as exact equations. This can occur if there is an underlying conserved quantity, for example energy, and appropriate manipulations are made on the original equation of motion. This is illustrated in the next example.

Example 1.9

A mass m slides on a frictionless parabolic wire in a vertical plane. The horizontal displacement is x, and the shape of the wire is $y = (\alpha/2)x^2$. The equation of motion in the x variable is given as

$$m(1 + \alpha^2 x^2)\frac{d^2x}{dt^2} + m\alpha^2 x\left(\frac{dx}{dt}\right)^2 + mg\alpha x = 0$$

Use $u = dx/dt$, recast the equation such that u is the independent variable and x is the dependent variable, and solve for the horizontal velocity u as a function of the horizontal displacement x.

Solution

In this problem it is useful to define the velocity as a dependent variable, $u = dx/dt$, and then recast the equation so that x takes the role of the independent variable. To do this, multiply the differential equation by dx/dt, as

$$m(1 + \alpha^2 x^2)\frac{d^2x}{dt^2}\frac{dx}{dt} + m\alpha^2 x\left(\frac{dx}{dt}\right)^2\frac{dx}{dt} + mg\alpha x\frac{dx}{dt} = 0$$

and then selectively apply $u = dx/dt$, which allows us to write

$$m(1+\alpha^2x^2)u\frac{du}{dt} + m\alpha^2xu^2\frac{dx}{dt} + mg\alpha x\frac{dx}{dt} = 0$$

Divide by dx/dt so that

$$m(1+\alpha^2x^2)u\frac{du}{dx} + m\alpha^2xu^2 + mg\alpha x = 0$$

which is in the form of

$$N(x,u)\frac{du}{dx} + M(x,u) = 0$$

Check the conditions of exactness:

$$\frac{\partial N}{\partial x} = 2m\alpha^2x = \frac{\partial M}{\partial u}$$

and therefore the system is exact, namely there is a solution in the form $\phi(x,u) = C$, where $\partial\phi/\partial x = M(x,u) = m\alpha^2xu^2 + mg\alpha x$ and $\partial\phi/\partial u = N(x,u) = m(1+\alpha^2x^2)u$. From the latter,

$$\phi = \int N(x,u)du + f(x) = \frac{m}{2}(1+\alpha^2x^2)u^2 + f(x)$$

Differentiating,

$$\frac{\partial\phi}{\partial x} = m\alpha^2xu^2 + \frac{df}{dx}$$

Comparing to $M(x, u) = m\alpha^2xu^2 + mg\alpha x$, it is clear that $df/dx = mg\alpha x$, and therefore $f(x) = 1/2(mg\alpha x^2 + C)$. Thus the solution is

$$\phi = \frac{m}{2}(1+\alpha^2x^2)u^2 + \frac{1}{2}mg\alpha x^2 = \frac{E}{2}$$

where $E = -C$ is the integration constant. In this problem, E physically represents twice the total energy.

We can solve for the velocity $u(x)$ as a function of displacement as

$$u(x) = \pm\sqrt{\frac{E - mg\alpha x^2}{m(1+\alpha^2x^2)}}$$

which is valid if $mg\alpha x^2 \le E$.

Comments: We started with a second-order differential equation, and converted it to an exact first-order differential equation. The original differential equation expressed the unknown displacement as a function of time. The exact form of the differential equation involved velocity as a function of displacement. The explicit time dependence was lost in this process. The time dependence, in theory, can be retained by writing $u(x) = dx/dt$ as a separable differential equation, although integration may not always be analytically tractable.

In dynamics, first-order linear differential equations might also occur. Consider a high-speed rotor spinning in a retainer bearing. The planar case has a configuration in which a spinning puck is sliding inside a rigid circular retainer, where the puck can revolve around the confines of the circle while sliding with a constant coefficient of friction, μ, and being acted on by gravity. We assume that the puck maintains sliding contact, and the model is valid until sliding contact is violated. We let x be the angular position of the contact point, and $u = dx/dt$ be its angular velocity. The equation of motion can be manipulated to cast x as the independent coordinate to write

$$\frac{du}{dx} = -\frac{\alpha \sin x}{u} - \mu u \tag{1.4.21}$$

where α is a parameter.

We can make this equation linear if we let $v = u^2/2$. Then

$$\frac{dv}{dx} = u\frac{du}{dx} = u\left(-\frac{\alpha \sin x}{u} - \mu u\right) \tag{1.4.22}$$

Hence we have

$$\frac{dv}{dx} + 2\mu v = -\alpha \sin x \tag{1.4.23}$$

which is a first-order linear equation.

Example 1.10

For the retainer bearing problem, use an integrating factor to solve for the angular velocity u as a function of the angular position x.

Solution

We work with Eq. 1.4.23 and first solve for v, keeping in mind that the quantity of interest, u, is related to v through $v = u^2/2$.

In Eq. 1.4.23 we establish the integrating factor as

$$F(x) = e^{\int 2\mu dx}$$

which is multiplied into the differential equation to obtain

$$e^{\int 2\mu dx}\frac{dv}{dx} + 2\mu e^{\int 2\mu dx} v = -\alpha e^{\int 2\mu dx} \sin x$$

We recognize that this expression can be written as

$$\frac{d}{dx}\left[e^{\int 2\mu dx} v\right] = -\alpha e^{\int 2\mu dx} \sin x$$

which can be separated and integrated as

$$\int d\left[e^{\int 2\mu dx}v\right] = \int -\alpha e^{\int 2\mu dx} \sin x \, dx$$

The left-hand side is simply $\int d\left[e^{\int 2\mu dx}v\right] = e^{\int 2\mu dx}v = e^{2\mu x}v$, and the right hand side is integrated by parts, resulting in

$$e^{2\mu x}v = -\frac{\alpha}{1+4\mu^2}e^{2\mu x}(2\mu \sin x - \cos x) + C$$

where C is an integration constant. Solving for v leads to

$$v = -\frac{\alpha}{1+4\mu^2}(2\mu \sin x - \cos x) + Ce^{-2\mu x}$$

Recalling that $v = u^2/2$, we finally obtain the general solution,

$$u = \pm\sqrt{-\frac{2\alpha}{1+4\mu^2}(2\mu \sin x - \cos x) + Ce^{-2\mu x}}$$

valid while the radicand is positive, and while slip is maintained. The integration constant can be specified by the initial condition $u(x_0) = u_0$.

1.5 LINEAR DIFFERENTIAL EQUATIONS

Many differential equations that describe physical phenomena are linear differential equations. The coefficients of the various derivatives may or may not be constants and the equations may or may not be homogeneous. If the coefficients are constants and the equation is homogeneous the solution can be written in terms of exponential functions; such second-order equations will be considered in the following articles. In this article we will discuss the general solution of the linear differential equation. We shall illustrate using a second-order equation.

The general linear second-order differential equation is written in standard form as

$$\frac{d^2u}{dx^2} + f(x)\frac{du}{dx} + g(x)u = h(x) \tag{1.5.1}$$

The associated homogeneous equation is written by supressing the term not containing the dependent variable or its derivatives; it is

$$\frac{d^2u}{dx^2} + f(x)\frac{du}{dx} + g(x)u = 0 \tag{1.5.2}$$

It has the homogeneous solution $u_h(x)$ given by

$$u_h(x) = c_1u_1(x) + c_2u_2(x) \tag{1.5.3}$$

where $u_1(x)$ and $u_2(x)$ are the two independent solutions of Eq. 1.5.2. *Superimposing solutions to obtain a general solution is only possible if the differential equation is linear.*

The $h(x)$ on the right of Eq. 1.5.1 requires that a *particular solution* $u_p(x)$ be added to the solution $u_h(x)$ of the homogeneous equation to give the general solution to Eq. 1.5.1 as

$$u(x) = c_1u_1(x) + c_2u_2(x) + u_p(x) \tag{1.5.4}$$

The particular solution can be found by various methods. It is often found by inspection. What is required is that when $u_p(x)$ is substituted for $u(x)$ in Eq. 1.5.1, equality will result. In Section 1.8 one such method will be presented.

The above is also true of first-order, *linear* equations. For first-order equations the solution would be of the form

$$u(x) = c_1u_1(x) + u_p(x) \tag{1.5.5}$$

Equation 1.3.23 has this form if $u_1(x) = \exp\,(-\int f\,dx)$.

A nonlinear second-order equation can be solved analytically only if the equation is separable. Numerical methods will be presented in Chapter 8 that can be used to solve both linear and nonlinear equations.

Example 1.11

Solve the equation

$$\frac{du}{dx} + 2xu = 2x^2 + 1$$

using (a) the integrating factor method, and (b) the method of this section.

Solution

a) Using the integrating factor method, we find that

$$F(x) = e^{\int 2x\,dx} = e^{x^2}$$

The solution provided by Eq. 1.3.23 is then

$$u(x) = e^{-x^2}\left[\int (2x^2+1)e^{x^2}\,dx + C\right]$$

The integration is easily carried out if we integrate by parts and recognize that

$$\int e^{x^2}(2x\,dx) = e^{x^2}$$

The resulting solution is

$$u(x) = x + Ce^{-x^2}$$

b) Now, let us use the method of this section. The associated homogeneous equation is

$$\frac{du}{dx} + 2xu = 0$$

It is put in the separable form

$$\frac{du}{u} = -2x\,dx$$

Integrating once yields the general solution

$$\ln u = -x^2 + \ln C$$

or, equivalently,

$$u = Ce^{-x^2}$$

The particular solution is found by inspection to be

$$u_p = x$$

This is easily verified by substituting back into the original differential equation. Finally, the solution is

$$u(x) = Ce^{-x^2} + x$$

which agrees with that obtained using an integrating factor.

1.6 HOMOGENEOUS SECOND-ORDER LINEAR EQUATIONS WITH CONSTANT COEFFICIENTS

We will focus our attention on second-order differential equations with constant coefficients; the homogeneous equation is written in standard form as

$$\frac{d^2u}{dx^2} + a\frac{du}{dx} + bu = 0 \tag{1.6.1}$$

It possesses exponential solutions. Hence, we assume

$$u = e^{mx} \tag{1.6.2}$$

When substituted into Eq. 1.6.1, we find that

$$e^{mx}(m^2 + am + b) = 0 \tag{1.6.3}$$

Thus, $u = e^{mx}$ is a solution of Eq. 1.6.1 if

$$m^2 + am + b = 0 \tag{1.6.4}$$

This is the *characteristic equation*. It has the two roots

$$m_1 = -\frac{a}{2} + \frac{1}{2}\sqrt{a^2 - 4b}, \qquad m_2 = -\frac{a}{2} - \frac{1}{2}\sqrt{a^2 - 4b} \tag{1.6.5}$$

$u_1(x)$ and $u_2(x)$ are the two independent solutions mentioned in Section 1.5.

It then follows that

$$u_1 = e^{m_1 x}, \qquad u_2 = e^{m_2 x} \tag{1.6.6}$$

are solutions of Eq. 1.6.1. The general solution is

$$u(x) = c_1 e^{m_1 x} + c_2 e^{m_2 x} \tag{1.6.7}$$

If $(a^2 - 4b) < 0$, then the roots are written as, using $i = \sqrt{-1}$,

$$m_1 = -\frac{a}{2} + \frac{i}{2}\sqrt{4b - a^2}, \qquad m_2 = -\frac{a}{2} - \frac{i}{2}\sqrt{4b - a^2} \tag{1.6.8}$$

Curiously, Euler's formula implies $e^{i\pi} + 1 = 0$, called Euler's identity, involving the five most fundamental quantities!

Recalling* from Euler's formula that

$$e^{i\theta} = \cos\theta + i\sin\theta \tag{1.6.9}$$

the general solution (1.6.7) can be put in the form

$$u(x) = e^{-(a/2)x}\left[A\cos\left(\frac{1}{2}\sqrt{4b - a^2}\,x\right) + B\sin\left(\frac{1}{2}\sqrt{4b - a^2}\,x\right)\right] \tag{1.6.10}$$

It, of course, is acceptable to write the sin term first with the coefficient A and the cos term second with the coefficient B.

If $(a^2 - 4b) = 0$, $m_1 = m_2$ and a double root occurs. For this case the solution is not that given in Eq. 1.6.7. One independent solution is $u_1 = e^{mx}$. Let us assume a second solution of the form

$$u_2 = v(x)u_1 \tag{1.6.11}$$

Substitute into Eq. 1.6.1 and we have

$$\left(\frac{d^2u_1}{dx^2} + a\frac{du_1}{dx} + bu_1\right)v + \left(2\frac{du_1}{dx} + au_1\right)\frac{dv}{dx} + u_1\frac{d^2v}{dx^2} = 0 \tag{1.6.12}$$

The first quantity in parentheses is zero, since $u_1(x)$ is a solution to the differential equation Eq. 1.6.1. The second quantity in parentheses is

$$2\frac{du_1}{dx} + au_1 = (2m + a)e^{mx} = \left[2\left(-\frac{a}{2}\right) + a\right]e^{mx} = 0 \tag{1.6.13}$$

using $u_1 = e^{mx}$ and $m = -a/2$. This leaves the final term in Eq. 1.6.12 as zero, demanding that

$$\frac{d^2v}{dx^2} = 0 \tag{1.6.14}$$

*Equation 1.6.9 should have been used in a former mathematics course. It can be verified by expanding the quantities in power series. See Eqs. 2.1.6.

which provides us with the solution

$$v = x \tag{1.6.15}$$

The second independent solution is then

$$u_2 = xe^{mx} \tag{1.6.16}$$

For the condition of a double root, the general solution has been shown to be

$$u(x) = (c_1 + c_2 x)e^{mx} \tag{1.6.17}$$

Initial or boundary conditions are used to evaluate the arbitrary constants in the solution above.

The technique above can also be used for solving differential equations with constant coefficients of order greater than 2. The substitution $u = e^{mx}$ leads to a characteristic equation which is solved for the various roots. The solution follows as above.

Example 1.12

Determine the general solution of the differential equation

$$\frac{d^2u}{dx^2} + 5\frac{du}{dx} + 6u = 0$$

Solution

We assume that the solution has the form $u(x) = e^{mx}$. Substitute this into the differential equation and find the characteristic equation to be

$$m^2 + 5m + 6 = 0$$

This is factored into

$$(m + 3)(m + 2) = 0$$

The roots are obviously

$$m_1 = -3, \qquad m_2 = -2$$

The two independent solutions are then

$$u_1(x) = e^{-3x}, \qquad u_2(x) = e^{-2x}$$

These solutions are superimposed to yield the general solution

$$u(x) = c_1 e^{-3x} + c_2 e^{-2x}$$

Example 1.13

Find the specific solution of the differential equation

$$\frac{d^2u}{dx^2} + 6\frac{du}{dx} + 9u = 0, \qquad \text{if} \quad u(0) = 2 \quad \text{and} \quad \frac{du}{dx}(0) = 0$$

Solution
Assume a solution of the form $u(x) = e^{mx}$. The characteristic equation

$$m^2 + 6m + 9 = 0$$

yields the roots

$$m_1 = -3, \qquad m_2 = -3$$

The roots are identical; therefore, the general solution is (see Eq. 1.6.17)

$$u(x) = c_1e^{-3x} + c_2xe^{-3x}$$

This solution must satisfy the imposed conditions. Using $u(0) = 2$, we have

$$2 = c_1$$

Differentiating, the expression for $u(x)$ gives

$$\frac{du}{dx} = (c_1 + xc_2)(-3e^{-3x}) + e^{-3x}c_2$$

Then, setting $du/dx(0) = 0$, we have

$$0 = -3c_1 + c_2$$

Hence,

$$c_2 = 6$$

The specific solution is then

$$u(x) = 2(1 + 3x)e^{-3x}$$

Example 1.14

Find the general solution of the differential equation

$$\frac{d^2u}{dx^2} + 2\frac{du}{dx} + 5u = 0$$

Solution
The assumed solution $u(x) = e^{mx}$ leads to the characteristic equation

$$m^2 + 2m + 5 = 0$$

The roots to this equation are

$$m_1 = -1 + 2i, \qquad m_2 = -1 - 2i$$

The general solution is then

$$u(x) = c_1 e^{(-1+2i)x} + c_2 e^{(-1-2i)x}$$

To obtain a more preferred expression for the solution, rewrite the equation above as

$$u(x) = e^{-x}\left(c_1 e^{2ix} + c_2 e^{-2ix}\right)$$

Using the relationship $e^{ix} = \cos x + i \sin x$, there results

$$\begin{aligned} u(x) &= e^{-x}[c_1(\cos 2x + i \sin 2x) + c_2(\cos 2x - i \sin 2x)] \\ &= e^{-x}(A \cos 2x + B \sin 2x) \end{aligned}$$

where

$$A = c_1 + c_2, \qquad B = (c_1 - c_2)i$$

1.7 SPRING–MASS SYSTEM—FREE MOTION

There are many physical phenomena that are described with linear, second-order, homogeneous differential equations. We wish to discuss one such phenomenon, the free motion of a spring–mass system, as an illustrative example. We shall restrict ourselves to systems with 1 degree of freedom; that is, only one independent variable is needed to describe the motion. Systems requiring more than one independent variable, such as a system with several masses and springs, lead to simultaneous ordinary differential equations and will not be considered.

Consider the simple spring–mass system shown in Fig. 1.2. We shall make the following assumptions:

1. The mass M, measured in kilograms, is constrained to move in the vertical direction only.
2. The viscous damping C, with units of kilograms per second, is proportional to the velocity dy/dt. For relatively small velocities this is usually acceptable; however, for large velocities the damping is more nearly proportional to the square of the velocity, which would lead to a nonlinear equation.
3. The force in the spring is Kd, where d is the distance measured in meters from the unstretched position. The spring modulus K, with units of newtons per meter (N/m), is assumed constant.
4. The mass of the spring is negligible compared with the mass M.
5. No external forces act on the system.

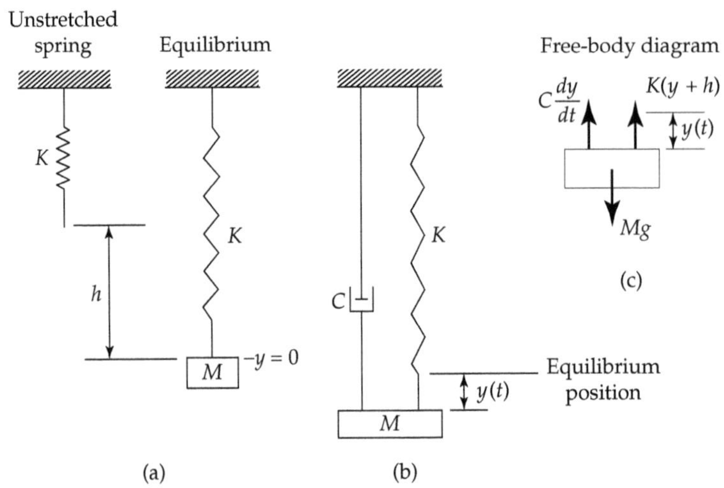

FIGURE 1.2 Spring–mass.

Newton's second law is used to describe the motion of the lumped mass. It states that the sum of the forces acting on a body in any particular direction equals the mass of the body multiplied by the acceleration of the body in that direction. This is written as

$$\sum F_y = Ma_y \tag{1.7.1}$$

for the y direction. Consider that the mass is suspended from an unstretched spring, as shown in Fig. 1.2a. The spring will then deflect a distance h, where h is found from the relationship

$$Mg = hK \tag{1.7.2}$$

which is a simple statement that for static equilibrium the weight must equal the force in the spring. The weight is the mass times the local acceleration of gravity. At this stretched position we attach a viscous damper, a dashpot, and allow the mass to undergo a displacement y about the equilibrium position. A free-body diagram of the mass is shown in Fig. 1.2c. Applying Newton's second law, we have, with the positive direction downward,

$$Mg - C\frac{dy}{dt} - K(y+h) = M\frac{d^2y}{dt^2} \tag{1.7.3}$$

Using Eq. 1.7.2, this simplifies to

$$M\frac{d^2y}{dt^2} + C\frac{dy}{dt} + Ky = 0 \tag{1.7.4}$$

This is a second-order, linear, homogeneous, ordinary differential equation with constant coefficients. Let us first consider the situation where the viscous damping coefficient C is sufficiently small that the viscous damping term may be neglected.

1.7.1 Undamped Motion

For the case where C is sufficiently small, it may be acceptable, especially for small time spans, to neglect the damping. If this is done, the differential equation that describes the motion is

$$M\frac{d^2y}{dt^2} + Ky = 0 \tag{1.7.5}$$

We assume a solution of the form e^{mt}, which leads to the characteristic equation

$$m^2 + \frac{K}{M} = 0 \tag{1.7.6}$$

The two roots are

$$m_1 = \sqrt{\frac{K}{M}}\,i, \qquad m_2 = -\sqrt{\frac{K}{M}}\,i \tag{1.7.7}$$

The solution is then

$$y(t) = Ae^{\sqrt{K/M}\,it} + Be^{-\sqrt{K/M}\,it} \tag{1.7.8}$$

or, equivalently, using $e^{i\theta} = \cos\theta + i\sin\theta$ and $e^{-i\theta} = \cos\theta - i\sin\theta$

$$y(t) = c_1\cos\sqrt{\frac{K}{M}}\,t + c_2\sin\sqrt{\frac{K}{M}}\,t \tag{1.7.9}$$

where $A + B = c_1$ and $i(A - B) = c_2$. The mass will undergo its first complete cycle as t goes from zero to $2\pi/\sqrt{K/M}$. Thus, one cycle is completed in $2\pi/\sqrt{K/M}$ seconds. The number of cycles per second, the *frequency*, is then $\sqrt{K/M}/2\pi$. The *angular frequency* ω_0 (this is often simply referred to as the "frequency") is given by

$$\omega_0 = \sqrt{\frac{K}{M}} \tag{1.7.10}$$

The solution is then written in the preferred form,

$$y(t) = c_1\cos\omega_0 t + c_2\sin\omega_0 t \tag{1.7.11}$$

This is the motion of the undamped mass. It is often referred to as a *harmonic oscillator*. It is important to note that the sum of the sine and cosine terms in Eq. 1.7.11 can be written as, using $\cos(\alpha + \beta) = \cos\alpha\cos\beta - \sin\alpha\sin\beta$,

$$y(t) = \Delta\cos(\omega_0 t - \delta) \tag{1.7.12}$$

where the *amplitude* Δ is related to c_1 and c_2 by $\Delta = \sqrt{c_1^2 + c_2^2}$ and $\tan\delta = c_2/c_1$. In this form Δ and δ are the arbitrary constants.

Two initial conditions, the initial displacement and velocity, are necessary to determine the two arbitrary coefficients in the above equations. For a zero initial velocity the motion would be as sketched in Fig. 1.3.

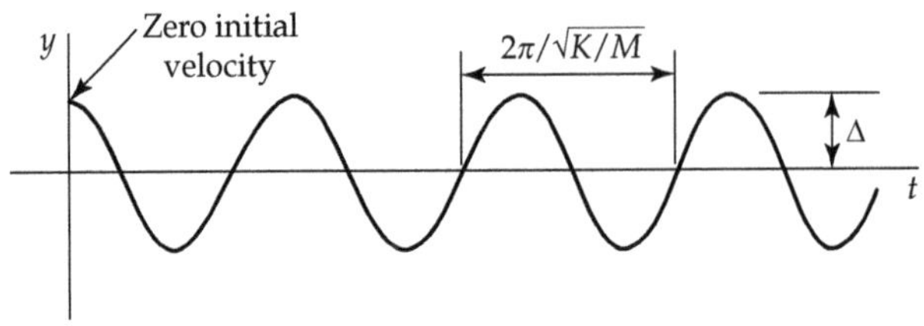

FIGURE 1.3 Harmonic oscillation.

Example 1.15

A 5-kg mass hangs from a spring with spring constant 20 N/m and is given a velocity of 4 m/s from its equilibrium position. If the damping is negligibly small, determine a) the angular frequency, b) the time required for one complete oscillation, and c) the maximum displacement of the mass.

Solution

a) The angular frequency of oscillation is given by Eq. 1.7.10. It is

$$\omega_0 = \sqrt{\frac{K}{M}} = \sqrt{\frac{20 \text{ N/m}}{5\text{N}\cdot\text{s}^2/\text{m}}} = \underline{2 \text{ rad/s}}$$

using $\text{N} = \text{kg}\cdot\text{m/s}^2$ from Newton's 2nd law.

b) The time Δt required for a complete oscillation is related to the frequency remembering that there are 2π radians per cycle. So,

$$\Delta t = \frac{2\pi}{\omega_0} = \frac{2\pi}{2} = \underline{3.142 \text{ s}}$$

Note: In general we provide at most 4 significant digits in a numerical answer. Most problems include a physical quantity such as mass or stiffness, which is seldom if ever known to 5 significant digits.

c) The maximum displacement is Δ in Eq. 1.7.12. Let us apply the two initial conditions to Eq. 1.7.11: $y(t) = c_1 \cos \omega_0 t + c_2 \sin \omega_0 t$. The initial conditions provide

$$y(0) = 0 \;\rightarrow\; c_1 = 0$$
$$\frac{dy}{dt}(0) = 4 \;\rightarrow\; 4 = c_2\omega_0 \;\rightarrow\; c_2 = 2$$

recognizing that $\frac{d \sin \omega_0 t}{dt} = \omega_0 \cos \omega_0 t$, $\sin 0 = 0$, and $\cos 0 = 1$. Refer to Eq. 1.7.12 to find

$$\Delta = \sqrt{c_1^2 + c_2^2} = c_2 = \underline{2 \text{ m}}$$

1.7.2 Damped Motion

Let us now include the viscous damping term in the equation. This would be necessary for long time spans, since viscous damping is always present, however small; or for short time periods, in which the damping coefficient C is not sufficiently small. The describing equation is

$$M\frac{d^2y}{dt^2} + C\frac{dy}{dt} + Ky = 0 \tag{1.7.13}$$

Assuming a solution of the form e^{mt}, the characteristic equation, is

$$Mm^2 + Cm + K = 0 \tag{1.7.14}$$

The two roots of this equation are

$$m_1 = -\frac{C}{2M} + \frac{1}{2M}\sqrt{C^2 - 4MK}, \qquad m_2 = -\frac{C}{2M} - \frac{1}{2M}\sqrt{C^2 - 4MK} \tag{1.7.15}$$

The solution is then, for $m_1 \neq m_2$, written as

$$y(t) = Ae^{-\frac{C}{2M}t + \frac{t}{2M}\sqrt{C^2-4MK}} + Be^{-\frac{C}{2M}t - \frac{t}{2M}\sqrt{C^2-4MK}} \tag{1.7.16}$$

or, equivalently,

$$y(t) = e^{-(C/2M)t}\left[Ae^{\sqrt{C^2-4MK}(t/2M)} + Be^{-\sqrt{C^2-4MK}(t/2M)}\right] \tag{1.7.17}$$

This solution obviously takes on three different forms, depending on the magnitude of the damping. The three cases are:

Case 1.	*Overdamped*	$C^2 - 4KM > 0.$	m_1 and m_2 are real.
Case 2.	*Critically damped*	$C^2 - 4KM = 0.$	$m_1 = m_2.$
Case 3.	*Underdamped*	$C^2 - 4KM < 0.$	m_1 and m_2 are complex.

Let us investigate each case separately.

Case 1. *Overdamped.* For this case the damping is so large that $C^2 > 4KM$. The roots m_1 and m_2 are real and the solution is best presented as in Eq. 1.7.17. Several overdamped motions are shown in Fig. 1.4. For large time the solution approaches $y = 0$.

Case 2. *Critically damped.* For this case the damping is just $C = 2\sqrt{KM}$. There is a double root of the characteristic equation, so the solution is (see Eq. 1.6.17)

$$y(t) = Ae^{mt} + Bte^{mt} \tag{1.7.18}$$

For the spring–mass system this becomes

$$y(t) = e^{-(C/2M)t}[A + Bt] \tag{1.7.19}$$

A sketch of the solution is not unlike that of the overdamped case. It is shown in Fig. 1.5.

Case 3. *Underdamped.* The most interesting of the three cases is that of underdamped motion. If $C^2 - 4KM$ is negative, we may write Eq. 1.7.17 as

$$y(t) = e^{-(C/2M)t}\left[Ae^{i\sqrt{4KM-C^2}(t/2M)} + Be^{-i\sqrt{4KM-C^2}(t/2M)}\right] \tag{1.7.20}$$

since $\sqrt{-1} = i$. Note the "i" before each radical Eq. 1.7.20.

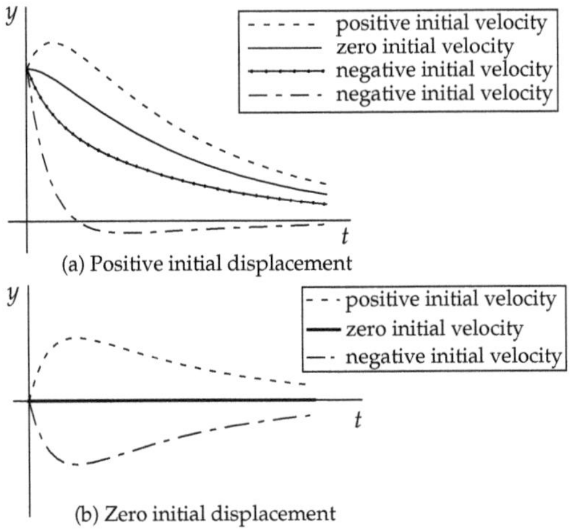

FIGURE 1.4 Overdamped motion.

This is expressed in the equivalent form, using $e^{i\theta} = \cos\theta + i\sin\theta$ and $e^{-i\theta} = \cos\theta - i\sin\theta$,

$$y(t) = e^{-(C/2M)t}\left[c_1 \cos\left(\sqrt{4KM - C^2}\,\frac{t}{2M}\right) + c_2 \sin\left(\sqrt{4KM - C^2}\,\frac{t}{2M}\right)\right] \tag{1.7.21}$$

The motion is an oscillating motion with a decreasing amplitude with time. The frequency of damped oscillation is $\sqrt{4KM - C^2}/4\pi M$ and approaches that of the undamped case as $C \to 0$. Equation 1.7.21 can be written in a form from which a sketch can more easily be made. It is

$$y(t) = Ae^{-(C/2M)t}\cos(\Omega t - \delta) \tag{1.7.22}$$

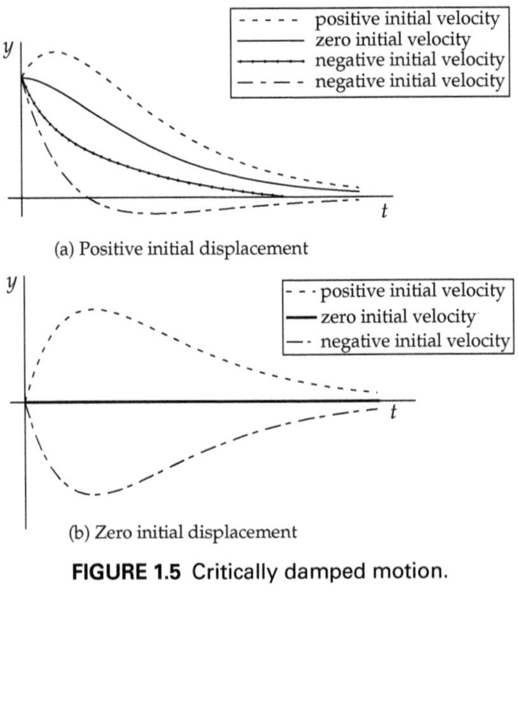

FIGURE 1.5 Critically damped motion.

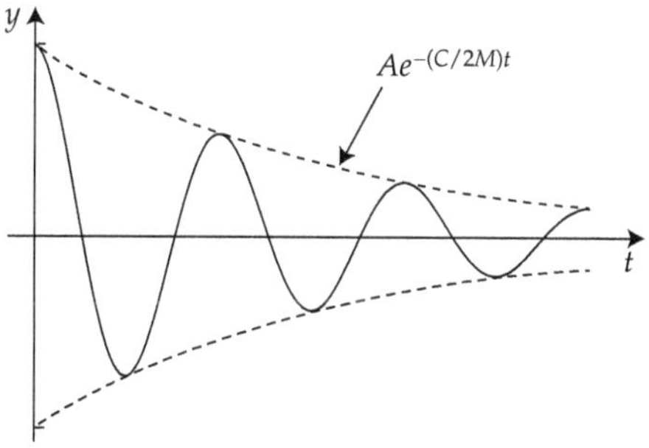

FIGURE 1.6 Underdamped motion.

where

$$\Omega = \frac{\sqrt{4KM - C^2}}{2M}, \qquad \tan\delta = \frac{c_2}{c_1}, \qquad A = \sqrt{c_1^2 + c_2^2} \tag{1.7.23}$$

The underdamped motion is sketched in Fig. 1.6, for an initial displacement with initial zero velocity. The motion damps out for large time.

The ratio of successive maximum displacements y_n is a quantity of particular interest for underdamped oscillations. We will show in Example 1.13 that this ratio is given by

$$\frac{y_n}{y_{n+1}} = e^{\pi C/\Omega M} \tag{1.7.24}$$

It is a constant for a particular underdamped motion for all time. The logarithm of this ratio is called the *logarithmic decrement D*, defined by

$$D = \ln\frac{y_n}{y_{n+1}} = \frac{\pi C}{\Omega M} \tag{1.7.25}$$

Note that the *logarithmic decrement*, *D*, is dimensionless.

Returning to the definition of Ω, this is expressed as

$$D = \frac{2\pi C}{\sqrt{4KM - C^2}} \tag{1.7.26}$$

In terms of the *critical damping*,* $C_c = 2\sqrt{KM}$, this is

$$D = \frac{2\pi C}{\sqrt{C_c^2 - C^2}} \tag{1.7.27}$$

or, alternatively,

$$\frac{C}{C_c} = \frac{D}{\sqrt{D^2 + 4\pi^2}} \tag{1.7.28}$$

Since y_n and y_{n+1} are easily measured, the logarithmic decrement D can be evaluated quite simply. This allows a quick method for determining the fraction of the critical damping that exists in a particular system.

---

*If $C = C_c$ there is no oscillation since $D = \infty$ (undefined) implying that $y_n = 0$ in Eq. 1.7.25.

Example 1.16

A damped 5-kg mass hangs from a spring with spring constant 20 N/m and is given a velocity of 4 m/s from its equilibrium position. Determine the solution $y(t)$ if the damping coefficient is a) 25 kg/s, b) 20 kg/s, and c) 12 kg/s.

Solution

a) First, make the calculation $\sqrt{C^2 - 4KM} = \sqrt{25^2 - 4 \times 20 \times 5} = \sqrt{225} = 15$. This is overdamped and the solution is given by Eq. 1.7.17:

$$\begin{aligned}
y(t) &= e^{-(C/2M)t}\left[Ae^{\sqrt{C^2-4MK}(t/2M)} + Be^{-\sqrt{C^2-4MK}(t/2M)}\right] \\
&= e^{-(25/2\times5)t}\left[Ae^{\sqrt{15}(t/2\times5)} + Be^{-\sqrt{15}(t/2\times5)}\right] \\
&= e^{-2.5t}\left[Ae^{0.3873t} + Be^{-0.3873t}\right]
\end{aligned}$$

The initial conditions $y(0) = 0$ and $dy/dt(0) = 4$ provide the following:

$$\begin{aligned}
0 &= A + B \\
\frac{dy}{dt} &= -2.5e^{-2.5t}\left[Ae^{0.3873t} + Be^{-0.3873t}\right] + e^{-2.5t}\left[0.3873Ae^{0.3873t} - 0.3873Be^{-0.3873t}\right] \\
4 &= -2.5(A + B) + 0.3873(A - B) = -2.113A - 2.887B
\end{aligned}$$

A simultaneous solution provides

$$A = 5.17 \text{ and } B = -5.17$$

The solution is then

$$y(t) = 5.17e^{-2.5t}\left[e^{0.3873t} - e^{-0.3873t}\right]$$

b) The quantity of interest is $\sqrt{C^2 - 4KM} = \sqrt{20^2 - 4 \times 20 \times 5} = 0$. This is critically damped so the solution is given by Eq. 1.7.19:

$$\begin{aligned}
y(t) = e^{-(C/2M)t}[A + Bt] &= e^{-(20/2\times5)}[A + Bt] \\
&= e^{-2t}[A + Bt]
\end{aligned}$$

The initial conditions $y(0) = 0$ and $dy/dt(0) = 4$ provide the following:

$$\begin{aligned}
0 &= A \\
\frac{dy}{dt} &= -2e^{-2t}[A + Bt] + Be^{-2t} \\
4 &= -2A + B, \qquad \therefore B = 4
\end{aligned}$$

The solution is then

$$y(t) = 4te^{-2t}$$

c) For this situation $\sqrt{C^2 - 4KM} = \sqrt{12^2 - 4 \times 20 \times 5} = 16i$, indicating an underdamped motion. There are three forms of the solution. First, select the solution given by Eq. 1.7.20:

$$y(t) = e^{-(C/2M)t}[Ae^{i\sqrt{4MK-C^2}(t/2M)} + Be^{-i\sqrt{4MK-C^2}(t/2M)}]$$
$$= e^{-2.5t}[Ae^{i0.3673t} + Be^{-i0.3673t}]$$

The initial conditions are now imposed resulting in a simultaneous solution for A and B:

$$A = -5.164i \text{ and } B = 5.164i$$

Finally, the solution is written, in the rather undesirable form, as

$$y(t) = 5.164ie^{-2.5t}[e^{-i0.3673t} - e^{i0.3673t}] \tag{1}$$

Using the identities $e^{i\theta} = \cos\theta + i\sin\theta$ and $e^{-i\theta} = \cos\theta - i\sin\theta$, the above equation takes the form

$$y(t) = 5.164ie^{-2.5t}[\cos 0.3673t + i\sin 0.3673t - \cos 0.3673t + i\sin 0.3673t]$$
$$= -5.164e^{-2.5t}\sin 0.3673t \tag{2}$$

If it is desired to put the solution in the form of Eq. 1.7.22, we would use $\cos(\alpha + \pi) = \cos\alpha\cos\pi - \sin\alpha\sin\pi = -\sin\alpha$ since $\cos\pi = 0$. The solution then takes the form

$$y(t) = 5.164e^{-2.5t}\cos(0.3673t + \pi) \tag{3}$$

The above solutions (1), (2), and (3) are equivalent. For this example, solution (2) would be the most desired form for this underdamped spring–mass system.

Example 1.17

Determine the ratio of successive local maximum displacements y_n for the free motion of an underdamped oscillation, and the logarithmic decrement for Example 1.12c.

Solution

The displacement function for an underdamped spring–mass system is

$$y(t) = Ae^{-(C/2M)t}\cos(\Omega t - \delta)$$

To find the local maximum displacement we set $dy/dt = 0$ and solve for the particular t that yields this condition. Differentiating, we have

$$\frac{dy}{dt} = -\left[\frac{C}{2M}\cos(\Omega t - \delta) + \Omega\sin(\Omega t - \delta)\right]Ae^{-(C/2M)t} = 0$$

This gives

$$\tan(\Omega t - \delta) = -\frac{C}{2M\Omega}$$

or, more generally,

$$\tan^{-1}\left(-\frac{C}{2M\Omega}\right) + n\pi = \Omega t - \delta$$

The time at which a local maximum occurs in the amplitude is given by

$$t = \frac{\delta}{\Omega} - \frac{1}{\Omega}\tan^{-1}\frac{C}{2M\Omega} + \frac{n\pi}{\Omega}$$

where $n = 0$ represents the first maximum, $n = 2$ the second maximum, and so on. For $n = 1$, a minimum would result. We are interested in the ratio y_n/y_{n+1}. If we let

$$B = \frac{\delta}{\Omega} - \frac{1}{\Omega}\tan^{-1}\frac{C}{2M\Omega}$$

this ratio becomes

$$\frac{y_n}{y_{n+2}} = \frac{Ae^{-\frac{C}{2M}\left(B+\frac{n\pi}{\Omega}\right)}\cos\left[\Omega\left(B+\frac{n\pi}{\Omega}\right)-\delta\right]}{Ae^{-\frac{C}{2M}\left[B+\frac{n+2}{\Omega}\pi\right]}\cos\left[\Omega\left(B+\frac{n+2}{\Omega}\pi\right)-\delta\right]}$$

By periodicity, $\cos\theta = \cos(\theta + 2\pi)$ and so cancellation takes place.

$$= e^{\pi C/\Omega M}\frac{\cos[B\Omega + n\pi - \delta]}{\cos[B\Omega + n\pi - \delta + 2\pi]} = e^{\pi C/\Omega M}$$

Hence, we see that the ratio of successive maximum amplitudes is dependent only on M, K, and C and is independent of time. It is constant for a particular spring–mass system.

The logarithmic decrement D is found to be

$$D = \ln\frac{y_n}{y_{n+2}} = \frac{\pi C}{\Omega M} = \frac{2\pi CM}{\sqrt{4KM - C^2 M}}$$

$$= \frac{2\pi \times 12}{\sqrt{4 \times 20 \times 5 - 12^2}} = \underline{4.712}$$

1.7.3 The Electrical Circuit Analog

We now consider the solution to Eq. 1.4.4 for the case $dv/dt = 0$. By comparing Eq. 1.4.4 with Eq. 1.7.4, we see that we can interchange the spring–mass system parameters with the circuit parameters as follows:

Spring–Mass		Series Circuit
M	→	L
C	→	R
K	→	$1/C$

The solutions that we have just considered for $y(t)$ may then be taken as solutions for $i(t)$.

Thus, for the undamped circuit, we have $R = 0$, and there is no dissipation of electrical energy. The current in this case is given by (see Eq. 1.7.11)

$$i(t) = c_1 \cos \omega_0 t + c_2 \sin \omega_0 t \tag{1.7.29}$$

where

$$\omega_0 = \sqrt{\frac{1}{LC}} \tag{1.7.30}$$

This value is typically very large for electrical circuits, since both L and C are usually quite small.

For the damped circuit the solution for $i(t)$ may be deduced from Eq. 1.7.17 to be

$$i(t) = e^{-(R/2L)t}[Ae^{\sqrt{R^2-4L/C}(t/2L)} + Be^{-\sqrt{R^2-4L/C}(t/2L)}] \tag{1.7.31}$$

Now the damping criteria become

Case 1. *Overdamped* $R^2 - \dfrac{4L}{C} > 0$

Case 2. *Critically damped* $R^2 - \dfrac{4L}{C} = 0$

Case 3. *Underdamped* $R^2 - \dfrac{4L}{C} < 0$

Example 1.18

Use Kirchhoff's second law to establish the differential equation for the parallel electrical circuit shown in Fig 1.7. Give the appropriate analogies with the spring–mass system and write the solution to the homogeneous equation.

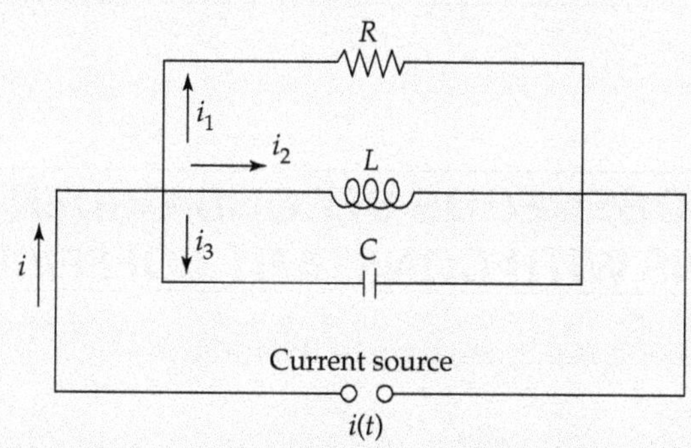

FIGURE 1.7

Solution

Kirchhoff's second law states that the current flowing to a point in a circuit must equal the current flowing away from the point. This demands that

$$i(t) = i_1 + i_2 + i_3$$

Use the observed relationships of current to impressed voltage for the components of our circuit,

$$\text{current flowing through a resistor} = \frac{v}{R} = i_1$$

$$\text{current flowing through a capacitor} = C\frac{dv}{dt} = i_3$$

$$\text{current flowing through an inductor} = \frac{1}{L}\int v\,dt = i_2$$

The equation above becomes

$$i(t) = \frac{v}{R} + C\frac{dv}{dt} + \frac{1}{L}\int v\,dt$$

If we differentiate our expression for $i(t)$, we find the differential equation to be

$$C\frac{d^2v}{dt^2} + \frac{1}{R}\frac{dv}{dt} + \frac{v}{L} = -\frac{di}{dt}$$

The analogy with the spring–mass system is

$$M \to C$$

$$C \to \frac{1}{R}$$

$$K \to \frac{1}{L}$$

Following Eq. 1.7.17, the solution to the homogeneous equation ($di/dt = 0$) is

$$v(t) = e^{-(t/2CR)}\left[Ae^{\sqrt{\frac{1}{R^2}-\frac{4C}{L}}\frac{t}{2C}} + Be^{-\sqrt{\frac{1}{R^2}-\frac{4C}{L}}\frac{t}{2C}}\right]$$

1.8 NONHOMOGENEOUS SECOND-ORDER LINEAR EQUATIONS WITH CONSTANT COEFFICIENTS

The solution of the second-order equation of the form

$$\frac{d^2u}{dx^2} + a\frac{du}{dx} + bu = h(x) \tag{1.8.1}$$

is found by adding the particular solution $u_p(x)$ to the solution $u_h(x)$ of the homogeneous equation

$$\frac{d^2u}{dx^2} + a\frac{du}{dx} + bu = 0 \qquad (1.8.2)$$

The solution of the homogeneous equation was presented in the preceding section; therefore, we now focus on finding $u_p(x)$. One approach may be taken by using the *method of undetermined coefficients*. Three common types of functions which are terms often found in $h(x)$ are listed below. Let us present the form of $u_p(x)$ for each.

1. $h(x)$ in Eq. 1.8.1 is a polynomial function (10, x, $x^2 - 4$, etc.). Choose $u_p(x)$ to be a polynomial of the same order but with undetermined coefficients. For $h(x) = x^2 - 4$ we would choose $u_p(x) = Ax^2 + Bx + C$.
2. $h(x)$ is an exponential function, Ae^{kx}, and k is not a root of the characteristic equation (see Eq. 1.6.4). Choose $u_p(x) = Ce^{kx}$. If k is a single root of the characteristic equation, choose $u_p(x) = Cxe^{kx}$, and if k is a double root, choose $u_p(x) = Cx^2e^{kx}$.
3. $h(x)$ is a sine or cosine function (e.g., $C \cos kx$), and ik is not a root of the characteristic equation. Choose $u_p(x) = A \cos kx + B \sin kx$. If ik is a single root of the characteristic equation, choose $u_p(x) = Ax \cos kx + Bx \sin kx$.

Should $h(x)$ include a combination of the above functions, as in Example 1.17, the particular solution would be found by superimposing the appropriate particular solutions listed above. For functions $h(x)$ that are not listed above, the particular solution would be found using some other technique. For periodic functions, a method using Fourier series may be used as presented in Section 1.10. Variation of parameters, presented in Section 1.12, may also be attempted for nonperiodic functions.

Example 1.19

Find the general solution of the differential equation

$$\frac{d^2u}{dx^2} + u = x^2$$

Solution

The solution of the homogeneous equation

$$\frac{d^2u}{dx^2} + u = 0$$

is found to be

$$u_h(x) = c_1 \cos x + c_2 \sin x$$

A particular solution is assumed to have the form

$$u_p(x) = Ax^2 + Bx + C$$

This is substituted into the original differential equation to give

$$2A + Ax^2 + Bx + C = x^2$$

Equating coefficients of the various powers of x, we have

$$x^0: \quad 2A + C = 0$$
$$x^1: \quad B = 0$$
$$x^2: \quad A = 1$$

These equations are solved simultaneously to give the particular solution as

$$u_p(x) = x^2 - 2$$

Finally, the general solution is

$$\begin{aligned} u(x) &= u_h(x) + u_p(x) \\ &= c_1 \cos x + c_2 \sin x + x^2 - 2 \end{aligned}$$

Example 1.20

Find the general solution of the differential equation

$$\frac{d^2u}{dx^2} + 4u = 2\sin 2x$$

Solution

The solution of the homogeneous equation is

$$u_h(x) = c_1 \cos 2x + c_2 \sin 2x$$

One root of the characteristic equation is $2i$; hence, we assume the particular solution to be

$$u_p(x) = Ax\cos 2x + Bx\sin 2x$$

Substitute this into the original differential equation:

$$\begin{aligned} &-2A \sin 2x + 2B\cos 2x - 2A \sin 2x + 2B\cos 2x - 4Ax\cos 2x \\ &- 4Bx \sin 2x + 4Ax \cos 2x + 4Bx \sin 2x = 2\sin 2x \end{aligned}$$

Equating coefficients yields

$$\sin 2x: \quad -2A - 2A = 2$$
$$\cos 2x: \quad 2B + 2B = 0$$
$$x\sin 2x: \quad -4B + 4B = 0$$
$$x\cos 2x: \quad -4A + 4A = 0$$

These equations require that $A = -1/2$ and $B = 0$. Thus,

$$u_p(x) = -\frac{1}{2}x\cos 2x$$

The general solution is then

$$\begin{aligned} u(x) &= u_h(x) + u_p(x) \\ &= c_1 \cos 2x + c_2 \sin 2x - \frac{1}{2}x\cos 2x \end{aligned}$$

Example 1.21

Find a particular solution of the differential equation

$$\frac{d^2u}{dx^2} + \frac{du}{dx} + 2u = 4e^x + 2x^2$$

Solution

Since $k = 1$ is not a root of $m^2 + m + 2 = 0$, assume the particular solution to have the form

$$u_p(x) = Ae^x + Bx^2 + Cx + D$$

Substitute this into the given differential equation and there results

$$Ae^x + 2B + Ae^x + 2Bx + C + 2Ae^x + 2Bx^2 + 2Cx + 2D = 4e^x + 2x^2$$

Equating the various coefficients yields

$$\begin{aligned} e^x:&\quad A + A + 2A = 4 \\ x^0:&\quad 2B + C + 2D = 0 \\ x^1:&\quad 2B + 2C = 0 \\ x^2:&\quad 2B = 2 \end{aligned}$$

From the equations above we find $A = 1$, $B = 1$, $C = -1$, and $D = -1/2$. Thus,

$$u_p(x) = e^x + x^2 - x - \frac{1}{2}$$

1.9 SPRING–MASS SYSTEM—FORCED MOTION

The spring–mass system shown in Fig. 1.2 is acted upon by a force $F(t)$, a *forcing function*, as shown in Fig. 1.8. The equation describing this motion is again found by applying Newton's second law to the mass M. We have

$$F(t) + Mg - K(y + h) - C\frac{dy}{dt} = M\frac{d^2y}{dt^2} \qquad (1.9.1)$$

where $y + h$ is the spring deformation and where h is as defined in Fig. 1.2, so that $Mg = Kh$. The equation above becomes

$$M\frac{d^2y}{dt^2} + C\frac{dy}{dt} + Ky = F(t) \qquad (1.9.2)$$

It is a nonhomogeneous equation and can be solved by the techniques introduced in the preceding section, depending on the form of $F(t)$.

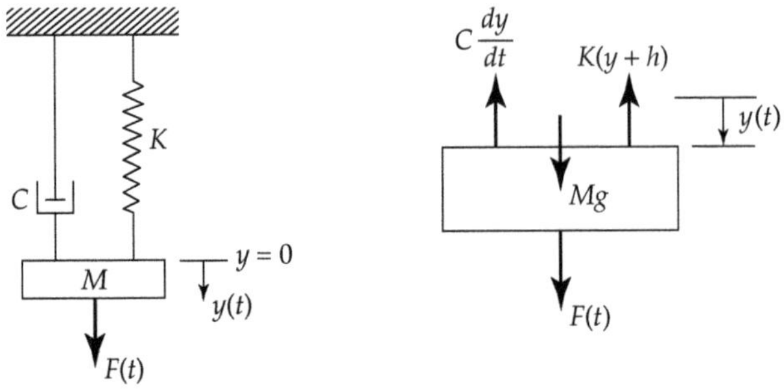

FIGURE 1.8 Spring–mass system with a forcing function.

We shall discuss the form of the solution for a sinusoidal forcing function,

$$F(t) = F_0 \cos \omega t \tag{1.9.3}$$

The particular solution has the form

$$y_p(t) = A\cos\omega t + B\sin\omega t \tag{1.9.4}$$

Substitute into Eq. 1.9.2 to obtain

$$\begin{aligned}[(K - M\omega^2)A + \omega CB]\cos\omega t + [(K - M\omega^2)B - \omega CA]\sin\omega t \\ = F_0 \cos\omega t\end{aligned} \tag{1.9.5}$$

Equating coefficients of $\cos\omega t$ and $\sin\omega t$ results in

$$\begin{aligned}(K - M\omega^2)A + \omega CB &= F_0 \\ -\omega CA + (K - M\omega^2)B &= 0\end{aligned} \tag{1.9.6}$$

A simultaneous solution yields

$$\begin{aligned}A &= F_0 \frac{K - M\omega^2}{(K - M\omega^2)^2 + \omega^2C^2} \\ B &= F_0 \frac{\omega C}{(K - M\omega^2)^2 + \omega^2C^2}\end{aligned} \tag{1.9.7}$$

The particular solution is then

$$y_p(t) = \frac{(K - M\omega^2)F_0}{(K - M\omega^2)^2 + \omega^2C^2}\left[\cos\omega t + \frac{\omega C}{K - M\omega^2}\sin\omega t\right] \tag{1.9.8}$$

This is added to the homogeneous solution presented in Section 1.7 to form the general solution

$$\begin{aligned}y(t) &= e^{-(C/2M)t}\left[Ae^{\sqrt{C^2-4MK}(t/2M)} + Be^{-\sqrt{C^2-4MK}(t/2M)}\right] \\ &\quad + \frac{(K - M\omega^2)F_0}{(K - M\omega^2)^2 + \omega^2C^2}\left[\cos\omega t + \frac{\omega C}{K - M\omega^2}\sin\omega t\right]\end{aligned} \tag{1.9.9}$$

Let us now discuss this solution in some detail, but first an example.

Example 1.22

A damped 5-kg mass hanging from a spring with spring constant 20 N/m and a damping coefficient of 25 kg/s. It starts from rest at its equilibrium position. Determine the displacement of the mass after 0.2 seconds, i.e., $y(0.2)$, if the forcing function is 50 sin $2t$.

Solution

We must determine the solution of the differential equation (1.9.2) and then apply the initial conditions $y(0) = 0$ and $dy/dt(0) = 0$. First, make the calculation $\sqrt{C^2 - 4KM} = \sqrt{25^2 - 4\times 20\times 5} = \sqrt{225} = 15$. This is overdamped according to Case 1 on Page 29 and the solution to the homogeneous differential equation is given by Eq. 1.7.17:

$$\begin{aligned} y_h(t) &= e^{-(C/2M)t}\left[c_1 e^{\sqrt{C^2-4MK}(t/2M)} + c_2 e^{-\sqrt{C^2-4MK}(t/2M)}\right] \\ &= e^{-(25/2\times 5)t}\left[c_1 e^{\sqrt{15}(t/2\times 5)} c_2 B e^{-\sqrt{15}(t/2\times 5)}\right] \\ &= e^{-2.5t}\left[c_1 e^{0.3873t} + c_2 e^{-0.3873t}\right] \end{aligned}$$

Next, the particular solution must be found. It is assumed to take the form $y_p(t) = A\sin 2t + B\cos 2t$. Substitute this assumed solution into Eq. 1.9.2:

$$\begin{aligned} &5(-4A\sin 2t - 4B\cos 2t) + 25(2A\cos 2t - 2B\sin 2t) + 20(A\sin 2t + B\cos 2t) = 5\sin 2t \\ &\therefore A = 0 \text{ and } B = -1 \end{aligned}$$

The solution of Eq. 1.9.2 for $y(t)$ is then

$$y(t) = y_h(t) + y_p(t) = e^{-2.5t}\left[c_1 e^{0.3873t} + c_2 e^{-0.3873t}\right] - \cos 2t$$

The initial conditions $y(0) = 0$ and $dy/dt(0) = 0$ for this mass initially at rest are now applied to the above solution:

$$\begin{aligned} &y(0) = 0 = [c_1 + c_2] - 1 \\ &\frac{dy}{dt}(0) = 0 = -2.5[c_1 + c_2] + [0.3873c_1 - 0.3873c_2] - 0 = -2.113c_1 - 2.887c_2 \end{aligned}$$

A simultaneous solution yields $c_1 = -1.73$ and $c_2 = 2.73$. The specific solution is found to be

$$y(t) = e^{-2.5t}[-1.73e^{0.3873t} + 2.73e^{-0.3873t}] - \cos 2t$$

This solution allows us to determine $y(0.2)$. It is

$$\begin{aligned} y(0.2) &= e^{-0.5}[-1.73e^{0.0775} + 2.73e^{-0.0775}] - \cos 0.4 \\ &= 0.6065[-1.867 + 2.526] - 0.9211 = \underline{-0.521\text{ m}} \end{aligned}$$

1.9.1 Resonance

An interesting and very important phenomenon is observed in the solution (1.9.9) if we let the damping coefficient C, which is often very small, be zero. The general solution is then [see Eq. 1.7.11 and let $C = 0$ in Eq. 1.9.8]

$$y(t) = c_1 \cos \omega_0 t + c_2 \sin \omega_0 t + \frac{F_0}{M(\omega_0^2 - \omega^2)} \cos \omega t \tag{1.9.10}$$

where $\omega_0 = \sqrt{K/M}$ and $\omega_0/2\pi$ is the natural frequency of the free oscillation (the oscillation that occurs with no forcing function). Consider the condition $\omega \to \omega_0$; that is, the frequency of the forcing function approaches the natural frequency. We observe from Eq. 1.9.10 that the amplitude of the particular solution becomes unbounded as $\omega \to \omega_0$. This condition is referred to as *resonance*. The amplitude, of course, does not become unbounded in a physical situation; the damping term may limit the amplitude, the physical situation may change for large amplitude, or failure may occur. The latter must be guarded against in the design of oscillating systems. Soldiers break step on bridges so that resonance will not occur. The spectacular failure of the Tacoma Narrows bridge provided a very impressive example of resonant failure (Google "Tacoma Narrows bridge" for details). One must be extremely careful to make the natural frequency of oscillating systems different, if at all possible, from the frequency of any probable forcing function.

If $\omega = \omega_0$, Eq. 1.9.10 is, of course, not a solution to the differential equation with no damping since the coefficeient of $\cos \omega t$ is undefined. For that case $i\omega_0$ is a root of the characteristic equation

$$m^2 + \omega_0^2 = 0 \tag{1.9.11}$$

of the undamped spring–mass system. The particular solution takes the form

$$y_p(t) = t(A \cos \omega_0 t + B \sin \omega_0 t) \tag{1.9.12}$$

By substituting into the differential equation

$$\frac{d^2 y}{dt^2} + \omega_0^2 y = \frac{F_0}{M} \cos \omega_0 t \tag{1.9.13}$$

we find the particular solution to be

$$y_p(t) = \frac{F_0}{2M\omega_0} t \sin \omega_0 t \tag{1.9.14}$$

As time t becomes large, the amplitude becomes large and will be limited by either damping, a changed physical condition, or failure. The particular solution $y_p(t)$ for resonance is shown in Fig. 1.9.

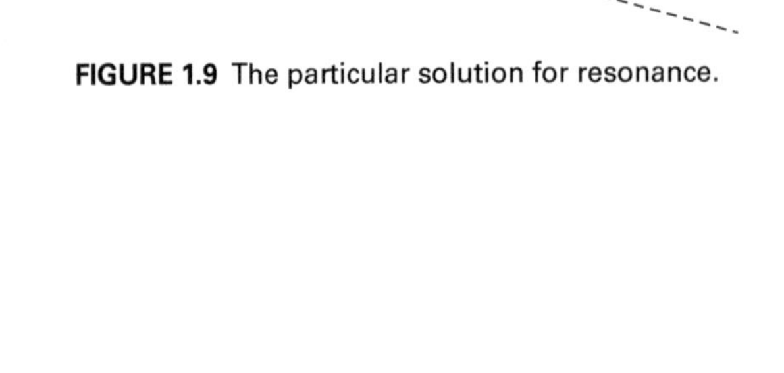

FIGURE 1.9 The particular solution for resonance.

Example 1.23

A damped 5-kg mass hanging from a spring with spring constant 20 N/m and negligible damping starts from rest at its equilibrium position. Determine the displacement of the mass after 0.5π s and 4.5π s if the forcing function is $5\sin 2t$.

Solution

The solution of the homogeneous differential equation $\ddot{y} + 4y = 0$ (we have used a dot to indicate a derivative with respect to time, so a double dot indicates a second derivative) is $y_h(t) = c_1 \cos 2t + c_2 \sin 2t$. Since the input frequency equals the natural frequency (2 rad/s), this represents resonance. The particular solution $y_p(t) = At \cos 2t + Bt \sin 2t$ is found by substituting into the governing differential equation $5\ddot{y} + 20y = 5 \sin 2t$. The general solution, using the particular solution (1.9.14), is then

$$\begin{aligned} y(t) = y_h(t) + y_p(t) &= c_1 \cos 2t + c_2 \sin 2t + \frac{F_0}{2M\omega_0} t \sin \omega_0 t \\ &= c_1 \cos 2t + c_2 \sin 2t + 0.25t \sin 2t \end{aligned}$$

To find the specific solution, the initial conditions $y(0) = 0$ and $\dot{y}(0) = 0$ for this mass initially at rest are now applied to the above solution:

$$\left.\begin{aligned} y(0) &= 0 = c_1 \\ \dot{y}(0) &= 0 = 2c_2 \end{aligned}\right\} \quad \therefore c_1 = c_2 = 0$$

So, the specific solution and the displacement at $t = 0.5\pi$ and 4.5π seconds (the first and third maximums in Fig. 1) are

$$\begin{aligned} y(t) &= 0.25t \sin 2t \\ y(0.5\pi) &= 0.25 \times 0.5\pi \sin 0.5\pi = \underline{0.393 \text{ m}} \\ y(4.5\pi) &= 0.25 \times 4.5\pi \sin 4.5\pi = \underline{3.53 \text{ m}} \end{aligned}$$

Note the substantial increase in amplitude of the mass between 1.57 s and 14.1 s, about a factor of 9.

1.9.2 Near Resonance

Another phenomenon occurs when the forcing frequency is approximately equal to the natural frequency; that is, the quantity $\omega_0 - \omega$ is small. Let us consider a situation for which $dy/dt(0) = 0$ and $y(0) = 0$ in the underdamped case. The arbitrary constants in Eq. 1.9.10 are then

$$c_2 = 0, \qquad c_1 = -\frac{F_0}{M(\omega_0^2 - \omega^2)} \tag{1.9.15}$$

The specific solution then becomes

$$y(t) = \frac{F_0}{M(\omega_0^2 - \omega^2)}[\cos\omega t - \cos\omega_0 t] \tag{1.9.16}$$

With the use of a trigonometric identity, this can be put in the form*

$$y(t) = \frac{2F_0}{M(\omega_0^2 - \omega^2)}\sin\left[(\omega_0 + \omega)\frac{t}{2}\right]\sin\left[(\omega_0 - \omega)\frac{t}{2}\right] \tag{1.9.17}$$

The quantity $\omega_0 - \omega$ is small; thus, the period of the sine wave $\sin[(\omega_0 - \omega)(t/2)]$ is large compared to the period of $\sin[(\omega_0 + \omega)(t/2)]$.

For $\omega_0 \cong \omega$, we can write

$$\frac{\omega_0 + \omega}{2} \cong \omega, \qquad \frac{\omega_0 - \omega}{2} = \varepsilon \tag{1.9.18}$$

where ε is small. Then the near-resonance solution (1.9.17) is expressed as

$$y(t) = \left[\frac{2F_0 \sin\varepsilon t}{M(\omega_0^2 - \omega^2)}\right]\sin\omega t \tag{1.9.19}$$

where the quantity in brackets is the slowly varying amplitude. A plot of $y(t)$ is sketched in Fig. 1.10. The larger wave-length wave appears as a "beat" and can often be heard when two sound waves are of approximately the same frequency.

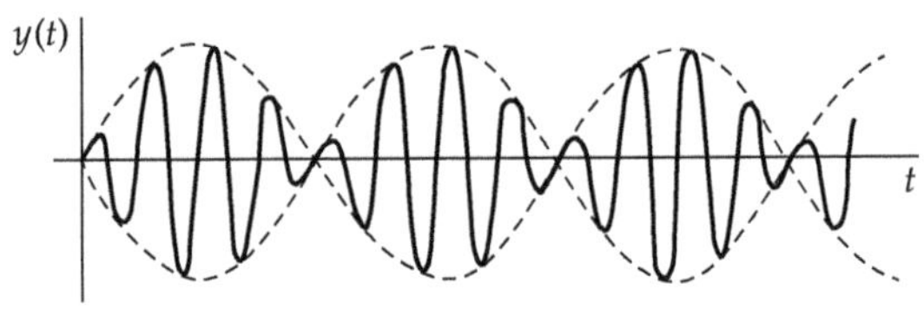

FIGURE 1.10 Near resonance—"beats."

Example 1.24

A 5-kg mass, which hangs from a spring with spring constant 20 N/m and negligible damping, starts from rest at its equilibrium position. Determine the frequency of the "beat" and the maximum displacement of the mass, shown in Fig. 1.10, if the forcing function is $0.5\sin 2.02t$. Also, calculate the time needed for one complete cycle of the natural oscillations and one complete cycle of the "beat."

*This is accomplished by writing $\cos\omega_0 t = \cos[((\omega + \omega_0)/2)t + ((\omega - \omega_0)/2)t]$ and $\cos\omega_0 t = \cos[((\omega + \omega_0)/2)t - ((\omega - \omega_0)/2)t]$ and then using the trigonometric identity $\cos(\alpha + \beta) = \cos\alpha\cos\beta - \sin\alpha\sin\beta$.

Solution

This is a near-resonance situation since the input frequency of the forcing function is quite close to the natural frequency of the undamped system. The solution is given by Eq. 1.9.19 with $\omega = \sqrt{K/M} = \sqrt{20/5} = 2$ rad/s:

$$y(t) = \frac{2F_0}{M(\omega_0^2 - \omega^2)} \sin \varepsilon t \sin \omega t$$

In the above equation, the frequency of the "beat" of Fig. 1.9 is ε, given in Eq. 1.9.18. Hence, it is

Note that the beats occur twice per period of sin εt. So the frequency of observed bursts is actually 2ε.

$$\varepsilon = \frac{\omega_0 - \omega}{2} = \frac{2.02 - 2}{2} = \underline{0.01 \text{ rad/s}}$$

The maximum displacement is given by, letting both sine functions equal 1,

$$y_{\max} = \frac{2F_0}{M(\omega_0^2 - \omega^2)} = \frac{2 \times 0.5}{5 \times (2.02^2 - 2^2)} = \underline{2.49 \text{ m}}$$

The time required to complete one cycle is

$$\Delta t = \frac{2\pi \text{ rad/cycle}}{2 \text{ rad/s}} = \underline{3.146 \text{ s/cycle}}$$

The frequency of the "beat" is 0.01 rad/s. So, the time to complete one cycle of the "beat" is

$$\Delta t_{\text{beat}} = \frac{2\pi \text{ rad/cycle}}{0.01 \text{ rad/s}} = \underline{6.292 \text{ s/cycle}}$$

1.9.3 Forced Oscillations with Damping

The homogeneous solution (1.7.17)

$$y_h(t) = e^{-(C/2M)t}\left[Ae^{\sqrt{C^2-4MK}(t/2M)} + Be^{-\sqrt{C^2-4MK}(t/2M)}\right] \tag{1.9.20}$$

for damped oscillations includes a factor $e^{-(C/2M)t}$ which is approximately zero after a sufficiently long time. Thus, the general solution $y(t)$ tends to the particular solution $y_p(t)$ after a long time; hence, $y_p(t)$ is called the *steady-state solution*. For short times the homogeneous solution must be included and $y(t) = y_h(t) + y_p(t)$ is the *transient solution*.

With damping included, the amplitude of the particular solution is not unbounded as $\omega \to \omega_0$, but it can still become large. The condition of resonance can be approached for the case of small damping. Hence, even with some damping, the condition $\omega = \omega_0$ is to be avoided, if at all possible. The Tacoma Narrows bridge obviously had some damping.

We are normally interested in the amplitude of the particular solution. To better display the amplitude for the input $F_0 \cos \omega t$, write Eq. 1.9.8 in the equivalent form*

$$y_p(t) = \frac{F_0}{\sqrt{M^2(\omega_0^2 - \omega^2)^2 + \omega^2 C^2}} \cos(\omega t - \alpha) \tag{1.9.21}$$

where we have used $\omega_0^2 = K/M$. The angle α is called the *phase angle* or *phase lag*. The amplitude Δ of the oscillation is

$$\Delta = \frac{F_0}{\sqrt{M^2(\omega_0^2 - \omega^2)^2 + \omega^2 C^2}} \tag{1.9.22}$$

We can find the maximum amplitude by setting $d\Delta/d\omega = 0$. Do this and find that the maximum amplitude occurs when

$$\omega^2 = \omega_0^2 - \frac{C^2}{2M^2} \tag{1.9.23}$$

Note that for sufficiently large damping, $C^2 > 2M^2\omega_0^2$, there is no value of ω that represents a maximum for the amplitude. However, if $C^2 < 2M^2\omega_0^2$, then the maximum occurs at the value of ω as given by Eq. 1.9.23. Substituting this into Eq. 1.9.22 gives the maximum amplitude

$$\Delta_{\text{max}} = \frac{2F_0 M}{C\sqrt{4M^2\omega_0^2 - C^2}} \tag{1.9.24}$$

The amplitude given by Eq. 1.9.22 is sketched in Fig. 1.11 as a function of ω. Large amplitudes due to resonance can thus be avoided by a sufficient amount of damping.

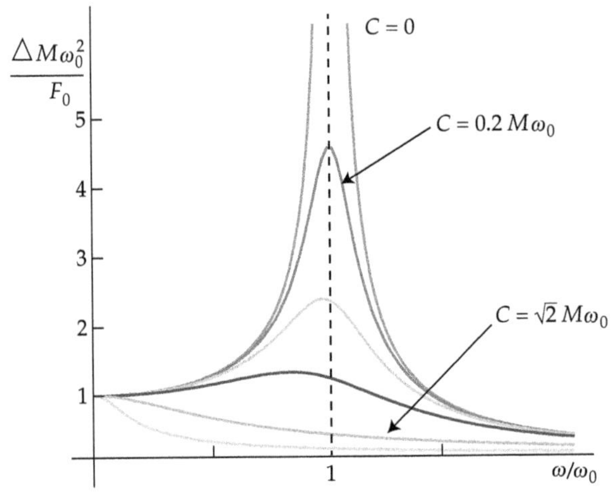

FIGURE 1.11 Amplitude as a function of ω for various degrees of damping.

*Equation 1.9.4 can be written in the equivalent form $y_p(t) = \Delta \cos(\omega t - \alpha)$.

Example 1.25

The ratio of sucessive maximum displacements for a particular spring–mass system for which $K = 100$ N/m and $M = 4$ kg is found to be 0.8 when the system undergoes free motion. If a forcing function $F = 10 \cos \omega t$ is imposed on the system, determine the maximum amplitude (over all forcing frequencies, ω) of the steady-state motion.

Solution

Damping causes the amplitude of the free motion to decrease with time. The logarithmic decrement is found to be (see Eq. 1.7.25)

$$D = \ln \frac{y_n}{y_{n+2}} = \ln \frac{1}{0.8} = 0.223$$

The damping is then calculated from Eq. 1.7.28. It is

$$C = C_c \frac{D}{\sqrt{D^2 + 4\pi^2}} = 2\sqrt{KM}\frac{D}{\sqrt{D^2 + 4\pi^2}}$$
$$= 2\sqrt{100 \times 4}\frac{0.223}{\sqrt{0.223^2 + 4\pi^2}} = 1.42 \text{ kg/s}$$

The natural frequency of the undamped system is

$$\omega_0 = \sqrt{\frac{K}{M}} = \sqrt{\frac{100}{4}} = 5 \text{ rad/s}$$

The maximum deflection has been expressed by Eq. 1.9.24. It is now calculated to be

$$\Delta_{\max} = \frac{2F_0 M}{C\sqrt{4M^2\omega_0^2 - C^2}}$$
$$= \frac{2 \times 10 \times 4}{1.42\sqrt{4 \times 4^2 \times 5^2 - 1.42^2}} = \underline{1.41 \text{ m}}$$

Example 1.26

For the network shown in Fig. 1.12, using Kirchhoff's laws, determine the currents $i_1(t)$ and $i_2(t)$, assuming all currents to be zero at $t = 0$.

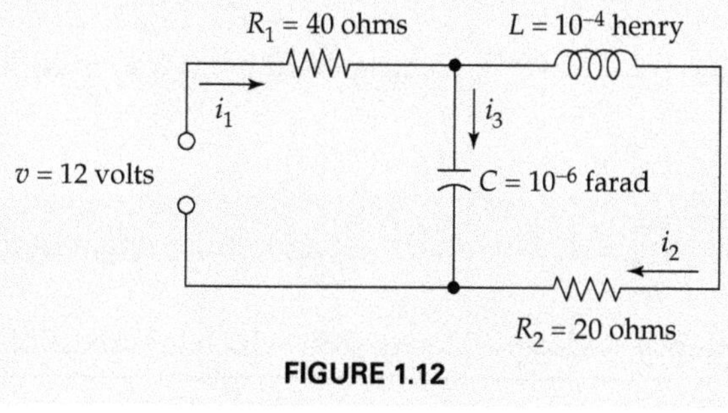

FIGURE 1.12

Solution

Using Kirchhoff's first law on the circuit on the left, we find that (see Eq. 1.4.3)

$$40i_1 + \frac{q}{10^{-6}} = 12 \tag{1}$$

where q is the charge on the capacitor. For the circuit around the outside of the network, we have

$$40i_1 + 10^{-4}\frac{di_2}{dt} + 20i_2 = 12 \tag{2}$$

Kirchhoff's second law requires that

$$i_1 = i_2 + i_3 \tag{3}$$

Using the relationship

$$i_3 = \frac{dq}{dt} \tag{4}$$

and the initial conditions, that $i_1 = i_2 = i_3 = 0$ at $t = 0$, we can solve the set of equations above. To do this, substitute (4) and (1) into (3). This gives

$$\frac{1}{40}(12 - 10^6 q) - \frac{dq}{dt} = i_2$$

Substituting this and (1) into (2) results in

$$10^{-4}\frac{d^2q}{dt^2} + 22.5\frac{dq}{dt} + 1.5 \times 10^6 q = 6$$

The appropriate initial conditions can be found from (1) and (4) to be $q = 12 \times 10^{-6}$ and $dq/dt = 0$ at $t = 0$. Solving the equation above, using the methods of this chapter, gives the charge as

$$q(t) = e^{-1.12\times 10^5 t}[c_1 \cos 48\,200t + c_2 \sin 48\,200t] + 4 \times 10^{-6}$$

The initial conditions allow the constants to be evaluated. They are

$$c_1 = 8 \times 10^{-6}, \qquad c_2 = 0.468 \times 10^{-6}$$

The current $i_1(t)$ is then found using (1) to be

$$i_1(t) = 0.2 - e^{-1.12\times 10^5 t}[0.2\cos 48\,200t + 0.468 \sin 48\,200t]$$

The current $i_2(t)$ is found by using (4) and (3). It is

$$i_2(t) = 0.2 + e^{-1.12\times 10^5 t}[-0.2\cos 48\,200t + 2.02 \sin 48\,200t]$$

Note the high frequency and rapid decay rate, which is typical of electrical circuits.

1.10 PERIODIC INPUT FUNCTIONS—FOURIER SERIES

We have learned thus far that nonhomogeneous differential equations with constant coefficients, containing sinusoidal input functions (e.g., $A \sin\omega t$) can be solved quite easily for any input frequency. There are many examples, however, of periodic input functions that are not sinusoidal but may be as shown in Fig. 1.13. The voltage input to a circuit or the force on a spring–mass system may be periodic but possess discontinuities such as those of Fig. 1.13. The object of this section is to present a technique for solving such problems. Other more complicated inputs may, of course, also act on a system.

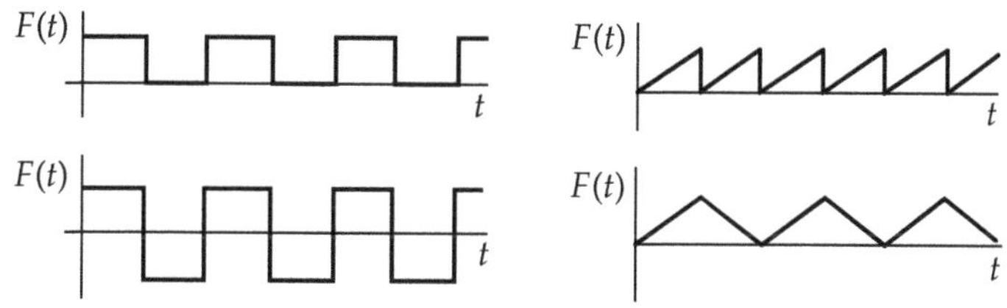

FIGURE 1.13 Various period input functions.

We will apply the fundamental *Fourier theorem*, which states that:

If a function $f(t)$ is a bounded periodic function of period $2T$, and if in any period it has a finite maxima and minima and a finite number of discontinuities, then the Fourier series

$$f(t) = \frac{a_0}{2} + \sum_{n=1}^{\infty}\left[a_n \cos\frac{n\pi t}{T} + b_n \sin\frac{n\pi t}{T}\right] \tag{1.10.1}$$

where

$$\begin{aligned} a_0 &= \frac{1}{T}\int_{-T}^{T} f(t)dt, \quad a_n = \frac{1}{T}\int_{-T}^{T} f(t)\cos\frac{n\pi t}{T}dt \\ b_n &= \frac{1}{T}\int_{-T}^{T} f(t)\sin\frac{n\pi t}{T}dt, \quad n = 1, 2, 3, \cdots \end{aligned} \tag{1.10.2}$$

More generally, one can integrate from t_0 to $t_0 + 2T$ if a specific choice of t_0 is convenient. Here, $t_0 = -T$.

converges to $f(t)$ at all points where $f(t)$ is continuous, and converges to the average of the right- and left-hand limits of $f(t)$ at each point where $f(t)$ is discontinuous.

To show that the expression above for the coefficients a_n is indeed true, let us multiply Eq. 1.10.1 by $\cos(m\pi t/T)dt$, where m is a fixed positive integer, and integrate from $-T$ to T. We have

$$\begin{aligned} \int_{-T}^{T} f(t)\cos\frac{m\pi t}{T}dt &= \frac{a_0}{2}\int_{-T}^{T}\cos\frac{m\pi t}{T}dt \\ &+ \sum_{n=1}^{\infty}\left[a_n\int_{-T}^{T}\cos\frac{n\pi t}{T}\cos\frac{m\pi t}{T}dt + b_n\int_{-T}^{T}\sin\frac{n\pi t}{T}\cos\frac{m\pi t}{T}dt\right] \end{aligned} \tag{1.10.3}$$

Consider the three integrals on the right in this equation. We have*

$$\int_{-T}^{T} \cos\frac{m\pi t}{T}dt = \frac{T}{m\pi}\sin\frac{m\pi t}{T}\Bigg|_{-T}^{T} = 0$$

$$\int_{-T}^{T} \sin\frac{n\pi t}{T}\cos\frac{m\pi t}{dt}dt = \frac{1}{2}\int_{-T}^{T}\left[\sin(n+m)\frac{\pi t}{T} - \sin(n-m)\frac{\pi t}{T}\right]dt$$

$$= \frac{1}{2}\left[-\frac{T}{\pi(n+m)}\cos(n+m)\frac{\pi t}{T}\Bigg|_{-T}^{T} + \frac{T}{\pi(n-m)}\cos(n-m)\frac{\pi t}{T}\Bigg|_{-T}^{T}\right] = 0 \qquad (1.10.4)$$

$$\int_{-T}^{T} \cos\frac{n\pi t}{T}\cos\frac{m\pi t}{T}dt = \frac{1}{2}\int_{-T}^{T}\left[\cos(n+m)\frac{\pi t}{T} + \cos(n-m)\frac{\pi t}{T}\right]dt$$

$$= \frac{1}{2}\left[\frac{T}{\pi(n+m)}\sin(n+m)\frac{\pi t}{T}\Bigg|_{-T}^{T} + \frac{T}{\pi(n-m)}\sin(n-m)\frac{\pi t}{T}\Bigg|_{-T}^{T}\right]$$

This last integral is also zero if $n \neq m$. However, when $n = m$ it has a nonzero value. It is found by simply letting $n = m$ in the integral on the right to obtain a value of T. Thus, we see that the right-hand side of Eq. 1.10.3 is equal to $a_m T$. Our result is

$$a_m = \frac{1}{T}\int_{-T}^{T} f(t)\cos\frac{m\pi t}{T}dt, \quad m = 1, 2, 3, \cdots \qquad (1.10.5)$$

Multiplying Eq. 1.10.1 by $\sin(m\pi t/T)dt$, integrating from $-T$ to T, and following a procedure similar to the one above, we can show that

$$b_m = \frac{1}{T}\int_{-T}^{T} f(t)\sin\frac{m\pi t}{T}dt, \qquad m = 1, 2, 3, \cdots \qquad (1.10.6)$$

To complete this discussion we must verify the expression for a_0. If we simply multiply Eq. 1.10.1 by dt and integrate from $-T$ to T, there results

$$\int_{-T}^{T} f(t)dt = \int_{-T}^{T}\frac{a_0}{2}dt + \sum_{n=1}^{\infty}\left[a_n\int_{-T}^{T}\cos\frac{n\pi t}{T}dt + b_n\int_{-T}^{T}\sin\frac{n\pi t}{T}dt\right]$$

$$= \frac{a_0}{2}(2T) + \sum_{n=1}^{\infty}\left[a_n\frac{T}{n\pi}\sin\frac{n\pi t}{T}\Bigg|_{-T}^{T} - b_n\frac{T}{n\pi}\cos\frac{n\pi t}{T}\Bigg|_{-T}^{T}\right] = a_0 T \qquad (1.10.7)$$

This results in

$$a_0 = \frac{1}{T}\int_{-T}^{T} f(t)dt \qquad (1.10.8)$$

The expressions for the coefficients in the Fourier series have thus been verified.

The class of functions that can be represented by the Fourier series is quite large. The function may consist of a number of disjointed pieces of various curves, each represented by a different equation, and still possess a Fourier series expansion. We shall consider two special classes of functions, following some examples.

---

*We shall make use of the trigonometric identities, $\sin\alpha\,\cos\beta = 1/2\,[\sin(\alpha+\beta) - \sin(\alpha-\beta)]$ and $\cos\alpha\cos\beta = 1/2\,[\cos(\alpha+\beta) + \cos(\alpha-\beta)]$.

Example 1.27

Write the Fourier series representation of the periodic function $f(t)$ of Fig. 1.14 if in one period

$$f(t) = t, \qquad -\pi < t < \pi$$

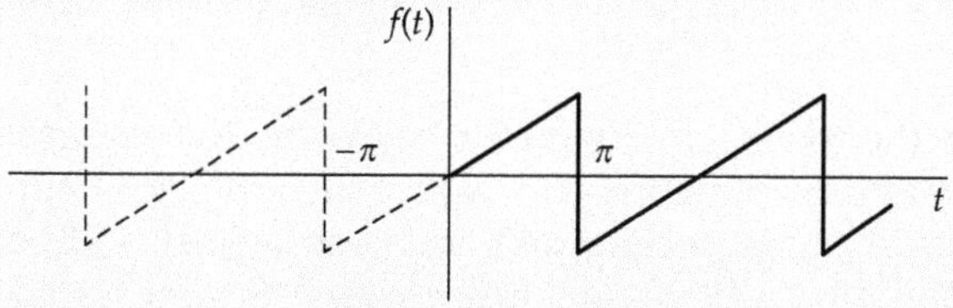

FIGURE 1.14

Solution

For this example, $T = \pi$. Formula (1.10.2) for the a_n provides us with

$$\begin{aligned} a_0 &= \frac{1}{\pi}\int_{-\pi}^{\pi} f(t)dt \\ &= \frac{1}{\pi}\int_{-\pi}^{\pi} t\,dt = \left.\frac{t^2}{2\pi}\right|_{-\pi}^{\pi} = 0 \\ a_n &= \frac{1}{\pi}\int_{-\pi}^{\pi} f(t)\cos nt\,dt, \qquad n = 1, 2, 3, \cdots \\ &= \frac{1}{\pi}\int_{-\pi}^{\pi} t\cos nt\,dt = \frac{1}{\pi}\left[\frac{t}{n}\sin nt + \frac{1}{n^2}\cos nt\right]_{-\pi}^{\pi} = 0 \end{aligned}$$

recognizing that $\cos n\pi = \cos(-n\pi)$ and $\sin n\pi = -\sin(-n\pi) = 0$. Also, in performing the integration, we have integrated by parts.* For b_n we have

$$\begin{aligned} b_n &= \frac{1}{\pi}\int_{-\pi}^{\pi} f(t)\sin nt\,dt, \qquad n = 1, 2, 3, \cdots \\ &= \frac{1}{\pi}\int_{-\pi}^{\pi} t\sin nt\,dt = \frac{1}{\pi}\left[-\frac{t}{n}\cos nt + \frac{1}{n^2}\sin nt\right]_{-\pi}^{\pi} = -\frac{2}{n}\cos n\pi \end{aligned}$$

The Fourier series representation has only sine terms. It is given by

$$f(t) = -2\sum_{n=1}^{\infty}\frac{(-1)^n}{n}\sin nt$$

---

*To integrate $\int_{-\pi}^{\pi} t\cos t\,dt$, integrate by parts; that is, let $u = t$ and $dv = \cos t\ dt$. Then $du = dt$ and $v = \sin t$. Thus,

$$\int_{-\pi}^{\pi} t\cos t\,dt = t\sin t\Big|_{-\pi}^{\pi} - \int_{-\pi}^{\pi}\sin t\,dt = 0 + [\cos\pi - \cos(-\pi)] = -1-(-1) = 0$$

where we have used $\cos n\pi = (-1)^n$. Writing out several terms, we have

$$f(t) = -2\left[-\sin t + \frac{1}{2}\sin 2t - \frac{1}{3}\sin 3t + \cdots\right]$$
$$= 2\sin t - \sin 2t + \frac{2}{3}\sin 3t - \cdots$$

Note the sketch in Fig. 1.15, showing the increasing accuracy with which the terms approximate the $f(t)$. Notice also the close approximation using three terms. Obviously, using a computer and keeping, say, 15 terms, a remarkably good approximation can result using Fourier series. (See the next example.)

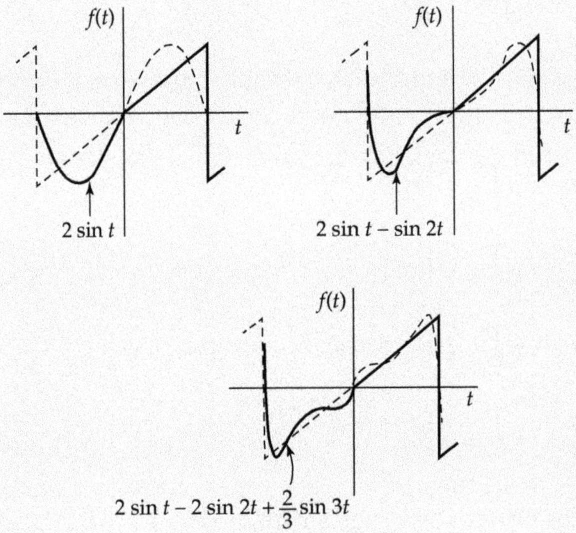

FIGURE 1.15

Example 1.28

The preceding example illustrated a Fourier series expansion for a function $f(t)$ that was continuous throughout each period. Let us now find the Fourier series expansion for the periodic function $f(t)$ of Fig. 1.16 if in one period

$$f(t) = \begin{cases} 0, & -\pi < t < 0 \\ t, & 0 < t < \pi \end{cases}$$

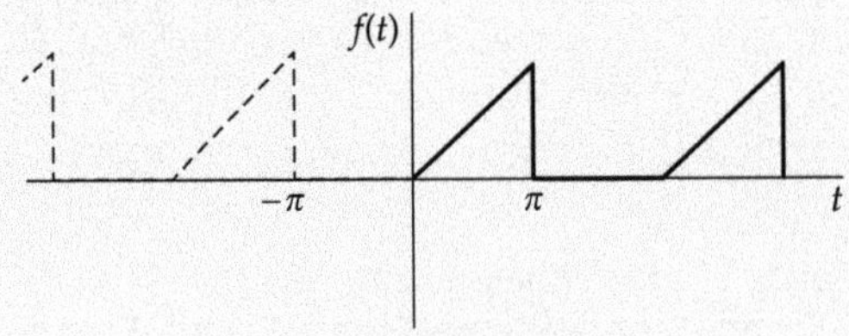

FIGURE 1.16

Solution

The period is again 2π; thus, $T = \pi$. The Fourier coefficients are given by

$$a_0 = \frac{1}{\pi}\int_{-\pi}^{\pi} f(t)\,dt = \frac{1}{\pi}\int_{-\pi}^{\pi} t\,dt = \frac{\pi}{2}$$

$$a_n = \frac{1}{\pi}\int_{-\pi}^{\pi} f(t)\cos nt\,dt = \frac{1}{\pi}\int_{-\pi}^{0} 0\cos nt\,dt + \frac{1}{\pi}\int_{0}^{\pi} t\cos nt\,dt$$

$$= \frac{1}{\pi}\left[\frac{t}{n}\sin nt + \frac{1}{n^2}\cos nt\right]_0^{\pi} = \frac{1}{\pi n^2}(\cos n\pi - 1), \qquad n = 1, 2, 3, \cdots$$

$$b_n = \frac{1}{\pi}\int_{-\pi}^{\pi} f(t)\sin nt\,dt = \frac{1}{\pi}\int_{-\pi}^{0} 0\sin nt\,dt + \frac{1}{\pi}\int_{0}^{\pi} t\sin nt\,dt$$

$$= \frac{1}{\pi}\left[-\frac{t}{n}\cos nt + \frac{1}{n^2}\sin nt\right]_0^{\pi} = -\frac{1}{n}\cos n\pi, \qquad n = 1, 2, 3, \cdots$$

The Fourier series representation is, then, using $\cos n\pi = (-1)^n$,

$$f(t) = \frac{\pi}{4} + \sum_{n=1}^{\infty}\left[\frac{(-1)^n - 1}{\pi n^2}\cos nt - \frac{(-1)^n}{n}\sin nt\right]$$

$$= \frac{\pi}{4} - \frac{2}{\pi}\cos t - \frac{2}{9\pi}\cos 3t + \cdots + \sin t - \frac{1}{2}\sin 2t + \frac{1}{3}\sin 3t + \cdots$$

A computer program was written with n equal to 5, 10, and 20, respectively, in the equation above. The results are plotted on the graph of Fig. 1.17.

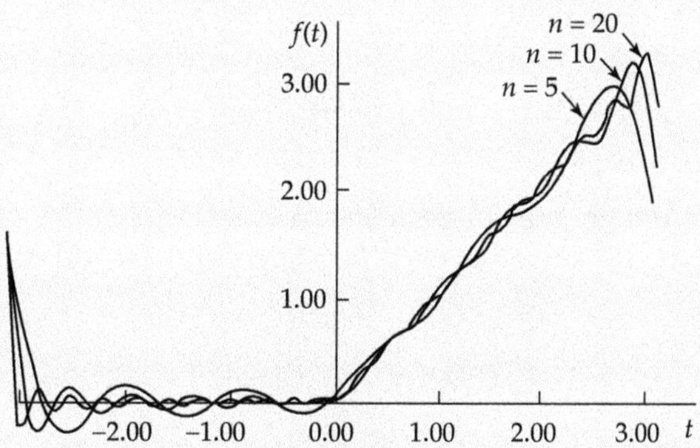

FIGURE 1.17

1.10.1 Even and Odd Functions

The Fourier series expansion can be accomplished with less effort if we recognize whether a function is even or odd. Recall that an even function requires that

$$f(-t) = f(t) \tag{1.10.9}$$

resulting in a symmetric graph with respect to the vertical axis. For an odd function,

$$f(-t) = -f(t) \tag{1.10.10}$$

The cosine is an even function, whereas the sine is an odd function.

Even and odd functions are displayed in Fig. 1.18. Observe that for an even function, the slope is zero at $t = 0$, whereas for an odd function, the function is zero at $t = 0$ (for a discontinuous function, its average would be zero). Two additional properties of even and odd functions must be noted.

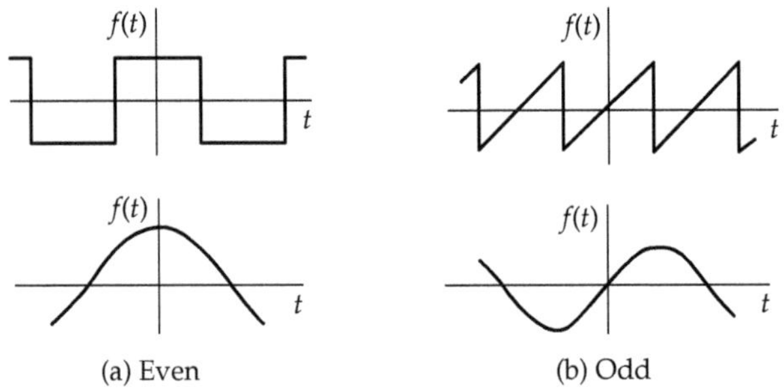

FIGURE 1.18 Even and odd functions.

First, if $f(t)$ is an even function,

$$\int_{-T}^{T} f(t)dt = 2\int_{0}^{T} f(t)dt \tag{1.10.11}$$

For an odd function,

$$\int_{-T}^{T} g(t)dt = 0 \tag{1.10.12}$$

as can be observed from net area under the curves in Fig. 1.18.

Second, the product $h = fg$ of an even function f and an odd function g is odd. This is shown by

$$h(-t) = f(-t)g(-t) = f(t)[-g(t)] = -h(t) \tag{1.10.13}$$

Conversely, the product of two even functions, or two odd functions, is an even function.

Hence, an *even function* $f(t)$ has the Fourier series expansion

$$f(t) = \frac{a_0}{2} + \sum_{n=1}^{\infty} a_n \cos\frac{n\pi t}{T} \tag{1.10.14}$$

where

$$a_0 = \frac{2}{T}\int_{0}^{T} f(t)dt, \quad a_n = \frac{2}{T}\int_{0}^{T} f(t)\cos\frac{n\pi t}{T}dt, \qquad n = 1, 2, 3, \cdots \tag{1.10.15}$$

For an *odd function* $f(t)$, the Fourier series takes the form

$$f(t) = \sum_{n=1}^{\infty} b_n \sin\frac{n\pi t}{T} \qquad (1.10.16)$$

where

$$b_n = \frac{2}{T}\int_0^T f(t)\sin\frac{n\pi t}{T}dt, \qquad n = 1, 2, 3, \cdots \qquad (1.10.17)$$

Observe that a function $f(t)$ is not intrinsically even or odd. Consider the function displayed in Fig. 1.19. It is even if the vertical axis is located as shown in part (a), it is odd if the vertical is located as in part (b), and it is neither even nor odd in part (c). If the $t = 0$ location is unimportant, we may expand a function in terms of cosines only (an even function), or in terms of sines only (an odd function), depending on where $t = 0$ is located.

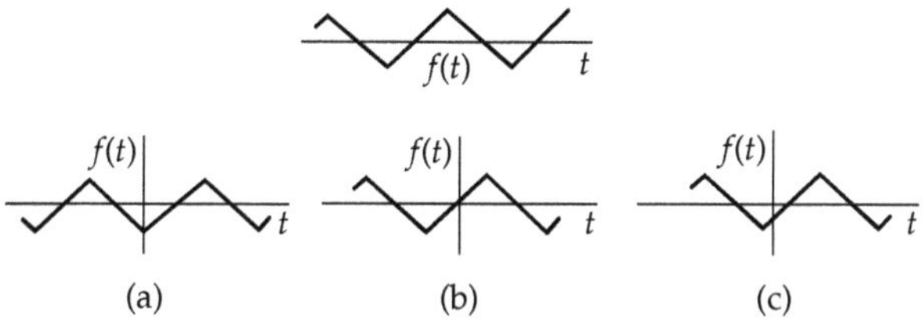

FIGURE 1.19 A function $f(t)$.

Example 1.29

The periodic forcing function shown in Fig. 1.20 acts on a spring–mass system. Find a sine-series representation by considering the function to be odd, and a cosine-series representation by considering the function to be even.

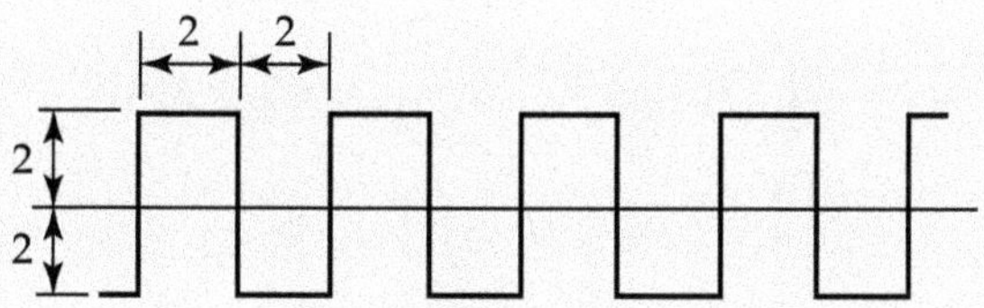

FIGURE 1.20

Solution

If the $t = 0$ location is selected as shown in Fig. 1.21, the resulting odd function can be written, for one period, as

$$f(t) = \begin{cases} -2, & -2 < t < 0 \\ 2, & 0 < t < 2 \end{cases}$$

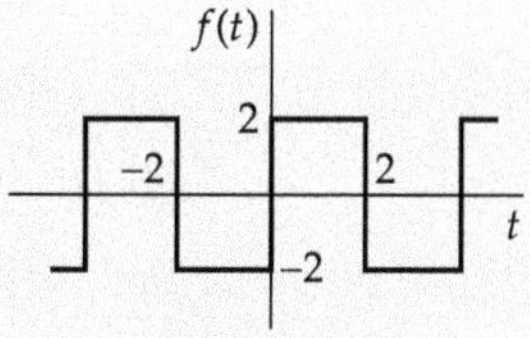

FIGURE 1.21

For an odd function we know that

$$a_n = 0$$

Hence, we are left with the task of finding b_n. We have, using $T = 2$,

$$b_n = \frac{2}{T}\int_0^T f(t)\sin\frac{n\pi t}{T}dt, \quad n = 1, 2, 3, \cdots$$
$$= \frac{2}{2}\int_0^2 2\sin\frac{n\pi t}{2}dt = -\frac{4}{n\pi}\cos\frac{n\pi t}{2}\bigg|_0^2 = -\frac{4}{n\pi}(\cos n\pi - 1)$$

The Fourier sine series is, then, again substituting $\cos n\pi = (-1)^n$,

$$f(t) = \sum_{n=1}^{\infty}\frac{4\left[1-(-1)^n\right]}{n\pi}\sin\frac{n\pi t}{2}$$
$$= \frac{8}{\pi}\sin\frac{\pi t}{2} - \frac{8}{3\pi}\sin\frac{3\pi t}{2} + \frac{8}{5\pi}\sin\frac{5\pi t}{2} - \cdots$$

If we select the $t = 0$ location as displayed in Fig. 1.22, an even function results. Over one period it is

$$f(t) = \begin{cases} -2, & -2 < t < -1 \\ 2, & -1 < t < 1 \\ -2, & 1 < t < 2 \end{cases}$$

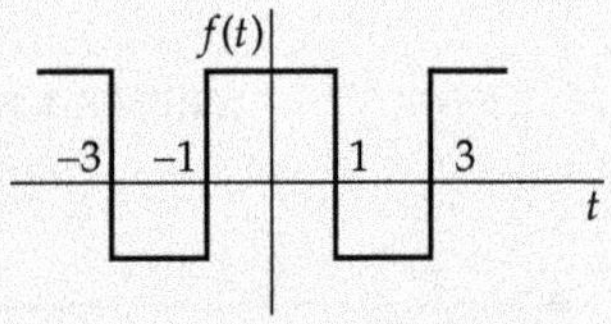

FIGURE 1.22

For an even function we know that

$$b_n = 0$$

The coefficients a_n are found from

$$a_n = \frac{2}{T}\int_0^T f(t)\cos\frac{n\pi t}{T}dt, \qquad n = 1,2,3,\cdots$$
$$= \frac{2}{2}\left[\int_0^1 2\cos\frac{n\pi t}{2}dt + \int_0^2 (-2)\cos\frac{n\pi t}{2}dt\right]$$
$$= \frac{4}{n\pi}\sin\frac{n\pi t}{2}\bigg|_0^1 - \frac{4}{n\pi}\sin\frac{n\pi t}{2}\bigg|_1^2 = \frac{8}{n\pi}\sin\frac{n\pi}{2}$$

The result for $n = 0$ is found from

$$a_0 = \frac{2}{T}\int_0^T f(t)dt$$
$$= \frac{2}{2}\left[\int_0^1 2dt + \int_0^2 (-2)dt\right] = 2 - 2 = 0$$

Finally, the Fourier cosine series is

$$f(t) = \sum_{n=1}^{\infty} \frac{8}{n\pi} \sin\frac{n\pi}{2} \cos\frac{n\pi t}{2}$$
$$= \frac{8}{\pi}\cos\frac{\pi t}{2} - \frac{8}{3\pi}\cos\frac{3\pi t}{2} + \frac{8}{5\pi}\cos\frac{5\pi t}{2} + \cdots$$

1.10.2 Half-Range Expansions

In modeling some physical phenomena it is necessary that we consider the values of a function only in the interval 0 to T. This is especially true when solving partial differential equations, as we shall do in Chapter 6. There is no condition of periodicity in the function, since there is no interest in the function outside the interval 0 to T. Consequently, we can extend the function arbitrarily to include the interval $-T$ to 0. Consider the function $f(t)$ shown in Fig. 1.23a. If we extend it as in part (b), an even function results; an extension as in part (c) would provide an odd function. A Fourier series expansion consisting of cosines for the function of part (b) would converge to the given $f(t)$ in the interval 0 to T, as would the Fourier series expansion consisting of sine terms for the function of part (c). In other words, we could expand the $f(t)$ of Fig. 1.23a as an even function using Eqs. 1.10.14 and 1.10.15, or as an odd function using Eqs. 1.10.16 and 1.10.17. Such series expansions are known as *half-range expansions*. An example will illustrate such expansions.

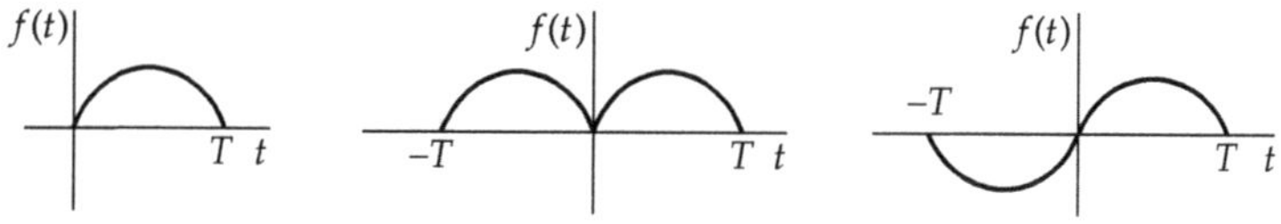

(a) $f(t)$ (b) Even function (c) Odd function

FIGURE 1.23 Extension of a function.

Example 1.30

A function $f(t)$ is defined only over the range $0 < t < 4$ is displayed in Fig. 1.24.

$$f(t) = \begin{cases} t, & 0 < t < 2 \\ 4 - t, & 2 < t < 4 \end{cases}$$

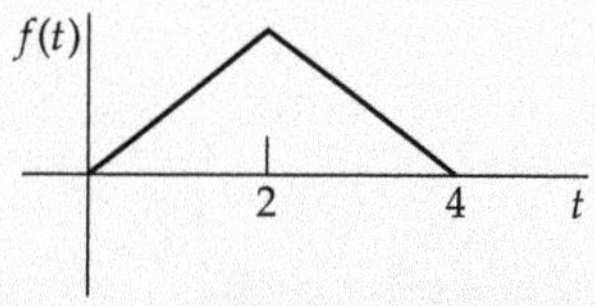

FIGURE 1.24

Find the half-range cosine and sine expansions of $f(t)$.

Solution

A half-range cosine expansion would be found by forming a symmetric extension to $f(t)$. The b_n of the Fourier series would be zero. The coefficients a_n are, from Eq. 1.10.15,

$$\begin{aligned} a_n &= \frac{2}{T}\int_0^T f(t)\cos\frac{n\pi t}{T}dt, \qquad n = 1, 2, 3, \ldots \\ &= \frac{2}{4}\int_0^2 t\cos\frac{n\pi t}{4}dt + \frac{2}{4}\int_2^4 (4-t)\cos\frac{n\pi t}{4}dt \\ &= \frac{1}{2}\left[\frac{4t}{n\pi}\sin\frac{n\pi t}{4} + \frac{16}{\pi^2 n^2}\cos\frac{n\pi t}{4}\right]_0^2 + \frac{1}{2}\left[\frac{16}{n\pi}\sin\frac{n\pi t}{4}\right]_2^4 \\ &\qquad -\frac{1}{2}\left[\frac{4t}{n\pi}\sin\frac{n\pi t}{4} + \frac{16}{n^2\pi^2}\cos\frac{n\pi t}{4}\right]_2^4 \\ &= -\frac{8}{n^2\pi^2}\left[1 + \cos n\pi - 2\cos\frac{n\pi}{2}\right] \end{aligned}$$

For $n = 0$ the coefficient a_0 is

$$a_0 = \frac{1}{2}\int_0^2 t\,dt + \frac{1}{2}\int_2^4 (4-t)dt = 2$$

The half-range cosine expansion is then

$$\begin{aligned} f(t) &= 1 + \sum_{n=1}^{\infty}\frac{8}{n^2\pi^2}\left(2\cos\frac{n\pi}{2} - \cos n\pi - 1\right)\cos\frac{n\pi t}{4} \\ &= 1 - \frac{8}{\pi^2}\left[\cos\frac{\pi t}{2} + \frac{1}{9}\cos\frac{3\pi t}{2} + \cdots\right] \end{aligned}$$

It is an even periodic extension that graphs as in Fig. 1.25.

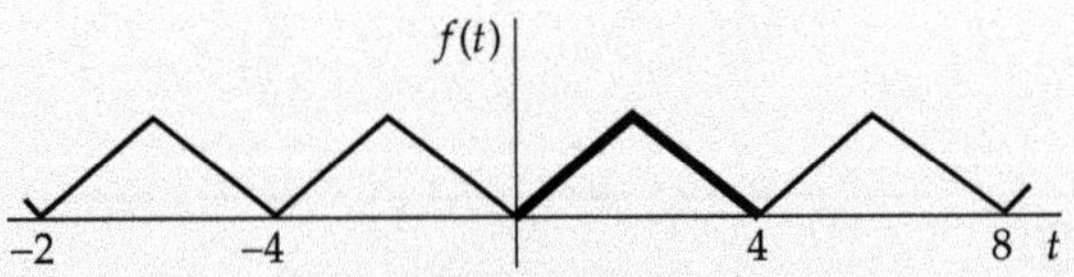

FIGURE 1.25

For the half-range sine expansion of $f(t)$, all a_n would be zero. The coefficients b_n are, from Eq. 1.10.17,

$$\begin{aligned} b_n &= \frac{2}{T}\int_0^T f(t)\sin\frac{n\pi t}{T}dt, \qquad n = 1, 2, 3, \ldots \\ &= \frac{2}{4}\int_0^2 t\sin\frac{n\pi t}{4}dt + \frac{2}{4}\int_2^4 (4-t)\sin\frac{n\pi t}{4}dt = \frac{8}{n^2\pi^2}\sin\frac{n\pi}{2} \end{aligned}$$

The half-range sine expansion is then

$$f(t) = \sum_{n=1}^{\infty} \frac{8}{n^2\pi^2} \sin\frac{n\pi}{2} \sin\frac{n\pi t}{4}$$
$$= \frac{8}{\pi^2}\left[\sin\frac{\pi t}{4} - \frac{1}{9}\sin\frac{3\pi t}{4} + \frac{1}{25}\sin\frac{5\pi t}{4} - \cdots\right]$$

This odd periodic extension appears as in Fig 1.26.

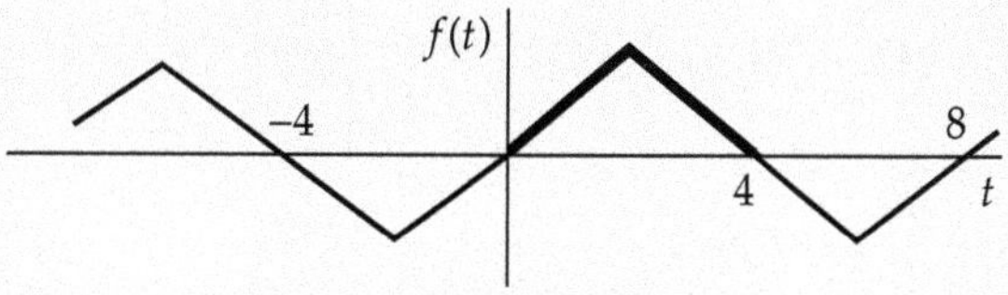

FIGURE 1.26

Both series would provide us with a good approximation to $f(t)$ in the interval $0 < t < 4$ if a sufficient number of terms are retained in each series. One would expect the accuracy of the sine series to be better than that of the cosine series for a given number of terms, since fewer discontinuities exist in the sine series. This is generally the case; if we make the extension smooth, greater accuracy results for a particular number of terms.

1.10.3 Forced Oscillations

We shall now consider an important application involving an external force acting on a spring–mass system. The differential equation describing this motion is

$$M\frac{d^2y}{dt^2} + C\frac{dy}{dt} + Ky = F(t) \tag{1.10.18}$$

If the input function $F(t)$ is a sine or cosine function, the steady-state solution is a harmonic motion having the frequency of the input function. We will now see that if $F(t)$ is periodic with frequency ω but is not a sine or cosine function, then the steady-state solution to Eq. 1.10.18 will contain the input frequency ω and multiples of this frequency contained in the terms of a Fourier series expansion of $F(t)$. If one of these higher frequencies is close to the natural frequency of an underdamped system, then the particular term containing that frequency may play the dominant role in the system response. This is somewhat surprising, since the input frequency may be considerably lower than the natural frequency of the system; yet that input could lead to serious problems if it is not purely sinusoidal. This will be illustrated with an example.

Example 1.31

Consider the force $F(t)$ to act on the spring–mass system shown in Fig. 1.27. Determine the steady-state response to such a forcing function.

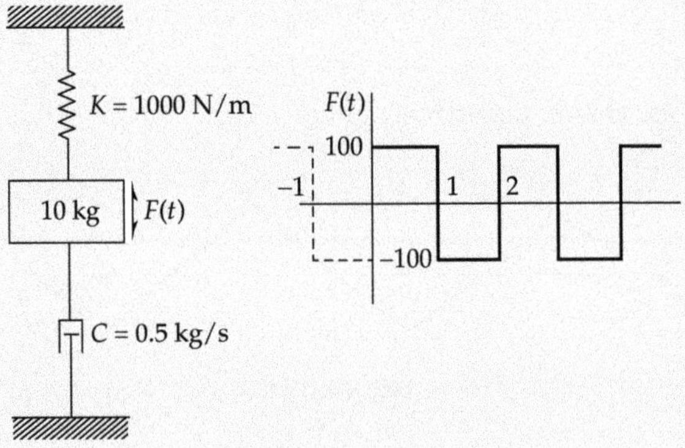

FIGURE 1.27

Solution

The coefficients in the Fourier series expansion to the odd forcing function $F(t)$ are (see Example 1.25)

$$a_n = 0$$

$$b_n = \frac{2}{1}\int_0^1 100 \sin\frac{n\pi t}{1}dt = -\frac{200}{n\pi}\cos n\pi t\bigg|_0^1 = -\frac{200}{n\pi}(\cos n\pi - 1), \qquad n = 1, 2, 3, \cdots$$

The Fourier series representation of $F(t)$ is then

$$\begin{aligned} F(t) &= \sum_{n=1}^{\infty}\frac{200}{n\pi}(1 - \cos n\pi)\sin n\pi t \\ &= \frac{400}{\pi}\sin \pi t - \frac{400}{3\pi}\sin 3\pi t + \frac{80}{\pi}\sin 5\pi t - \cdots \end{aligned}$$

The differential equation can be written as

$$10\frac{d^2y}{dt^2} + 0.5\frac{dy}{dt} + 1000y = \frac{400}{\pi}\sin \pi t - \frac{400}{3\pi}\sin 3\pi t + \frac{80}{\pi}\sin 5\pi t - \cdots$$

Because the differential equation is linear, we can first find the particular solution $(y_p)_1$ corresponding to the first term on the right, then $(y_p)_2$ corresponding to the second term, and so on. Finally, the steady-state solution is

This is an application of the principle of superposition.

$$y_p(t) = (y_p)_1 + (y_p)_2 + \cdots$$

Doing this for the three terms shown, using the methods developed in section 1.9.3, we have

$$
\begin{aligned}
(y_p)_1 &= 0.141 \sin \pi t - 2.5 \times 10^{-4} \cos \pi t \\
(y_p)_2 &= -0.376 \sin 3\pi t + 1.56 \times 10^{-3} \cos 3\pi t \\
(y_p)_3 &= -0.0174 \sin 5\pi t - 9.35 \times 10^{-5} \cos 5\pi t
\end{aligned}
$$

Actually, rather than solving the problem each time for each term, we could have found a $(y_p)_n$ corresponding to the term $[-(200/n\pi)(\cos n\pi - 1) \sin n\pi t]$ as a general function of n. Note the amplitude of the sine term in $(y_p)_2$. It obviously dominates the solution, as displayed in a sketch of $y_p(t)$ in Fig. 1.28. Yet it has an annular frequency of 3π rad/s,

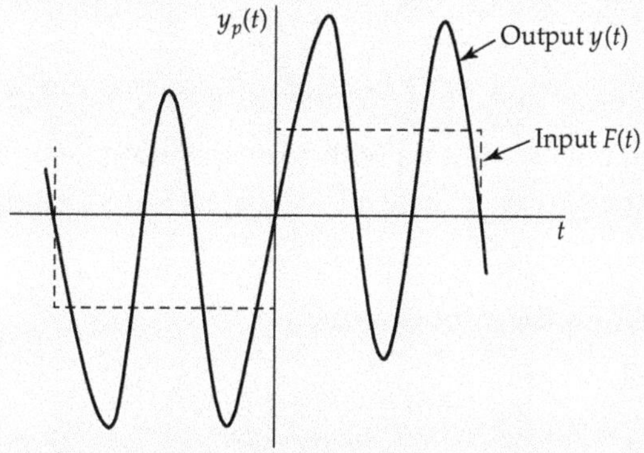

FIGURE 1.28

whereas the frequency of the input function was π rad/s. This happened since the natural frequency of the undamped system was 10 rad/s, very close to the frequency of the second sine term in the Fourier series expansion. Hence, it was this overtone that resonated with the system, and not the fundamental. Overtones may dominate the steady-state response for any underdamped system that is forced with a periodic function having a frequency smaller than the natural frequency of the system. The galloping power line is an example; when ice coats a power line, the wind can create shedding vortices with a shedding frequency that results in the power line undergoing an oscillation referred to as galloping. Google "galloping power lines" and watch the phenomenon.

1.11 THE CAUCHY EQUATION

In the preceding sections we have discussed differential equations with constant coefficients. In this section we shall present the solution to a class of second-order differential equations with variable coefficients. Such a class of equations is called the *Cauchy equation of order* 2.* It is

$$
x^2 \frac{d^2u}{dx^2} + ax\frac{du}{dx} + bu = 0 \tag{1.11.1}
$$

*Cauchy's equation is of nth order and can be written as

$$
x^n \frac{d^n u}{dx^n} + a_1 x^{n-1} \frac{d^{n-1} u}{dx^{n-1}} + \cdots + a_n u = f(x)
$$

The solution has the general form

$$u(x) = x^m \tag{1.11.2}$$

This is substituted into Eq. 1.11.1 to obtain

$$x^2 m(m-1)x^{m-2} + axmx^{m-1} + bx^m = 0 \tag{1.11.3}$$

or, equivalently,

$$[m(m-1) + am + b]x^m = 0 \tag{1.11.4}$$

By setting the quantity in brackets equal to zero, we can find two roots for m. This *auxiliary equation*, written as

$$m^2 + (a-1)m + b = 0 \tag{1.11.5}$$

yields the two unique roots m_1 and m_2 with corresponding independent solutions

$$u_1 = x^{m_1} \quad \text{and} \quad u_2 = x^{m_2} \tag{1.11.6}$$

The general solution, for unique roots, is then

$$u(x) = c_1 x^{m_1} + c_2 x^{m_2} \tag{1.11.7}$$

If a double root results from the auxiliary equation, that is, $m_1 = m_2$, then u_1 and u_2 are not independent and Eq. 1.11.7 is not the general solution. To find a second independent solution, assuming that $u_1 = x^m$ is one independent solution, we assume, as in Eq. 1.6.11,

$$u_2 = v(x)u_1 \tag{1.11.8}$$

Following the steps outlined in the equations following Eq. 1.6.11 we find that

$$v(x) = \ln x \tag{1.11.9}$$

The general solution, for double roots, is then

$$u(x) = (c_1 + c_2 \ln x)x^m \tag{1.11.10}$$

Two examples will illustrate.

Example 1.32

Find the general solution to the differential equation

$$x^2 \frac{d^2u}{dx^2} - 5x\frac{du}{dx} + 8u = 0$$

Solution

The auxiliary equation is

$$m^2 - 6m + 8 = 0$$

The two unique roots are

$$m_1 = 4, \qquad m_2 = 2$$

with corresponding independent solutions

$$u_1 = x^4, \qquad u_2 = x^2$$

The general solution is then

$$u(x) = c_1x^4 + c_2x^2$$

Example 1.33

Determine the specific solution to the differential equation

$$x^2\frac{d^2u}{dx^2} - 3x\frac{du}{dx} + 4u = 0$$

if the initial conditions are $u(1) = 2$ and $du/dx(1) = 8$.

Solution

The auxiliary equation is

$$m^2 - 4m + 4 = 0$$

A double root $m = 2$ occurs; thus, the general solution is (see Eq. 1.11.10)

$$u(x) = (c_1 + c_2 \ln x)x^2$$

To use the initial conditions, we must have du/dx. It is

$$\frac{du}{dx} = \left(\frac{c_2}{x}\right)x^2 + (c_1 + c_2 \ln x)2x$$

The initial conditions then give

$$2 = (c_1 + c_2 \overset{0}{\cancel{\ln 1}})1^2$$

$$8 = \left(\frac{c_2}{1}\right)1^2 + (c_1 + c_2 \overset{0}{\cancel{\ln 1}})2$$

These two equations result in

$$c_1 = 2, \qquad c_2 = 4$$

Finally, the specific solution is

$$u(x) = 2(1 + 2\ln x)x^2$$

1.12 VARIATION OF PARAMETERS

In Section 1.10 we discussed the particular solution associated with an input function that was periodic. In this section we shall present a method that will allow us to determine the particular solution associated with nonperiodic input functions that may differ from those of Section 1.8.

Consider the equation

$$\frac{d^2u}{dx^2} + f(x)\frac{du}{dx} + g(x)u = h(x) \tag{1.12.1}$$

The solution $u(x)$ is found by adding the particular solution $u_p(x)$ to the solution of the homogeneous equation, to obtain

$$u(x) = c_1u_1(x) + c_2u_2(x) + u_p(x) \tag{1.12.2}$$

where $u_1(x)$ and $u_2(x)$ are solutions to the homogeneous equation

$$\frac{d^2u}{dx^2} + f(x)\frac{du}{dx} + g(x)u = 0 \tag{1.12.3}$$

To find the particular solution, assume the form

$$u_p(x) = v_1(x)u_1(x) + v_2(x)u_2(x) \tag{1.12.4}$$

Differentiate and obtain

$$\frac{du_p}{dx} = v_1\frac{du_1}{dx} + v_2\frac{du_2}{dx} + u_1\frac{dv_1}{dx} + u_2\frac{dv_2}{dx} \tag{1.12.5}$$

We will seek a solution such that

$$u_1\frac{dv_1}{dx} + u_2\frac{dv_2}{dx} = 0 \tag{1.12.6}$$

We are free to impose this one restriction on $v_1(x)$ and $v_2(x)$ without loss of generality. Thus, we have

$$\frac{du_p}{dx} = v_1\frac{du_1}{dx} + v_2\frac{du_2}{dx} \tag{1.12.7}$$

Differentiating this equation again results in

$$\frac{d^2u_p}{dx^2} = v_1\frac{d^2u_1}{dx^2} + v_2\frac{d^2u_2}{dx^2} + \frac{dv_1}{dx}\frac{du_1}{dx} + \frac{dv_2}{dx}\frac{du_2}{dx} \tag{1.12.8}$$

Substituting into Eq. 1.12.1, we find that

$$v_1\left(\frac{d^2u_1}{dx^2} + f\frac{du_1}{dx} + gu_1\right) + v_2\left(\frac{d^2u_2}{dx^2} + f\frac{du_2}{dx} + gu_2\right) + \frac{dv_1}{dx}\frac{du_1}{dx} + \frac{dv_2}{dx}\frac{du_2}{dx} = h(x) \tag{1.12.9}$$

The quantities in parentheses are both zero since u_1 and u_2 are solutions of the homogeneous equation. Hence,

$$\frac{dv_1}{dx}\frac{du_1}{dx}+\frac{dv_2}{dx}\frac{du_2}{dx}=h(x) \tag{1.12.10}$$

This equation and Eq. 1.12.6 are solved simultaneously to find

$$\frac{dv_1}{dx}=-\frac{u_2 h(x)}{u_1\dfrac{du_2}{dx}-u_2\dfrac{du_1}{dx}}, \qquad \frac{dv_2}{dx}=\frac{u_1 h(x)}{u_1\dfrac{du_2}{dx}-u_2\dfrac{du_1}{dx}} \tag{1.12 11}$$

The quantity in the denominator is the *Wronskian W* of $u_1(x)$ and $u_2(x)$,

$$W(x)=u_1\frac{du_2}{dx}-u_2\frac{du_1}{dx} \tag{1.12.12}$$

We can now integrate Eqs. 1.12.11 and obtain

$$v_1(x)=-\int\frac{u_2 h}{W}dx, \qquad v_2(x)=\int\frac{u_1 h}{W}dx \tag{1.12.13}$$

The particular solution is then

$$u_p(x)=-u_1\int\frac{u_2 h}{W}dx+u_2\int\frac{u_1 h}{W}dx \tag{1.12.14}$$

The general solution of the nonhomogeneous equation follows. This technique is referred to as the *method of variation of parameters*.

Example 1.34

The general solution of $(d^2u/dx^2)+u=x^2$ was found in Example 1.15. Find the particular solution to this equation using the method of variation of parameters.

Solution

The two independent homogeneous solutions are

$$u_1(x)=\sin x, \qquad u_2(x)=\cos x$$

The Wronskian is then

$$\begin{aligned}W(x)&=u_1\frac{du_2}{dx}-u_2\frac{du_1}{dx}\\&=-\sin^2 x-\cos^2 x=-1\end{aligned}$$

The particular solution is then found from Eq. 1.12.14 to be

$$u_p(x)=\sin x\int x^2\cos x\,dx-\cos x\int x^2\sin x\,dx$$

This is integrated by parts twice to give

$$u_p(x)=x^2-2$$

This, of course, agrees with the result of Example 1.15.

1.13 MISCELLANEOUS INFORMATION

There are several other relationships and techniques that are useful when solving differential equations. Occasionally, we are interested in changing variables from x and y to ξ and η. This *transformation* of variables is accomplished by expressing ξ and η in terms of x and y,

$$\xi = \xi(x, y), \qquad \eta = \eta(x, y) \tag{1.13.1}$$

then the derivatives are transformed by using

$$\begin{aligned} \frac{\partial f}{\partial x} &= \frac{\partial f}{\partial \xi}\frac{\partial \xi}{\partial x} + \frac{\partial f}{\partial \eta}\frac{\partial \eta}{\partial x} \\ \frac{\partial f}{\partial y} &= \frac{\partial f}{\partial \xi}\frac{\partial \xi}{\partial y} + \frac{\partial f}{\partial \eta}\frac{\partial \eta}{\partial y} \end{aligned} \tag{1.13.2}$$

Note that partial derivatives are necessary, since two independent variables are involved. Higher-order derivatives follow by reapplying the relationships above after the quantities, $\partial\xi/\partial x, \partial\xi/\partial y, \partial\eta/\partial x,$ and $\partial\eta/\partial y$ are expressed in terms of ξ and η.

The Taylor series is one of the most important relationships used in the solution of problems in the physical sciences. If one knows the value of a function and the values of the derivatives at a particular point, the Taylor series allows us to approximate the value of the function at a neighboring point. It is written as

$$f(x + \Delta x) = f(x) + \Delta x \frac{df}{dx} + \frac{(\Delta x)^2}{2!}\frac{d^2 f}{dx^2} + \frac{(\Delta x)^3}{3!}\frac{d^3 f}{dx^3} + \cdots \tag{1.13.3}$$

where the derivatives are evaluated at x.

If the function f depends on more than one variable, say (x, y, z, t), then, if x is allowed to vary and the other variables are held constant, we write the Taylor series as

$$f(x + \Delta x, y, z, t) = f(x, y, z, t) + \Delta x \frac{\partial f}{\partial x} + \frac{(\Delta x)^2}{2!}\frac{\partial^2 f}{\partial x^2} + \cdots \tag{1.13.4}$$

If Δx is sufficiently small, we may approximate a function at $x + \Delta x$ as

$$f(x + \Delta x) \cong f(x) + \Delta x \frac{df}{dx} \tag{1.13.5}$$

This approximation is used in many derivations of describing equations for various physical phenomena.

Suppose that the differential equation describing a certain physical phenomenon is

$$v\frac{d^4 f}{dx^4} + f\frac{d^2 f}{dx^2} + 3\left(\frac{df}{dx}\right)^2 = 0 \tag{1.13.6}$$

where v is a small quantity. The boundary conditions are given at $x = 0$ and $x = 0.01$. We then often *normalize* the equation so that the coefficients in the differential equation are all of order unity and so that the boundary condition at $x = 0.01$ occurs at an x value of order unity. Suppose that $v = 10^{-4}$; then we would choose a new independent variable $\eta = x/\sqrt{v}$ with which to normalize. This would lead to the normalized equation

$$\frac{d^4 f}{d\eta^4} + f\frac{d^2 f}{d\eta^2} + 3\left(\frac{df}{dy}\right)^2 = 0 \tag{1.13.7}$$

with boundary conditions at $\eta = 0$ and $\eta = 1$.

PROBLEMS

1.1 Classify completely each of the following equations.

(a) $\dfrac{d^2u}{dx^2} + \dfrac{du}{dx} - 10x^2 = 0$

(b) $\dfrac{d^2u}{dx^2} + u\dfrac{du}{dx} = 10x$

(c) $\dfrac{\partial^2 u}{\partial x \partial y} + \dfrac{\partial u}{\partial y} = u^2$

(d) $\sqrt{\dfrac{\partial u}{\partial x} + u} + \dfrac{\partial u}{\partial y} = 0$

(e) $\sin\dfrac{du}{dx} + u = 0$

(f) $\dfrac{d^2u}{dx^2} = \sqrt{u+10}$

1.2 The acceleration of an object is given by $a = d^2s/dt^2$, where s is the displacement. For a constant deceleration of 20 m/s^2, find the distance an object travels before coming to rest if the initial velocity is 100 m/s.

1.3 The deflection of a particular 10-m-long cantilever–cantilever beam with constant loading is found by solving the differential equation, $(d^4u/dx^4) = 0.006$. Find the maximum deflection of the beam. Each cantilever end requires both deflection and slope to be zero.

1.4 An object is dropped from a house roof 8 m above the ground. How long does it take to hit the ground neglecting any drag? The acceleration due to gravity is 9.81 m/s^2.

1.5 Is $u(x) = A\sin(ax) - B\cos(ax)$ a solution to $d^2u/dx^2 + a^2u = 0$? If $u(0) = 10$ and $u(\pi/2a) = 20$, determine the specific solution.

1.6 Determine the general solution for each of the following separable equations.

(a) $x(x+2)\dfrac{du}{dx} - u^2 = 0$

(b) $(x^3 + u^3) - xu^2\dfrac{du}{dx} = 0$

(c) $3u + (u+x)\dfrac{du}{dx} = 0$

1.7 Show that each of the following equations is exact and find the general solution.

(a) $(2+x^2)\dfrac{du}{dx} + 2xu = 0$

(b) $x^2 + 3u^2\dfrac{du}{dx} = 0$

(c) $\sin 2x\dfrac{du}{dx} + 2u\cos 2x = 0$

(d) $e^x\left(\dfrac{du}{dx} + u\right) = 0$

1.8 Find the specific solution for each of the following differential equations.

(a) $x\dfrac{du}{dx} - u + x^3 = 0; \quad u(1) = 0$

(b) $2xu\dfrac{du}{dx} + u^2 = 4; \quad u(1) = 0$

(c) $\dfrac{du}{dx} - u = \cos x; \quad u(0) = 2$

(d) $\dfrac{du}{dx} - u - xu^5 = 0; \quad u(0) = 1$

$\left(\textit{Hint: } \text{Use } v = \dfrac{1}{u^4}\right)$

1.9 A constant voltage of 12 volts is impressed on a series circuit composed of a 10-ohm resistor and a 10^{-4}-henry inductor. Determine the current after 2 micro-seconds if the current is zero at $t = 0$.

1.10 An exponentially increasing voltage of $0.2e^{2t}$ is impressed on a series circuit containing a 20-ohm resistor and a 10^{-3}-henry inductor. Calculate the resulting current as a function of time using $i = 0$ at $t = 0$.

1.11 A series circuit composed of a 50-ohm resistor and a 10^{-7}-farad capacitor is excited with the voltage 12 sin $2t$. What is the general expression for the charge on the capacitor? For the current?

1.12 A constant voltage of 12 volts is impressed on a series circuit containing a 200-ohm resistor and a 10^{-6}-farad capacitor. Determine the general expression for the charge. How long will it take before the capacitor is half-charged?

1.13 The initial concentration of salt in 10 m^3 of solution is 0.2 kg/m^3. Fresh water flows into the tank at the rate of 0.1 m^3/s until the volume is 20 m^3, at which time t_f the solution flows out at the same rate as it flows into the tank. Express the concentration C as a function of time. One function will express $C(t)$ for $t < t_f$ and another for $t > t_f$.

1.14 An average person takes 18 breaths per minute and each breath exhales 0.0016 m^3 of air containing 4% CO_2. At the start of a seminar with 300 participants, the room air contains 0.4% CO_2. The ventilation system delivers 10 m^3 of air per minute to the 1500-m^3 room. Find an expression for the concentration level of CO_2 in the room.

1.15 Determine an expression for the height of water in the funnel shown in Fig. 1.29. What time is necessary to drain the funnel?

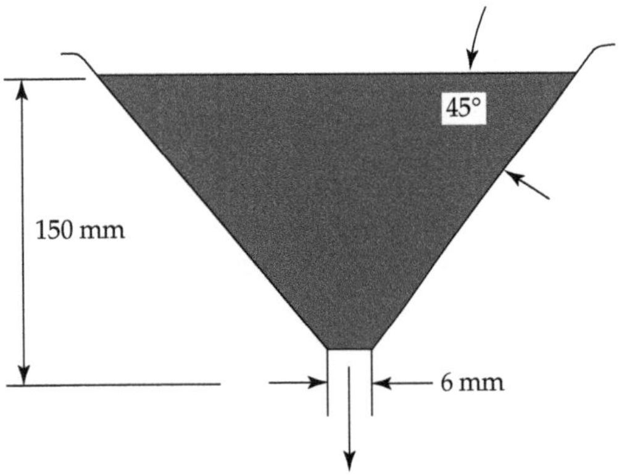

FIGURE 1.29

1.16 A square tank, 3 m on a side, is filled with water to a depth of 2 m. A vertical slot 6 mm wide from the top to the bottom allows the water to drain out. Determine the height h as a function of time and the time necessary for one half of the water to drain out.

1.17 A body falls from rest and is resisted by air drag. Determine the time necessary to reach a velocity of $V = 50$ m/s if the 100-kg body is resisted by a force equal to (a) 0.01 V and (b) 0.004 V^2. Check if the equation $M(dV/dt) = Mg - D$, where D is the drag force, describes the motion.

1.18 Calculate the velocity of escape from the earth for a rocket fired radially outward on the surface ($R \cong 6400$ km) of the earth. Use Newton's law of gravitation, which states that $dv/dt = k/r^2$, where, for the present problem, $k = -gR^2$. Also, to eliminate t, use $dt = dr/v$.

1.19 The rate in kilowatts (kW) at which heat is conducted in a solid is proportional to the area and the temperature gradient with the constant of proportionality being the thermal conductivity k (kW/m · °C). For a long, laterally insulated rod this takes the form $q = -kA(dT/dx)$. At the left end heat is transferred at the rate of 10 kW. Determine the temperature distribution in the rod if the right end at $x = 2$ m is held constant at 50°C. The cross-sectional area is 1200 mm^2 and $k = 100$ kW/m · °C.

1.20 An object at a temperature of 80°C to be cooled is placed in a refrigerator maintained at 5°C. It has been observed that the rate of temperature change of such an object is proportional to the surface area A and the difference between its temperature T and the temperature of the surrounding medium. Determine the time for the temperature to reach 8°C if the constant of proportionality $\alpha = 0.02$ $(s \cdot m^2)^{-1}$ and $A = 0.2$ m^2.

1.21 The evaporation rate of moisture from a sheet hung on a clothesline is proportional to the moisture content. If one half of the moisture is lost in the first 20 minutes, calculate the time necessary to evaporate 95% of the moisture.

1.22 An ant of mass m is in freefall under the action of a gravitational force, mg, and a resistance force, cv^2, where the gravitational acceleration $g = 9.81$ m/s^2 and the velocity v are both positive downward, c is a constant, and the resistance force opposes the velocity. If the ant drops from rest, find (a) the velocity as a function of time in terms of the parameters, and (b) if the terminal velocity is 6.4 km/hr, and the mass of the ant is $m = 0.0040$ grams, find the time it takes for the ant to reach 98% terminal velocity.

1.23 A squirrel was observed to fall out of a tree and then scamper away unharmed. Suppose the terminal velocity of a squirrel of mass $m = 0.30$ kg is 10.3 m/s. Starting at rest, it falls out of a tree. If the force of air resistance is modeled as cv^2, and the gravitational acceleration is $g = 9.81$ m/s^2, what will be the squirrel's velocity when it hits the ground 4 meters below?

1.24 Consider a simple pendulum of mass m and length l in the field of gravity. Suppose θ is the deflection angle from the downward position and the pendulum starts at rest from the angle A. The equation of motion and initial conditions are

$$ml^2\frac{d^2\theta}{dt^2} + mgl\sin\theta = 0$$

$$\theta(0) = A, \qquad \dot{\theta}(0) = 0$$

(a) Multiply the equation by $d\theta/dt$ and selectively replace $d\theta/dt = u$ to obtain a first order differential equation, and its boundary condition, with u as the dependent variable and θ as the independent variable.

(b) Solve for $u(\theta)$.

1.25 Consider a mass m and nonlinear restoring cubic spring force $f(x) = -\alpha mx^3$, where x is the displacement of the mass, and where $\alpha > 0$. The mass starts at $x = 0$ with initial velocity v. The equation of motion and initial conditions are

$$m\frac{d^2x}{dt^2} + \alpha mx^3 = 0$$

$$x(0) = 0, \qquad \dot{x}(0) = v$$

(a) Multiply the equation by $d\theta/dt$ and selectively replace $u = dx/dt$ to obtain a first-order differential equation, and its boundary condition, with u as the dependent variable and x as the independent variable.

(b) Solve for $u(x)$.

1.26 A particle moves freely on a sinusoidal frictionless track defined by $y = a\cos\gamma x$. If x measures the horizontal distance along the track, and $u = dx/dt$, then a differential equation that relates u to x is given as

$$mu(\ + a^2\gamma^2\cos^2\gamma x)\frac{du}{dx} +$$

$$(-ma^2\gamma^3u^2\cos\gamma x\sin\gamma x + mga\gamma\cos\gamma x) = 0$$

Determine that the equation is exact, and then solve for $u(x)$.

1.27 A rotor with time varying damping is modeled with a first-order linear differential equation of the form

$$\frac{d\Omega}{dt} + \frac{\alpha}{1+t}\Omega = T$$

where Ω is the angular speed of the rotor, and T is a constant applied torque. If the rotor starts at rest, such that $\Omega(0) = 0$, find the rotor speed $\Omega(t)$.

1.28 Show that Eq. 1.6.10 follows from substituting Eq. 1.6.9 and Eq. 1.6.8 into Eq. 1.6.7.

1.29 Find the general solution for each of the following differential equations.

(a) $\frac{d^2u}{dx^2} - \frac{du}{dx} - 6u = 0$

(b) $\frac{d^2u}{dx^2} - 9u = 0$

(c) $\frac{d^2u}{dx^2} + 9u = 0$

(d) $4\frac{d^2u}{dx^2} + \frac{du}{dx} = 0$

(e) $\frac{d^2u}{dx^2} - 4\frac{du}{dx} + 4u = 0$

(f) $\frac{d^2u}{dx^2} + 4\frac{du}{dx} + 4u = 0$

(g) $\frac{d^2u}{dx^2} - 4\frac{du}{dx} - 4u = 0$

(h) $\frac{d^2u}{dx^2} + 4\frac{du}{dx} - 4u = 0$

(i) $\frac{d^2u}{dx^2} - 4\frac{du}{dx} = 0$

(j) $\frac{d^2u}{dx^2} - 4\frac{du}{dx} + 8u = 0$

(k) $\frac{d^2u}{dx^2} + 2\frac{du}{dx} + 5u = 0$

(l) $2\frac{d^2u}{dx^2} + 6\frac{du}{dx} + 5u = 0$

1.30 Derive the differential equation that describes the motion of a mass M swinging from the end of a string of length L. Assume small angles. Find the general solution of the differential equation. Neglect any damping.

1.31 Determine the motion of a mass moving toward the origin with a force of attraction proportional to the distance from the origin. Assume that the 10-kg mass starts at rest at a distance of 10 m and that the constant of proportionality is 10 N/m. What will the speed of the mass be 5 m from the origin?

1.32 A spring–mass system has zero damping. Find the general solution and determine the frequency of oscillation if $M = 4$ kg and $K = 100$ N/m.

1.33 Calculate the time necessary for a 0.03-kg mass hanging from a spring with spring constant 0.5 N/m to undergo one complete oscillation.

1.34 A 4-kg mass is hanging from a spring with $K = 100$ N/m. Sketch, on the same plot, the two specific solutions found from (a) $y(0) = 0.5$ m, $dy/dt(0) = 0$, and (b) $y(0) = 0$, $dy/dt(0) = 10$ m/s. The coordinate y is measured from the equilibrium position.

1.35 A damped spring–mass system involves a mass of 4 kg, a spring with $K = 64$ N/m, and a dashpot with $C = 32$ kg/s. The mass is raised 1 m from its equilibrium position and released from rest. Sketch $y(t)$ for the first 2 s.

1.36 A damped spring–mass system is given an initial velocity of 50 m/s from the equilibrium position. Find $y(t)$ if $M = 4$ kg, $K = 64$ N/m, and $C = 40$ kg/s.

1.37 A body weighs 50 N and hangs from a spring with spring constant of 50 N/m. A dashpot is attached to the body. If the body is raised 2 m from its equilibrium position and released from rest, determine the solution if (a) $C = 17.7$ kg/s, and (b) $C = 40$ kg/s.

1.38 After a period of time a dashpot deteriorates, so the damping coefficient decreases. For Problem 1.35 sketch $y(t)$ if the damping coefficient is reduced to 20 kg/s.

1.39 Show that solution for the overdamped motion of a spring–mass system can be written as

$$y(t) = c_1 e^{-(C/2M)t} \sinh\left[\sqrt{C^2 - 4MK}\,\frac{t}{2M} + c^2\right].$$

Note: $\sinh x = \frac{e^x - e^{-x}}{2}$

1.40 A maximum occurs for the overdamped motion of curve 1 of Fig. 1.4b. Using the results of Problem 1.39, show that the maximum occurs when

$$t = \frac{2M}{\sqrt{C^2 - 4MK}} \tanh^{-1} \frac{\sqrt{C^2 - 4MK}}{C}$$

1.41 Using the results of the preceding two problems find an expression for y_{max} of curve ① of Fig. 1.4b if v_0 is the initial velocity.

1.42 Determine the time between consecutive maximum amplitudes for a spring–mass system in which $M = 30$ kg, $K = 2000$ N/m, and $C = 300$ kg/s.

1.43 Find the damping as a percentage of critical damping for the motion $y(t) = 2e^{-t}\sin t$. Also, find the time for the first maximum and sketch the curve.

1.44 Find the displacement $y(t)$for a mass of 5 kg hanging from a spring with $K = 100$ N/m if there is a dashpot attached having $C =$ 30 kg/s. The initial conditions are $y(0) = 1$ m and $dy/dt(0) = 0$. Express the solution in all three forms. Refer to Eqs. 1.7.20, 1.7.21, and 1.7.22.

1.45 An electrical circuit is composed of an inductor with $L = 10^{-3}$ henry, a capacitor with $C = 2 \times 10^{-5}$ farad, and a resistor. Determine the critical resistance that will just lead to an oscillatory current if the elements are connected (a) in series, and (b) in parallel.

1.46 The amplitudes of two successive maximum currents in a series circuit containing an inductor with $L = 10^{-4}$ henry and a capacitor with $C = 10^{-6}$ farad are measured to be 0.02 ampere (A) and 0.01 A. Determine the resistance and write the solution for $i(t)$ in the form of Eq. 1.7.22.

1.47 Determine the current $i(t)$ in a series circuit containing a resistor with $R = 20$ ohms, a capacitor with $C = 10^{-6}/2$ farad, and an inductor with $L = 10^{-3}$ henry. The initial conditions are $i(0) = 10$ amps and $(di/dt)(0) = 0$.

1.48 An input torque on a circular shaft is $T(t)$. It is resisted by a clamping torque proportional to the rate of angle change $d\theta/dt$ and an elastic torque proportional to the angle itself, the constants of proportionality being c and k, respectively. We have observed that the moment of inertia I times the angular acceleration $d^2\theta/dt^2$ equals the net torque. Write the appropriate differential equation and note the analogy with the spring–mass system.

1.49 Find the particular solution for each of the following differential equations.

(a) $$\frac{d^2u}{dx^2} + 2u = 2x$$

(b) $$\frac{d^2u}{dx^2} + \frac{du}{dx} + 2u = 2x$$

(c) $$\frac{d^2u}{dx^2} + u = e^{-x}$$

(d) $$\frac{d^2u}{dx^2} - u = e^{x}$$

(e) $$\frac{d^2u}{dx^2} + 10u = 5\sin x$$

(f) $$\frac{d^2u}{dx^2} + 9u = \cos 3x$$

(g) $$\frac{d^2u}{dx^2} + 4\frac{du}{dx} + 4u = e^{-2x}$$

(h) $$\frac{d^2u}{dx^2} + 9u = x^2 + \sin 3x$$

1.50 Find the general solution for each of the following differential equations.

(a) $$\frac{d^2u}{dx^2} + u = e^{2x}$$

(b) $$\frac{d^2u}{dx^2} + 4\frac{du}{dx} + 4u = x^2 + x + 4$$

(c) $$\frac{d^2u}{dx^2} + 9u = x^2 + \sin 2x$$

(d) $$\frac{d^2u}{dx^2} + 4u = \sin 2x$$

(e) $$\frac{d^2u}{dx^2} - 16u = e^{4x}$$

(f) $$\frac{d^2u}{dx^2} + 5\frac{du}{dx} + 6u = 3\sin 2x$$

1.51 Find the specific solution for each of the following initial-value problems.

(a) $\frac{d^2u}{dx^2} + 4\frac{du}{dx} = 4u = x^2,$

$$u(0) = 0, \quad \frac{du}{dx}(0) = \frac{1}{2}$$

(b) $\frac{d^2u}{dx^2} + 4u = 2\sin x, \quad u(0) = 1, \quad \frac{du}{dx}(0) = 0$

(c) $\frac{d^2u}{dx^2} + 4\frac{du}{dx} + 5u = x^2 + 5,$

$$u(0) = 0, \quad \frac{du}{dx}(0) = 0$$

(d) $\frac{d^2u}{dx^2} + 4u = 2\sin 2x, \quad u(0) = 0, \quad \frac{du}{dx}(0) = 0$

(e) $\frac{d^2u}{dx^2} + 6\frac{du}{dx} + 10u = \cos 2x,$

$$u(0) = 0, \quad \frac{du}{dx}(0) = 0$$

(f) $\frac{d^2u}{dx^2} - 16u = 2e^{4x}, \quad u(0) = 0, \quad \frac{du}{dx}(0) = 0$

1.52 Find the solution for $M(d^2y/dt^2) + C(dy/dt) + Mg = 0$. Show that this represents the motion of a body rising with drag proportional to velocity.

1.53 For Problem 1.52, assume that the initial velocity is 100 m/s upward, $C = 0.4$ kg/s, and $M = 2$ kg. How high will the body rise?

1.54 For the body of Problem 1.53, calculate the time required for the body to rise to the maximum height and compare this to the time it takes for the body to fall back to the original position.

1.55 A body weighing 100 N is dropped from rest. The drag is assumed to be proportional to the first power of the velocity with the constant of proportionality being 0.5. Approximate the time necessary for the body to attain terminal velocity. Define terminal velocity to be equal to 0.99 V_∞, where V_∞ is the velocity attained as $t \to \infty$.

1.56 Find the general solution to the equation $M(d^2y/dt^2) + Ky = F_0 \cos \omega t$ and verify Eq. 1.9.10 by letting $\omega_0 = \sqrt{K/M}$.

1.57 A 2-kg mass is suspended by a spring with $K = 32$ N/m. A force of 0.1 sin 4t is applied to the mass. Calculate the time required for failure to occur if the spring breaks when the amplitude of the oscillation exceeds 0.5 m. The motion starts from rest and damping is neglected.

1.58 A 20-N weight is suspended by a frictionless spring with $K = 98$ N/m. A force of 2 cos 7t acts on the weight. Calculate the frequency of the "beat" and find the maximum amplitude of the motion, which starts from rest.

1.59 Use the sinusoidal forcing function as $F_0 \sin \omega t$, and with $C = 0$ show that the amplitude $F_0/M(\omega_0^2 - \omega^2)$ of the particular solution remains unchanged for the spring–mass system.

1.60 Show that the particular solution given by Eq. 1.9.14 for $\omega = \omega_0$ follows from the appropriate equations for $F(t) = F_0 \cos \omega_0 t$.

1.61 Find the steady-state solution for each of the following differential equations.

(a) $\frac{d^2y}{dt^2} + \frac{dy}{dt} + 4y = 2\sin 2t$

(b) $\frac{d^2y}{dt^2} + 2\frac{dy}{dt} + y = \cos 3t$

(c) $\frac{d^2y}{dt^2} + \frac{dy}{dt} + y = 2\sin t + \cos t$

(d) $\frac{d^2y}{dt^2} + 0.1\frac{dy}{dt} + 2y = 2\sin 2t$

(e) $\frac{d^2y}{dt^2} + 2\frac{dy}{dt} + 5y = \sin t - 2\cos 3t$

(f) $\frac{d^2y}{dt^2} + \frac{dy}{dt} + 2y = \cos t - \sin 2t$

1.62 Determine the transient solution for each of the following differential equations.

(a) $\frac{d^2y}{dt^2} + 5\frac{dy}{dt} + 4y = \cos 2t$

(b) $\frac{d^2y}{dt^2}+7\frac{dy}{dt}+10y=2\sin t-\cos 2t$

(c) $\frac{d^2y}{dt^2}+4\frac{dy}{dt}+4y=4\sin t$

(d) $\frac{d^2y}{dt^2}+0.1\frac{dy}{dt}+2y=\cos 2t$

1.63 Solve for the specific solution to each of the following initial-value problems.

(a) $\frac{d^2y}{dt^2}+5\frac{dy}{dt}+6y=52\cos 2t,$

$$y(0)=0,\ \frac{dy}{dt}(0)=0$$

(b) $\frac{d^2y}{dt^2}+2\frac{dy}{dt}+y=2\sin t,$

$$y(0)=0,\ \frac{dy}{dt}(0)=0$$

(c) $\frac{d^2y}{dt^2}+2\frac{dy}{dt}+10y=26\sin 2t,$

$$y(0)=1,\ \frac{dy}{dt}(0)=0$$

(d) $\frac{d^2y}{dt^2}+0.1\frac{dy}{dt}+2y=20.1\cos t,$

$$y(0)=0,\ \frac{dy}{dt}(0)=10$$

(e) $\frac{d^2y}{dt^2}+3\frac{dy}{dt}+2y=10\sin t,$

$$y(0)=0,\ \frac{dy}{dt}(0)=0$$

(f) $\frac{d^2y}{dt^2}+0.02\frac{dy}{dt}+16y=2\sin 4t,$

$$y(0)=0,\ \frac{dy}{dt}(0)=0$$

1.64 The motion of a 3-kg mass, hanging from a spring with $K=12$ N/m, is damped with a dashpot with $C=5$ kg/s. (a) Show that Eq. 1.9.22 gives the amplitude of the steady-state solution if $F(t)=F_0\sin\omega t$. (b) Determine the phase lag and amplitude of the steady-state solution if a force $F=20\sin 2t$ acts on the mass.

1.65 For Problem 1.64 let the forcing function be $F(t)=20\sin\omega t$. Calculate the maximum possible amplitude of the steady-state solution and the associated forcing-function frequency.

1.66 A forcing function $F=10\sin 2t$ is to be imposed on a spring–mass system with $M=2$ kg and $K=8$ N/m. Determine the damping coefficient necessary to limit the amplitude of the resulting motion to 2 m.

1.67 A constant voltage of 12 volts is impressed on a series circuit containing elements with $R=30$ ohms, $L=10^{-4}$ henry, and $C=10^{-6}$ farad. Determine expressions for both the charge on the capacitor and the current if $q=i=0$ at $t=0$.

1.68 A series circuit is composed of elements with $R=60$ ohms, $L=10^{-3}$ henry, and $C=10^{-5}$ farad. Find an expression for the steady-state current if a voltage of $120\cos 120\pi t$ is applied at $t=0$.

1.69 A circuit is composed of elements with $R=80$ ohms, $L=10^{-4}$ henry, and $C=10^{-6}$ farad connected in parallel. The capacitor has an initial charge of 10^{-4} columb. There is no current flowing through the capacitor at $t=0$. What is the current flowing through the resistor at $t=10^{-4}$ s?

1.70 The circuit of Problem 1.69 is suddenly subjected to a current source of $2\cos 200t$. Find the steady-state voltage across the elements.

1.71 The inductor and the capacitor are interchanged in Example 1.26. Determine the resulting current $i_2(t)$ flowing through R_2. Also, find the steady-state charge on the capacitor.

1.72 Write the Fourier series representation for each of the periodic functions shown below and in Fig. 1.30. One period is defined for each. Express the answer as a series using the summation symbol.

(a) $f(t)=\begin{cases}-t, & -\pi<t<0\\ t, & 0<t<\pi\end{cases}$

(b) $f(t)=t^2,\quad -\pi<t<\pi$

(c) $f(t) = \cos\dfrac{t}{2}, \quad -\pi < t < \pi$

(d) $f(t) = t + 2\pi, \quad -2\pi < t < 2\pi$

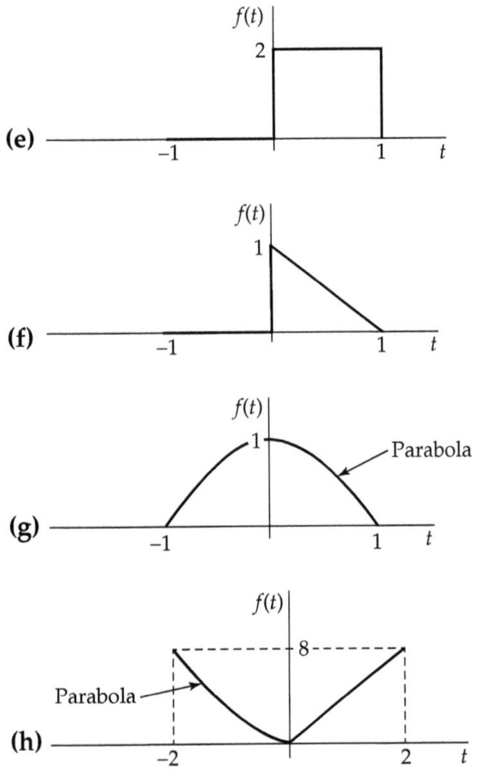

FIGURE 1.30

1.73 In Problem 1.72, (a) which of the functions are even, (b) which of the functions are odd, (c) which of the functions could be made even by shifting the vertical axis, and (d) which of the functions could be made odd by shifting the vertical axis?

1.74 Expand each of the periodic functions shown below and in Fig. 1.31 in a Fourier sine series and a Fourier cosine series.

(a) $f(t) = 4t, \quad 0 < t < \pi$

(b) $f(t) = \begin{cases} 10, & 0 < t < \pi \\ 0, & \pi < t < 2\pi \end{cases}$

(c) $f(t) = \sin t, \quad 0 < t < \pi$

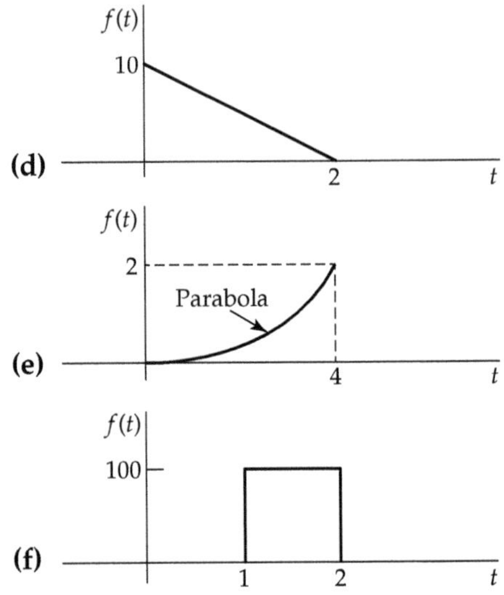

FIGURE 1.31

1.75 Rework Example 1.30 for a more general function. Let the two zero points of $f(t)$ be at $t = 0$ and $t = T$. Let the maximum of $f(t)$ at $t = T/2$ be K.

1.76 Find a half-range cosine expansion and a half-range sine expansion for the function $f(t) = t - t^2$ for $0 < t < 1$. Which expansion would be the more accurate for an equal number of terms? Write the first three terms in each series.

1.77 Find half-range sine expansion of

$$f(t) = \begin{cases} t, & 0 < t < 2 \\ 2, & 2 < t < 4 \end{cases}$$

Make a sketch of the first three terms in the series.

1.78 Find the particular solution associated with the system displayed in Fig. 1.32. Which term dominates the solution?

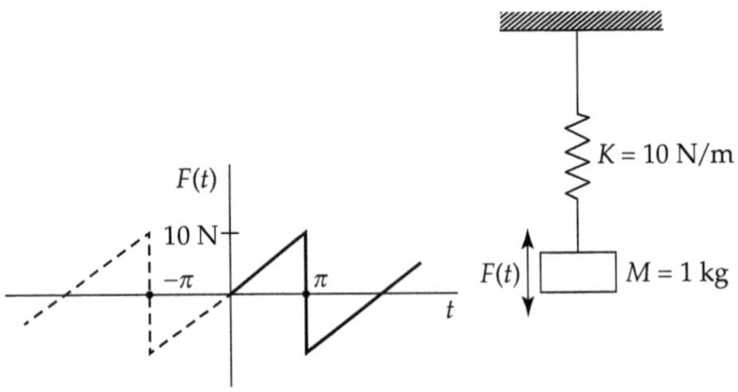

FIGURE 1.32

1.79 What is the steady-state response of the mass to the forcing function of Fig. 1.33?

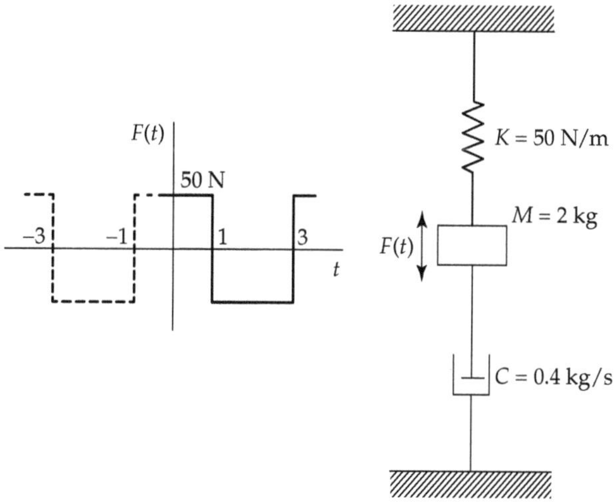

FIGURE 1.33

1.80 Determine the steady-state current in the circuit of Fig. 1.34.

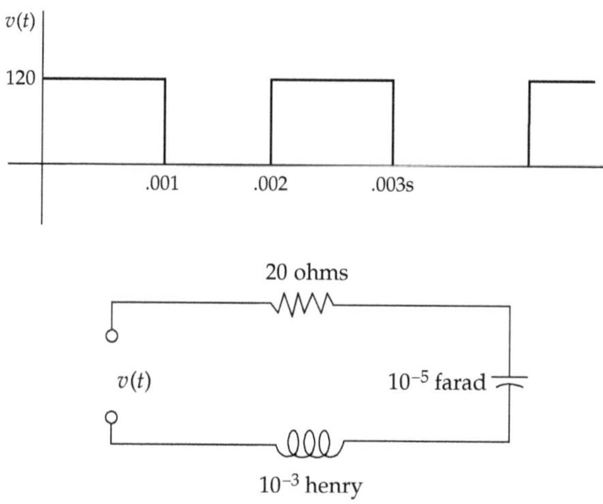

FIGURE 1.34

1.81 Determine the general solution for each of the following differential equations.

(a) $x^2 = \dfrac{d^2u}{dx^2} + 7x\dfrac{du}{dx} + 8u = 0$

(b) $x^2\dfrac{d^2u}{ax^2} + 9x\dfrac{du}{dx} + 12u = 0$

(c) $x^2\dfrac{d^2u}{dx^2} - 12u = 24x$

(d) $x^2\dfrac{d^2u}{dx^2} + 2x\dfrac{du}{dx} - 12u = 24$

1.82 Show that Eq. 1.11.9 follows from the equations of Section 1.11 using the suggestion following Eq. 1.11.8.

1.83 Find the particular solution for each of the following differential equations.

(a) $\dfrac{d^2y}{dt^2} + y = t\sin t$

(b) $\dfrac{d^2y}{dt^2} + 5\dfrac{dy}{dt} + 4y = te^t$

(c) $\dfrac{d^2y}{dt^2} + 4\dfrac{dy}{dt} + 4y = te^{-2t}$

1.84 Express the partial differential equation $\partial^2u/\partial t^2 = a^2(\partial^2u/\partial x^2)$ in terms of ξ and η, where $\xi = x - at$ and $\eta = x + at$.

1.85 Develop series expressions for e^x, $\sin x$, and $\cos x$. Use the expressions to verify that $\sin x = (e^{ix} - e^{-ix})/2i$ and $\cos x = (e^{ix} + e^{-ix})/2$. (*Hint:* Expand in a Taylor series about $x = 0$.)

1.86 Find an approximate expression for the quantity $[\sin(\alpha + \Delta\alpha) - \sin\alpha]$ if $\Delta\alpha$ is small.

2 Power-Series Methods

Outline

2.1 Power Series
2.2 Linear Differential Equations with Variable Coefficients
2.3 Legendre's Equation
2.4 The Method of Frobenius
2.4.1 Distinct Roots Not Differing by an Integer
2.4.2 Double Roots
2.4.3 Roots Differing by an Integer
2.5 Bessel's Equation
Problems

2.1 POWER SERIES

We have studied linear differential equations with constant coefficients and have solved such equations using exponential functions. In general, a linear differential equation with variable coefficients cannot be solved by the use of exponential functions. We did, however, solve a special equation with variable coefficients, the Cauchy equation, by assuming a simple power solution. A more general method will be presented in this chapter that utilizes power series to obtain a solution. A *power series* is the sum of the infinite number of terms of the form

$$b_0 + b_1(x-a) + b_2(x-a)^2 + b_3(x-a)^3 + \cdots = \sum_{n=0}^{\infty} b_n(x-a)^n \tag{2.1.1}$$

where $a, b_0, b_1, b_2, \ldots$ are constants. A power series does not include terms with negative powers.

Assume that a solution $u(x)$ can be represented by the power series

$$u(x) = b_0 + b_1(x-a) + b_2(x-a)^2 + \cdots = \sum_{n=0}^{\infty} b_n(x-a)^n \tag{2.1.2}$$

The derivatives are found by differentiating term by term to obtain

$$\frac{du}{dx} = b_1 + 2b_2(x-a) + 3b_3(x-a)^2 + \cdots = \sum_{n=1}^{\infty} nb_n(x-a)^{n-1} \tag{2.1.3}$$

M. C. Potter and B. F. Feeny, *Mathematical Methods for Engineering and Science*,
https://doi.org/10.1007/978-3-031-26151-0_2

$$\frac{d^2u}{dx^2} = 2b_2 + 6b_3(x-a) + 12b_4(x-a)^2 + \cdots = \sum_{n=2}^{\infty} n(n-1)b_n(x-a)^{n-2} \qquad (2.1.4)$$

and similarly for derivatives of higher order. If the coefficients in the differential equation can be expressed in terms of powers of $(x-a)$, the expressions above may be substituted into the differential equation and an equation of the following form results:

$$k_0 + k_1(x-a) + k_2(x-a)^2 + \cdots = 0 \qquad (2.1.5)$$

If this equation is valid *for all* x within an interval of interest, the coefficients k_0, k_1, $k_2, \ldots$ must all be zero. From these equations the b_n's in Eq. 2.1.2 can be calculated. If the series converges, often several terms are all that are necessary for a solution to approximate $u(x)$.

To reduce a differential equation to the form (2.1.5), it is necessary to express the coefficients in the differential equation in terms of $(x-a)$. We often choose $a=0$ and expand in a series about $x=0$. Several elementary functions that may appear as coefficients are expanded in powers of x as

$$\begin{aligned}
\frac{1}{1-x} &= 1 + x + x^2 + \cdots \\
e^x &= 1 + x + \frac{x^2}{2!} + \cdots \\
\sin x &= x - \frac{x^3}{3!} + \frac{x^5}{5!} - \cdots \\
\cos x &= 1 - \frac{x^2}{2!} + \frac{x^4}{4!} - \cdots \\
\ln(1+x) &= x - \frac{x^2}{2} + \frac{x^3}{3} - \cdots
\end{aligned} \qquad (2.1.6)$$

These expressions are easily derived by using a Taylor series expansion (see Eq. 1.13.3) about $x=0$.

There are several properties of a series that we will consider before we look at some examples illustrating this procedure. The sum s_m of the first m terms in a power series is

$$s_m = b_0 + b_1(x-a) + \cdots + b_m(x-a)^m \qquad (2.1.7)$$

and is called the mth *partial sum* of the series. The sum R_m of the remaining terms is

$$R_m(x) = b_{m+1}(x-a)^{m+1} + b_{m+2}(x-a)^{m+2} + \cdots \qquad (2.1.8)$$

and is the *remainder.* A power series *converges* if $R_m \to 0$ as $m \to \infty$; otherwise, it *diverges.* Since we wish to approximate a solution with a finite, and usually small, number of terms it is necessary that the power series representing the solution converges.

There is usually an interval over which the power series converges with the midpoint at $x=a$; that is, the series converges if

$$|x-a| < R \qquad (2.1.9)$$

where R is the *radius of convergence*. This radius is given by

$$\frac{1}{R} = \lim_{n\to\infty}\left|\frac{b_{n+1}}{b_n}\right| \tag{2.1.10}$$

This formula will not be developed here.

A function $f(x)$ is *analytic* at the point $x = a$ if it can be expressed as a power series $\sum_{n=0}^{\infty} b_n(x-a)^n$ with $R > 0$. This property of a function $f(x)$ is of great importance in the solution of differential equations. If the functions $f(x)$, $g(x)$, and $h(x)$ in the differential equation

$$\frac{d^2u}{dx^2} + f(x)\frac{du}{dx} + g(x)u = h(x) \tag{2.1.11}$$

are analytic at the point $x = a$, the solution can be represented by a power series with a finite radius of convergence; that is,

$$u(x) = \sum_{n=0}^{\infty} b_n(x-a)^n \tag{2.1.12}$$

with $R > 0$; the point $x = a$ is called an *ordinary point*. If $f(x)$, $g(x)$, or $h(x)$ is not analytic at $x = a$, the point $x = a$ is said to be a *singular point*. If a function is singular at $x = a$, that function becomes infinite as $x \to a$. The function $1/x$ is singular at $x = 0$, the function $1/(x-1)$ is singular at $x = 1$, and $1/(x^2-1)$ is singular at $x = \pm 1$. These three functions are analytic at all points other than those singular points.

Another more convenient and somewhat simpler method of determining the radius of convergence is to locate all the singular points on the complex plane. We shall consider x to be a complex variable with real and imaginary parts. As an example, consider the function $x/[(x^2+9)(x-6)]$. It has singularities (the function becomes infinite) at the following points: $x = 6, 3i, -3i$. The singular points are plotted in Fig. 2.1a. If we expand about the origin, the radius of convergence is established by drawing a circle, with center at the origin, passing through the nearest singular point, as shown in Fig. 2.1b. This gives $R = 3$, a rather surprising result, since the first singular point on the x axis is at $x = 6$. The singularity at $x = 3i$ prevents the series from converging for $x \geq 3$. If we expand about $x = 5$, that is, in powers of $(x-5)$, the nearest singularity would be located at (6, 0). This would give a radius of convergence of $R = 1$ and the series would converge for $4 < x < 6$.

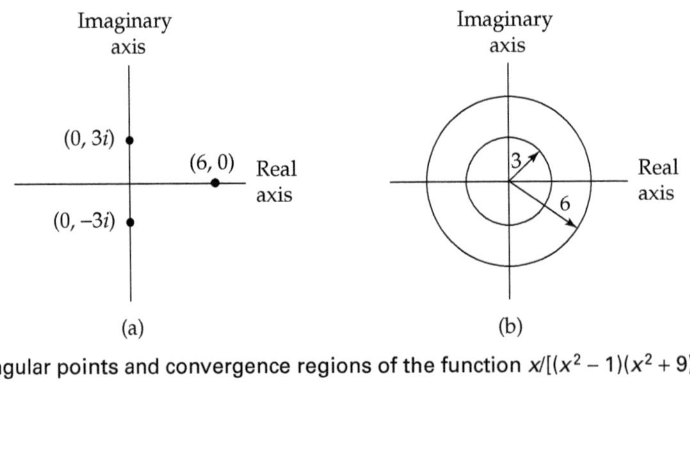

FIGURE 2.1 Singular points and convergence regions of the function $x/[(x^2-1)(x^2+9)(x-6)]$.

Example 2.1

Derive the series expansion for $\sin x$ given in Eq. 2.1.6 using a Taylor series expansion.

Solution

Recall that a Taylor series, Eq. 1.13.3, is written as

$$f(x+\Delta x) = f(x) + \Delta x \frac{df}{dx} + \frac{(\Delta x)^2}{2!}\frac{d^2 f}{dx^2} + \frac{(\Delta x)^3}{3!}\frac{d^3 f}{dx^3} + \cdots$$

where the derivatives are evaluated at x. Let us expand about $x = 0$. Then the Taylor series is

$$f(\Delta x) = f(0) + \Delta x \left.\frac{df}{dx}\right|_{x=0} + \frac{(\Delta x)^2}{2!}\left.\frac{d^2 f}{dx^2}\right|_{x=0} + \cdots$$

Since we are expanding about the origin, Δx is simply the distance from the origin; that is, we can replace Δx with x, giving us

$$f(x) = f(0) + x \left.\frac{df}{dx}\right|_{x=0} + \frac{x^2}{2!}\left.\frac{d^2 f}{dx^2}\right|_{x=0} + \cdots$$

Finally, let $f(x) = \sin x$, and we have

$$\begin{aligned}\sin x &= \overset{0}{\cancel{\sin 0}} + x\cos 0 - \frac{x^2}{2!}\overset{0}{\cancel{\sin 0}} - \frac{x^3}{3!}\cos 0 + \cdots \\ &= x - \frac{x^3}{3!} + \frac{x^5}{5!} - \cdots\end{aligned}$$

Example 2.2

By using the expansions of Eqs. 2.1.6, find a series expansion for (a) $1/(x^2-4)$, and (b) $\tan^{-1} x$.

Solution

a) First we factor the given function,

$$\frac{1}{x^2-4} = \frac{1}{x-2}\cdot\frac{1}{x+2}$$

Next, we write the fractions in the form

$$\frac{1}{x-2} = -\frac{1}{2-x} = -\frac{1}{2}\left(\frac{1}{1-x/2}\right)$$

$$\frac{1}{x+2} = \frac{1}{2+x} = \frac{1}{2}\left(\frac{1}{1+x/2}\right)$$

Now, we use the first expansion of Eqs. 2.1.6, replacing x with $x/2$ in the first expansion and x with $(-x/2)$ in the second expansion. There results

$$\frac{1}{x-2} = -\frac{1}{2}\left[1+\frac{x}{2}+\left(\frac{x}{2}\right)^2+\left(\frac{x}{2}\right)^3+\cdots\right]$$

$$= -\frac{1}{2}-\frac{x}{4}-\frac{x^2}{8}-\frac{x^3}{16}-\cdots$$

$$\frac{1}{x+2} = \frac{1}{2}\left[1+\left(-\frac{x}{2}\right)+\left(-\frac{x}{2}\right)^2+\left(-\frac{x}{2}\right)^3+\cdots\right]$$

$$= \frac{1}{2}-\frac{x}{4}+\frac{x^2}{8}-\frac{x^3}{16}+\cdots$$

Finally, we multiply these two series to obtain

$$\frac{1}{x^2-4} = -\left(\frac{1}{2}+\frac{x}{4}+\frac{x^2}{8}+\cdots\right)\left(\frac{1}{2}-\frac{x}{4}+\frac{x^2}{8}+\cdots\right)$$

$$= -\left(\frac{1}{4}+\frac{x^2}{16}+\frac{x^4}{64}+\cdots\right)$$

b) We recognize that the function $\tan^{-1}x$ is

$$\tan^{-1}x = \int_0^x \frac{dt}{1+t^2}$$

The series expansion for $1/(1+t^2)$ is found, as in part (a), to be

$$\frac{1}{1+t^2} = \frac{1}{1-(-t^2)} = 1+(-t^2)+(-t^2)^2+(-t^2)^3+\cdots$$

$$= 1-t^2+t^4-t^6+\cdots$$

It then follows that

$$\tan^{-1}x = \int_0^x (1-t^2+t^4-t^6+\cdots)dt$$

$$= x-\frac{x^3}{3}+\frac{x^5}{5}-\cdots$$

where we have integrated term by term.

2.2 LINEAR DIFFERENTIAL EQUATIONS WITH VARIABLE COEFFICIENTS

Let us illustrate the solution of differential equations with variable coefficients using the power-series method by solving the equation

$$(1-x^2)\frac{d^2u}{dx^2}+u = 0 \tag{2.2.1}$$

The coefficient $g(x) = 1/(1 - x^2)$ in Eq. 2.1.11 is singular at $x = \pm 1$ and analytic at all other points. Hence, we can find a solution in the form of a power series by expanding about $x = 0$ that will converge for $-1 < x < 1$. Assume a solution in the form

$$u(x) = \sum_{n=0}^{\infty} b_n x^n = b_0 + b_1 x + b_2 x^2 + \cdots \tag{2.2.2}$$

Then*

$$\frac{d^2u}{dx^2} = 2b_2 + 6b_3 x + 12b_4 x^2 + \cdots \tag{2.2.3}$$

Substituting into Eq. 2.2.1 gives

$$2b_2 + 6b_3 x + 12b_4 x^2 + \cdots - 2b_2 x^2 - 6b_3 x^3 - \cdots + b_0 + b_1 x + b_2 x^2 + \cdots = 0 \tag{2.2.4}$$

Collect the terms as powers of x:

$$(2b_2 + b_0) + (6b_3 + b_1)x + (12b_4 - b_2)x^2 + \cdots = 0 \tag{2.2.5}$$

Set each coefficient equal to zero. Several coefficients are found to be

$$\begin{aligned} b_2 &= -\frac{b_0}{2} \\ b_3 &= -\frac{b_1}{6} \\ b_4 &= \frac{b_2}{12} = -\frac{b_0}{24} \\ b_5 &= \frac{b_3}{4} = -\frac{b_1}{24} \end{aligned} \tag{2.2.6}$$

All the coefficients can be expressed in terms of the two arbitrary coefficients b_0 and b_1. The solution can now be written as

$$u(x) = b_0\left(1 - \frac{x^2}{2} - \frac{x^4}{24} + \cdots\right) + b_1\left(x - \frac{x^3}{6} - \frac{x^5}{24} + \cdots\right) \tag{2.2.7}$$

For values of x in the interval $-1 < x < 1$, the series will converge. Generally, only a small number of terms is necessary for a good approximation to $u(x)$.

The solution above is of the form

$$u(x) = b_0 u_1(x) + b_1 u_2(x) \tag{2.2.8}$$

where

$$u_1(x) = 1 - \frac{x^2}{2} - \frac{x^4}{24} + \cdots, \qquad u_2(x) = x - \frac{x^3}{6} - \frac{x^5}{24} + \cdots \tag{2.2.9}$$

*Note that we usually write three terms of a series followed by three dots; this is standard procedure.

The functions $u_1(x)$ and $u_2(x)$ are two independent solutions of the second-order linear differential equation. Two conditions must be given to determine the two arbitrary coefficients b_0 and b_1.

A more convenient technique can be illustrated by solving the preceding problem again. The summation notation is used instead of writing out the expanded series. The second derivative can be written as (see Eq. 2.2.3)

$$\frac{d^2u}{dx^2} = \sum_{n=1}^{\infty} n(n-1)b_n x^{n-2} \tag{2.2.10}$$

Note that we start the series with $n = 2$, since the first two terms for $n = 0$ and 1 are zero. The differential equation is expressed in terms of the power series as

$$\sum_{n=2}^{\infty} n(n-1)b_n x^{n-2} - \sum_{n=2}^{\infty} n(n-1)b_n x^n + \sum_{n=0}^{\infty} b_n x^n = 0 \tag{2.2.11}$$

In the first series let $n - 2 = m$; then, when $n = 2$ and $m = 0$ we have

$$\sum_{m=0}^{\infty} (m+2)(m+1)b_{m+2}x^m - \sum_{n=2}^{\infty} n(n-1)b_n x^n + \sum_{n=0}^{\infty} b_n x^n = 0 \tag{2.2.12}$$

Now, replace m with n in the first series* and factor out x^n, to obtain

$$2b_2 + 6b_3 x + b_0 + b_1 x + \sum_{n=2}^{\infty} \{(n+2)(n+1)b_{n+2} - [n(n-1)-1]b_n\} x^n = 0 \tag{2.2.13}$$

The first four terms are the $n = 0$ and $n = 1$ terms from the two series that start with $n = 0$. Equating coefficients of x^0 and x to zero there results

$$b_2 = -\frac{b_0}{2}, \qquad b_3 = -\frac{b_1}{6} \tag{2.2.14}$$

and

$$b_{n+2} = \frac{n(n-1)-1}{(n+2)(n+1)} b_n, \qquad n = 2, 3, 4, \cdots \tag{2.2.15}$$

This is a *recurrence relation* which relates the coefficients of higher-order terms to coefficients of lower-order terms. From it we find that

$$b_4 = \frac{b_2}{12}, \qquad b_5 = \frac{b_3}{4}, \text{ etc.} \tag{2.2.16}$$

The solution can then be written as in Eq. 2.2.7.

*We can do this since m is a "dummy" index—it does not appear anywhere in the expanded series. This is quite similar to the x in $\int_0^1 x\,dx$. It is a dummy variable since we could replace it with y and the result would be the same.

The latter method is preferred, since the recurrence relation is part of the solution. Once the recurrence relation is obtained, the coefficients can be determined more easily. This is especially helpful when using a computer.

Example 2.3

Find the solution to the differential equation

$$\frac{d^2u}{dx^2} + 9u = 0$$

using a power-series solution.

Solution

Assume the solution to be the power series

$$u(x) = b_0 + b_1 x + b_2 x^2 + \cdots$$

The second derivative is

$$\frac{d^2u}{dx^2} = 2b_2 + 6b_3 x + 12b_4 x^2 + \cdots$$

Substitute these back into the differential equation to get

$$2b_2 + 6b_3 x + 12b_4 x^2 + \cdots + 9b_0 + 9b_1 x + 9b_2 x^2 + \cdots = 0$$

Collect the terms as powers of x:

$$2b_2 + 9b_0 + (6b_3 + 9b_1)x + (12b_4 + 9b_2)x^2 + \cdots = 0$$

Now, for this equation to be satisfied for all x we demand that every coefficient of each power of x be zero. That is,

$$\begin{aligned} x^0&: \quad 2b_2 + 9b_0 = 0 \\ x^1&: \quad 6b_3 + 9b_1 = 0 \\ x^2&: \quad 12b_4 + 9b_2 = 0 \\ x^3&: \quad 20b_5 + 9b_3 = 0 \end{aligned}$$

Solving these in terms of b_0 and b_1 results in

$$b_2 = -\frac{9}{2}b_0, \quad b_4 = -\frac{3}{4}b_2 = \frac{27}{8}b_0, \quad b_3 = -\frac{3}{2}b_1,$$

$$b_5 = -\frac{9}{20}b_3 = \frac{27}{40}b_1, \cdots$$

Substituting back into the original series gives the solution as

$$u(x) = b_0 + b_1 x + \left(-\frac{9}{2}b_0\right)x^2 + \left(-\frac{3}{2}b_1\right)x^3 + \left(\frac{27}{8}b_0\right)x^4 + \left(\frac{27}{40}b_1\right)x^5 + \cdots$$
$$= b_0\left[1 - \frac{9}{2}x^2 + \frac{27}{8}x^4 + \cdots\right] + b_1\left[x - \frac{3}{2}x^3 + \frac{27}{40}x^5 + \cdots\right]$$

This is the power-series solution. The second-order differential equation again yields two arbitrary coefficients, as it must.

We can put this solution in a more recognizable form, as follows:

$$u(x) = b_0\left[1 - \frac{(3x)^2}{2!} + \frac{(3x)^4}{4!} - \cdots\right] + \frac{b_1}{3}\left[3x - \frac{(3x)^3}{3!} + \frac{(3x)^5}{5!} - \cdots\right]$$
$$= A\sin 3x + B\cos 3x$$

where $A = b_0$ and $B = b_1/3$. This is, of course, the solution that we would expect using the methods of Chapter 1. It is not always possible, though, to put the power-series solution in a form that is recognizable as elementary functions. The solution is usually left as a power series.

Example 2.4

Using the power-series method, solve the differential equation

$$\frac{d^3u}{dx^3} + \frac{d^2u}{dx^2} + \frac{du}{dx} - u = 0$$

Solution

Assume the power-series solution

$$u(x) = \sum_{n=0}^{\infty} b_n x^n$$

Substitute into the differential equation to find

$$\sum_{n=3}^{\infty} n(n-1)(n-2)b_n x^{n-3} + \sum_{n=2}^{\infty} n(n-1)b_n x^{n-2} - \sum_{n=1}^{\infty} nb_n x^{n-1} - \sum_{n=0}^{\infty} b_n x^n = 0$$

Letting $m = n - 3$ in the first series, $p = n - 2$ in the second series, and $q = n - 1$ in the third series, we have

$$\sum_{m=0}^{\infty} (m+3)(m+2)(m+1)b_{m+3}x^m + \sum_{p=0}^{\infty} (p+2)(p+1)b_{p+2}x^p$$
$$-\sum_{q=0}^{\infty} (q+1)b_{q+1}x^q - \sum_{n=0}^{\infty} b_n x^n = 0$$

Letting all the dummy indexes be n, we can write this as

$$\sum_{n=0}^{\infty}[(n+3)(n+2)(n+1)b_{n+3}+(n+2)(n+1)b_{n+2}-(n+1)b_{n+1}-b_n]x^n=0$$

Setting the quantity in brackets equal to zero gives the recurrence relation,

$$b_{n+3}=\frac{(n+2)(n+1)b_{n+2}-(n+1)b_{n+1}-b_n}{(n+3)(n+2)(n+1)}, \qquad n=0,1,2,\cdots$$

Several coefficients are

$$\begin{aligned} b_3 &= \frac{1}{6}(2b_2-b_1-b_0) \\ b_4 &= \frac{1}{24}(6b_3-2b_2-b_1)=-\frac{1}{24}(2b_1-b_0) \\ b_5 &= \frac{1}{60}(12b_4-3b_3-b_2)=-\frac{1}{120}(4b_2+b_1-2b_0) \end{aligned}$$

The solution is

$$\begin{aligned} u(x) &= b_0+b_1x+b_2x^2+\frac{1}{6}(2b_2-b_1-b_0)x^3-\frac{1}{24}(2b_1-b_0)x^4 \\ &\qquad -\frac{1}{120}(4b_2+b_1-2b_0)x^5+\cdots \\ &= b_0\left[1-\frac{x^3}{6}+\frac{x^4}{24}+\cdots\right]+b_1\left[x-\frac{x^3}{6}+\frac{x^4}{24}+\cdots\right] \\ &\qquad +b_2\left[x^2+\frac{x^3}{3}-\frac{x^5}{30}+\cdots\right] \end{aligned}$$

Note the three arbitrary coefficients and the standardized form for presenting the solution.

Example 2.5

Solve the differential equation

$$\frac{d^2u}{dx^2}+x^2u=x$$

assuming a series solution.

Solution

Assume that

$$u(x)=\sum_{n=0}^{\infty}b_nx^n$$

Substitute into the given differential equation and find

$$\sum_{n=2}^{\infty} n(n-1)b_n x^{n-2} + x^2 \sum_{n=0}^{\infty} b_n x^n = x$$

Let $n-2=m$ in the first series and multiply the x^2 times the second series; then

$$\sum_{m=0}^{\infty} (m+2)(m+1)b_{m+2}x^m + \sum_{n=0}^{\infty} b_n x^{2+n} = x$$

Now, let $n+2=m$ in the second series. We have

$$\sum_{m=0}^{\infty} (m+2)(m+1)b_{m+2}x^m + \sum_{m=2}^{\infty} b_{m-2}x^m = x$$

The first series starts at $m=0$, but the second starts at $m=2$. Thus, we must extract the first two terms from the first series. There results, letting $m=n$,

$$2b_2 + 6b_3 x + \sum_{n=2}^{\infty} (n+2)(n+1)b_{n+2}x^n + \sum_{n=2}^{\infty} b_{n-2}x^n = x$$

Now we can combine the two series, resulting in

$$2b_2 + 6b_3 x + \sum_{n=2}^{\infty} [(n+2)(n+1)b_{n+2} + b_{n-2}]x^n = x$$

Equating coefficients of the various powers of x gives

$$x^0: \quad 2b_2 = 0 \qquad \therefore \quad b_2 = 0$$

$$x^1: \quad 6b_3 = 1 \qquad \therefore \quad b_3 = \frac{1}{6}$$

$$x^n: \quad (n+2)(n+1)b_{n+2} + b_{n-2} = 0$$

$$\therefore \quad b_{n+2} = -\frac{b_{n-2}}{(n+2)(n+1)}, \qquad n = 2, 3, 4, \ldots$$

The recursion formula above allows us to write

$$b_4 = -\frac{b_0}{12}, \qquad b_8 = -\frac{b_4}{56} = \frac{b_0}{672}$$

$$b_5 = -\frac{b_1}{20}, \qquad b_9 = -\frac{b_5}{72} = -\frac{b_1}{1440}$$

$$b_6 = -\frac{b_2}{30} = 0, \qquad b_{10} = -\frac{b_6}{90} = 0$$

$$b_7 = -\frac{b_3}{42} = -\frac{1}{252}, \qquad b_{11} = -\frac{b_7}{110} = \frac{1}{27720}$$

The solution is found by substituting into

$$
\begin{aligned}
u(x) &= b_0 + b_1 x + b_2 x^2 + \cdots \\
&= b_0 + b_1 x - \frac{x^3}{6} - \frac{b_0}{12}x^4 - \frac{b_1}{20}x^5 - \frac{x^7}{252} + \frac{b_0}{672}x^8 + \frac{b_1}{1440}x^9 + \frac{x^{11}}{27720} + \cdots \\
&= b_0\left(1 - \frac{x^4}{12} + \frac{x^8}{672} + \cdots\right) + b_1\left(x - \frac{x^5}{20} + \frac{x^9}{1440} + \cdots\right) \\
&\qquad + \frac{x^3}{6} - \frac{x^7}{252} + \frac{x^{11}}{27720} + \cdots
\end{aligned}
$$

This solution is of the form

$$u(x) = b_0 u_1(x) + b_1 u_2(x) + u_p(x)$$

Since we treated a nonhomogeneous equation directly, the solution included the particular solution.

where $u_p(x)$ results because the differential equation was nonhomogeneous. The arbitrary constants b_0 and b_1 would be evaluated from given conditions. It was necessary to determine so many coefficients so that three terms in each series could be specified.

Example 2.6

Solve the differential equation

$$x\frac{d^2u}{dx^2} + u = 0$$

expanding about the point $x = 1$. Find the specific solution if $u(1) = 2$ and $du/dx(1) = 2$, and find an approximate value for $u(x)$ at $x = 1.5$.

Solution

If put in the form of Eq. 2.1.11, we note that $g(x) = 1/x$, which is singular at $x = 0$. Hence, we cannot expand with $a = 0$ so we choose the point $x = 1$ to expand about. To do this we must express the coefficients of the given differential equation in terms of $(x - 1)$. This is accomplished as follows:

$$[(x-1)+1]\frac{d^2u}{dx^2} + u = 0$$

Assume the power-series solution

$$u(x) = \sum_{n=0}^{\infty} b_n (x-1)^n$$

Substitute into the given equation and obtain

$$\sum_{n=2}^{\infty} n(n-1)b_n(x-1)^{n-2} + \sum_{n=2}^{\infty} n(n-1)b_n(x-1)^{n-1} + \sum_{n=0}^{\infty} b_n(x-1)^n = 0$$

Let $m = n - 2$ in the first series and $p = n - 1$ in the second series. There results, again letting $m = n$ and $p = n$,

$$\sum_{n=0}^{\infty}(n+2)(n+1)b_{n+2}(x-1)^n + \sum_{n=1}^{\infty}(n+1)nb_{n+1}(x-1)^n + \sum_{n=0}^{\infty}b_n(x-1)^n = 0$$

We can collect all terms under one summation symbol after we extract the extra terms in the first and last series. This gives us

$$b_0 + 2b_2 + \sum_{n=1}^{\infty}[(n+2)(n+1)b_{n+2} + (n+1)nb_{n+1} + b_n](x-1)^n = 0$$

This equation yields the results

$$b_2 = -\frac{1}{2}b_0$$

$$b_{n+2} = -\frac{n(n+1)b_{n+1} + b_n}{(n+2)(n+1)}, \qquad n = 1, 2, 3, \cdots$$

Several additional coefficients are then

$$b_3 = -\frac{1}{6}(2b_2 + b_1) = -\frac{1}{6}(b_1 - b_0)$$

$$b_4 = -\frac{1}{12}(6b_3 + b_2) = \frac{1}{24}(2b_1 - b_0)$$

$$b_5 = -\frac{1}{20}(12b_4 + b_3) = -\frac{1}{120}(5b_1 - 2b_0)$$

The solution is now written as

$$u(x) = b_0 + b_1(x-1) + \left(-\frac{b_0}{2}\right)(x-1)^2 + \left(-\frac{1}{6}\right)(b_1 - b_0)(x-1)^3 + \cdots$$

$$= b_0\left[1 - \frac{(x-1)^2}{2} + \frac{(x-1)^3}{6} - \cdots\right] + b_1\left[(x-1) - \frac{(x-1)^3}{6} + \frac{(x-1)^4}{12} - \cdots\right]$$

To evaluate b_0 and b_1 we use the conditions given:

$$u(1) = 2 = b_0[1 + 0 + 0 + \cdots] + b_1[0 + 0 + 0 + \cdots]$$

$$\frac{du}{dx}(1) = 2 = b_0[0 + 0 + 0 + \cdots] + b_1[1 + 0 + 0 + \cdots]$$

Thus,

$$b_0 = 2, \qquad b_1 = 2$$

The solution becomes

$$u(x) = 2x - (x-1)^2 + \frac{(x-1)^4}{12} - \frac{1}{20}(x-1)^5 + \cdots$$

At $x = 1.5$ the dependent variable $u(x)$ takes the approximate value,

$$u(1.5) = 3 - 0.25 + 0.0052 - 0.00156 + \cdots \cong \underline{2.754}$$

2.3 LEGENDRE'S EQUATION

A differential equation that attracts much attention in the solution of a number of physical problems is *Legendre's equation*,

$$(1-x^2)\frac{d^2u}{dx^2} - 2x\frac{du}{dx} + \lambda(\lambda+1)u = 0 \tag{2.3.1}$$

It is encountered most often when modeling a phenomenon in spherical coordinates. The parameter λ is a nonnegative, real constant. Legendre's equation is written in standard form as

$$\frac{d^2u}{dx^2} - \frac{2x}{1-x^2}\frac{du}{dx} + \frac{\lambda(\lambda+1)}{1-x^2}u = 0 \tag{2.3.2}$$

The solution for a negative value of λ, say λ_n, where $\lambda_n \le -1$, is the same as that for $\lambda = -(\lambda_n + 1)$; hence, it is sufficient to consider only nonnegative values.

The variable coefficients can be expressed as a power series about the origin and thus are analytic at $x = 0$. They are not analytic at $x = \pm 1$. Let us find the power-series solution of Legendre's equation valid for $-1 < x < 1$.

Assume a power-series solution

$$u(x) = \sum_{n=0}^{\infty} b_n x^n \tag{2.3.3}$$

Substitute into Eq. 2.3.1 and let $\lambda(\lambda+1) = k$. Then

$$(1-x^2)\sum_{n=2}^{\infty} n(n-1)b_n x^{n-2} - 2x\sum_{n=1}^{\infty} nb_n x^{n-1} + k\sum_{n=0}^{\infty} b_n x^n = 0 \tag{2.3.4}$$

This can be written as

$$\sum_{n=2}^{\infty} n(n-1)b_n x^{n-2} - \sum_{n=2}^{\infty} n(n-1)b_n x^n - \sum_{n=1}^{\infty} 2nb_n x^n + \sum_{n=0}^{\infty} kb_n x^n = 0 \tag{2.3.5}$$

The first sum can be rewritten as

$$\sum_{n=2}^{\infty} n(n-1)b_n x^{n-2} = \sum_{n=0}^{\infty} (n+2)(n+1)b_{n+2} x^n \tag{2.3.6}$$

Then, extracting the terms for $n = 0$ and $n = 1$, Eq. 2.3.5 becomes

$$\sum_{n=2}^{\infty}\{(n+2)(n+1)b_{n+2} - [n(n-1) + 2n - k]b_n\}x^n + 2b_2 + kb_0 \\ + (6b_3 - 2b_1 + kb_1)x = 0 \qquad (2.3.7)$$

Equating coefficients of like powers of x to zero, we find that

$$\begin{aligned} b_2 &= -\frac{k}{2}b_0 \\ b_3 &= \frac{2-k}{6}b_1 \\ b_{n+2} &= \frac{n^2 + n - k}{(n+2)(n+1)}b_n, \qquad n = 2, 3, 4, \cdots \end{aligned} \qquad (2.3.8)$$

Substituting $\lambda(\lambda + 1) = k$ back into the coefficients, we have

$$b_{n+2} = \frac{(n-\lambda)(n+\lambda+1)}{(n+2)(n+1)}b_n, \qquad n = 2, 3, 4, \cdots \qquad (2.3.9)$$

There are two arbitrary coefficients b_0 and b_1. The coefficients with even subscripts can be expressed in terms of b_0 and those with odd subscripts in terms of b_1. The solution can then be written as

$$u(x) = b_0 u_1(x) + b_1 u_2(x) \qquad (2.3.10)$$

where

$$u_1(x) = 1 - \frac{\lambda(\lambda+1)}{2!}x^2 + \frac{(\lambda-2)\lambda(\lambda+1)(\lambda+3)}{4!}x^4 + \cdots \qquad (2.3.11)$$

and

$$u_2(x) = x - \frac{(\lambda-1)(\lambda+2)}{3!}x^3 + \frac{(\lambda-3)(\lambda-1)(\lambda+2)(\lambda+4)}{5!}x^5 + \cdots \qquad (2.3.12)$$

are the two independent solutions.

Let us investigate these solutions for various positive integer values of λ. If λ is an even integer,

$$\begin{aligned} \lambda &= 0, \qquad u_1(x) = 1 \\ \lambda &= 2, \qquad u_1(x) = 1 - 3x^2 \\ \lambda &= 4, \qquad u_1(x) = 1 - 10x^2 + \frac{35}{3}x^4, \qquad \text{etc.} \end{aligned} \qquad (2.3.13)$$

All the higher-power terms contain factors that are zero. Thus, only polynomials result. For odd integers,

$$\begin{aligned} \lambda &= 1, \qquad u_2(x) = x \\ \lambda &= 3, \qquad u_2(x) = x - \frac{5}{3}x^3 \\ \lambda &= 5, \qquad u_2(x) = x - \frac{14}{3}x^3 + \frac{21}{5}x^5, \qquad \text{etc.} \end{aligned} \qquad (2.3.14)$$

The polynomials above represent independent solutions to Legendre's equation for the various λ's indicated; that is, if $\lambda = 5$, one independent solution is $x - \frac{14}{3}x^3 + \frac{21}{5}x^5$. Obviously, if $u_1(x)$ is a solution to the differential equation, then $Cu_1(x)$, where C is a constant, is also a solution. We shall choose the constant C such that the polynomials above all have the value unity at $x = 1$. If we do that, the polynomials are called *Legendre polynomials*. Several are

$$\begin{aligned} P_0(x) &= 1, & P_1(x) &= x \\ P_2(x) &= \frac{1}{2}(3x^2 - 1), & P_3(x) &= \frac{1}{2}(5x^3 - 3x) \\ P_4(x) &= \frac{1}{8}(35x^4 - 30x^2 + 3), & P_5(x) &= \frac{1}{8}(63x^5 - 70x^3 + 15x) \end{aligned} \tag{2.3.15}$$

We can write Legendre polynomials in the general form

$$P_\lambda(x) = \sum_{n=0}^{N} (-1)^n \frac{(2\lambda - 2n)!}{2^\lambda n!(\lambda - n)!(\lambda - 2n)!} x^{\lambda - 2n} \tag{2.3.16}$$

where $N = \lambda/2$ if λ is even and $N = (\lambda - 1)/2$ if λ is odd. Some Legendre polynomials are sketched in Fig. 2.2.

When λ is an even integer, $u_2(x)$ has the form of an infinite series, and when λ is an odd integer, $u_1(x)$ is expressed as an infinite series. *Legendre's functions of the second kind* are multiples of the infinite series defined by

$$Q_\lambda(x) = \begin{cases} u_1(1)u_2(x), & \lambda \text{ even} \\ -u_2(1)u_1(x), & \lambda \text{ odd} \end{cases} \tag{2.3.17}$$

The general solution of Legendre's equation is now written as

$$u(x) = c_1 P_\lambda(x) + c_2 Q_\lambda(x) \tag{2.3.18}$$

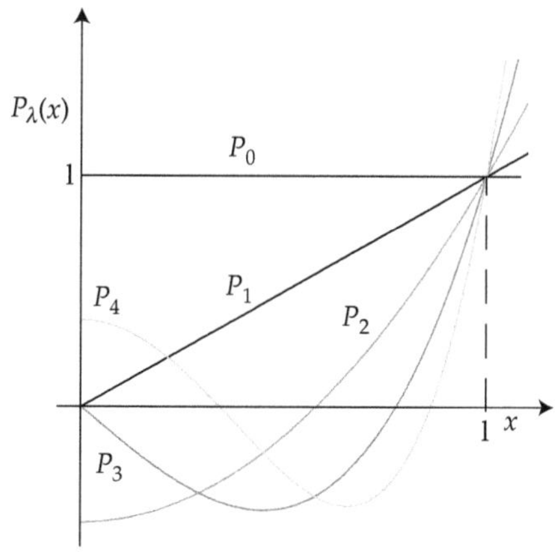

FIGURE 2.2 Legendre polynomials.

Several Legendre functions of the second kind can be shown, by involved manipulation, to be

$$\begin{aligned}
Q_0(x) &= \frac{1}{2}\ln\frac{1+x}{1-x} \\
Q_1(x) &= xQ_0(x) - 1 \\
Q_2(x) &= P_2(x)Q_0(x) - \frac{3}{2}x \\
Q_3(x) &= P_3(x)Q_0(x) - \frac{5}{2}x^2 + \frac{2}{3} \\
Q_4(x) &= P_4(x)Q_0(x) - \frac{35}{8}x^3 + \frac{55}{24}x \\
Q_5(x) &= P_5(x)Q_0(x) - \frac{63}{8}x^4 + \frac{49}{8}x^2 - \frac{8}{15}
\end{aligned} \tag{2.3.19}$$

Note that all the functions are singular at the point $x = 1$, since $Q_0(x) \to \infty$ as $x \to 1$, and thus the functions above are valid only for $|x| < 1$.

If we make the change of variables $x = \cos\phi$, we transform Legendre's equation (2.3.1) into

$$\frac{d^2u}{d\phi^2} + \cot\phi\frac{du}{d\phi} + \lambda(\lambda+1)u = 0 \tag{2.3.20}$$

or, equivalently,

$$\frac{1}{\sin\phi}\frac{d}{d\phi}\left(\sin\phi\frac{du}{d\phi}\right) + \lambda(\lambda+1)u = 0 \tag{2.3.21}$$

Legendre's equations of this form arise in various physical problems in which spherical coordinates are used.

For example, the treatment of electrical potential on a sphere in Section 6.7 leads to Legendre's equation.

Example 2.7

Find the specific solution to the differential equation

$$(1-x^2)\frac{d^2u}{dx^2} - 2x\frac{du}{dx} + 12u = 0 \qquad \text{if } \frac{du}{dx}(0) = 4$$

and the function $u(x)$ is well behaved at $x = 1$. This latter condition is often imposed in physical situations.

Solution

We note that the given differential equation is Legendre's equation with λ determined from

$$\lambda(\lambda+1) = 12$$

This can be written as

$$(\lambda+4)(\lambda-3) = 0$$

giving

$$\lambda = -4,\ 3$$

We choose the positive root and write the general solution as

$$u(x) = c_1 P_3(x) + c_2 Q_3(x)$$

If the function is to be well behaved at $x = 1$, we must let $c_2 = 0$, since $Q_3(1)$ is not defined. The other condition gives

$$4 = c_1 \frac{dP_3}{dx}(0) = -\frac{3}{2}c_1 \quad \text{or} \quad c_1 = -\frac{8}{3}$$

The solution is then

$$u(x) = -\frac{4}{3}(5x^3 - 3x)$$

2.4 THE METHOD OF FROBENIUS

There are second-order differential equations that appear in physical applications which have coefficients that cannot be expressed in power series at the point* $x = 0$; the origin is a singular point. The method illustrated in the previous section cannot be used to obtain a solution valid about $x = 0$ for such equations. Consider the second-order equation

$$\frac{d^2u}{dx^2} + \frac{f(x)}{x}\frac{du}{dx} + \frac{g(x)}{x^2}u = 0 \tag{2.4.1}$$

where $f(x)$ and $g(x)$ are analytic at $x = 0$. A differential equation of this form can be solved by the method of Frobenius. The solution is of the form

$$u(x) = x^r \sum_{n=0}^{\infty} a_n x^n \tag{2.4.2}$$

where r may be complex and $a_0 \neq 0$.

To solve Eq. 2.4.1 we write it in a more convenient form:

$$x^2 \frac{d^2u}{dx^2} + xf(x)\frac{du}{dx} + g(x)u = 0 \tag{2.4.3}$$

and express $f(x)$ and $g(x)$ as

$$f(x) = \sum_{n=0}^{\infty} b_n x^n, \qquad g(x) = \sum_{n=0}^{\infty} c_n x^n \tag{2.4.4}$$

*In general, the singular point may exist at $x = a$. We would simply define a new variable $x - a = t$, with the result that the singular point exists at the origin $t = 0$.

The derivatives du/dx and d^2u/dx^2 are

$$\frac{du}{dx} = \sum_{n=0}^{\infty}(n+r)a_n x^{n+r-1}, \qquad \frac{d^2u}{dx^2} = \sum_{n=0}^{\infty}(n+r-1)(n+r)a_n x^{n+r-2} \tag{2.4.5}$$

Substitution of the series expressions into Eq. 2.4.3 yields, in expanded form,

$$\begin{aligned} r(r-1)a_0x^r + (r+1)ra_1x^{r+1} + \cdots + (b_0 + b_1x + \cdots)\big[ra_0x^r \\ + (r+1)a_1x^{r+1} + \cdots\big] + (c_0 + c_1x + \cdots)a_0x^r \\ + a_1x^{r+1} + \cdots) = 0 \end{aligned} \tag{2.4.6}$$

Sorting out the coefficient of x^r and setting the coefficient to zero gives

$$r(r-1) + b_0r + c_0 = 0 \tag{2.4.7}$$

which is called the *indicial equation.* It has two roots, which (1) may be distinct and not differ by an integer, (2) may both be equal (a double root), and (3) may differ by an integer. We shall discuss each case separately.

2.4.1 Distinct Roots Not Differing by an Integer

Let the two roots of the indicial equation be r_1 and r_2. For the root r_1 a solution is

$$u_1(x) = \sum_{n=0}^{\infty} a_n x^{n+r_1} = (a_0 + a_1x + a_2x^2 + \cdots)x^{r_1} \tag{2.4.8}$$

For the root r_2 another solution is

$$u_2(x) = \sum_{n=0}^{\infty} d_n x^{n+r_2} = (d_0 + d_1x + d_2x^2 + \cdots)x^{r_2} \tag{2.4.9}$$

where, in general, the a_n's and d_n's will be different. The general solution is then

$$u(x) = Au_1(x) + Bu_2(x) \tag{2.4.10}$$

An example follows.

Example 2.8

Find the general solution, valid near the origin, of the differential equation

$$8x^2\frac{d^2u}{dx^2} + 6x\frac{du}{dx} + (x-1)u = 0$$

Solution

We recognize that the given equation is singular at the origin and has the same form as Eq. 2.4.3. Thus, we assume that

$$u(x) = \sum_{n=0}^{\infty} a_n x^{n+r}$$

This is substituted into the differential equation to obtain

$$\sum_{n=0}^{\infty} 8(n+r)(n+r-1)a_n x^{n+r} + \sum_{n=0}^{\infty} 6(n+r)a_n x^{n+r} + \sum_{n=0}^{\infty} a_n x^{n+r+1} - \sum_{n=0}^{\infty} a_n x^{n+r} = 0$$

In the third term we let $n + 1 = m$, so that $n + r + 1 = m + r$, and the series starts at $m = 1$. Extracting the first term from each of the series starting at $n = 0$, the equation above becomes

$$a_0[8r(r-1) + 6r - 1]x^r + \sum_{n=1}^{\infty} \{[8(n+r)(n+r-1) + 6(n+r) - 1]a_n + a_{n-1}\} x^{n+r} = 0$$

The indicial equation is then the coefficient of the x^r term; it is

$$8r(r-1) + 6r - 1 = 0$$

The recurrence relation is found by setting the quantity in braces to zero, resulting in

$$a_n = -\frac{a_{n-1}}{8(n+r)(n+r-1) + 6(n+r) - 1}, \qquad n = 1, 2, 3, \cdots$$

The two roots of the indicial equation are

$$r_1 = \frac{1}{2}, \qquad r_2 = -\frac{1}{4}$$

For the first root, $r_1 = \frac{1}{2}$, the recurrence relation gives

$$a_n = -\frac{a_{n-1}}{2n(4n+3)}, \qquad n = 1, 2, 3, \cdots$$

Several coefficients are

$$a_1 = -\frac{a_0}{14}, \quad a_2 = -\frac{a_1}{44} = \frac{a_0}{616}, \quad a_3 = -\frac{a_2}{90} = -\frac{a_0}{55440}$$

The first independent solution is then

$$u_1(x) = a_0 \left[x^{1/2} - \frac{x^{3/2}}{14} + \frac{x^{5/2}}{616} - \frac{x^{7/2}}{55440} + \cdots \right]$$

The second independent solution is found by using the root $r_2 = -1/4$. The recurrence relation provides us with, letting the coefficients be denoted by d_n,

$$d_n = -\frac{d_{n-1}}{2n(4n-3)}, \qquad n = 1, 2, 3, \cdots$$

Several coefficients in this second solution are

$$d_1 = -\frac{d_0}{2}, \quad d_2 = -\frac{d_1}{20} = \frac{d_0}{40}, \quad d_3 = -\frac{d_2}{54} = -\frac{d_0}{2160}$$

The second independent solution is then

$$u_2(x) = d_0\left[x^{-1/4} - \frac{x^{3/4}}{2} + \frac{x^{7/4}}{40} - \frac{x^{11/4}}{2160} + \cdots\right]$$

Finally, the general solution is

$$u(x) = a_0 x^{1/2}\left[1 - \frac{x}{14} + \frac{x^2}{616} + \cdots\right] + d_0 x^{-1/4}\left[1 - \frac{x}{2} + \frac{x^2}{40} - \cdots\right]$$

If the solution is to have a finite value at $x = 0$, the coefficient d_0 would be set to zero. This would be a condition imposed in a particular problem. Another condition would be sufficient to determine a_0.

2.4.2 Double Roots

The first solution is obtained as before and is

$$u_1(x) = \sum_{n=0}^{\infty} a_n x^{n+r} = (a_0 + a_1 x + a_2 x^2 + \cdots)x^r \tag{2.4.11}$$

The roots to the indicial equation are

$$r = \frac{1}{2}\left[1 - b_0 \pm \sqrt{(1-b_0)^2 - 4c_0}\right] \tag{2.4.12}$$

The roots are equal if $(1 - b_0)^2 = 4c_0$. Then the root is $r = (1 - b_0)/2$. The second solution is then assumed to have the form

$$u_2(x) = v(x)u_1(x) \tag{2.4.13}$$

The derivatives are

$$\frac{du_2}{dx} = v\frac{du_1}{dx} + u_1\frac{dv}{dx}, \qquad \frac{d^2u_2}{dx^2} = v\frac{d^2u_1}{dx^2} + 2\frac{dv}{dx}\frac{du_1}{dx} + u_1\frac{d^2v}{dx^2} \tag{2.4.14}$$

Substitute into the differential equation (2.4.3) and obtain

$$x^2\left(\frac{d^2v}{dx^2}u_1 + 2\frac{dv}{dx}\frac{du_1}{dx} + v\frac{d^2u_1}{dx^2}\right) + xf\left(\frac{dv}{dx}u_1 + v\frac{du_1}{dx}\right) + gvu_1 = 0 \tag{2.4.15}$$

Note that the quantity

$$v\left(x^2\frac{d^2u_1}{dx^2} + xf\frac{du_1}{dx} + gu_1\right) = 0 \tag{2.4.16}$$

since $u_1(x)$ is a solution to the differential equation given by Eq. 2.4.3. Thus,

$$u_1 x^2 \frac{d^2 v}{dx^2} + \left(2x^2 \frac{du_1}{dx} + xfu_1\right)\frac{dv}{dx} = 0 \tag{2.4.17}$$

Dividing by $x^2 u_1$, we find that

$$\frac{d^2 v}{dx^2} + \left[\frac{2(du_1/dx)}{u_1} + \frac{f}{x}\right]\frac{dv}{dx} = 0 \tag{2.4.18}$$

In terms of their respective series, we may write

$$\frac{f}{x} = \frac{b_0}{x} + b_1 + b_2 x + \cdots \tag{2.4.19}$$

$$\begin{aligned}
\frac{du_1/dx}{u_1} &= \frac{(a_1 + 2a_2 x + \cdots)x^r + (a_0 + a_1 x + \cdots)rx^{r-1}}{(a_0 + a_1 x + a_2 x^2 + \cdots)x^r} \\
&= \frac{1}{x}\frac{ra_0 + (r+1)a_1 x + \cdots}{a_0 + a_1 x + \cdots} \\
&= \frac{r}{x}\left(1 + \frac{a_1}{a_0}x + \cdots\right)
\end{aligned} \tag{2.4.20}$$

Insert this result into Eq. 2.4.18 and there results

$$\frac{d^2 v}{dx^2} + \left[\frac{2r + b_0}{x} + \left(\frac{ra_1}{a_0} + b_2\right) + \cdots\right]\frac{dv}{dx} = 0 \tag{2.4.21}$$

Using $r = (1 - b_0)/2$, we find that

$$\frac{d^2 v}{dx^2} + \left\{\frac{1}{x} + \left[(1 - b_0)\frac{a_1}{a_0} + b_1\right] + \cdots\right\}\frac{dv}{dx} = 0 \tag{2.4.22}$$

or

$$\frac{d(dv/dx)}{dv/dx} = \left[-\frac{1}{x} + k_0 + k_1 x + \cdots\right]dx \tag{2.4.23}$$

where the k's are constants. The result above can be integrated to yield

$$\ln\frac{dv}{dx} = -\ln x + k_0 x + \frac{k_1}{2}x^2 + \cdots$$

or

$$\frac{dv}{dx} = e^{-\ln x + k_0 x + \cdots} = \frac{1}{x}e^{k_0 x + (k_1/2)x^2 + \cdots} \tag{2.4.24}$$

Integrate again, using $e^{k_0 x + (k_1/2)x^2 + \cdots} = 1 + k_0 x + \frac{1}{2}\left(k_0^2 + k_1\right)x^2 + \cdots$ (see Eqs. 2.1.6), and we have

$$v(x) = \ln x + k_0 x + \frac{1}{4}\left(k_0^2 + k_1\right)x^2 + \cdots \tag{2.4.25}$$

Finally,

$$u_2(x) = u_1(x)\left[\ln x + k_0 x + \frac{1}{4}\left(k_0^2 + k_1\right)x^2 + \cdots\right]$$
$$= u_1(x)\ln x + (a_0 + a_1 x + \cdots)\left[k_0 + \frac{1}{4}\left(k_0^2 + k_1\right)x + \cdots\right]x^{r+1}$$
$$= u_1(x)\ln x + x^{r+1}\sum_{n=0}^{\infty} A_n x^n \quad (2.4.26)$$

The general solution is then

$$u(x) = Au_1(x) + Bu_2(x) \quad (2.4.27)$$

If double roots occur, it is important to note that the solution contains the logarithmic term.

Example 2.9

Find the general solution to the differential equation

$$x(1-x)\frac{d^2u}{dx^2} + \frac{du}{dx} - u = 0$$

valid about the origin.

Solution

We observe that the differential equation can be written as

$$\frac{d^2u}{dx^2} + \frac{1/(1-x)}{x}\frac{du}{dx} + \frac{x/(1-x)}{x^2}u = 0$$

so that

$$f(x) = \frac{1}{1-x} = 1 + x + x^2 + \cdots$$
$$g(x) = \frac{x}{1-x} = x + x^2 + x^3 + \cdots$$

Frobenius's method can be used since $f(x)$ and $g(x)$ are analytic at $x = 0$. Assume the solution to have the form

$$u(x) = \sum_{n=0}^{\infty} a_n x^{n+r}$$

Insert this into the original differential equation and find that

$$\sum_{n=0}^{\infty}(n+r)(n+r-1)a_n x^{n+r-1} - \sum_{n=0}^{\infty}(n+r)(n+r-1)a_n x^{n+r}$$
$$+\sum_{n=0}^{\infty}(n+r)a_n x^{n+r-1} - \sum_{n=0}^{\infty} a_n x^{n+r} = 0$$

Letting $n - 1 = m$ in the first and third series and extracting the first term in each of these two series gives

$$-[r(r-1)+r]a_0x^{r-1} + \sum_{n=0}^{\infty}\{[(n+r+1)(n+r)+n+r+1]a_{n+1} \\ -[(n+r)(n+r-1)+1]a_n\}x^{n+r} = 0$$

The coefficient of the term with the lowest exponent is set equal to zero, giving

$$r^2 = 0$$

This yields the double root $r_1 = r_2 = 0$. Next, we set the coefficient in braces equal to zero and let $r = 0$, with the result that

$$a_{n+1} = \frac{n^2 - n + 1}{(n+1)^2}a_n, \qquad n = 0, 1, 2, \cdots$$

Several coefficients are determined to be

$$a_1 = a_0, \quad a_2 = \frac{a_1}{4} = \frac{a_0}{4}, \quad a_3 = \frac{a_2}{3} = \frac{a_0}{12}$$

The first independent solution is thus*

$$u_1(x) = a_0\left[1 + x + \frac{x^2}{4} + \frac{x^3}{12} + \cdots\right]$$

The second solution is, following the procedure outlined in the preceding section,

$$u_2(x) = v(x)u_1(x)$$

This is substituted into the differential equation to obtain

$$x(1-x)\left[u_1\frac{d^2v}{dx^2} + 2\frac{du_1}{dx}\frac{dv}{dx} + v\frac{d^2u_1}{dx^2}\right] + u_1\frac{dv}{dx} + v\frac{du_1}{dx} - vu_1 = 0$$

Since $u_1(x)$ satisfies the original differential equation, the equation above reduces to

$$x(1-x)\left(u_1\frac{d^2v}{dx^2} + 2\frac{du_1}{dx}\frac{dv}{dx}\right) + u_1\frac{dv}{dx} = 0$$

If we substitute the expression for $u_1(x)$ into the result above (we can set $a_0 = 1$ with no loss of generality), we have

$$\frac{d(dv/dx)}{dv/dx} = \frac{\left(1 + x + \frac{x^2}{4} + \frac{x^3}{12} + \cdots\right) + 2x(1-x)\left(1 + \frac{x}{2} + \frac{x^2}{4} + \cdots\right)}{x(x-1)\left(1 + x + \frac{x^2}{4} + \frac{x^3}{12} + \cdots\right)}dx$$

$$= -\left(\frac{1}{x} + 3 + 2x^2 + \cdots\right)dx$$

*Note that we could let $a_0 = 1$, since the independent solutions are multiplied by an arbitrary constant when the general solution is formed; that is, we could absorb the a_0 in A of Eq. 2.4.27. This is often done.

Integrate once to obtain

$$\ln\frac{dv}{dx} = -\ln x - 3x - \frac{2}{3}x^3 + \cdots$$

or

$$\begin{aligned}\frac{dv}{dx} &= e^{-\ln x - 3x - (2/3)x^3 + \cdots} \\ &= \frac{1}{x}e^{-3x-(2/3)x^3+\cdots} \\ &= \frac{1}{x}\left[1 - 3x + \frac{9}{2}x^2 - \frac{29}{3}x^3 + \cdots\right]\end{aligned}$$

Integrating once again results in

$$v(x) = \ln x - 3x + \frac{9}{4}x^2 - \frac{29}{9}x^3 + \cdots$$

The general solution is finally

$$u(x) = \left(1 + x\frac{x^2}{4} + \frac{x^3}{12} + \cdots\right) \times \left[A + B\left(\ln x - 3x + \frac{9}{4}x^2 - \frac{29}{9}x^3 + \cdots\right)\right]$$

Rather than follow the procedure outlined above we could have substituted the final expression in Eq. 2.4.26 for $u_2(x)$ into the differential equation and solved for the coefficients A_n. This is often done.

2.4.3 Roots Differing by an Integer

If the roots differ by an integer, one independent solution is

$$u_1(x) = \sum_{n=0}^{\infty} a_n x^{n+r_1} \tag{2.4.28}$$

where r_1 is one root of the indicial equation. The other root is $r_2 = r_1 - p$ (p is a positive integer) and may result in a solution that is not independent if we follow the procedure of Section 2.4.1.* We again assume that

$$u_2(x) = v(x)u_1(x) \tag{2.4.29}$$

Follow the procedure outlined in Section 2.4.2 and find that

$$\frac{d^2v}{dx^2} + \left[\frac{2r_1 + b_0}{x} + \left(\frac{r_1 a_1}{a_0} + b_1\right) + \cdots\right]\frac{dv}{dx} = 0 \tag{2.4.30}$$

*Occasionally, the process is easier if we find $u_1(x)$ by using the smaller root. The technique presented here will, however, always work.

The difference of the two roots $r_1 - r_2$ is (see Eq. 2.4.12)

$$r_1 - r_2 = \sqrt{(1-b_0)^2 - 4c_0} = p \tag{2.4.31}$$

so

$$2r_1 = 1 - b_0 + p \tag{2.4.32}$$

Then Eq. 2.4.30 becomes, with $s_0 = r_1 a_1/a_0 + b_1$,

$$\frac{d^2v}{dx^2} + \left[\frac{1+p}{x} - s_0 - s_1 x + \cdots\right]\frac{dv}{dx} = 0 \tag{2.4.33}$$

Integrating, we obtain

$$\ln\left(\frac{dv}{dx}\right) = -(1+p)\ln x + s_0 x + s_1\frac{x^2}{2} + \cdots \tag{2.4.34}$$

or

$$\frac{dv}{dx} = x^{-(p+1)}e^{s_0x+\cdots} = x^{-(p+1)}[1 + k_0 x + k_1 x^2 + \cdots] \tag{2.4.35}$$

where $k_0 = s_0, k_1 = \frac{1}{2}(s_0^2 + s_1), \ldots$. Multiplying out the result above, we have

$$\frac{dv}{dx} = \frac{1}{x^{p+1}} + \frac{k_0}{x^p} + \frac{k_1}{x^{p-1}} + \cdots + \frac{k_{p-1}}{x} + k_p + k_{p+1}x + \cdots \tag{2.4.36}$$

Finally,

$$v(x) = -\frac{1}{px^p} - \frac{k_0}{(p-1)x^{p-1}} - \cdots + k_{p-1}\ln x + k_p x + \cdots \tag{2.4.37}$$

Returning to Eq. 2.4.29 the second independent solution is

$$\begin{aligned} u_2(x) &= u_1(x)v(x) \\ &= (a_0 + a_1 x + \cdots)x^{r_1}\left[-\frac{1}{px^p} - \frac{k_0}{(p-1)x^{p-1}} - \cdots \right. \\ &\qquad \left. + k_{p-1}\ln x + k_p x + \cdots\right] \\ &= k_{p-1}u_1(x)\ln x + (a_0 + a_1 x + \cdots)x^{r_1 - p} \\ &\qquad \times\left[-\frac{1}{p} - \frac{k_0 x}{p-1} - \cdots + k_p x^{p+1} + \cdots\right] \\ &= ku_1(x)\ln x + x^{r_1}\sum_{n=0}^{\infty} B_n x^n \end{aligned} \tag{2.4.38}$$

where k is a constant. For the case of roots differing by an integer, the logarithmic part of $u_2(x)$ may disappear if $k = 0$.

Example 2.10

Find the general solution, valid near the origin, to the differential equation

$$x\frac{d^2u}{dx^2} + u = 0$$

Solution

This equation is of the proper form to apply the method of Frobenius, since $f(x) = 0$ and $g(x) = x$. Assume, as before, that

$$u(x) = \sum_{n=0}^{\infty} a_n x^{n+r}$$

Substitute into the differential equation, to find

$$\sum_{n=0}^{\infty} (n+r)(n+r-1)a_n x^{n+r-1} + \sum_{n=0}^{\infty} a_n x^{n+r} = 0$$

This gives

$$[(r-1)r]a_0 x^{r-1} + \sum_{n=0}^{\infty}[(n+r+1)(n+r)a_{n+1} + a_n]x^{n+r} = 0$$

The indicial equation gives

$$r_1 = 1, \qquad r_2 = 0$$

The difference of the roots is an integer. Now, we set the quantity in brackets in the summation above equal to zero, letting $r = 1$, and obtain

$$a_{n+1} = -\frac{a_n}{(n+2)(n+1)}, \qquad n = 0, 1, 2, \cdots$$

This gives several roots as

$$a_1 = -\frac{a_0}{2}, \qquad a_2 = -\frac{a_1}{6} = \frac{a_0}{12}, \qquad a_3 = -\frac{a_2}{12} = -\frac{a_0}{144}$$

The first independent solution is then

We can use $r = 0$ to obtain the same result for $u_1(x)$.

$$u_1(x) = a_0\left[x - \frac{x^2}{2} + \frac{x^3}{12} - \frac{x^4}{144} + \cdots\right]$$

Next, we assume the second solution to be

$$u_2(x) = u_1(x)v(x)$$

This is substituted into the differential equation, to obtain

$$2\left(1 - x + \frac{x^2}{4} - \frac{x^3}{36} + \cdots\right)\frac{dv}{dx} + \left(x - \frac{x^2}{2} + \frac{x^3}{3} - \frac{x^4}{144} + \cdots\right)\frac{d^2v}{dx^2} = 0$$

which can be put in the form

$$\frac{d(dv/dx)}{dv/dx} = \left(-\frac{2}{x} + 1 + \frac{x}{6} + \frac{x^2}{24} + \cdots\right)dx$$

Integrating once gives

$$\ln\frac{dv}{dx} = -2\ln x + x + \frac{x^2}{12} + \frac{x^3}{72} + \cdots$$

or, equivalently,

$$\begin{aligned}\frac{dv}{dx} &= e^{-2\ln x}\left[e^{x+x^2/12+x^3/72+\cdots}\right] \\ &= \frac{1}{x^2}\left[1 + x + \frac{7}{12}x^2 + \frac{19}{72}x^3 + \cdots\right]\end{aligned}$$

Integrating again results in

$$v(x) = -\frac{1}{x} + \ln x + \frac{7}{12}x + \frac{19}{144}x^2 + \cdots$$

Using $u_2(x) = u_1(x)v(x)$, we have the general solution as

$$u(x) = \left[x - \frac{x^2}{2} + \frac{x^3}{12} - \frac{x^4}{144} + \cdots\right] \times \left[A + B\left(-\frac{1}{x} + \ln x + \frac{7}{12}x + \frac{19}{144}x^2 + \cdots\right)\right]$$

Note the $\ln x$ term in this solution. It does not always happen that the $\ln x$ term appears in the solution when the roots to the indicial equation differ by an integer. The series expansion for dv/dx may not include the term k/x, which leads to the $\ln x$ term.

2.5 BESSEL'S EQUATION

An equation that appears in a variety of problems from diverse areas is *Bessel's differential equation,*

$$x^2\frac{d^2u}{dx^2} + x\frac{du}{dx} + (x^2 - \lambda^2)u = 0 \tag{2.5.1}$$

It is most often encountered when solving problems while using cylindrical coordinates. The parameter λ is assumed for convenience to be real and nonnegative. We shall use Frobenius's method, since there is a singularity at $x = 0$ and we wish to find a series expansion in powers of x. Assume a solution of the form

$$u(x) = \sum_{n=0}^{\infty} a_n x^{n+r} \tag{2.5.2}$$

Substitution of $u(x)$ and its derivatives into Bessel's equation yields

$$\sum_{n=0}^{\infty}(n+r)(n+r-1)a_n x^{n+r} + \sum_{n=0}^{\infty}(n+r)a_n x^{n+r} + \sum_{n=0}^{\infty} a_n x^{n+r+2} - \sum_{n=0}^{\infty}\lambda^2 a_n x^{n+r} = 0 \tag{2.5.3}$$

Changing the third summation so that the exponent on x is $n + r$, we have

$$\sum_{n=0}^{\infty}[(n+r)(n+r-1)+(n+r)-\lambda^2]a_n x^{n+r} + \sum_{n=2}^{\infty} a_{n-2} x^{n+r} = 0 \tag{2.5.4}$$

Writing out the first two terms on the first summation gives

$$[r(r-1)+r-\lambda^2]a_0 x^r + [(1+r)r+(1+r)-\lambda^2]a_1 x^{1+r} + \sum_{n=2}^{\infty}\{[(n+r)(n+r-1)+(n+r)-\lambda^2]a_n + a_{n-2}\}x^{n+r} = 0 \tag{2.5.5}$$

Equating coefficients of like powers of x to zero gives

$$(r^2-\lambda^2)a_0 = 0 \tag{2.5.6}$$

$$(r^2+2r+1-\lambda^2)a_1 = 0 \tag{2.5.7}$$

$$[(n+r)^2-\lambda^2]a_n + a_{n-2} = 0 \tag{2.5.8}$$

Equation 2.5.6 requires that

$$r^2-\lambda^2 = 0 \tag{2.5.9}$$

since $a_0 \neq 0$ according to the method of Frobenius. The indicial equation above has roots $r_1 = \lambda$ and $r_2 = -\lambda$.

Next, we shall find $u_1(x)$ corresponding to $r_1 = \lambda$. Equation 2.5.7 gives $a_1 = 0$, since the quantity in parentheses is not zero. From the recursion relation (2.5.8), we find that $a_3 = a_5 = a_7 = \cdots = 0$. All the coefficients with an odd subscript vanish. For the coefficients with an even subscript, we find that

$$\begin{aligned} a_2 &= -\frac{a_0}{2^2(\lambda+1)} \\ a_4 &= -\frac{a_2}{2^2\cdot 2(\lambda+2)} = \frac{a_0}{2^4\cdot 2(\lambda+1)(\lambda+2)} \\ a_6 &= -\frac{a_4}{2^2\cdot 3(\lambda+3)} = -\frac{a_0}{2^6\cdot 3\cdot 2(\lambda+1)(\lambda+2)(\lambda+3)}, \quad \text{etc.} \end{aligned} \tag{2.5.10}$$

In general, we can relate the coefficients with even subscripts to the arbitrary coefficient a_0 by the equation

$$a_{2n} = \frac{(-1)^n a_0}{2^{2n} n!(\lambda+1)(\lambda+2)\ldots(\lambda+n)}, \quad n = 0, 1, 2, \cdots \tag{2.5.11}$$

Because a_0 is arbitrary, it is customary to normalize the a_n's by letting

$$a_0 = \frac{1}{2^\lambda \Gamma(\lambda+1)} \tag{2.5.12}$$

where the *gamma function* $\Gamma(\lambda + 1)$ is defined by

$$\Gamma(\lambda+1) = \int_0^\infty e^{-t} t^\lambda dt \tag{2.5.13}$$

Because of the common use of the gamma function, we shall present some important properties of the function and then return to our discussion of the solution to Eq. 2.5.1.

To observe some of these properties, let us integrate the integral in Eq. 2.5.13 by parts:

$$\int_0^\infty e^{-t} t^\lambda dt = -e^{-t} t^\lambda \Big|_0^\infty + \lambda \int_0^\infty t^{\lambda-1} e^{-t} dt \tag{2.5.14}$$

$$\begin{pmatrix} u = t^\lambda, & dv = e^{-t}dt \\ du = \lambda t^{\lambda-1}dt, & v = -e^{-t} \end{pmatrix}$$

The quantity $e^{-t}t^\lambda$ vanishes at $t = \infty$ and $t = 0$. Thus, we have

$$\Gamma(\lambda+1) = \lambda \int_0^\infty e^{-t} t^{\lambda-1} dt \tag{2.5.15}$$

The last integral is simply $\Gamma(\lambda)$. Thus, we have the important property,

$$\Gamma(\lambda+1) = \lambda \Gamma(\lambda) \tag{2.5.16}$$

If we let $\lambda = 0$ in Eq. 2.5.13, there results

$$\begin{aligned} \Gamma(1) &= \int_0^\infty e^{-t} dt \\ &= -e^{-t}\Big|_0^\infty = 1 \end{aligned} \tag{2.5.17}$$

Using Eq. 2.5.16, there follows

$$\begin{aligned} \Gamma(2) &= 1 \cdot \Gamma(1) = 1 \\ \Gamma(3) &= 2 \cdot \Gamma(2) = 2! \\ \Gamma(4) &= 3 \cdot \Gamma(3) = 3! \end{aligned} \tag{2.5.18}$$

The equations above represent another important property of the gamma function if λ is a positive integer,

$$\Gamma(\lambda+1) = \lambda! \tag{2.5.19}$$

It is not necessary, however, that λ be a positive integer for the gamma function to have a value. It can take on any numerical value. It is interesting to note, though, that $\Gamma(0) = \infty$. This is observed from Eq. 2.5.16 by letting $\lambda = 0$. Then, it follows that $\Gamma(-1) = -\Gamma(0) = -\infty$, $\Gamma(-2) = -\frac{1}{2}\Gamma(-1) = \infty$, etc. The gamma function of all negative integers is undefined. This is shown in Fig. 2.3. Because of the recurrence relation of Eq. 2.5.16, the gamma function is usually tabulated only for $1 \le \lambda \le 2$. This is given in Table A2 of the Appendix. An example will illustrate the use of the gamma functions in performing integrations.

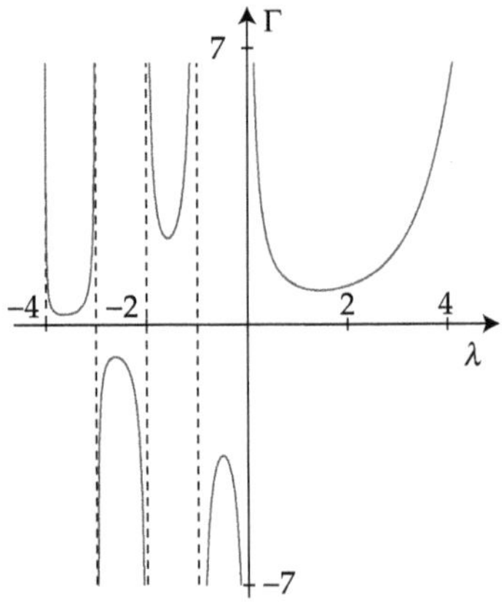

FIGURE 2.3 The gamma function.

Now, let us return to the problem of finding two independent solutions of Eq. 2.5.1. With the introduction of the normalizing factor (2.5.12) we have

$$a_{2n} = \frac{(-1)^n}{2^{2n+\lambda} n!\Gamma(\lambda+n+1)}, \qquad n = 0, 1, 2, \cdots \tag{2.5.20}$$

where we have used

$$\Gamma(\lambda+n+1) = (\lambda+n)(\lambda+n-1)\cdots(\lambda+1)\Gamma(\lambda+1) \tag{2.5.21}$$

By substituting the coefficients above into our series solution (2.5.2) (replace n with $2n$), we have found one independent solution of Bessel's equation to be

$$J_\lambda(x) = \sum_{n=0}^{\infty} \frac{(-1)^n x^{2n+\lambda}}{2^{2n+\lambda} n!\Gamma(\lambda+n+1)} \tag{2.5.22}$$

Bessel functions of the first kind can describe shapes of synchronous vibrations of circular membranes.

where $J_\lambda(x)$ is called the *Bessel function of the first kind* of order λ. The series converges for all values of x, since there are no singular points other than $x = 0$; this results in an infinite radius of convergence. Sketches of $J_0(x)$ and $J_1(x)$ are shown in Fig. 2.4. Tables giving the numerical values of $J_0(x)$ and $J_1(x)$ for $0 < x < 15$ are located in the Appendix.

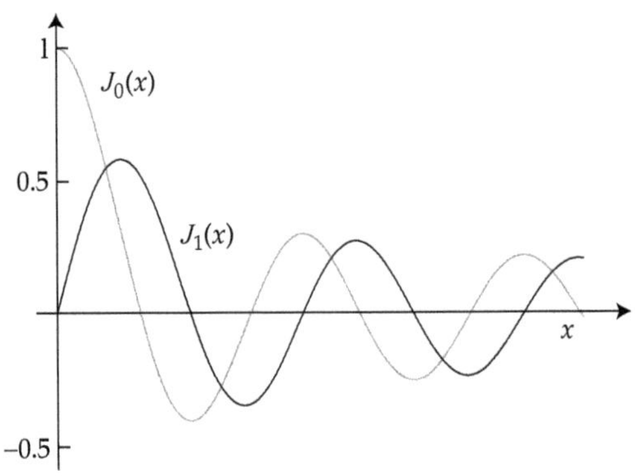

FIGURE 2.4 Bessel functions of the first kind.

The solution corresponding to $r_2 = -\lambda$ is found simply by replacing λ with $(-\lambda)$. This can be verified by following the steps leading to the expression for $J_\lambda(x)$. Hence, if λ is not an integer, the solution

$$J_{-\lambda}(x) = x^{-\lambda} \sum_{n=0}^{\infty} \frac{(-1)^n x^{2n}}{2^{2n-\lambda}\Gamma(n - \lambda + 1)} \tag{2.5.23}$$

is a second independent solution. It is singular at $x = 0$. The general solution is then

$$u(x) = AJ_\lambda(x) + BJ_{-\lambda}(x) \tag{2.5.24}$$

If λ is zero or an integer, $J_{-\lambda}(x)$ is not independent but can be shown to be related to $J_\lambda(x)$ by the relation

$$J_{-\lambda}(x) = (-1)^\lambda J_\lambda(x) \tag{2.5.25}$$

Hence, a second independent solution must be found before the general solution can be formed. For this case we follow the method of Frobenius and assume a solution of the form

$$u_2(x) = v(x)J_\lambda(x) \tag{2.5.26}$$

Substitute into Bessel's equation and we find, after some rearranging,

$$x^2 J_\lambda \frac{d^2v}{dx^2} + \left[2x^2 \frac{dJ_\lambda}{dx} + xJ_\lambda\right]\frac{dv}{dx} + \left[x^2 \frac{d^2J_\lambda}{dx^2} + x\frac{dJ_\lambda}{dx} + (x^2 - \lambda^2)J_\lambda\right]v = 0 \tag{2.5.27}$$

The group of terms in the last bracket vanishes since $J_\lambda(x)$ is a solution of Bessel's equation. We are left with

$$x^2 J_\lambda \frac{d^2v}{dx^2} + \left[2x^2 \frac{dJ_\lambda}{dx} + xJ_\lambda\right]\frac{dv}{dx} = 0 \tag{2.5.28}$$

or, equivalently,

$$\frac{d(dv/dx)}{dv/dx} + \frac{2}{J_\lambda}dJ_\lambda + \frac{dx}{x} = 0 \tag{2.5.29}$$

This can be integrated to yield*

$$\ln\left(\frac{dv}{dx}\right) + 2\ln J_\lambda + \ln x = 0 \tag{2.5.30}$$

---

*Note that we are not concerned with the constant of integration. It can be absorbed into the arbitrary constant of the general solution. We could add a term ln C to Eq. 2.5.30 with no change in the solution.

or, in a more usable form,

$$\frac{dv}{dx} = \frac{1}{xJ_\lambda^2} \tag{2.5.31}$$

Finally,

$$v = \int \frac{dx}{xJ_\lambda^2} \tag{2.5.32}$$

The second independent solution is then

$$u_2(x) = J_\lambda(x) \int \frac{dx}{xJ_\lambda^2} \tag{2.5.33}$$

Letting $\lambda = 0$, we can expand $J_0(x)$ in its series form and integrate to obtain

$$u_2(x) = J_0(x)\left[\ln x + \frac{x^2}{4} + \frac{5x^4}{128} + \cdots\right] \tag{2.5.34}$$

Letting $\lambda = 1$, we have

$$u_2(x) = J_1(x)\left[-\frac{2}{x^2} + \ln x + \frac{7x^2}{96} + \cdots\right] \tag{2.5.35}$$

Similarly, the second independent solution can be formed for λ's of larger values.

It is customary to adjust the function of Eq. 2.5.33 by a suitable factor so that numerical tabulations for various values of λ will be uniform. This is done and the second independent solution can be expressed in the series form

$$\begin{aligned} Y_\lambda(x) = \frac{2}{\pi} J_\lambda(x)\left[\ln\frac{x}{2} + \gamma\right] &+ \frac{x^1}{\pi}\sum_{n=0}^{\infty} \frac{(-1)^{n-1}(h_n + h_1)x^{2n}}{2^{2n+\lambda}\, n!(n+\lambda)!} \\ &- \frac{x^{-\lambda}}{\pi}\sum_{n=0}^{\lambda-1} \frac{(\lambda - n - 1)!\, x^{2n}}{2^{2n-\lambda}\, n!} \end{aligned} \tag{2.5.36}$$

where γ is *Euler's constant*, $\gamma = 0.57721566490$, and

$$h_n = 1 + \frac{1}{2} + \frac{1}{3} + \cdots + \frac{1}{n} \tag{2.5.37}$$

The quantity $Y_\lambda(x)$ is called *Bessel's function of the second kind* of order λ. The general solution is then

$$u(x) = AJ_\lambda(x) + BY_\lambda(x) \tag{2.5.38}$$

Bessel functions of the second kind can be used to describe vibrations of a hanging cable.

Graphs of $Y_0(x)$ and $Y_1(x)$ are shown in Fig. 2.5. Since $Y_0(x)$ and $Y_1(x)$ are not defined at $x = 0$, that is, $Y_0(0) = Y_1(0) = -\infty$, the solution (2.5.38) for a problem with a finite boundary condition at $x = 0$ requires that $B = 0$; the solution would then only involve Bessel functions of the first kind.

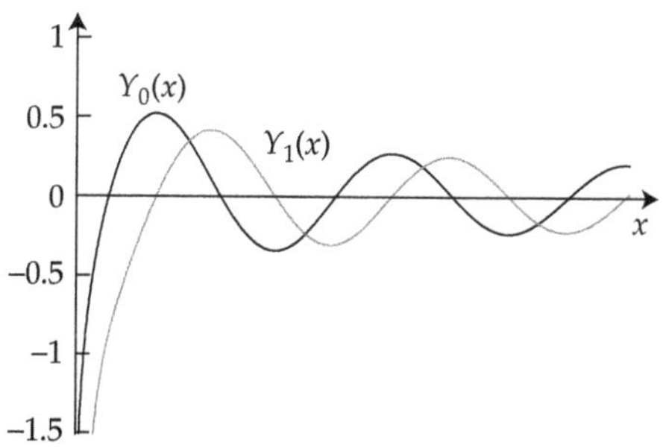

FIGURE 2.5 Bessel functions of the second kind.

In the manipulation of Bessel functions a number of helpful identities are used. We will now present some of the most important of these identities. Let us first show that

$$\frac{d}{dx}[x^{\lambda+1}J_{\lambda+1}(x)] = x^{\lambda+1}J_\lambda(x) \tag{2.5.39}$$

The series expansion (2.5.22) gives

$$x^{\lambda+1}J_{\lambda+1}(x) = \sum_{n=0}^{\infty} \frac{(-1)^n x^{2n+2\lambda+2}}{2^{2n+\lambda+1} n!\Gamma(\lambda+n+2)} \tag{2.5.40}$$

This is differentiated to yield

$$\begin{aligned} \frac{d}{dx}[x^{\lambda+1}J_{\lambda+1}(x)] &= \sum_{n=0}^{\infty} \frac{(-1)^n (2n+2\lambda+2) x^{2n+2\lambda+1}}{2^{2n+\lambda+1} n!\Gamma(\lambda+n+2)} \\ &= \sum_{n=0}^{\infty} \frac{(-1)^n 2(n+\lambda+1) x^{2n+2\lambda+1}}{2\cdot 2^{2n+\lambda} n!(\lambda+n+1)\Gamma(\lambda+n+1)} \\ &= x^{\lambda+1}\sum_{n=0}^{\infty} \frac{(-1)^n x^{2n+\lambda}}{2^{2n+\lambda} n!\Gamma(\lambda+n+1)} \\ &= x^{\lambda+1}J_\lambda(x) \end{aligned} \tag{2.5.41}$$

This proves the relationship (2.5.39). Following this procedure, we can show that

$$\frac{d}{dx}[x^{-\lambda}J_\lambda(x)] = -x^{-\lambda}J_{\lambda+1}(x) \tag{2.5.42}$$

From the two identities above we can perform the indicated differentiation on the left-hand sides and arrive at

$$\begin{aligned} x^{\lambda+1}\frac{dJ_{\lambda+1}}{dx} + (\lambda+1)x^\lambda J_{\lambda+1} &= x^{\lambda+1}J_\lambda \\ x^{-\lambda}\frac{dJ_\lambda}{dx} - \lambda x^{-\lambda-1}J_\lambda &= x^{-\lambda}J_{\lambda+1} \end{aligned} \tag{2.5.43}$$

Let us multiply the first equation above by $x^{-\lambda-1}$ and the second by x^λ. There results

$$\begin{aligned} \frac{dJ_{\lambda+1}}{dx} + \frac{\lambda+1}{x}J_{\lambda+1} &= J_\lambda \\ \frac{dJ_\lambda}{dx} - \frac{\lambda}{x}J_\lambda &= -J_{\lambda+1} \end{aligned} \tag{2.5.44}$$

If we now replace $\lambda + 1$ with λ in the first equation, we have

$$\frac{dJ_\lambda}{dx} + \frac{\lambda}{x} J_\lambda = J_{\lambda-1} \tag{2.5.45}$$

This equation can be added to the second equation of (2.5.44) to obtain

$$\frac{dJ_\lambda}{dx} = \frac{1}{2}(J_{\lambda-1} - J_{\lambda-1}) \tag{2.5.46}$$

Equation 2.5.45 can also be subtracted from the second equation of (2.5.44) to obtain the important recurrence relation

$$J_{\lambda+1}(x) = \frac{2\lambda}{x} J_\lambda(x) - J_{\lambda-1}(x) \tag{2.5.47}$$

This allows us to express Bessel functions of higher order in terms of Bessel functions of lower order. This is the reason that tables only give $J_0(x)$ and $J_1(x)$ as entries. All higher-order Bessel functions can be related to $J_0(x)$ and $J_1(x)$. By rewriting Eq. 2.5.47, we can also relate Bessel functions of higher negative order to $J_0(x)$ and $J_1(x)$. We would use

$$J_{\lambda-1}(x) = \frac{2\lambda}{x} J_\lambda(x) - J_{\lambda+1}(x) \tag{2.5.48}$$

In concluding this section, let us express the differentiation identities (2.5.39) and (2.5.42) as integration identities. By integrating once we would have

$$\begin{aligned} \int x^{\lambda+1} J_\lambda(x) dx &= x^{\lambda+1} J_{\lambda+1}(x) + C \\ \int x^{-\lambda} J_{\lambda+1}(x) dx &= -x^{-\lambda} J_\lambda(x) + C \end{aligned} \tag{2.5.49}$$

These formulas are used when integrating Bessel functions.

Example 2.11

Evaluate the integral $\int_0^\infty x^{5/4} e^{-\sqrt{x}} dx$.

Solution

The gamma functions are quite useful in evaluating integrals of this type. To make the exponent to the exponential function equal to $-t$, we let

$$x = t^2, \qquad dx = 2t\,dt$$

Then the integral becomes (the limits remain unchanged)

$$\int_0^\infty x^{5/4} e^{-\sqrt{x}} dx = 2\int_0^\infty t^{7/2} e^{-t}\, dt$$

By using Eq. 2.5.13, we have

$$2\int_0^\infty e^{-t} t^{7/2} dt = 2\Gamma\left(\frac{9}{2}\right)$$

The recurrence relation (2.5.16) gives

$$\Gamma\left(\frac{9}{2}\right) = \frac{7}{2}\cdot\frac{5}{2}\cdot\frac{3}{2}\Gamma\left(\frac{3}{2}\right) = \frac{105}{8}\Gamma\left(\frac{3}{2}\right)$$

From the tabulated values in the Appendix for the gamma function, we have

$$\Gamma\left(\frac{3}{2}\right) = 0.886$$

Finally, the value of the integral is

$$\int_0^\infty x^{5/4} e^{-\sqrt{x}} dx = 2 \times \frac{105}{8} \times 0.886 = \underline{23.3}$$

Example 2.12

Find numerical values for the quantities $J_4(3)$ and $J_{-4}(3)$ using the recurrence relations.

Solution

We use the recurrence relation (2.5.47) to write $J_4(3)$ in terms of $J_0(3)$ and $J_1(3)$. It gives

$$\begin{aligned} J_4(3) &= \frac{2\cdot 3}{3} J_3(3) - J_2(3) \\ &= 2\left[\frac{2\cdot 2}{3} J_2(3) - J_1(3)\right] - J_2(3) \\ &= \frac{5}{3}\left[\frac{2}{3} J_1(3) - J_0(3)\right] - 2J_1(3) \\ &= -\frac{8}{9} J_1(3) - \frac{5}{3} J_0(3) \\ &= -\frac{8}{9}\times 0.339 - \frac{5}{3}\times(-0.260) = \underline{0.132} \end{aligned}$$

Using the tables in the Appendix, we see that $J_0(3) = -0.26005$ and $J_1(3) = 0.33906$.

Now, to find a value for $J_{-4}(3)$ we use Eq. 2.5.48 to get

$$\begin{aligned} J_{-4}(3) &= \frac{2(-3)}{3} J_{-3}(3) - J_{-2}(3) \\ &= -2\left[\frac{2(-2)}{3} J_{-2}(3) - J_{-1}(3)\right] - J_{-2}(3) \\ &= \frac{5}{3}\left[\frac{2(-1)}{3} J_{-1}(3) - J_0(3)\right] + 2J_{-1}(3) \\ &= \frac{8}{9}[-J_1(3)] - \frac{5}{3} J_0(3) = \underline{0.132} \end{aligned}$$

We see that $J_4(x) = J_{-4}(x)$, which was shown to be the case in Eq. 2.5.25.

Example 2.13

Integrals involving Bessel functions are often encountered in the solution of physically motivated problems. Determine an expression for

$$\int x^2 J_2(x)dx$$

Solution

To use the second integration formula of (2.5.49), we put the integral in the form

$$\int x^2 J_2(x)dx = \int x^3[x^{-1}J_2(x)]dx$$

Integrate by parts:

$$\begin{aligned} u &= x^3, \quad & dv &= x^{-1}J_2(x)dx \\ du &= 3x^2, \quad & v &= -x^{-1}J_1(x) \end{aligned}$$

Then

$$\int x^2 J_2(x)dx = -x^2 J_1(x) + 3\int xJ_1(x)dx$$

Again we integrate by parts:

$$\begin{aligned} u &= x, \quad & dv &= J_1(x)dx \\ du &= dx, \quad & v &= -J_0(x) \end{aligned}$$

There results

$$\int x^2 J_2(x)dx = -x^2 J_1(x) - 3xJ_0(x) + 3\int J_0(x)dx$$

The last integral, $\int J_0(x)dx$, cannot be evaluated using our integration formulas. Because it often appears when integrating Bessel functions, it has been tabulated, although we will not include it in this work. However, we must recognize when we arrive at $\int J_0(x)dx$, our integration is complete. In general, whenever we integrate $\int x^n J_m(x)dx$ and $n + m$ is even and positive, the integral $\int J_0(x)dx$ will appear.

PROBLEMS

2.1 Derive a power-series expansion of each of the following functions by expanding in a Taylor series about $x = 0$. Recall that the Taylor series expanded about the origin can be written as

$$f(x) = f(0) + x\left.\frac{df}{dx}\right|_{x=0} + \left.\frac{x^2}{2!}\frac{d^2f}{dx^2}\right|_{x=0} + \left.\frac{x^3}{3!}\frac{d^3f}{dx^3}\right|_{x=0} + \cdots$$

(a) $\dfrac{1}{1-x}$ **(b)** e^x
(c) $\sin x$ **(d)** $\cos x$
(e) $\ln x$ **(f)** $\ln(1+x)$

2.2 Find a series expansion for each of the following functions.

(a) $\dfrac{1}{1+x}$ **(b)** $\dfrac{1}{x+2}$
(c) $\dfrac{1}{x^2+3x+2}$ **(d)** $\dfrac{7}{x^2-x-12}$
(e) e^{2x+1} **(f)** e^{-x^2}
(g) $\sin x^2$ **(h)** $\tan x$
(i) $\ln\dfrac{x+1}{2}$ **(j)** $\ln\dfrac{4-x^2}{4}$
(k) $\dfrac{e^x}{x+4}$ **(l)** $e^{-x}\sin x$

2.3 Find a series expansion for each of the following integrals by first expanding the integrand.

(a) $\displaystyle\int_0^x \frac{dt}{1+t}$ **(b)** $\displaystyle\int_0^x \frac{dt}{4-t^2}$
(c) $\displaystyle\int_0^x \frac{t\,dt}{1+t^2}$ **(d)** $\displaystyle\int \sin^2 x\,dx$
(e) $\displaystyle\int \tan x\,dx$ **(f)** $\displaystyle\int \sin x\cos x\,dx$

2.4 The function $(x^2-1)/[(x-4)(x^2+1)]$ is to be expanded in a power series about (a) the origin, (b) the point $x = 1$, and (c) the point $x = 2$. Determine the radius of convergence for each expansion.

2.5 For each of the following equations, list all singular points and determine the radius of convergence if we expand about the origin.

(a) $\dfrac{d^2u}{dx^2} + (x^2-1)u = x^2$

(b) $(x^2-1)\dfrac{d^2u}{dx^2} + u = x^2$

(c) $x(x^2+4)\dfrac{d^2u}{dx^2} + x\dfrac{du}{dx} = 0$

(d) $\dfrac{d^2u}{dx^2} + xu = \dfrac{1}{1-x}$

(e) $\dfrac{d^2u}{dx^2} + \dfrac{x-1}{x+1}\dfrac{du}{dx} + u = 0$

(f) $\cos x\dfrac{d^2u}{dx^2} + u = \sin x$

2.6 Determine the radius of convergence for each of the following series.

(a) $\displaystyle\sum_{n=0}^{\infty} x^n$ **(b)** $\displaystyle\sum_{n=0}^{\infty} \frac{1}{n!}x^n$
(c) $\displaystyle\sum_{n=0}^{\infty} \frac{n(n-1)}{2^n}x^n$ **(d)** $\displaystyle\sum_{n=0}^{\infty} 2^n x^n$
(e) $\displaystyle\sum_{n=0}^{\infty} \frac{1}{n!}(x-2)^n$ **(f)** $\displaystyle\sum_{n=0}^{\infty} \frac{(-1)^n}{(2n)!}(x-1)^n$

2.7 Solve each of the following differential equations for the general solution using the power-series method by expanding about $x = 0$. Note the radius of convergence for each solution.

(a) $\dfrac{du}{dx} + u = 0$

(b) $\dfrac{du}{dx} + u = x^2$

(c) $(1-x)\dfrac{du}{dx} + u = x$

(d) $x\dfrac{du}{dx} + x^2u = \sin x$

(e) $\dfrac{d^2u}{dx^2} - 4u = 0$

(f) $(x^2-1)\dfrac{d^2u}{dx^2} - 4u = 0$

(g) $\dfrac{d^2u}{dx^2} + 2\dfrac{du}{dx} + u = x^2$

(h) $\dfrac{d^2u}{dx^2} + 5\dfrac{du}{dx} + 6u = x^2 + 2\sin x$

2.8 Find the specific solution to each of the following differential equations by expanding about $x = 0$. State the limits of convergence for each series.

(a) $x\dfrac{du}{dx} + u\sin x = 0, \qquad u(0) = 1$

(b) $(4 - x^2)\dfrac{d^2u}{dx^2} + u = x^2 + 2x,$

$$u(0) = 0, \quad \frac{du}{dx}(0) = 0$$

(c) $\dfrac{d^2u}{dx^2} + (1 - x)u = 4x,$

$$u(0) = 1, \quad \frac{du}{dx}(0) = 0$$

(d) $\dfrac{d^2u}{dx^2} - x^2\dfrac{du}{dx} + u\sin x = 4\cos x,$

$$u(0) = 0, \quad \frac{du}{dx}(0) = 1$$

2.9 Solve $(1 - x)df/dx - f = 2x$ using a power-series expansion. Let $f = 6$ for $x = 0$, and expand about $x = 0$. Obtain five terms in the series and compare with the exact solution for values of $x = 0, 1/4, 1/2, 1,$ and 2.

2.10 The solution to $(1 - x)df/dx - f = 2x$ is desired in the interval from $x = 1$ to $x = 2$. Expand about $x = 2$ and determine the value of $f(x)$ at $x = 1.9$ if $f(2) = 1$. Compare with the exact solution.

2.11 Find the general solution to each of the following differential equations by expanding about the point specified.

(a) $(x - 2)\dfrac{d^2u}{dx^2} + u = 0 \quad \text{about } x = 1$

(b) $x^2\dfrac{d^2u}{dx^2} + u = 0 \quad \text{about } x = 1$

(c) $\dfrac{d^2u}{dx^2} + xu = x^2 \quad \text{about } x = 2$

2.12 Solve the differential equation $(d^2u/dx^2) + x^2u = 2x$ using the power-series method if $u(0) = 4$ and $du/dx(0) = -2$. Find an approximate value for $u(x)$ at $x = 2$.

2.13 Solve the differential equation $x^2(d^2u/dx^2) + 4u = 0$ by expanding about the point $x = 2$. Find an approximate value for $u(3)$ if $u(2) = 2$ and $du/dx(2) = 4$.

2.14 If $x(d^2u/dx^2) + (x - 1)u = 0$ find approximate values for $u(x)$ at $x = 1$ and at $x = 3$. We know that $u(2) = 10$ and $du/dx(2) = 0$.

2.15 Verify by substitution that the Legendre polynomials of Eqs. 2.3.15 satisfy Legendre's equation.

2.16 Write an expression for $P_8(x)$.

2.17 Show that

(a) $P_\lambda(-x) = (-1)^\lambda P_\lambda(x)$

(b) $\dfrac{dP_\lambda}{dx}(-x) = (-1)^{\lambda+1}\dfrac{dP_\lambda}{dx}(x)$

2.18 Verify that the formula

$$P_\lambda(x) = \frac{1}{2^\lambda \lambda!}\frac{d^\lambda}{dx^\lambda}(x^2 - 1)^\lambda$$

yields the first four Legendre polynomials. This is known as *Rodrigues's formula* and can be used for all Legendre polynomials with λ a positive integer.

2.19 Verify the formulas

$$\frac{dP_{\lambda+1}}{dx} - \frac{dP_{\lambda-1}}{dx} = (2\lambda + 1)P_\lambda$$

$$\int_x^1 P_\lambda(x)dx = \frac{1}{2\lambda + 1}[P_{\lambda-1}(x) - P_{\lambda+1}(x)]$$

for $\lambda = 2$ and $\lambda = 4$.

2.20 Determine the general solution for each of the following differential equations valid near the origin.

(a) $(1 - x^2)\dfrac{d^2u}{dx^2} - 2x\dfrac{du}{dx} + 12u = 0$

(b) $(1 - x^2)\dfrac{d^2u}{dx^2} - 2x\dfrac{du}{dx} + 6u = x^2$

(c) $4(1 - x^2)\dfrac{d^2u}{dx^2} - 8x\dfrac{du}{dx} + 3u = 0$

(d) $\dfrac{1}{\sin\phi}\dfrac{d}{d\phi}\left(\sin\phi\dfrac{du}{d\phi}\right) + 6u = 0$

(*Hint:* Let $x = \cos\phi$)

2.21 Find the specific solution to the differential equation

$$(1 - x^2)\frac{d^2u}{dx^2} - 2x\frac{du}{dx} + 20u = 14x^2$$

At $x = 0$, $u = 3$ and the function has a finite value at $x = 1$.

2.22 Find the general solution, valid in the vicinity of the origin, for each of the following differential equations (roots not differing by an integer).

(a) $2x\dfrac{d^2u}{dx^2}+(1-x)\dfrac{du}{dx}+u=0$

(b) $16x\dfrac{d^2u}{dx^2}+3(1+1/x)u=0$

(c) $2x(1-x)\dfrac{d^2u}{dx^2}+\dfrac{du}{dx}-u=0$

(d) $2x\dfrac{d^2u}{dx^2}+(1+4x)\dfrac{du}{dx}+u=0$

(e) $4x^2(1-x)\dfrac{d^2u}{dx^2}-x\dfrac{du}{dx}+(1-x)u=0$

(f) $2x^2\dfrac{d^2u}{dx^2}-7x\dfrac{du}{dx}+(x-10)=0$

(g) $2x^2\dfrac{d^2u}{dx^2}+x(x-1)\dfrac{du}{dx}+u=0$

(h) $2x^2\dfrac{d^2u}{dx^2}+x^2\dfrac{du}{dx}-u=0$

2.23 Determine the general solution for each of the following differential equations by expanding in a series about the origin (equal roots).

(a) $x\dfrac{d^2u}{dx^2}+\dfrac{du}{dx}+u=0$

(b) $x(1-x)\dfrac{d^2u}{dx^2}+\dfrac{du}{dx}+u=0$

(c) $x^2\dfrac{d^2u}{dx^2}-3x\dfrac{du}{dx}+(4-x)u=0$

2.24 Solve each of the following differential equations for the general solution valid about $x=0$ (roots differing by an integer).

(a) $x\dfrac{d^2u}{dx^2}-u=0$

(b) $x^2\dfrac{d^2u}{dx^2}+x\dfrac{du}{dx}+(x^2-1)u=0$

(c) $4x^2\dfrac{d^2u}{dx^2}-4x(1-x)\dfrac{du}{dx}+3u=0$

2.25 Solve each of the following differential equations by expanding about the point $x=1$.

(a) Problem 2.20c
(b) Problem 2.22f
(c) Problem 2.23b

2.26 Evaluate each of the following integrals.

(a) $\int_0^\infty \sqrt{x}e^{-x}dx$

(b) $\int_0^\infty x^2e^{-x^2}dx$

(c) $\int_0^\infty x^3e^{-x}dx$

(d) $\int_0^\infty x^{-4}e^{-\sqrt{x}}dx$

(e) $\int_0^\infty \dfrac{1}{\sqrt{x}}e^{-x^3}dx$

(f) $\int_0^\infty (1-x)^3e^{-\sqrt{x}}dx$

(g) $\int_1^\infty x^2e^{1-x}dx$

(h) $\int_0^\infty x^3e^{-x^{1/3}}dx$

2.27 Write out the first four terms in the expansion for (a) $J_0(x)$, and (b) $J_1(x)$.

2.28 From the expansions in Problem 2.27, calculate $J_0(2)$ and $J_1(2)$ to four decimal places. Compare with the tabulated values in the Appendix.

2.29 If we were interested in $J_0(x)$ and $J_1(x)$ for small x only (say for $x<0.1$), what algebraic expressions could be used to approximate $J_0(x)$ and $J_1(x)$? Using the expressions, find $J_0(0.1)$ and $J_1(0.1)$ and compare with the tabulated values in the Appendix.

2.30 Write the general solution for each of the following differential equations.

(a) $x^2\dfrac{d^2u}{dx^2}+x\dfrac{du}{dx}+(x^2-1)u=0$

(b) $x\dfrac{d^2u}{dx^2}+\dfrac{du}{dx}+xu=0$

(c) $4x^2\dfrac{d^2u}{dx^2}+4x\dfrac{du}{dx}+(4x^2-1)u=0$

(d) $x^2\dfrac{d^2u}{dx^2}+x\dfrac{du}{dx}+(4x^2-1)u=0$ (let $2x=y$)

(e) $x\dfrac{d^2u}{dx^2}+\dfrac{du}{dx}+u=0$ (let $x=y^2/4$)

(f) $\dfrac{d^2u}{dx^2}+u=0$ (let $u=\sqrt{x}v$)

2.31 Evaluate each of the following terms.

(a) $J_3(2)$ **(b)** $J_5(5)$

(c) $\dfrac{dJ_0}{dx}$ at $x=2$ **(d)** $\dfrac{dJ_2}{dx}$ at $x=4$

(e) $\dfrac{dJ_1}{dx}$ at $x=1$ **(f)** $\dfrac{dJ_3}{dx}$ at $x=1$

2.32 Find an expression in terms of $J_1(x)$ and $J_0(x)$ for each of the following integrals.

(a) $\int x^3J_2(x)dx$ **(b)** $\int xJ_2(x)dx$

(c) $\int \dfrac{J_4(x)}{x}dx$ **(d)** $\int xJ_1(x)dx$

(e) $\int x^3J_1(x)dx$ **(f)** $\int \dfrac{J_3(x)}{x}dx$

3 Laplace Transforms

Outline

3.1 Introduction
3.2 The Laplace Transform
3.3 Laplace Transforms of Derivatives and Integrals
3.4 Derivatives and Integrals of Laplace Transforms
3.5 Laplace Transforms of Periodic Functions
3.6 Inverse Transforms—Partial Fractions
3.6.1 Unrepeated Linear Factor $(s-a)$
3.6.2 Repeated Linear Factor $(s-a)^m$
3.6.3 Unrepeated Quadratic Factor $[(s-a)^2+b^2]$
3.6.4 Repeated Quadratic Factor $[(s-a)^2+b^2]^2$
3.7 Solution of Differential Equations
Problems
Table 3.1 Laplace Transforms

3.1 INTRODUCTION

The solution of a linear, ordinary differential equation with constant coefficients may be obtained by using the Laplace transformation. It is particularly useful in solving nonhomogeneous equations that result when modeling systems involving suddenly applied impulsive inputs, or discontinuous, periodic input functions, such as was done with the Fourier series in Chapter 1. It is not necessary, however, when using Laplace transforms that a homogeneous solution and a particular solution be added together to form the general solution. In fact, we do not find a general solution when using Laplace transforms. The initial conditions must be given and with them we obtain the specific solution to the nonhomogeneous equation directly, with no additional steps. This makes the technique quite attractive.

Another attractive feature of using Laplace transforms to solve a differential equation is that the transformed equation becomes an algebraic equation. The algebraic equation is then used to determine the solution to the differential equation. The technique requires only algebraic manipulation and should prove to be less difficult than the procedure using Fourier series.

The general technique of solving a differential equation using Laplace transforms involves finding the transform of each term in the equation, solving the resulting algebraic equation in terms of the new transformed variable, then finally solving the inverse transform to retrieve the original variables. We shall follow that order in this chapter. Let us first find the Laplace transform of the various quantities that may occur in our differential equations.

M. C. Potter and B. F. Feeny, *Mathematical Methods for Engineering and Science*,
https://doi.org/10.1007/978-3-031-26151-0_3

3.2 THE LAPLACE TRANSFORM

Let the function $f(t)$ be the dependent variable of an ordinary differential equation that we wish to solve. Multiply $f(t)$ by e^{-st} and integrate with respect to t from 0 to infinity. The independent variable t integrates out and there remains a function of s, say $F(s)$. This is expressed as

$$F(s) = \int_0^\infty f(t)e^{-st}dt \tag{3.2.1}$$

The function $F(s)$ is called the *Laplace transform* of the function $f(t)$. We will return often to this definition of the Laplace transform. It is usually written as

$$\mathcal{L}(f) = F(s) = \int_0^\infty f(t)e^{-st}dt \tag{3.2.2}$$

where the operator script $\mathcal{L}$ denotes the Laplace transform. We shall consistently use a lowercase letter to represent a function and its capital to denote its Laplace transform; that is, $Y(s)$ would denote the Laplace transform of $y(t)$. The *inverse Laplace transform* will be denoted by $\mathcal{L}^{-1}$, resulting in

$$f(t) = \mathcal{L}^{-1}(F) \tag{3.2.3}$$

The Laplace transform exists even if there are discontinuities in the function $f(t)$; however, it must be sectionally continuous so that the integral in Eq. 3.2.1 exists. It could be a function as sketched in Fig. 3.1. Note that the right-hand limit at a discontinuity, t_3 for example, will be designated $f(t_3^+)$ and the left-hand limit will be $f(t_3^-)$. There are functions, though, which do not possess a Laplace transform. The function e^{t^2} is such a function. If this function is substituted into the integral of Eq. 3.2.1, the integral does not exist; no $F(s)$ could be found. This, however, is an unusual function not often encountered in the solution of real problems. By far the majority of functions representing some physical quantity will possess a Laplace transform.

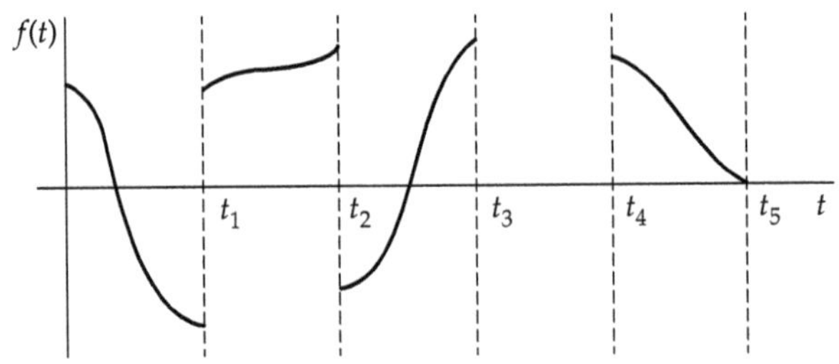

FIGURE 3.1 Sectionally continuous function.

Before considering some examples that demonstrate how the Laplace transforms of various functions are found, let us consider three very important properties of the Laplace transform. First, the Laplace transform operator $\mathcal{L}$ is a *linear operator*. This is expressed as

$$\mathcal{L}[af(t) + bg(t)] = a\mathcal{L}(f) + b\mathcal{L}(g) \tag{3.2.4}$$

where a and b are constants. To verify that this is true, we simply substitute the quantity $[af(t) + bg(t)]$ into the definition for the Laplace transform, obtaining

$$\begin{aligned}\mathcal{L}[af(t) + bg(t)] &= \int_0^\infty [af(t) + bg(t)]e^{-st}dt \\ &= a\int_0^\infty f(t)e^{-st}dt + b\int_0^\infty g(t)e^{-st}dt \\ &= a\mathcal{L}(f) + b\mathcal{L}(g) \end{aligned} \tag{3.2.5}$$

The second property is often called the *first shifting property*. It is expressed as

$$\mathcal{L}[e^{at}f(t)] = F(s-a) \tag{3.2.6}$$

where $F(s)$ is the Laplace transform of $f(t)$. This is shown to be true by using $e^{at}f(t)$ in place of $f(t)$ in Eq. 3.2.2; there results

$$\begin{aligned}\mathcal{L}[e^{at}f(t)] &= \int_0^\infty e^{at}f(t)e^{-st}dt \\ &= \int_0^\infty f(t)e^{-(s-a)t}dt \end{aligned} \tag{3.2.7}$$

Now, let $s - a = \boldsymbol{s}$. Then we have

$$\begin{aligned}\mathcal{L}[e^{at}f(t)] &= \int_0^\infty f(t)e^{-\boldsymbol{s}t}dt \\ &= F(\boldsymbol{s}) = F(s-a) \end{aligned} \tag{3.2.8}$$

The third property is the *second shifting property*. It is stated as follows: If the Laplace transform of $f(t)$ is known to be

$$\mathcal{L}(f) = F(s) \tag{3.2.9}$$

and if

$$g(t) = \begin{cases} f(t-a), & t > a \\ 0, & t < a \end{cases} \tag{3.2.10}$$

then the Laplace transform of $g(t)$ is

$$\mathcal{L}(g) = e^{-as}F(s) \tag{3.2.11}$$

To show this result, the Laplace transform of $g(t)$ given by Eq. 3.2.10 is

$$\begin{aligned}\mathcal{L}(g) &= \int_0^\infty g(t)e^{-st}dt \\ &= \int_0^a \cancelto{0}{0\cdot e^{-st}}dt + \int_a^\infty f(t-a)e^{-st}dt \end{aligned} \tag{3.2.12}$$

Make the substitution $\tau = t - a$. Then $d\tau = dt$ and we have

$$\begin{aligned}\mathcal{L}(g) &= \int_0^\infty f(\tau)e^{-s(\tau+a)}d\tau \\ &= e^{-as}\int_0^\infty f(\tau)e^{-s\tau}d\tau \\ &= e^{-as}F(s) \end{aligned} \tag{3.2.13}$$

and the second shifting property is verified.

The three properties above should simplify the task of finding the Laplace transform of a particular function $f(t)$, or the inverse transform of $F(s)$. This will be illustrated in the following examples. Table 3.1, which gives the Laplace transform of a variety of functions, may be found at the end of the chapter, following the Problems.

Example 3.1

Find the Laplace transform of the unit step function $u_0(t)$ shown in Fig. 3.2.

$$u_0(t) = \begin{cases} 1, & t > 0 \\ 0, & t < 0 \end{cases}$$

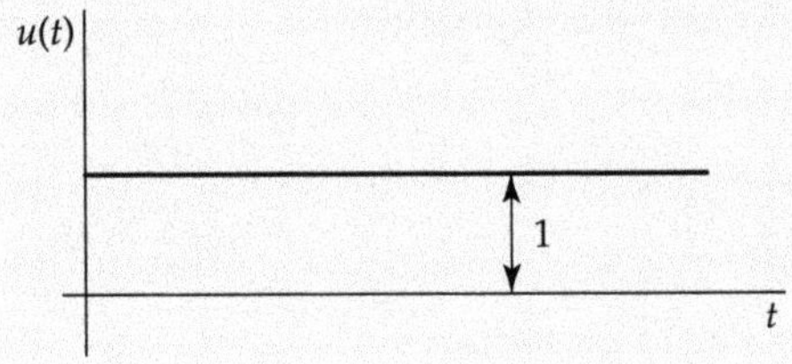

FIGURE 3.2

Solution
Using the definition of the Laplace transform, we have

$$\mathcal{L}(u_0) = \int_0^\infty u_0(t)e^{-st}dt$$
$$= \int_0^\infty e^{-st}dt = -\frac{1}{s}e^{-st}\Big|_0^\infty = \frac{1}{s}$$

This will also be used as the Laplace transform of unity, that is, $\mathcal{L}(1) = 1/s$, since the integration occurs between zero and infinity, as above.

Example 3.2

Use the first shifting property and find the Laplace transform of e^{at}.

Solution
Equation 3.2.6 provides us with

$$\mathcal{L}(e^{at}) = \mathcal{L}(e^{at} \cdot 1) = F(s - a)$$

where the transform of unity is, from Example 3.1,

$$\mathcal{L}(1) = F(s) = \frac{1}{s}$$

We simply substitute $s - a$ for s and obtain

$$\mathcal{L}(e^{at}) = \frac{1}{s-a}$$

Example 3.3

Use the second shifting property and find the Laplace transform of the unit step function $u_a(t)$ defined in Fig. 3.3.

$$u_a(t) = \begin{cases} 1, & t > a \\ 0, & t < a \end{cases}$$

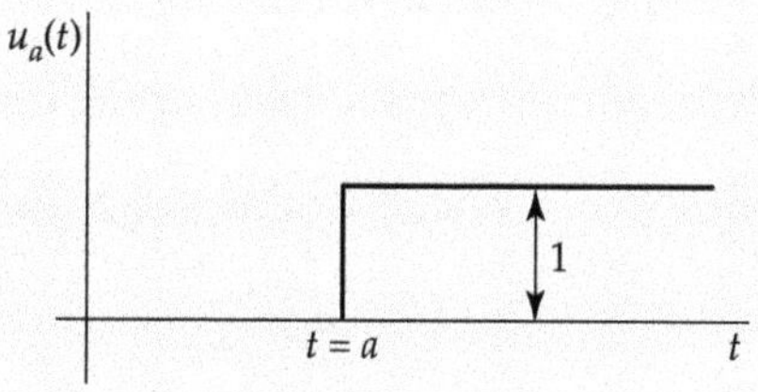

FIGURE 3.3

Check the result by using the definition of the Laplace transform.

Solution

Using the second shifting theorem given by Eq. 3.2.11, there results

$$\begin{aligned} \mathcal{L}(u_a) &= e^{-as}F(s) \\ &= \frac{1}{s}e^{-at} \end{aligned}$$

where $F(s)$ is the Laplace transform of unity given in Example 3.1.

To check the result above, we use the definition of the Laplace transform:

$$\begin{aligned} \mathcal{L}(u_a) &= \int_0^\infty u_a(t)e^{-st}dt \\ &= \cancel{\int_0^a 0 \cdot e^{-st}dt}^{\,0} + \int_a^\infty 1e^{-st}dt \\ &= -\frac{1}{s}e^{-st}\Big|_a^\infty = \frac{1}{s}e^{-as} \end{aligned}$$

This, of course, checks the result obtained with the second shifting theorem.

Example 3.4

Determine the Laplace transform of $\sin \omega t$ and $\cos \omega t$ by using

$$e^{i\theta} = \cos\theta + i\sin\theta$$

the first shifting property, and the linearity property.

Solution

The first shifting property allows us to write (see Example 3.2)

$$\begin{aligned}\mathcal{L}(e^{l\omega t}) &= \frac{1}{s - i\omega} \\ &= \frac{1}{s-i\omega}\frac{s+i\omega}{s+i\omega} = \frac{s+i\omega}{s^2+\omega^2} = \frac{s}{s^2+\omega^2} + i\frac{\omega}{s^2+\omega^2}\end{aligned}$$

Using the linearity property expressed by Eq. 3.2.4, we have

$$\mathcal{L}(e^{l\omega t}) = \mathcal{L}(\cos\omega t + i\sin\omega t) = \mathcal{L}(\cos\omega t) + i\mathcal{L}(\sin\omega t)$$

Equating the real and imaginary parts of the two equations above results in

$$\mathcal{L}(\sin\omega t) = \frac{\omega}{s^2+\omega^2}$$

$$\mathcal{L}(\cos\omega t) = \frac{s}{s^2+\omega^2}$$

These two Laplace transforms could have been obtained by substituting directly into Eq. 3.2.2, each of which would have required integrating by parts twice.

Example 3.5

Find the Laplace transform of t^k.

Solution

The Laplace transform of t^k is given by

$$\mathcal{L}(t^k) = \int_a^\infty t^k e^{-st} dt$$

To integrate this, we make the substitution

$$\eta = st, \quad dt = \frac{d\eta}{s}$$

There then results

$$\begin{aligned}\mathcal{L}(t^k) = \int_0^\infty t^k e^{-st} dt &= \frac{1}{s^{k+1}} \int_0^\infty \eta^k e^{-\eta} d\eta \\ &= \frac{1}{s^{k+1}}\Gamma(k+1)\end{aligned}$$

where the gamma function is $\Gamma(k+1) = \int_0^\infty e^{-t}t^\lambda dt$, as defined by Eq. 2.5.13. If k is an integer, say $k = n$, then Eq. 2.5.19 gives

$$\Gamma(n+1) = n!$$

Our result is

$$\mathcal{L}(t^n) = \frac{n!}{s^{n+1}}$$

Example 3.6

Use the linearity property and find the Laplace transform of $\cosh \omega t$.

Solution

The $\cosh \omega t$ can be written as

$$\cosh \omega t = \frac{1}{2}(e^{\omega t} + e^{-\omega t})$$

The Laplace transform is then

$$\mathcal{L}(\cosh \omega t) = \mathcal{L}\left(\frac{1}{2}e^{\omega t} + \frac{1}{2}e^{-\omega t}\right)$$

$$= \frac{1}{2}\mathcal{L}(e^{\omega t}) + \frac{1}{2}\mathcal{L}(e^{-\omega t})$$

Using the results of Example 3.2, we have

$$\mathcal{L}(\cosh \omega t) = \frac{1}{2(s-\omega)} + \frac{1}{2(s+\omega)}$$

$$= \frac{s}{s^2 - \omega^2}$$

Example 3.7

Find the Laplace transform of the function defined in Fig. 3.4, using the result of Example 3.3.

$$f(t) = \begin{cases} 1, & t < a \\ A, & a < t < b \\ 0, & b < t \end{cases}$$

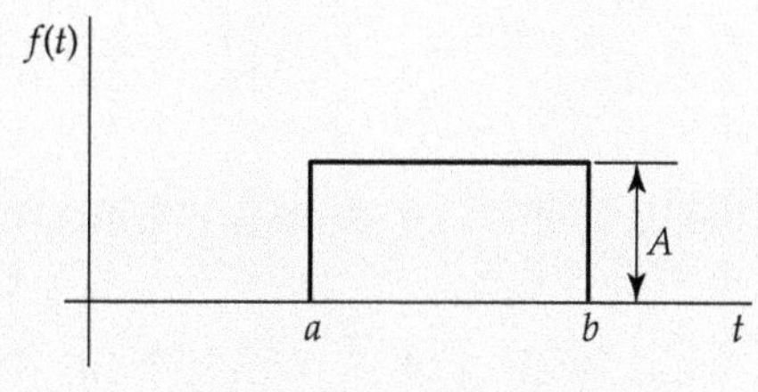

FIGURE 3.4

Solution

The function $f(t)$ can be written in terms of the unit step function as

$$f(t) = Au_a(t) - Au_b(t)$$

The Laplace transform is, from Example 3.3,

$$\mathcal{L}(f) = \frac{A}{s}e^{-as} - \frac{A}{s}e^{-bs}$$
$$= \frac{A}{s}[e^{-as} - e^{-bs}]$$

Example 3.8

An extension of the function shown in Example 3.7 is the function shown in Fig. 3.5. If $\epsilon \to 0$, the *unit impulse function* results. It is often denoted by $\delta_0(t)$. It has an area of unity, its height approaches ∞ as its base approaches zero. Find $\mathcal{L}(f)$ for the unit impulse function if it occurs at (a) $t = 0$ as shown, and (b) at $t = a$.

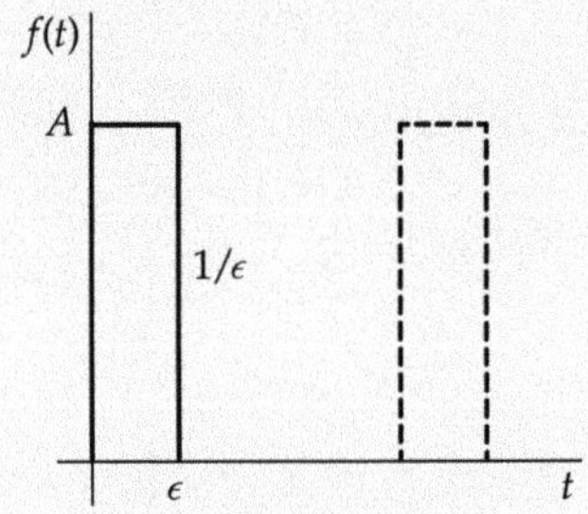

FIGURE 3.5

Solution

a) Let us use the results of Example 3.7. With the function $f(t)$ shown with the solid line, the Laplace transform would be, using $A = 1/\epsilon$,

$$\mathcal{L}(f) = \frac{1}{\epsilon s}[1 - e^{-\epsilon s}]$$

To find the limit as $\epsilon \to 0$, expand $e^{-\epsilon s}$ in a series. This gives

$$e^{-\epsilon s} = 1 - \epsilon s + \frac{\epsilon^2 s^2}{2!} - \frac{\epsilon^3 s^3}{3!} + \cdots$$

Hence,

$$\frac{1-e^{-\epsilon s}}{\epsilon s} = 1 - \frac{\epsilon s}{2!} + \frac{\epsilon^2 s^2}{3!} - \cdots$$

As $\epsilon \to 0$, the expression above approaches unity. Thus,

$$\mathcal{L}(\delta_0) = 1$$

b) If the impulse function occurs at a time $t = a$, it is denoted by $\delta_a(t)$. Then, using the second shifting property, we have

$$\mathcal{L}(\delta_a) = e^{-as}$$

Examples of the use of the impulse function would be a concentrated load $P\delta_a(x)$ located at $x = a$, an electrical potential $V\delta_a(t)$ applied instantaneously to a circuit at $t = a$, or an impact excitation to a vibration system at $t = a$.

Example 3.9

Find the Laplace transform of

$$f(t) = \begin{cases} 0, & 0 < t < 1 \\ t^2, & 1 < t < 2 \\ 0, & 2 < t \end{cases}$$

Solution

The function $f(t)$ is written in terms of the unit step function as

$$f(t) = u_1(t)t^2 - u_2(t)t^2$$

We cannot apply the second shifting property with $f(t)$ in this form since, according to Eq. 3.2.10, we must have for the first term a function of $(t-1)$ and for the second term a function of $(t-2)$. The function $f(t)$ is thus rewritten as follows:

$$f(t) = u_1(t)[(t-1)^2 + 2(t-1) + 1] + u_2(t)[(t-2)^2 + 4(t-2) + 4]$$

Now, we can apply the second shifting property to the result above, to obtain,

$$\mathcal{L}(f) = \mathcal{L}\{u_1(t)[(t-1)^2 + 2(t-1) + 1]\} + \mathcal{L}\{u_2(t)[(t-2)^2 + 4(t-2) + 4]\}$$

For the first set of braces $f_1(t) = t^2 + 2t + 1$ shifted by 1, and for the second set of braces $f_2(t) = t^2 + 4t + 4$ shifted by 2. The result is

$$\mathcal{L}(f) = e^{-s}\left[\frac{2}{s^3} + \frac{2}{s^2} + \frac{1}{s}\right] + e^{-2s}\left[\frac{2}{s^3} + \frac{4}{s^2} + \frac{4}{s}\right]$$

The Laplace transforms of the terms in the braces were obtained by using linearity and Table 3.1.

Note that, in general, $f(t)$ is not the function given in the statement of the problem, which in this case was $f(t) = t^2$.

Example 3.10

A section of the square-wave function is as shown in Fig. 3.6. Determine its Laplace transform.

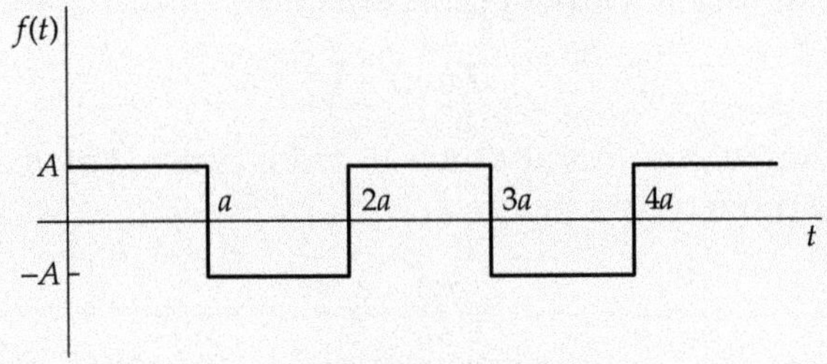

FIGURE 3.6

Solution

The function $f(t)$ can be represented using the unit step function. It is

$$f(t) = Au_0(t) - 2Au_a(t) + 2Au_{2a}(t) - 2Au_{3a}(t) + \cdots$$

The Laplace transform of the above is, referring to Example 3.3,

$$\begin{aligned}\mathcal{L}(f) &= A\left[\frac{1}{s} - \frac{2}{s}e^{-as} + \frac{2}{s}e^{-2as} - \frac{2}{s}e^{-3as} + \cdots\right] \\ &= \frac{A}{s}[1 - 2e^{-as}(1 - e^{-as} + e^{-2as} - \cdots)]\end{aligned}$$

Letting $e^{-as} = \eta$, we have

$$\mathcal{L}(f) = \frac{A}{s}[1 - 2\eta(1 - \eta + \eta^2 - \eta^3 + \cdots)]$$

The quantity in parentheses is recognized as the series expansion for $1/(1 + \eta)$, see Eqs. 2.1.6. Hence, we can write

$$\mathcal{L}(f) = \frac{A}{s}\left[1 - \frac{2e^{-as}}{1 + e^{-as}}\right]$$

This can be put in the form

$$\mathcal{L}(f) = \frac{A}{s}\frac{1 - e^{-as}}{1 + e^{-as}} = \frac{A}{s}\frac{e^{as/2} - e^{-as/2}}{e^{as/2} + e^{-as/2}}\frac{e^{-as/2}}{e^{-as/2}} = \frac{A}{s}\frac{e^{as/2} - e^{-as/2}}{e^{as/2} + e^{-as/2}}$$

This form is recognized* as

$$\mathcal{L}(f) = \frac{A}{s}\tanh\frac{as}{2}$$

* $\tanh x = \sinh x/\cosh x = (e^x - e^{-x})/(e^x + e^{-x})$

Example 3.11

Use the Laplace transforms from Table 3.1 and find $f(t)$ when $F(s)$ is given by

a) $\dfrac{2s}{s^2+4}$

b) $\dfrac{6s}{s^2+4s+13}$

c) $\dfrac{4e^{-2s}}{s^2-16}$

Solution

a) The Laplace transform of $\cos \omega t$ is

$$\mathcal{L}(\cos \omega t) = \frac{s}{s^2+\omega^2}$$

Then,

$$\mathcal{L}(2\cos 2t) = 2\mathcal{L}(\cos 2t) = \frac{2s}{s^2+2^2}$$

Thus, if $F(s) = 2s/(s^2+4)$, then $f(t)$ is given by

$$f(t) = \mathcal{L}^{-1}\left(\frac{2s}{s^2+4}\right) = 2\cos 2t$$

b) Let us write the given $F(s)$ as (this is suggested by the term $4s$ in the denominator)

$$F(s) = \frac{6s}{s^2+4s+13} = \frac{6(s+2)-12}{(s+2)^2+9} = \frac{6(s+2)}{(s+2)^2+9} - \frac{12}{(s+2)^2+9}$$

Using the first shifting property, Eq. 3.2.6, we can write

$$\mathcal{L}(e^{-2t}\cos 3t) = \frac{s+2}{(s+2)^2+9}$$

$$\mathcal{L}(e^{-2t}\sin 3t) = \frac{3}{(s+2)^2+9}$$

It then follows that

$$\mathcal{L}^{-1}\left(\frac{6s}{s^2+4s+13}\right) = 6e^{-2t}\cos 3t - 4e^{-2t}\sin 3t$$

or we have

$$f(t) = 2e^{-2t}(3\cos 3t - 2\sin 3t)$$

c) The second shifting property suggests that we write

$$\frac{4e^{-2s}}{s^2-16} = e^{-2s}\left[\frac{4}{s^2-16}\right]$$

and find the $f(t)$ associated with the quantity in brackets; that is,

$$\mathcal{L}(\sinh 4t) = \frac{4}{s^2-16}$$

or

$$\mathcal{L}^{-1}\left(\frac{4}{s^2-16}\right) = \sinh 4t$$

Finally, there results, using Eq. 3.2.10,

$$f(t) = \begin{cases} \sinh 4(t-2), & t > 2 \\ 0, & t < 2 \end{cases}$$

In terms of the unit step function, this can be written as

$$f(t) = u_2(t)\sinh 4(t-2)$$

3.3 LAPLACE TRANSFORMS OF DERIVATIVES AND INTEGRALS

The operations of differentiation and integration are significantly simplified when using Laplace transforms. Differentiation results when the Laplace transform of a function is multiplied by the transformed variable s and integration corresponds to dividing by s, as we shall see.

Let us consider a function $f(t)$ that is continuous and possesses* a derivative $f'(t)$ that may be sectionally continuous. An example of such a function is sketched in Fig. 3.7. We shall not allow discontinuities in the function $f(t)$, although we will discuss such a function subsequently. The Laplace transform of a derivative is defined to be

$$\mathcal{L}(f') = \int_0^\infty f'(t)e^{-st}dt \tag{3.3.1}$$

This can be integrated by parts if we let

Recall $\int u dv = uv - \int v du$.

$$\begin{aligned} u &= e^{-st}, & dv &= f'(t)dt = df \\ du &= -se^{-st}dt, & v &= f \end{aligned} \tag{3.3.2}$$

*It is common to let $df/dt = f'$ or $df/dt = \dot{f}$. We select $df/dt = f'$.

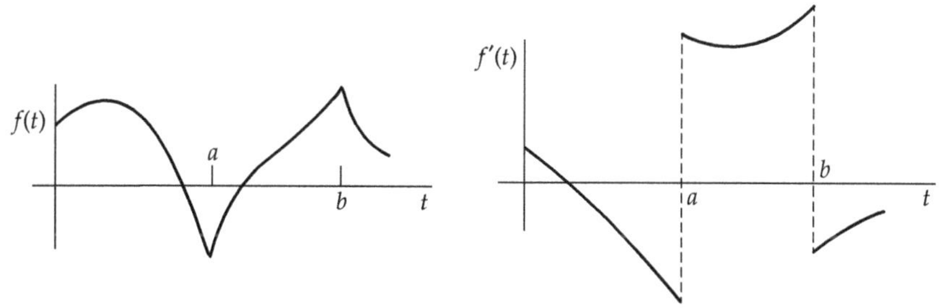

FIGURE 3.7 Continuous function possessing a sectionally continuous derivative.

Then

$$\mathcal{L}(f') = f(t)e^{-st}\Big|_0^\infty + s\int_0^\infty f(t)e^{-st}dt \tag{3.3.3}$$

Assuming that the quantity fe^{-st} vanishes at the upper limit, that is if the real part of s is greater than zero, this is written as

$$\begin{aligned}\mathcal{L}(f') &= -f(0) + s\int_0^\infty f(t)e^{-st}dt \\ &= s\mathcal{L}(f) - f(0)\end{aligned} \tag{3.3.4}$$

This result can be easily extended to the second-order derivative; however, we must demand that the first derivative $f(t)$ be continuous. Then, with the use of Eq. 3.3.4, we have

$$\begin{aligned}\mathcal{L}(f'') &= s\mathcal{L}(f') - f'(0) \\ &= s[s\mathcal{L}(f) - f(0)] - f'(0) \\ &= s^2\mathcal{L}(f) - sf(0) - f'(0)\end{aligned} \tag{3.3.5}$$

Note that the initial conditions must be known when finding the Laplace transforms of the derivatives.

Higher-order derivatives naturally follow giving us the relationship,

$$\mathcal{L}(f^{(n)}) = s^n\mathcal{L}(f) - s^{n-1}f(0) - s^{n-2}f'(0) - \cdots - f^{(n-1)}(0) \tag{3.3.6}$$

where all the functions $f(t), f'(t), \ldots, f^{(n-1)}(t)$ are continuous, with the quantities $f^{(n-1)}e^{-st}$ vanishing at infinity; the quantity $f^{(n)}(t)$ may be sectionally continuous.

Now, let us find the Laplace transform of a function possessing a discontinuity. Consider the function $f(t)$ to have one discontinuity at $t = a$, with $f(a^+)$ the right-hand limit and $f(a^-)$ the left-hand limit as shown in Fig. 3.8. The Laplace transform of the first derivative is then

$$\mathcal{L}(f') = \int_0^a f'(t)e^{-st}dt + \int_a^\infty f'(t)e^{-st}dt \tag{3.3.7}$$

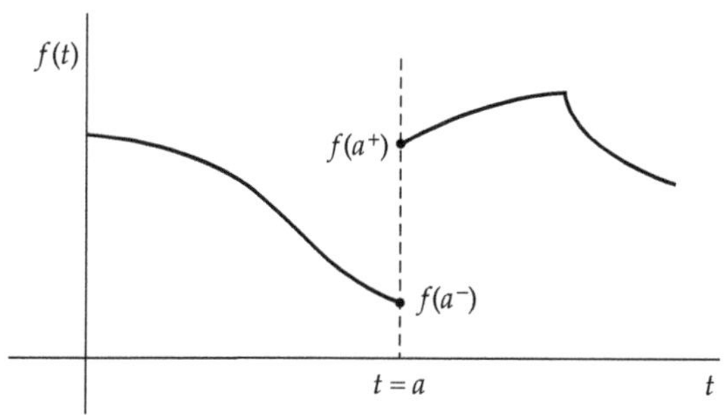

FIGURE 3.8 Function $f(t)$ with one discontinuity.

Integrating by parts allows us to write

$$\mathcal{L}(f') = f(t)e^{-st}\Big|_0^{a-} + s\int_0^{a^-} f(t)e^{-st}dt + f(t)e^{-st}\Big|_{a^+}^{\infty} + s\int_{a^+}^{\infty} f(t)e^{-st}dt$$
$$= f(a^-)e^{-as} - f(0) + s\int_0^{a^-} f(t)e^{-st}dt - f(a^+)e^{-as} + s\int_{a^+}^{\infty} f(t)e^{-st}dt \qquad (3.3.8)$$

The two integrals above can be combined, since there is no contribution to the integral between $t = a^-$ and $t = a^+$. We then have

$$\mathcal{L}(f') = s\int_0^{\infty} f(t)e^{-st}dt - f(0) - [f(a^+) - f(a^-)]e^{-as}$$
$$= s\mathcal{L}(f) - f(0) - [f(a^+) - f(a^-)]e^{-as} \qquad (3.3.9)$$

If two discontinuities exist in $f(t)$, the second discontinuity would be accounted for by adding the appropriate terms to the equation above.

We shall now find the Laplace transform of an integral quantity. Let the integral quantity be given by

$$g(t) = \int_0^t f(\tau)d\tau \qquad (3.3.10)$$

where the dummy variable of integration is arbitrarily chosen as τ; the variable t occurs only as the upper limit. The first derivative is then*

$$g'(t) = f(t) \qquad (3.3.11)$$

We also note that $g(0) = 0$. Now, applying Eq. 3.3.4, we have

$$\mathcal{L}(g') = s\mathcal{L}(g) - \overset{0}{\cancel{g(0)}} \qquad (3.3.12)$$

or, using Eq. 3.3.11, this can be written as

$$\mathcal{L}(g) = \frac{\mathcal{L}(g')}{s} = \frac{1}{s}\mathcal{L}(f) \qquad (3.3.13)$$

Written explicitly in terms of the integral, this is

$$\mathcal{L}\left(\int_0^t f(\tau)d\tau\right) = \frac{1}{s}\mathcal{L}(f) \qquad (3.3.14)$$

These transforms of derivatives and integrals are obviously necessary when solving differential equations or integrodifferential equations. They will also, however, find application in obtaining the Laplace transforms of various functions and the inverse transforms. Before we turn to the solution of differential equations, let us illustrate the latter use.

---

*Liebnitz's rule of differentiating an integral is

$$\frac{d}{dt}\int_{a(t)}^{b(t)} f(\tau, t)d\tau = \frac{db}{dt}f(b, t) - \frac{da}{dt}f(a, t) + \int_a^b \frac{\partial f}{\partial t}d\tau$$

Example 3.12

Find the Laplace transform of $f(t) = t^2$. Use the transform of a derivative.

Solution

We can use the Laplace transform of the third derivative obtained in Eq. 3.3.6. From the given function we have $f(0) = 0$, $f'(0) = 0$, and $f''(0) = 2$; this allows us to write

$$\mathcal{L}(f''') = s^3\mathcal{L}(f) - s^2\cancelto{0}{f(0)} - s\cancelto{0}{f'(0)} - f''(0)$$

or, recognizing that $f''' = 0$, we have

$$\mathcal{L}(0) = s^3\mathcal{L}(f) - 2 = 0$$

since $\mathcal{L}(0) = 0$. This results in

$$\mathcal{L}(t^2) = \frac{2}{s^3}$$

Example 3.13

Using the transform of a derivative, find the Laplace transform of $f(t) = t \sin t$ assuming that $\mathcal{L}(\cos t)$ is known.

Solution

The first and second derivatives of $f(t)$ are

$$f'(t) = t\cos t + \sin t$$
$$f''(t) = 2\cos t - t\sin t$$

The transform of a second derivative is

$$\mathcal{L}(f'') = s^2\mathcal{L}(f) - s\cancelto{0}{f(0)} - \cancelto{0}{f'(0)}$$

where we have used $f(0) = 0$ and $f'(0) = 0$. Thus, Eq. 3.3.5 gives

$$\mathcal{L}(2\cos t - t\sin t) = s^2\mathcal{L}(t\sin t)$$

This can be written as

$$2\mathcal{L}(\cos t) - \mathcal{L}(t\sin t) = s^2\mathcal{L}(t\sin t)$$

or, using Table 3.1,

$$(s^2 + 1)\mathcal{L}(t\sin t) = 2\mathcal{L}(\cos t)$$
$$= \frac{2s}{s^2 + 1}$$

Finally, we have

$$\mathcal{L}(t\sin t) = \frac{2s}{(s^2 + 1)^2}$$

Note: This approach is easier than integrating $\mathcal{L}(f)$ directly.

Example 3.14

Find $f(t)$ if $F(s) = 8/(s^2+4)^2$ by using $\mathcal{L}(t \sin 2t) = 2s/(s^2+4)^2$.

Solution

We write the given transform as

$$\mathcal{L}(f) = F(s) = \frac{1}{s}\frac{8s}{(s^2+4)^2}$$

Equation 3.3.14 allows us to write

$$\mathcal{L}\left(\int_0^t 4\tau \sin 2\tau \, d\tau\right) = \frac{1}{s}\frac{8s}{(s^2+4)^2}$$

Hence,

$$f(t) = 4\int_0^\tau \tau \sin 2\tau \, d\tau$$

This is integrated by parts if we let

$$u = \tau, \quad dv = \sin 2\tau \, d\tau$$

$$du = d\tau, \quad v = -\frac{1}{2}\cos 2\tau$$

There results

$$\begin{aligned} f(t) &= -2t \cos 2t + 2\int_0^t \cos 2\tau \, d\tau \\ &= -2t \cos 2t + \sin 2t \end{aligned}$$

3.4 DERIVATIVES AND INTEGRALS OF LAPLACE TRANSFORMS

The problem of determining the Laplace transform of a particular function or the function corresponding to a particular transform can often be simplified by either differentiating or integrating a Laplace transform. First, let us find the Laplace transform of the quantity $tf(t)$. It is, by definition,

$$\mathcal{L}(tf) = \int_0^\infty tf(t)e^{-st}dt \tag{3.4.1}$$

Using Liebnitz's rule of differentiating an integral (see the footnote associated with Eq. 3.3.11) we can differentiate Eq. 3.2.2 and obtain

$$\begin{aligned} F'(s) &= \frac{d}{ds}\int_0^\infty f(t)e^{-st}dt = \int_0^\infty f(t)\frac{\partial}{\partial s}(e^{-st})dt \\ &= -\int_0^\infty tf(t)e^{-st}dt \end{aligned} \tag{3.4.2}$$

Comparing this with Eq. 3.4.1, there follows

$$\mathcal{L}(tf) = -F'(s) \tag{3.4.3}$$

The second derivative is

$$F''(s) = \int_0^\infty f(t)\frac{\partial^2}{\partial s^2}(e^{-st})dt = \int_0^\infty t^2 f(t)e^{-st}dt \tag{3.4.4}$$

or

$$\mathcal{L}(t^2 f) = F''(s) \tag{3.4.5}$$

In general, this is written as

$$\mathcal{L}(t^n f) = (-1)^n F^{(n)}(s) \tag{3.4.6}$$

Next, we will find the Laplace transform of $f(t)/t$. Let

$$f(t) = tg(t) \tag{3.4.7}$$

Then, using Eq. 3.4.3, the Laplace transform of the equation above is

$$\begin{aligned} F(s) &= \mathcal{L}(f) \\ &= \mathcal{L}(tg) = -G'(s) \end{aligned} \tag{3.4.8}$$

This is written as

$$-dG = F(s)\,ds \tag{3.4.9}$$

Integrate as follows:

$$-\int_0^{G(s)} dG = \int_\infty^s F(\mathfrak{s})d\mathfrak{s} \tag{3.4.10}$$

where we assume that $G(s) \to 0$ as $s \to \infty$. The dummy variable of integration is written arbitrarily as $\mathfrak{s}$. We then have

In the examples we have considered, the Laplace transform functions indeed approach zero as $s \to \infty$.

$$G(s) = \int_s^\infty F(\mathfrak{s})d\mathfrak{s} \tag{3.4.11}$$

where the limits of integration have been interchanged to remove the negative sign. Finally, referring to Eq. 3.4.7, we see that

$$\begin{aligned} \mathcal{L}(f/t) &= \mathcal{L}(g) = G(s) \\ &= \int_s^\infty F(\mathfrak{s})d\mathfrak{s} \end{aligned} \tag{3.4.12}$$

The use of the expressions above for the derivatives and integral of a Laplace transform will be demonstrated in the following examples.

Example 3.15

Differentiate the Laplace transform of $f(t) = \sin \omega t$, thereby determining $\mathcal{L}(t \sin \omega t)$. Use $\mathcal{L}(\sin \omega t) = \omega/(s^2 + \omega^2)$.

Solution

Equation 3.4.3 allows us to write

$$\begin{aligned}\mathcal{L}(t \sin \omega t) &= -\frac{d}{ds}\mathcal{L}(\sin \omega t)\\ &= -\frac{d}{ds}\left(\frac{\omega}{s^2+\omega^2}\right)\\ &= \frac{2\omega s}{(s^2+\omega^2)^2}\end{aligned}$$

This transform was obviously much easier to obtain using Eq. 3.4.3 than the technique used in Example 3.13. It is also much easier than direct integration of $\mathcal{L}(f)$.

Example 3.16

Find the Laplace transform of $(e^{-t} - 1)/t$ using the transforms

$$\mathcal{L}(e^{-t}) = \frac{1}{s+1} \quad \text{and} \quad \mathcal{L}(1) = \frac{1}{s}$$

Solution

The Laplace transform of the function $f(t) = e^{-t} - 1$ is

$$\mathcal{L}(f) = \frac{1}{s+1} - \frac{1}{s}$$

Equation 3.4.12 gives us

$$\begin{aligned}\mathcal{L}(f/t) &= \int_s^\infty \left(\frac{1}{\mathcal{s}+1} - \frac{1}{\mathcal{s}}\right) d\mathcal{s}\\ &= \left[\ln(\mathcal{s}+1) - \ln \mathcal{s}\right]_s^\infty = \ln \left.\frac{\mathcal{s}+1}{\mathcal{s}}\right]_s^\infty = \ln \frac{s}{s+1}\end{aligned}$$

This problem could be reformulated to illustrate a function that had no Laplace transform. Consider the function $(e^{-t} - 2)/t$. The solution would have resulted in $\ln(s+1)/s^2\big|_s^\infty$. At the upper limit this quantity is not defined and thus $\mathcal{L}(f/t)$ does not exist.

Example 3.17

Determine the inverse Laplace transform of $\ln[s^2/(s^2+4)]$.

Solution

We know that if we differentiate $\ln[s^2/(s^2+4)]$ we may arrive at a recognizable fraction. Letting $G(s) = \ln[s^2/(s^2+4)] = \ln s^2 - \ln(s^2+4)$ we have (see Eq. 3.4.8)

$$\begin{aligned} F(s) = -G'(s) &= -\frac{2s}{s^2} + \frac{2s}{s^2+4} \\ &= -\frac{2}{s} + \frac{2s}{s^2+4} \end{aligned}$$

Now, the inverse transform of $F(s)$ is, referring to Table 3.1,

$$f(t) = -2 + 2\cos 2t$$

Finally, the desired inverse transform is

$$\mathcal{L}^{-1}\left(\ln\frac{s^2}{s^2+4}\right) = \frac{f(t)}{t} = -\frac{2}{t}(1 - \cos 2t)$$

3.5 LAPLACE TRANSFORMS OF PERIODIC FUNCTIONS

Before we turn to the solution of differential equations using Laplace transforms, we shall consider the problem of finding the transform of periodic functions that often exist as input functions in physical systems. The non-homogeneous part of differential equations would involve such periodic functions as was illustrated in Section 1.10. A periodic function is one that has the characteristic

$$\begin{aligned} f(t) &= f(t+a) = f(t+2a) = f(t+3a) \\ &= \cdots = f(t+na) = \cdots \end{aligned} \tag{3.5.1}$$

This is illustrated in Fig. 3.9. We can write the transform of $f(t)$ as the series of integrals

$$\begin{aligned} \mathcal{L}(f) &= \int_0^\infty f(t)e^{-st}dt \\ &= \int_0^a f(t)e^{-st}dt + \int_a^{2a} f(t)e^{-st}dt + \int_{2a}^{3a} f(t)e^{-st}dt + \cdots \end{aligned} \tag{3.5.2}$$

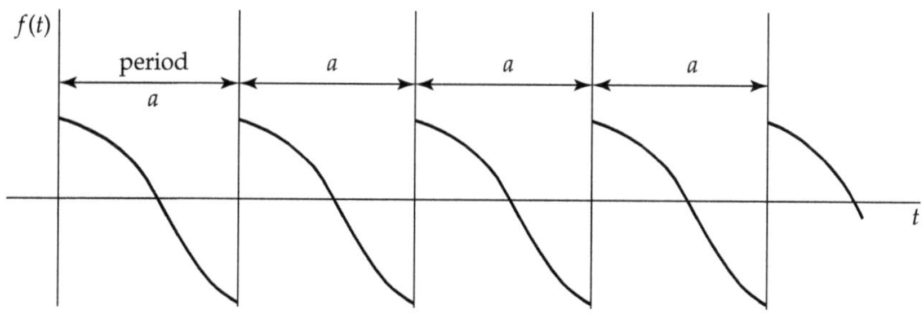

FIGURE 3.9 Periodic function.

In the second integral, let $t = \tau + a$; in the third integral, let $t = \tau + 2a$; in the fourth, let $t = \tau + 3a$, etc.; then the limits on each integral are 0 and a. There results

$$\mathcal{L}(f) = \int_0^a f(t)e^{-st}dt + \int_0^a f(\tau + a)e^{-s(\tau+a)}d\tau + \int_0^a f(\tau + 2a)e^{-s(\tau+2a)}d\tau + \cdots \quad (3.5.3)$$

The dummy variable of integration τ can be set equal to t, and with the use of Eq. 3.5.1 we have

$$\begin{aligned}\mathcal{L}(f) &= \int_0^a f(t)e^{-st}dt + e^{-as}\int_0^a f(t)e^{-st}dt + e^{-2as}\int_0^a f(t)e^{-st}dt + \cdots \\ &= [1 + e^{-as} + e^{-2as} + \cdots]\int_0^a f(t)e^{-st}dt \quad (3.5.4)\end{aligned}$$

Using the series expansion (see Eq. 2.1.6), $1/(1 - x) = 1 + x + x^2 + \cdots$, we can write the equation above as

$$\mathcal{L}(f) = \frac{1}{1 - e^{-as}}\int_0^a f(t)e^{-st}dt \quad (3.5.5)$$

provided $|e^{-as}| < 1$.

Example 3.18

Determine the Laplace transform of the square-wave function of Fig. 3.10. Compare with Example 3.10.

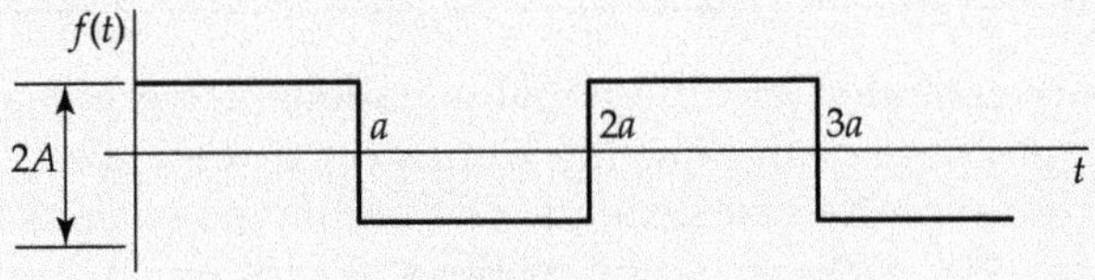

FIGURE 3.10

Solution

The function $f(t)$ is periodic with period $2a$. Using Eq. 3.5.5, we would have

$$\begin{aligned}\mathcal{L}(f) &= \frac{1}{1 - e^{-2as}}\int_0^{2a} f(t)e^{-st}dt \\ &= \frac{1}{1 - e^{-2as}}\left[\int_0^a Ae^{-st}dt + \int_a^{2a}(-A)e^{-st}dt\right] \\ &= \frac{1}{1 - e^{-2as}}\left[-\frac{A}{s}e^{-st}\Big|_0^a + \frac{A}{s}e^{-st}\Big|_a^{2a}\right] \\ &= \frac{1}{1 - e^{-2as}}\left[\frac{A}{s}(-e^{-as} + 1 + e^{-2as} - e^{-as})\right] \\ &= \frac{A}{s}\frac{1 - 2e^{-as} + e^{-2as}}{1 - e^{-2as}} = \frac{A}{s}\frac{(1 - e^{-as})(1 - e^{-as})}{(1 - e^{-as})(1 + e^{-as})} \\ &= \frac{A}{s}\frac{1 - e^{-as}}{1 + e^{-as}}\end{aligned}$$

This is the same result obtained in Example 3.10. It can be put in the more desired form,

$$\mathcal{L}(f) = \frac{A}{s}\tanh\frac{as}{2}$$

Example 3.19

Find the Laplace transform of the half-wave rectified sine wave shown in Fig. 3.11 with period 2π and an amplitude of 2.

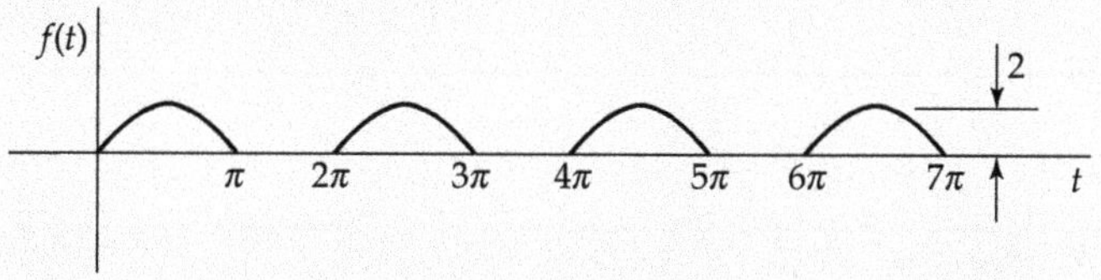

FIGURE 3.11

Solution

The function $f(t)$ is given by

$$f(t) = \begin{cases} \sin t, & 0 < t < \pi \\ 0, & \pi < t < 2\pi \end{cases}$$

Equation 3.5.5 provides us with

$$\begin{aligned} \mathcal{L}(f) &= \frac{1}{1-e^{-2\pi s}}\int_0^{2\pi} f(t)e^{-st}dt \\ &= \frac{1}{1-e^{-2\pi s}}\int_0^{\pi} \sin t\; e^{-st}dt \end{aligned}$$

Integrate by parts:

$$\begin{aligned} u &= \sin t, \quad & dv &= e^{-st}dt \\ du &= \cos t\, dt, \quad & v &= -\frac{1}{s}e^{-st} \end{aligned}$$

Then

$$\int_0^{\pi} \sin t\; e^{-st}dt = -\frac{1}{s}e^{-st}\overset{0}{\cancel{\sin t}}\Big|_0^{\pi} + \frac{1}{s}\int_0^{\pi} \cos t\; e^{-st}dt$$

Integrate by parts again:

$$\begin{aligned} u &= \cos t, \quad & dv &= e^{-st}dt \\ du &= -\sin t\, dt, \quad & v &= -\frac{1}{s}e^{-st} \end{aligned}$$

Again,

$$\int_0^{\pi} \sin t\, e^{-st} dt = \frac{1}{s}\left[-\frac{1}{s}e^{-st}\cos t\Big|_0^{\pi} - \frac{1}{s}\int_0^{\pi} \sin t\, e^{-st} dt\right]$$

This is rearranged to give

$$\int_0^{\pi} \sin t\, e^{-st} dt = \frac{s^2}{s^2+1}\left[\frac{1}{s^2}(e^{-s\pi}+1)\right] = \frac{1+e^{-\pi s}}{s^2+1}$$

Finally,

$$\mathcal{L}(f) = \frac{1+e^{-\pi s}}{(1-e^{-2\pi s})(s^2+1)} = \frac{1}{(1-e^{-\pi s})(s^2+1)}$$

Example 3.20

Find the Laplace transform of the periodic function of Fig. 3.12.

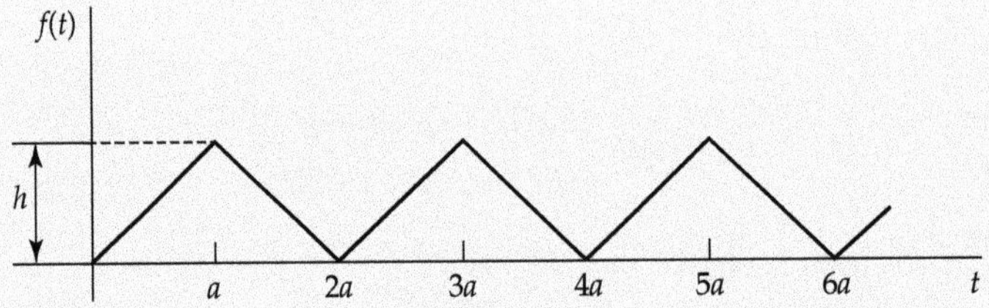

FIGURE 3.12

Solution
We could find the transform for the function by using Eq. 3.5.5 with

$$f(t) = \begin{cases} \dfrac{h}{a}t, & 0 < t < a \\ \dfrac{h}{a}(2a-t), & a < t < 2a \end{cases}$$

This would not be too difficult a task; but if we recognize that the $f(t)$ of this example is simply the integral of the square wave of Example 3.18 if we let $h = Aa$, then we can use Eq. 3.3.14 in the form

$$\mathcal{L}(f) = \mathcal{L}\left(\int_0^t g(\tau)d\tau\right) = \frac{1}{s}\mathcal{L}(g)$$

where $\mathcal{L}(g)$ is given in Example 3.18. There results

$$\mathcal{L}(f) = \frac{h}{as^2}\tanh\frac{as}{2}$$

This example illustrates that some transforms may be easier to find using the results of the preceding sections.

3.6 INVERSE TRANSFORMS—PARTIAL FRACTIONS

When solving differential equations using Laplace transforms we must frequently make use of partial fractions in finding the inverse of a transform. In this section we shall present a technique that will organize this procedure of finding the partial fractions.

3.6.1 Unrepeated Linear Factor $(s - a)$

Consider the ratio of two polynomials $P(s)$ and $Q(s)$ such that the order of $Q(s)$ is greater than the order of $P(s)$. Then the method of partial fractions allows us to write

$$F(s) = \frac{P(s)}{Q(s)} = \frac{A_1}{s - a_1} + \frac{A_2}{s - a_2} + \frac{A_3}{s - a_3} + \cdots + \frac{A_n}{s - a_n} \tag{3.6.1}$$

where it is assumed that $Q(s)$ can be factored into n factors with distinct roots $a_1, a_2, a_3, \ldots, a_n$. Let us attempt to find one of the coefficients, for example A_3. Multiply Eq. 3.6.1 by $(s - a_3)$ and let $s \to a_3$; there results

$$\lim_{s \to a_s} \frac{P(s)}{Q(s)}(s - a_3) = A_3 \tag{3.6.2}$$

since all other terms are multiplied by $(s - a_3)$, which goes to zero as $s \to a_3$.

Now, we may find the limit shown above. It is found as follows:

$$\lim_{s \to a_3} \frac{P(s)}{Q(s)}(s - a_3) = \lim_{s \to a_3} \left[P(s) \frac{s - a_3}{Q(s)} \right] = P(a_3) \frac{0}{0} \tag{3.6.3}$$

Because the quotient $0/0$ appears, we use L'Hospital's rule and differentiate both numerator and denominator with respect to s and then let $s \to a_3$. This yields

$$A_3 = P(a_3) \lim_{s \to a_3} \frac{1}{Q'(s)} = \frac{P(a_3)}{Q'(a_3)} \tag{3.6.4}$$

We could, of course, have chosen any coefficient; so, in general,

$$A_i = \frac{P(a_i)}{Q'(a_i)} \quad \text{or} \quad A_i = \frac{P(a_i)}{[Q(s)/(s - a_i)]_{s=a_i}} \tag{3.6.5}$$

This second formula is obtained from the limit in Eq. 3.6.2. With either of these formulas, the coefficients of the partial fractions are quite easily obtained.

3.6.2 Repeated Linear Factor $(s - a)^m$

If there are repeated roots in $Q(s)$, such as $(s - a)^m$, we would have the sum of partial fractions,

$$F(s) = \frac{P(s)}{Q(s)} = \frac{B_m}{(s - a_1)^m} + \cdots + \frac{B_2}{(s - a_1)^2} + \frac{B_1}{s - a_1} + \frac{A_2}{s - a_2} + \frac{A_3}{s - a_3} + \cdots \tag{3.6.6}$$

The A_i, the coefficients of the terms resulting from distinct roots, would be as given in Eq. 3.6.5. But the B_i would be given by

$$\begin{aligned} B_i &= \frac{1}{(m-i)!}\frac{d^{m-i}}{ds^{m-i}}\left[\frac{P(s)}{Q(s)/(s-a_1)^m}\right]_{s=a_1} \\ B_m &= \frac{P(a_1)}{[Q(s)/(s-a_1)^m]_{s=a_1}} \end{aligned} \tag{3.6.7}$$

3.6.3 Unrepeated Quadratic Factor $[(s-a)^2+b^2]$

Suppose that a quadratic factor appears in $Q(s)$, such as $[(s-a)^2+b^2]$. The transform $F(s)$ written in partial fractions would be

$$F(s) = \frac{P(s)}{Q(s)} = \frac{B_1 s + B_2}{(s-a)^2+b^2} + \frac{A_1}{s-a_1} + \frac{A_2}{s-a_2} + \cdots \tag{3.6.8}$$

where B_1 and B_2 are real constants. Now, multiply by the quadratic factor and let $s \to (a+ib)$. There results

$$B_1(a+ib) + B_2 = \left\{\frac{P(s)}{Q(s)/[(s-a)^2+b^2]}\right\}_{s=a+ib}. \tag{3.6.9}$$

The equation above is complex. The real part and the imaginary part allow both B_1 and B_2 to be calculated. The A_i would be given by Eq. 3.6.5.

3.6.4 Repeated Quadratic Factor $[(s-a)^2+b^2]^2$

If the square of a quadratic factor appears in $Q(s)$, the transform $F(s)$ would be expanded in partial fractions as

$$F(s) = \frac{P(s)}{Q(s)} = \frac{C_1 s + C_2}{[(s-a)^2+b^2]^2} + \frac{B_1 s + B_2}{(s-a)^2+b^2} + \frac{A_1}{s-a_1} + \frac{A_2}{s-a_2} + \cdots \tag{3.6.10}$$

The undetermined constants are obtained from the complex equations

$$C_1(a+ib) + C_2 = \left\{\frac{P(s)}{Q(s)/[(s-a)^2+b^2]^2}\right\}_{s=a+ib} \tag{3.6.11}$$

and

$$B_1(a+ib) + B_2 = \frac{d}{ds}\left\{\frac{P(s)}{Q(s)/[(s-a)^2+b^2]^2}\right\}_{s=a+ib}. \tag{3.6.12}$$

The A_i are again given by Eq. 3.6.5.

Example 3.21

Find $f(t)$ if the Laplace transform of $f(t)$ is given as

$$F(s) = \frac{s^3 + 3s^2 - 2s + 4}{s(s-1)(s-2)(s^2 + 4s + 3)}$$

Solution

The partial-fraction representation of $F(s)$ is

$$F(s) = \frac{A_1}{s} + \frac{A_2}{s-1} + \frac{A_3}{s-2} + \frac{A_4}{s+1} + \frac{A_5}{s+3}$$

The A_i will be found using the second formula of Eq. 3.6.5. For the $F(s)$ given, we have

$$\begin{aligned} P(s) &= s^3 + 3s^2 - 2s + 4 \\ Q(s) &= s(s-1)(s-2)(s^2 + 4s + 3) \\ &= s(s-1)(s-2)(s+3)(s+1) \end{aligned}$$

For the first root, $a_1 = 0$. Letting $s = 0$ in the expressions for $P(s)$ and $Q(s)/s$, there results

$$A_1 = \frac{P(0)}{[Q(s)/s]_{s=0}} = \frac{4}{6}$$

Proceeding, we have, with $a_2 = 1$, $a_3 = 2$, $a_4 = -1$, and $a_5 = -3$,

$$A_2 = \frac{P(1)}{[Q(s)/(s-1)]_{s=1}} = \frac{6}{-8}, \qquad A_3 = \frac{P(2)}{[Q(s)/(s-2)]_{s=2}} = \frac{20}{30}$$

$$A_4 = \frac{P(-1)}{[Q(s)/(s+1)]_{s=-1}} = \frac{8}{-12}, \qquad A_5 = \frac{P(-3)}{[Q(s)/(s+3)]_{s=-3}} = \frac{10}{120}$$

The partial-fraction representation is then

$$F(s) = \frac{\frac{2}{3}}{s} - \frac{\frac{3}{4}}{s-1} + \frac{\frac{2}{3}}{s-2} - \frac{\frac{2}{3}}{s+1} + \frac{\frac{1}{12}}{s+3}$$

Table 3.1 is consulted in finding $f(t)$. There results

$$f(t) = \frac{2}{3} - \frac{3}{4}e^{t} + \frac{2}{3}e^{2t} - \frac{2}{3}e^{-t} + \frac{1}{12}e^{-3t}$$

Example 3.22

Find $f(t)$ if the Laplace transform of $f(t)$ is

$$F(s) = \frac{s^2 - 1}{(s-2)^2(s^2+s-6)}$$

Solution

The denominator $Q(s)$ can be written as

$$Q(s) = (s-2)^3(s+3)$$

Thus, a triple root occurs and we use the partial-fraction expansion given by Eq. 3.6.6, that is,

$$F(s) = \frac{B_3}{(s-2)^3} + \frac{B_2}{(s-2)^2} + \frac{B_1}{s-2} + \frac{A_2}{s+3}$$

The constants B_i are determined from Eq. 3.6.7 to be

$$B_3 = \left[\frac{s^2-1}{s+3}\right]_{s=2} = \frac{3}{5}$$

$$B_2 = \frac{1}{1!}\frac{d}{ds}\left[\frac{s^2-1}{s+3}\right]_{s=2} = \left[\frac{s^2+6s+1}{(s+3)^2}\right]_{s=2} = \frac{17}{25}$$

$$B_1 = \frac{1}{2!}\frac{d^2}{ds^2}\left[\frac{s^2-1}{s+3}\right]_{s=2} = \frac{8}{125}$$

The constant A_2 is, using $a_2 = -3$,

$$A_2 = \frac{P(a_2)}{[Q(s)/(s+3)]_{s=a_2}} = -\frac{8}{125}$$

Hence we have

$$F(s) = \frac{\frac{3}{5}}{(s-2)^3} + \frac{\frac{17}{25}}{(s-2)^2} + \frac{\frac{8}{125}}{s-2} + \frac{-\frac{8}{125}}{s+3}$$

Table 3.1, at the end of the chapter, allows us to write $f(t)$ as

$$f(t) = \frac{3}{10}t^2e^{2t} + \frac{17}{25}te^{2t} + \frac{8}{125}e^{2t} - \frac{8}{125}e^{-3t}$$

Example 3.23

The Laplace transform of the displacement function $y(t)$ for a forced, frictionless, spring–mass system is found to be

$$Y(s) = \frac{\omega F_0/M}{(s^2+\omega_0^2)(s^2+\omega^2)}$$

for a particular set of initial conditions. Find $y(t)$.

Solution

The function $Y(s)$ can be written as

$$Y(s) = \frac{A_1 s + A_2}{s^2+\omega_0^2} + \frac{B_1 s + B_2}{s^2+\omega^2}$$

The functions $P(s)$ and $Q(s)$ are

$$P(s) = \frac{\omega F_0}{M}$$
$$Q(s) = (s^2+\omega_0^2)(s^2+\omega^2)$$

With the use of Eq. 3.6.9, we have

$$A_1(i\omega_0) + A_2 = \frac{\omega F_0/M}{(i\omega_0)^2+\omega^2} = \frac{\omega F_0/M}{\omega^2-\omega_0^2}$$
$$B_1(i\omega) + B_2 = \frac{\omega F_0/M}{(i\omega)^2+\omega_0^2} = -\frac{\omega F_0/M}{\omega^2-\omega_0^2}$$

where $a = 0$ and $b = \omega_0$ in the first equation; in the second equation $a = 0$ and $b = \omega$. Equating real and imaginary parts:

$$A_1 = 0, \qquad A_2 = \frac{\omega F_0/M}{\omega^2-\omega_0^2}$$
$$B_1 = 0, \qquad B_2 = -\frac{\omega F_0/M}{\omega^2-\omega_0^2}$$

The partial-fraction representation is then

$$Y(s) = \frac{\omega F_0/M}{\omega^2-\omega_0^2}\left[\frac{1}{s^2+\omega_0^2} - \frac{1}{s^2+\omega^2}\right]$$

Finally, using Table 3.1 we have

$$y(t) = \frac{\omega F_0/M}{\omega^2-\omega_0^2}\left[\frac{1}{\omega_0}\sin\omega_0 t - \frac{1}{\omega}\sin\omega t\right]$$

3.7 SOLUTION OF DIFFERENTIAL EQUATIONS

We are now in a position to solve linear ordinary differential equations with constant coefficients. The technique will be demonstrated with second-order equations, as was done in Chapter 1. The method is, however, applicable to any linear differential equation with constant coefficients. To solve a differential equation, we shall find the Laplace transform of each term of the differential equation, using the techniques presented in Sections 3.2 through 3.5. The resulting algebraic equation will then be organized into a form for which the inverse can be readily found. For nonhomogeneous equations this usually involves partial fractions as discussed in Section 3.6. Let us demonstrate the procedure using the equation

$$\frac{d^2y}{dt^2} + a\frac{dy}{dt} + by = r(t) \tag{3.7.1}$$

By taking the Laplace transform of both sides of Eq. 3.7.1, Eqs. 3.3.4 and 3.3.5 allow us to write this differential equation as the algebraic equation

$$s^2Y(s) - sy(0) - y'(0) + a[sY(s) - y(0)] + bY(s) = R(s) \tag{3.7.2}$$

where $Y(s) = \mathcal{L}(y)$ and $R(s) = \mathcal{L}(r)$. This algebraic equation is referred to as the *subsidiary equation* of the given differential equation. It can be rearranged in the form

$$Y(s) = \frac{(s+a)y(0) + y'(0)}{s^2 + as + b} + \frac{R(s)}{s^2 + as + b} \tag{3.7.3}$$

Note that the non zero initial conditions are responsible for the first term on the right and the nonhomogeneous part of the differential equation is responsible for the second term. To find the desired solution, our task is simply to find the inverse Laplace transform

$$y(t) = \mathcal{L}^{-1}(Y) \tag{3.7.4}$$

Let us illustrate with several examples.

Example 3.24

The differential equation that represents the damped harmonic motion of the spring-mass system shown in Fig. 3.13 is

$$\frac{d^2y}{dt^2} + 4\frac{dy}{dt} + 8y = 0$$

If the initial conditions are $y(0) = 2$, $dy/dt\,(0) = 0$, find its solution. For the derivation of this equation, see Section 1.7.

$y(t)$ is the displacement from the static equilibrium position.

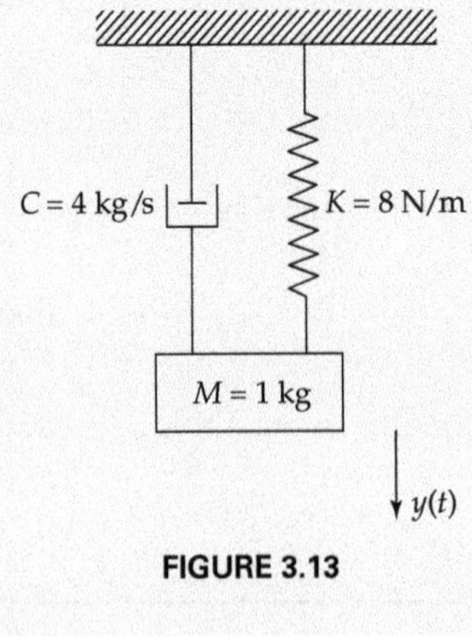

FIGURE 3.13

Solution

The subsidiary equation is found by taking the Laplace transform of the given differential equation:

$$s^2Y - sy(0) - \overset{0}{\cancel{\dot{y}}}(0) + 4[sY - y(0)] + 8Y = 0$$

This is rearranged and put in the form

$$Y(s) = \frac{2s+8}{s^2+4s+8}$$

To use Table 3.1 we write this as

$$\begin{aligned} Y(s) &= \frac{2(s+2)+4}{(s+2)^2+4} \\ &= \frac{2(s+2)}{(s+2)^2+4} + \frac{4}{(s+2)^2+4} \end{aligned}$$

The inverse transform is then found to be

$$\begin{aligned} y(t) &= e^{-2t}2\cos 2t + e^{-2t}2\sin 2t \\ &= 2e^{-2t}(\cos 2t + \sin 2t) \end{aligned}$$

Example 3.25

An inductor of 2 henrys and a capacitor of 0.02 farad is connected in series with an imposed voltage of 100 sin ωt volts, shown in Fig. 3.14. Determine the charge $q(t)$ on the capacitor as a function of ω if the initial charge on the capacitor and current in the circuit are zero.

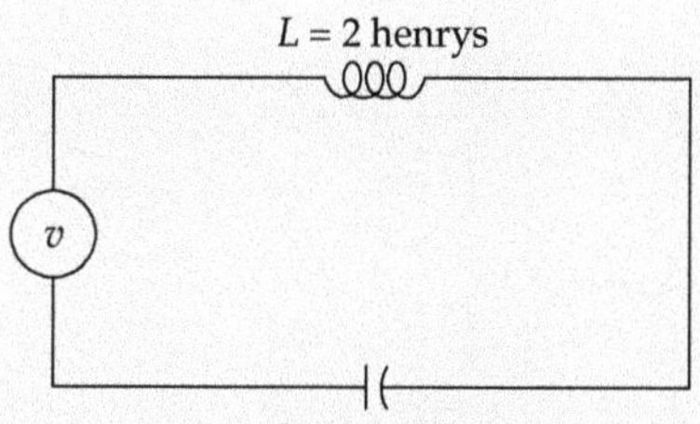

FIGURE 3.14

Solution

Kirchhoff's laws allow us to write (see Section 1.4)

$$2\frac{di}{dt} + \frac{q}{0.02} = 100\sin\omega t$$

where $i(t)$ is the current in the circuit. Using $i = dq/dt$, we have

$$2\frac{d^2q}{dt^2} + 50q = 100\sin\omega t$$

The Laplace transform of this equation is

$$2[s^2Q - \cancelto{0}{sq(0)} - \cancelto{0}{q'(0)}] + 50Q = \frac{100\omega}{s^2+\omega^2}$$

using $i(0) = q'(0) = 0$. The transform of $q(t)$ is then

$$Q(s) = \frac{50\omega}{(s^2+\omega^2)(s^2+25)}$$

The appropriate partial fractions are

$$Q(s) = \frac{A_1s + A_2}{s^2+\omega^2} + \frac{B_1s + B_2}{s^2+25}$$

The constants are found from (see Eq. 3.6.9)

$$A_1(i\omega) + A_2 = \frac{50\omega}{-\omega^2+25}, \qquad B_1(5i) + B_2 = \frac{50\omega}{-25+\omega^2}$$

They are

$$A_1 = 0, \quad A_2 = \frac{50\omega}{25-\omega^2}, \quad B_1 = 0, \quad B_2 = \frac{50\omega}{\omega^2-25}$$

Hence,

$$Q(s) = \frac{50}{25-\omega^2}\left[\frac{\omega}{s^2+\omega^2} - \frac{\omega}{s^2+25}\right]$$

The inverse Laplace transform is

$$q(t) = \frac{50}{25-\omega^2}[\sin\omega t - \sin 5t]$$

This solution is acceptable if $\omega \neq 5$ rad/s, and we observe that the amplitude becomes unbounded as $\omega \to 5$ rad/s.

If $\omega = 5$ rad/s, the initial expression of the Laplace transform becomes

$$Q(s) = \frac{250}{(s^2+25)^2} = \frac{C_1s + C_2}{(s^2+25)^2} + \frac{B_1s + B_2}{s^2+25}$$

Using Eqs. 3.6.11 and 3.6.12, we have

$$C_1(5i) + C_2 = 250, \qquad B_1(5i) + B_2 = \frac{d}{ds}(250) = 0$$

We have

$$C_1 = 0, \ \ C_2 = 250, \ \ B_1 = 0, \ \ B_2 = 0$$

Hence,

$$Q(s) = \frac{250}{(s^2 + 25)^2}$$

The inverse is

$$\begin{aligned} q(t) &= 250\left[\frac{1}{2 \times 5^3}(\sin 5t - 5t \cos 5t)\right] \\ &= \sin 5t - 5t \cos 5t \end{aligned}$$

Observe that the amplitude becomes unbounded as t gets large. This is *resonance*, a phenomenon that occurs in undampened oscillatory systems with input frequency equal to the natural frequency of the system. See Section 1.9.1 for a discussion of resonance.

Example 3.26

As an example of a differential equation that has boundary conditions at two locations, consider a beam loaded as shown in Fig. 3.15. The differential equation that describes the deflection $y(x)$ is

$$\frac{d^4y}{dx^4} = \frac{w}{EI}$$

with boundary conditions $y(0) = y''(0) = y(L) = y''(L) = 0$, where E is the elastic modulus and I is the cross-sectional second moment of area. Find $y(x)$.

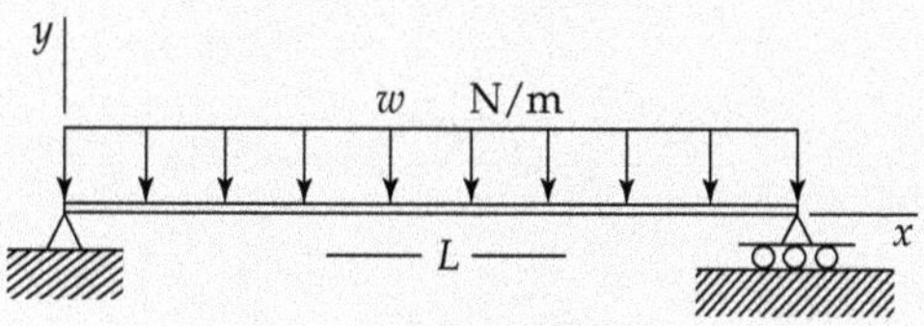

FIGURE 3.15

Solution

The Laplace transform of the differential equation is, according to Eq. 3.3.6,

$$s^4Y - s^3\overset{0}{\cancel{y(0)}} - s^2y'(0) - \overset{0}{\cancel{sy''(0)}} - y'''(0) = \frac{w}{EIs}$$

The two unknown initial conditions are replaced with

$$y'(0) = c_1 \quad \text{and} \quad y'''(0) = c_2$$

We then have

$$Y(s) = \frac{c_1}{s^2} + \frac{c_2}{s^4} + \frac{w}{EIs^5}$$

The inverse Laplace transform is

$$y(x) = c_1 x + c_2 \frac{x^3}{6} + \frac{w}{EI}\frac{x^4}{24}$$

The boundary conditions on the right end are now satisfied

$$y(L) = c_1 L + c_2 \frac{L^3}{6} + \frac{w}{EI}\frac{L_4}{24} = 0$$

$$y''(L) = c_2 L + \frac{w}{EI}\frac{L^2}{2} = 0$$

Hence,

$$c_2 = -\frac{wL}{2EI}, \qquad c_1 = -\frac{wL^3}{24EI}$$

Finally, the desired solution for the deflection of the beam is

$$y(x) = \frac{w}{24EI}\left[xL^3 - 2x^3L + x^4\right]$$

Example 3.27

Solve the differential equation

$$\frac{d^2y}{dt^2} + 0.02\frac{dy}{dt} + 25y = f(t)$$

which describes a slightly damped oscillating system where $f(t)$ is as shown in Fig. 3.16. Assume that the system starts from rest.

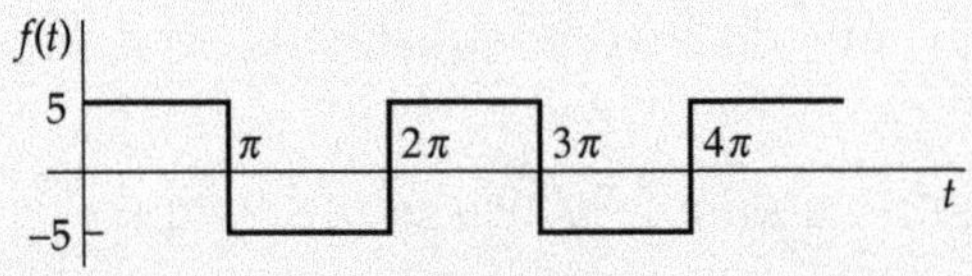

FIGURE 3.16

Solution

The subsidary equation is found by taking the Laplace transform of the given differential equation:

$$s^2Y - sy(0) - y'(0) + 0.02[sY - y(0)] + 25Y = F(s)$$

where $F(s)$ is given by (see Example 3.10)

$$F(s) = \frac{5}{s}[1 - 2e^{-\pi s} + 2e^{-2\pi s} - 2e^{-3\pi s} + \cdots]$$

Since the system starts from rest, $y(0) = y'(0) = 0$. The subsidiary equation becomes

$$\begin{aligned} Y(s) &= \frac{F(s)}{s^2 + 0.02s + 25} \\ &= \frac{5}{s(s^2 + 0.02s + 25)}[1 - 2e^{-\pi s} + 2e^{-2\pi s} - \cdots] \end{aligned}$$

Now let us find the inverse term by term. We must use*

$$\mathcal{L}^{-1}\left[\frac{5}{(s + 0.01)^2 + 5^2}\right] = e^{-0.01t}\sin 5t$$

With the use of Eq. 3.3.14, we have, for the first term (see Example 3.19 or Eq. 3.3.2 for integration by parts),

$$\begin{aligned} y_0(t) &= \mathcal{L}^{-1}\left[\frac{5}{s[(s + 0.01)^2 + 5^2]}\right] \\ &= \int_0^t e^{-0.01\tau}\sin 5\tau\, d\tau \\ &= 1 - \frac{1}{5}e^{-0.01t}[\cos 5t + 0.002\sin 5t] \end{aligned}$$

The inverse of the next term is found using the second shifting property (see Eq. 3.2.11):

$$\begin{aligned} y_1(t) &= \mathcal{L}^{-1}\left[e^{-\pi s}\frac{5}{s(s^2 + 0.02s + 25)}\right] \\ &= -2u_\pi(t)\left\{1 - \frac{1}{5}e^{-0.01(t-\pi)}[\cos 5(t - \pi) + 0.002\sin 5(t - \pi)]\right\} \\ &= -2u_\pi(t)\{1 + [1 - y_0(t)]e^{0.01\pi}\} \end{aligned}$$

where $u_\pi(t)$ is the unit step function and we have used $\cos(t - \pi) = -\cos t$ and $\sin(t - \pi) = -\sin t$. The third term provides us with

$$\begin{aligned} y_2(t) &= \mathcal{L}^{-1}\left[e^{-2\pi s}\frac{5}{s(s^2 + 0.02s + 25)}\right] \\ &= -2u_{2\pi}(t)\left\{1 - \frac{1}{5}e^{-0.01(t-2\pi)}[\cos 5(t - 2\pi) + 0.002\sin 5(t - 2\pi)]\right\} \\ &= -2u_{2\pi}(t)\left\{1 - [1 - y_0(t)]e^{0.02\pi}\right\} \end{aligned}$$

*We write $s^2 + 0.02s + 25 = (s + 0.01)^2 + 24.999 \cong (s + 0.01)^2 + 5^5$.

and so on. The solution $y(t)$ would be

$$y(t) = y_0(t) = 1 - \frac{1}{5}e^{-0.01t}[\cos 5t + 0.002\sin 5t], \qquad 0 < t < \pi$$

$$y(t) = y_0(t) + y_1(t)$$
$$= -1 - \frac{1}{5}e^{-0.01t}[\cos 5t + 0.002\sin 5t][1 + 2e^{0.01\pi}], \qquad \pi < t < 2\pi$$

$$y(t) = y_0(t) + y_1(t) + y_2(t)$$
$$= 1 - \frac{1}{5}e^{-0.01t}[\cos 5t + 0.002\sin 5t][1 + 2e^{0.01\pi} + 2e^{0.02\pi}],$$
$$2\pi < t < 3\pi$$

Now let us find the solution for large t, that is, for $n\pi < t < (n + 1)\pi$, with n large. Generalize the results above and obtain*

$$y(t) = y_0(t) + y_1(t) + \cdots + y_n(t), \qquad n\pi < t < (n+1)\pi$$
$$= (-1)^n - \frac{1}{5}e^{-0.01t}[\cos 5t + 0.002\sin 5t][1 + 2e^{0.01\pi} + 2e^{0.02\pi} + \cdots + 2e^{0.01n\pi}]$$
$$= (-1)^n + \frac{1}{5}e^{-0.01t}[\cos 5t + 0.002\sin 5t]$$
$$-\frac{2}{5}e^{-0.01t}[\cos 5t + 0.002\sin 5t][1 + 2e^{0.01\pi} + \cdots + 2e^{0.01n\pi}]$$
$$= (-1)^n + \frac{1}{5}e^{-0.01t}[\cos 5t + 0.002\sin 5t]$$
$$-\frac{2}{5}e^{-0.01t}[\cos 5t + 0.002\sin 5t]\frac{1 - e^{(n+1)0.01\pi}}{1 - e^{0.01\pi}}$$
$$= (-1)^n + \left[\frac{1}{5} - \frac{2}{5(1 - e^{0.01\pi})}\right]e^{-0.01t}[\cos 5t + 0.002\sin 5t]$$
$$+\frac{2e^{(n+1)0.01\pi - 0.01t}}{5(1 - e^{0.01\pi})}[\cos 5t + 0.002\sin 5t]$$

Then, letting t be large and in the interval $n\pi < t < (n + 1)\pi$, $e^{-0.01t} \to 0$ and we have

$$y(t) = (-1)^n + \frac{2e^{(n+1)0.01\pi - 0.01t}}{5(1 - e^{0.01\pi})}(\cos 5t + 0.002\sin 5t)$$
$$\cong (-1)^n - 12.5e^{0.01[(n+1)\pi - t]}\cos 5t$$

---

*In the manipulations we use

$$\frac{1}{1+x} = 1 + x + x^2 + x^3 + \cdots + x^n + x^{n+1} + \cdots$$
$$= 1 + x + x^2 + \cdots + x^n + x^{n+1}(1 + x + x + x^2 + \cdots)$$
$$= 1 + x + x^2 + \cdots + x^n + x^{n+1}/(1 - x)$$

Hence,

$$\frac{1 - x^{n+1}}{1 - x} = 1 + x + x^2 + \cdots + x^n$$

This is the steady-state response due to the square-wave input function shown in the example. One period is sketched in Fig. 3.17. The second half of the period is obtained by replacing n with $n+1$. Note that the input frequency of 1 rad/s results in a periodic response of 5 rad/s. Note also the large amplitude of the response, a resonance-type behavior. This is surprising, since the natural frequency of the system with no damping is 5 rad/s. This phenomenon occurs quite often when systems with little damping are subjected to nonsinusoidal periodic input functions.

The fifth harmonic of the Fourier series of the square-wave input is at resonance.

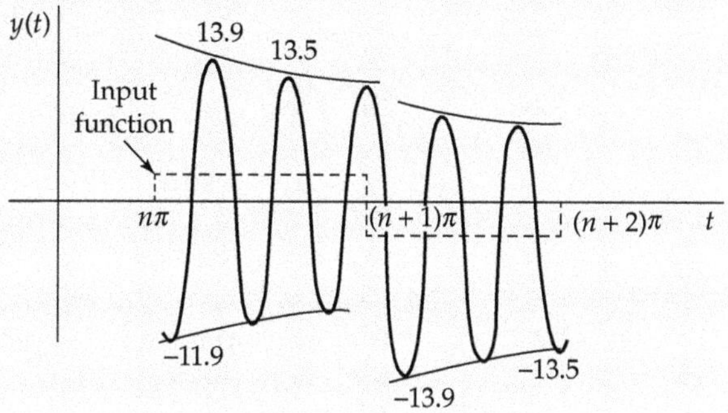

FIGURE 3.17 (For this sketch, n is considered even.)

PROBLEMS

3.1 Find the Laplace transform of each of the following functions and those in Fig. 3.18 by direct integration.

(a) $2t$ **(b)** $t-3$
(c) e^{3t} **(d)** $2\sin t$
(e) $\cos 4t$ **(f)** $t^{1/2}$
(g) $2t^{3/2}$ **(h)** $4t^2-3$
(i) $\sinh 2t$ **(j)** $(t-2)^2$
(k) $\cosh 4t$ **(l)** e^{2t-t}

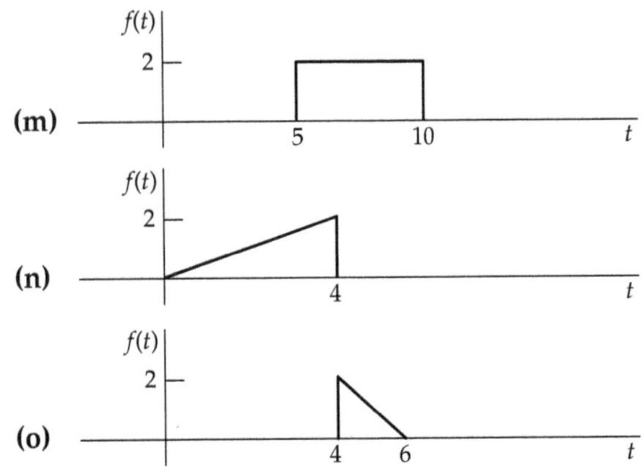

FIGURE 3.18

3.2 Use the first shifting property and Table 3.1 and find the Laplace transform of each of the following functions.

(a) $3te^{3t}$
(b) t^2e^{-t}
(c) $e^{-2}\cos 4t$
(d) $e^{2t}\sinh 2t$
(e) $3e^{-t}\sin 2t$
(f) $4e^{-2t}\cosh t$
(g) $e^{-t}(\cos 4t - 2\sin 4t)$
(h) $e^{-2t}(\sinh 2t + 3\cosh 2t)$
(i) $e^{-2t}(t^2+4t+5)$

3.3 Use the second shifting property and Table 3.1 and find the Laplace transform of each of the following functions. Sketch each function.

(a) $u_2(t)$
(b) $u_4(t)\sin \pi t$
(c) $f(t)=\begin{cases}0 & 0<t<2\\ 2t & 2<t<4\\ 0 & 4<t\end{cases}$
(d) $\dfrac{t}{2}-u_4(t)\dfrac{t}{2}$
(e) $u_4(t)(6-t)-u_6(t)(6-t)$
(f) $f(t)=\begin{cases}t & 0<t<2\\ 2 & 2<t\end{cases}$
(g) $f(t)=\begin{cases}\sin t & 0<t<2\pi\\ 0 & 2\pi<t\end{cases}$
(h) $f(t)=\begin{cases}\sin t & 0<t<\pi\\ \sin 2t & \pi<t\end{cases}$

3.4 Express each of the following hyperbolic functions in terms of exponential functions and, with the use of Table 3.1, find the Laplace transform.

(a) $2\cosh 2t\sin 2t$ **(b)** $2\sinh 3t\cos 2t$
(c) $4\cosh 2t\sinh 3t$ **(d)** $6\sinh t\cos t$
(e) $4\sinh 2t\sinh 4t$ **(f)** $2\cosh t\cos 2t$

3.5 Using Table 3.1, find the function $f(t)$ corresponding to each of the following Laplace transforms.

(a) $\dfrac{1}{s}\left(\dfrac{1}{s^2}+\dfrac{1}{s}-2\right)$ **(b)** $\dfrac{1}{s^2}\left(\dfrac{3}{s}+2\right)$

(c) $\dfrac{2s}{(s+3)^2}$ **(d)** $\dfrac{s}{(s+1)^3}$

(e) $\dfrac{1}{s(s+1)}$ **(f)** $\dfrac{1}{s^2(s-2)}$

(g) $\dfrac{1}{(s-2)(s+1)}$ **(h)** $\dfrac{1}{(s-1)(s+2)}$

(i) $\dfrac{2s}{(s-1)^2(s+1)}$ **(j)** $\dfrac{e^{-s}}{s+1}$

(k) $\dfrac{e^{-2s}}{s(s+1)^2}$ **(l)** $\dfrac{4}{s^2+2s+5}$

(m) $\dfrac{4s+3}{s^2+4s+13}$ **(n)** $\dfrac{2}{s^2-2s-3}$

(o) $\dfrac{3s+1}{s^2-4s-5}$ **(p)** $\dfrac{4se^{-2\pi s}}{s^2+2s+5}$

(q) $\dfrac{2}{(s^2-1)(s^2+1)}$ **(r)** $\dfrac{2s+3}{(s^2+4s+13)^2}$

3.6 If $\mathcal{L}[f(t)] = F(s)$, show that $\mathcal{L}[f(at)] = (1/a)F(s/a)$; use the definition of a Laplace transform. Then, if $\mathcal{L}[\cos t] = s/(s^2+1)$, find $\mathcal{L}(\cos 4t)$.

3.7 Show that $\mathcal{L}[J_0(t)] = (s^2+1)^{-1/2}$, where $J_0(t)$ is the Bessel function of order zero. Use a series expression for the Bessel function and the linearity property. Using the results of Problem 3.6, find $\mathcal{L}[J_0(2t)]$.

3.8 Expand $\sin\sqrt{t}$ in an infinite series and show that $\mathcal{L}(\sin\sqrt{t}) = (\sqrt{\pi}/2s^{3/2})e^{-1/4s}$.

3.9 **(a)** Write an expression for $\mathcal{L}(f^{(\text{iv})})$.

(b) Write an expression for $\mathcal{L}(f')$ if two discontinuities occur in $f(t)$, one at $t = a$ and the other at $t = b$.

3.10 Use Eq. 3.3.4 and find the Laplace transform of $f(t) = e^t$.

3.11 Use Eq. 3.3.5 and find the Laplace transform of each of the following

(a) $\sin \omega t$ **(b)** $\cos \omega t$
(c) $\sinh at$ **(d)** $\cosh at$
(e) e^{2t} **(f)** t

3.12 If $f(t) = \begin{cases} t, & 0 < t < 1 \\ 1, & 1 < t, \end{cases}$ find $\mathcal{L}(f)$. Also, find $\mathcal{L}(f')$. Is Eq. 3.3.4 verified for this $f(t)$?

3.13 If $f(t) = \begin{cases} t, & 0 < t < 1 \\ 0, & 1 < t, \end{cases}$ find $\mathcal{L}(f)$. Also, find $\mathcal{L}(f')$. Does Eq. 3.3.4 hold for this $f(t)$? Verify that Eq. 3.3.9 holds for this $f(t)$.

3.14 Using the equations for the Laplace transforms of derivatives from Section 3.3 and Table 3.1, find the transform of each of the following.

(a) te^t **(b)** $t \sin 2t$
(c) $t \cos t$ **(d)** $t^2 \sin t$
(e) $te^t \sin t$ **(f)** $(t^2+1)\cos 2t$
(g) $t \cosh 2t$ **(h)** t^2e^t
(i) $t \sinh 2t$

3.15 Find the function $f(t)$ corresponding to each of the following Laplace transforms.

(a) $\dfrac{1}{s^2+2s}$ **(b)** $\dfrac{2}{s^2-s}$
(c) $\dfrac{4}{s^3+4s}$ **(d)** $\dfrac{4}{s^4+4s^2}$
(e) $\dfrac{6}{s^3-9s}$ **(f)** $\dfrac{6}{s^4-9s^2}$
(g) $\dfrac{2}{s^4+2s^2}$ **(h)** $\dfrac{1}{s}\dfrac{s-1}{s+1}$
(i) $\dfrac{1}{s^2}\dfrac{s-1}{s^2+1}$

3.16 Determine the Laplace transform of each of the following functions using Table 3.1 and the equations of Section 3.4.

(a) $2t \sin 3t$ **(b)** $t \cos 2t$
(c) $t^2 \sin 2t$ **(d)** $t^2 \sinh t$
(e) $te^t \cos 2t$ **(f)** $t(e^t - e^{-t})$
(g) $t(e^t - e^{-2t})$ **(h)** $te^{-t} \sin t$
(i) $t^2e^{-t} \sin t$ **(j)** $t \cosh t$
(k) $\dfrac{2}{t}(1-\cos 2t)$ **(l)** $\dfrac{2}{t}(1-\cosh 2t)$
(m) $\dfrac{1}{t}(e^{2t}-e^{-2t})$ **(n)** $\dfrac{1}{t}(e^{2t}-1)$

3.17 Use Eq. 3.4.3 and find an expression for the Laplace transform of $f(t) = t^n e^{at}$ using $\mathcal{L}(e^{at}) = 1/(s-a)$.

3.18 Find the function $f(t)$ that corresponds to each of the following Laplace transforms using the equations of Section 3.4.

(a) $\dfrac{1}{(s+2)^2}$ **(b)** $\dfrac{4s}{(s^2+4)^2}$
(c) $\dfrac{s}{(s^2-4)^2}$ **(d)** $\ln\dfrac{s}{s-2}$
(e) $\ln\dfrac{s-2}{s+3}$ **(f)** $\ln\dfrac{s^2-4}{s^2+4}$
(g) $\ln\dfrac{s^2+1}{s^2+4}$ **(h)** $\ln\dfrac{s^2}{s^2+4}$
(i) $\ln\dfrac{s^2+4s+5}{s^2+2s+5}$

3.19 Determine the Laplace transform for each of the following periodic functions. The first period is stated. Also, sketch several periods of each function.

(a) $f(t) = \sin t,\ 0 < t < \pi$
(b) $f(t) = t,\quad 0 < t < 2$

(c) $f(t) = 2 - t, \quad 0 < t < 2$

(d) $f(t) = t - 2, \; 0 < t < 4$

(e) $f(t) = t^2, \quad 0 < t < \pi$

(f) $f(t) = \begin{cases} 1 & 0 < t < 2 \\ 0 & 2 < t < 4 \end{cases}$

(g) $f(t) = \begin{cases} t & 2 < t < 4 \\ 0 & 0 < t < 2 \end{cases}$

(h) $f(t) = \begin{cases} 2 - t & 0 < t < 1 \\ 0 & 1 < t < 2 \end{cases}$

(i) $f(t) = \begin{cases} 2 & 0 < t < 1 \\ 0 & 1 < t < 2 \\ -2 & 2 < t < 3 \\ 0 & 3 < t < 4 \end{cases}$

3.20 Find the function $f(t)$ corresponding to each of the following Laplace transforms.

(a) $\dfrac{120s}{(s-1)(s+2)(s^2-2s-3)}$

(b) $\dfrac{5s^2+20}{s(s-1)(s^2+5s+4)}$

(c) $\dfrac{s^3+2s}{(s^2+3s+2)(s^2+s-6)}$

(d) $\dfrac{s^2+2s+1}{(s-1)(s^2+2s-3)}$

(e) $\dfrac{8}{s^2(s-2)(s^2-4s+4)}$

(f) $\dfrac{s^2-3s+2}{s^2(s-1)^2(s-5s+4)}$

(g) $\dfrac{s^2-1}{(s^2+4)^2(s^2+1)}$

(h) $\dfrac{5}{(s^2+400)(s^2+441)}$

(i) $\dfrac{s-1}{(s+1)(s^2+4)}$

(j) $\dfrac{s^2+1}{(s+1)^2(s^2+4)}$

(k) $\dfrac{50}{(s^2+4)^2(s^2+1)}$

(l) $\dfrac{10}{(s^2+4)^2(s^2+1)^2}$

3.21 Determine the solution for each of the following initial-value problems.

(a) $\dfrac{d^2y}{dt^2} + 4y = 0, \quad y(0) = 0,\; y'(0) = 10$

(b) $\dfrac{d^2y}{dt^2} - 4y = 0, \quad y(0) = 2,\; y'(0) = 0$

(c) $\dfrac{d^2y}{dt^2} + y = 2, \quad y(0) = 0,\; y'(0) = 2$

(d) $\dfrac{d^2y}{dt^2} + 4y = 2\cos t, \quad y(0) = 0,\; y'(0) = 0$

(e) $\dfrac{d^2y}{dt^2} + 4y = 2\cos 2t, \quad y(0) = 0,\; y'(0) = 0$

(f) $\dfrac{d^2y}{dt^2} + y = e^t + 2, \quad y(0) = 0,\; y'(0) = 0$

(g) $\dfrac{d^2y}{dt^2} + 5\dfrac{dy}{dt} + 6y = 0,$

$y(0) = 0,\; y'(0) = 20$

(h) $\dfrac{d^2y}{dt^2} + 4\dfrac{dy}{dt} + 4y = 0,$

$y(0) = 1,\; y'(0) = 0$

(i) $\dfrac{d^2y}{dt^2} - 2\dfrac{dy}{dt} - 8y = 0,$

$y(0) = 1,\; y'(0) = 0$

(j) $\dfrac{d^2y}{dt^2} + 5\dfrac{dy}{dt} + 6y = 12,$

$y(0) = 0,\; y'(0) = 10$

(k) $\dfrac{d^2y}{dt^2} + 2\dfrac{dy}{dt} + y = 2t,$

$y(0) = 0,\; y'(0) = 0$

(l) $\dfrac{d^2y}{dt^2} + 4\dfrac{dy}{dt} + 4y = 4\sin 2t,$

$y(0) = 1,\; y'(0) = 0$

(m) $\dfrac{d^2y}{dt^2} + 4\dfrac{dy}{dt} + 104y = 2\cos 10t,$

$y(0) = 0,\; y'(0) = 0$

(n) $\dfrac{d^2y}{dt^2} + 2\dfrac{dy}{dt} + 101y = 5\sin 10t,$

$y(0) = 0,\; y'(0) = 20$

3.22 Solve for the displacement $y(t)$ of the mass in Fig. 3.19 if $y(0) = 0$, $y'(0) = 0$. Use a combination of the following friction coefficients and forcing functions.

(a) $C = 0$ kg/s	**(1)** $F(t) = 2$ N
(b) $C = 2$ kg/s	**(2)** $F(t) = 10 \sin 2t$
(c) $C = 24$ kg/s	**(3)** $F(t) = 10 \sin 6t$
(d) $C = 40$ kg/s	**(4)** $F(t) = 10[u_0(t) - u_{4\pi}(t)]$
	(5) $F(t) = 10e^{-0.2t}$
	(6) $F(t) = 100\delta_0(t)$

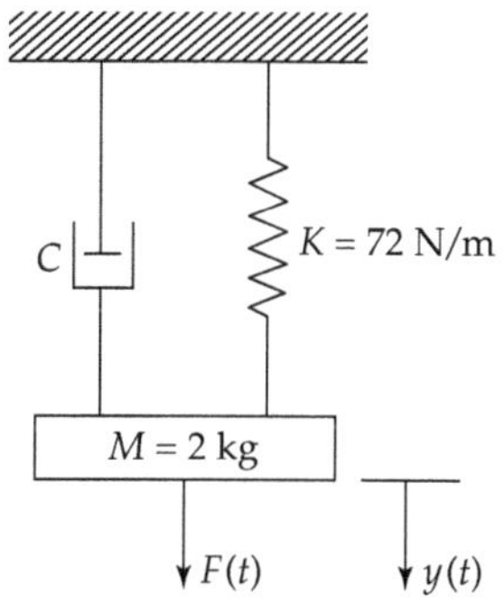

FIGURE 3.19

3.23 For a particular combination of the following resistances and input voltages in Fig. 3.20, calculate the current $i(t)$ if the circuit is quiescent at $t = 0$, that is, the initial charge on the capacitor $q(0) = 0$ and $i(0) = 0$. Sketch the solution.

(a) $R = 0$ ohms	**(1)** $v(t) = 10$ volts
(b) $R = 16$ ohms	**(2)** $v(t) = 10 \sin 20t$
(c) $R = 20$ ohms	**(3)** $v(t) = 5 \sin 10t$
(d) $R = 25$ ohms	**(4)** $v(t) = 10[u_0(t) - u_{2\pi}(t)]$
	(5) $v(t) = 10\ \delta_0(t)$
	(6) $v(t) = 20e^{-t}$

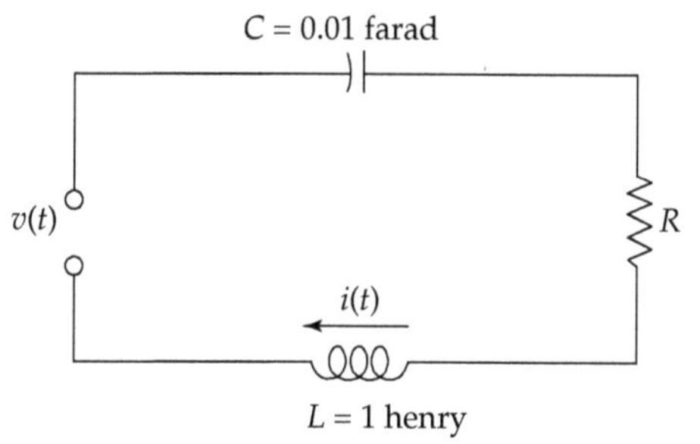

FIGURE 3.20

3.24 Calculate the response due to the input function $f(t)$ for one of the systems shown in parts (a) and (b) of Fig. 3.21. Assume each system to be quiescent at $t = 0$. The function $f(t)$ is given in parts (1) through (7).

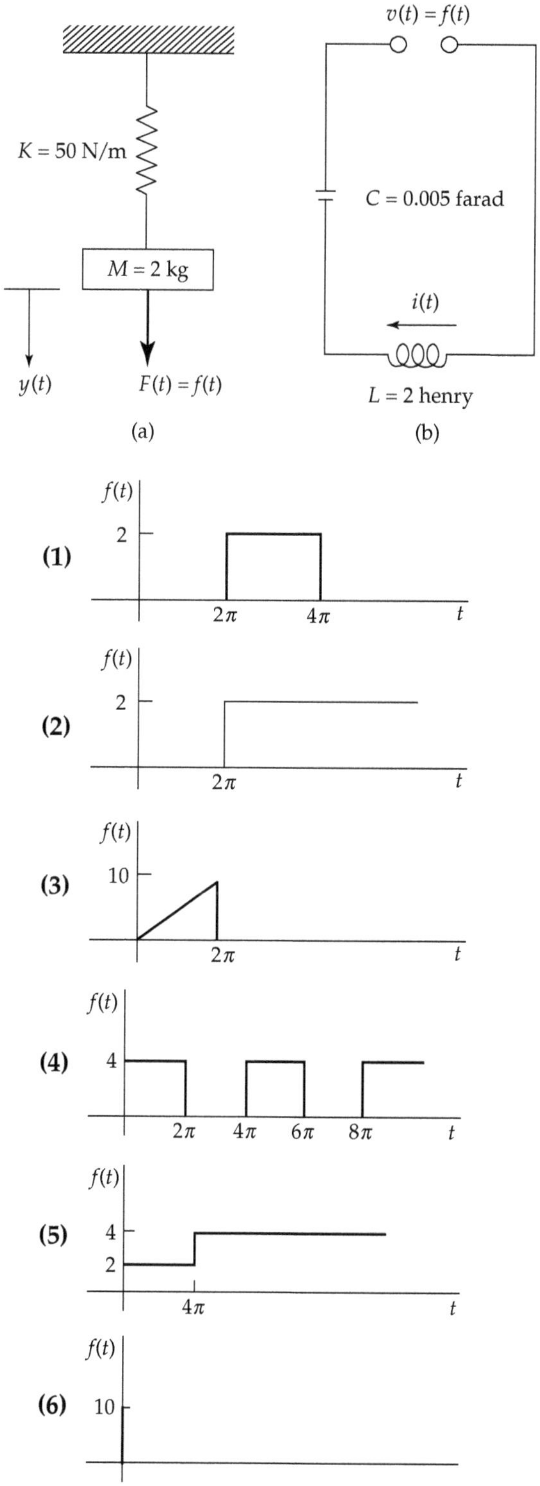

(7) 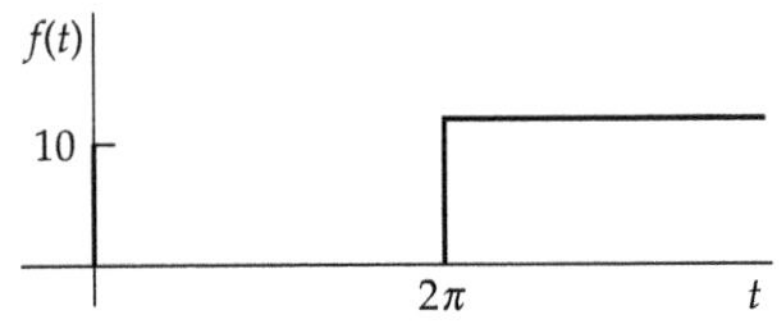

FIGURE 3.21

3.25 Determine the response function due to the input function for one of the systems shown in parts of (a) and (b) of Fig. 3.22. Each system is quiescent at $t = 0$. Use an input function $f(t)$ from Problem 3.24.

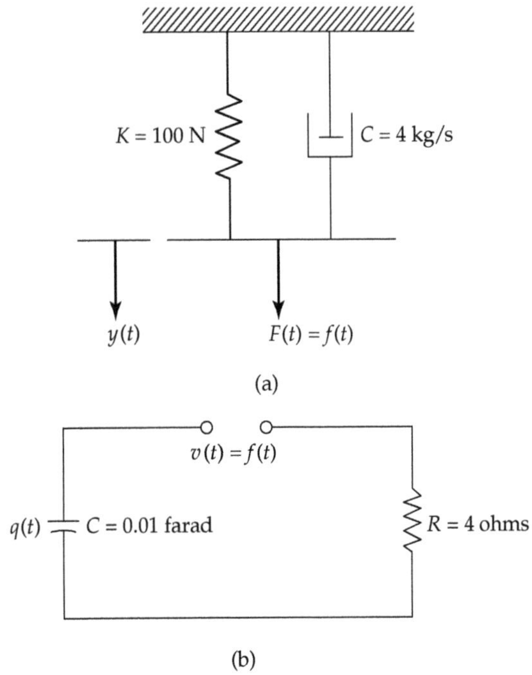

FIGURE 3.22

3.26 Find the deflection $y(x)$ of the beam shown in Fig. 3.23. The differential equation that describes the deflection is

$$\frac{d^4y}{dx^4} = \frac{w(x)}{EI}, \quad w(x) = P\delta_{L/2}(x) + w[u_0(x) - u_{L/2}(x)].$$

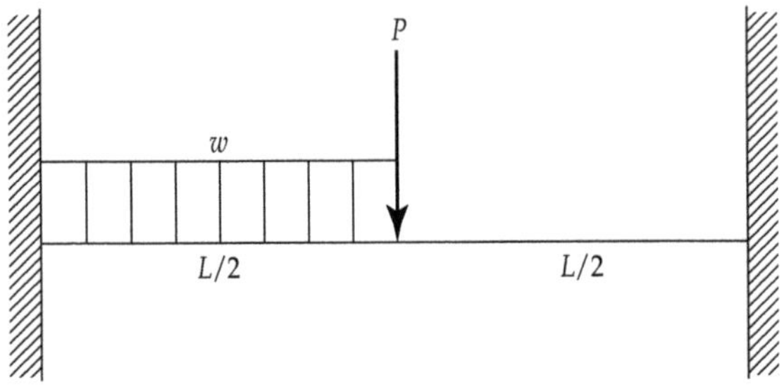

FIGURE 3.23

TABLE 3.1 Laplace Transforms

	$f(t)$	$F(s) = \mathcal{L}\{f(t)\}$
1	1	$1/s$
2	t	$1/s^2$
3	t^{n-1}	$(n-1)!/s^n \quad (n = 1, 2, \cdots)$
4	$t^{-1/2}$	$\sqrt{\pi/s}$
5	$t^{1/2}$	$\sqrt{\pi}/2s^{3/2}$
6	t^{k-1}	$\Gamma(k)/s^k \quad (k > 0)$
7	e^{at}	$\dfrac{1}{s-a}$
8	te^{at}	$\dfrac{1}{(s-a)^2}$
9	$t^{n-1}e^{at}$	$\dfrac{(n-1)!}{(s-a)^n} \quad (n = 1, 2, \cdots)$
10	$t^{k-1}e^{at}$	$\dfrac{\Gamma(k)}{(s-a)^k} \quad (k > 0)$
11	$e^{at} - e^{bt}$	$\dfrac{a-b}{(s-a)(s-b)} \quad (a \neq b)$
12	$ae^{at} - be^{bt}$	$\dfrac{(a-b)s}{(s-a)(s-b)} \quad (a \neq b)$
13	$\delta_0(t)$	1
14	$\delta_a(t)$	e^{-as}
15	$u_a(t)$	$\dfrac{1}{s}e^{-as}$
16	$\ln t$	$\dfrac{1}{s}\left(\ln\dfrac{1}{s} - 0.5772156\ldots\right)$
17	$\sin \omega t$	$\dfrac{\omega}{s^2 + \omega^2}$
18	$\cos \omega t$	$\dfrac{s}{s^2 + \omega^2}$
19	$\sinh at$	$\dfrac{a}{s^2 - a^2}$
20	$\cosh at$	$\dfrac{s}{s^2 - a^2}$
21	$e^{at} \sin \omega t$	$\dfrac{\omega}{(s-a)^2 + \omega^2}$
22	$e^{at} \cos \omega t$	$\dfrac{s-a}{(s-a)^2 + \omega^2}$

TABLE 3.1 Continued

	$f(t)$	$F(s) = \mathcal{L}\{f(t)\}$
23	$1 - \cos \omega t$	$\dfrac{\omega^2}{s(s^2 + \omega^2)}$
24	$\omega t - \sin \omega t$	$\dfrac{\omega^3}{s^2(s^2 + \omega^2)}$
25	$\sin \omega t - \omega t \cos \omega t$	$\dfrac{2\omega^3}{(s^2 + \omega^2)^2}$
26	$t \sin \omega t$	$\dfrac{2\omega s}{(s^2 + \omega^2)^2}$
27	$\sin \omega t + \omega t \cos \omega t$	$\dfrac{2\omega s^2}{(s^2 + \omega^2)^2}$
28	$\cos at - \cos bt$	$\dfrac{(b^2 - a^2)s}{(s^2 + a^2)(s^2 + b^2)} \quad (a^2 \neq b^2)$
29	$\sin at \cosh at - \cos at \sinh at$	$\dfrac{4a^3}{s^4 + 4a^4}$
30	$\sin at \sinh at$	$\dfrac{2a^2 s}{s^4 + 4a^4}$
31	$\sinh at - \sin at$	$\dfrac{2a^3}{s^4 - a^4}$
32	$\cosh at - \cos at$	$\dfrac{2a^2 s}{s^4 - a^4}$

4 Matrices and Determinants

Outline

4.1 Introduction
4.2 Matrices
4.3 Addition of Matrices
4.4 The Transpose and Some Special Matrices
4.5 Matrix Multiplication—Definition
4.6 Matrix Multiplication—Additional Properties
4.7 Determinants
4.8 The Adjoint and the Inverse Matrices
4.9 Solution of Simultaneous Linear Algebraic Equations
4.9.1 Nonhomogeneous Sets of Linear Algebraic Equations
4.9.2 Homogeneous Sets of Linear Algebraic Equations
4.9.3 Solutions to Sets of Linear Equations by MATLAB
4.10 Least-Squares Fit and the Pseudo Inverse
4.11 Eigenvalues and Eigenvectors
4.12 Eigenvalue Problems in Engineering
4.12.1 Moments of Inertia
4.12.2 Stress
4.12.3 Linear Dynamic Systems and Stability
Problems

4.1 INTRODUCTION

Problems in the real world can often be expressed with sufficient accuracy in terms of linear equations. Such equations may be algebraic equations, ordinary or partial differential equations, or integral equations. When solving a linear equation that models a particular phenomenon of interest, we must often, in the process, solve a set of simultaneous linear algebraic equations,

$$\begin{aligned} a_{11}x_1 + a_{12}x_2 + \cdots + a_{1n}x_n &= r_1 \\ a_{21}x_1 + a_{22}x_2 + \cdots + a_{2n}x_n &= r_2 \\ &\vdots \\ a_{m1}x_1 + a_{m2}x_2 + \cdots + a_{mn}x_n &= r_m \end{aligned} \tag{4.1.1}$$

where the coefficients a_{ij} are known constants. If the quantities r_i are also known constants, the variables x_i may be determined. The solution of such a set of linear algebraic equations has motivated the subject presented in this chapter: matrices and determinants.

M. C. Potter and B. F. Feeny, *Mathematical Methods for Engineering and Science*,
https://doi.org/10.1007/978-3-031-26151-0_4

This is a "mathematical shorthand" which facilitates manipulations that allow us to perform calculations with relative ease when compared with straightforward brute force. Our primary objective in this chapter will be to solve the set of equations (4.1.1). Before tackling this problem directly, we shall develop some of the concepts and operations that will be useful to this task.

4.2 MATRICES

A *matrix* is a rectangular array of numbers arranged into rows and columns. The coefficients of the variables x_i in Eqs. 4.1.1 form such a rectangular array. The matrix is displayed as

$$\begin{bmatrix} a_{11} & a_{12} & \cdots & a_{1n} \\ a_{21} & a_{22} & \cdots & a_{2n} \\ \vdots & \vdots & & \vdots \\ a_{m1} & a_{m2} & \cdots & a_{mn} \end{bmatrix} \tag{4.2.1}$$

This general matrix has m rows (the horizontal lines) and n columns (the vertical lines) and is said to be an *m by n matrix* (written $m \times n$). The numbers $a_{11}, a_{12}, \ldots, a_{mn}$ are called the *elements* of the matrix. It should be noted that the elements could be functions; the word "numbers" is meant to be general. An arbitrary element is represented by a_{ij}, where the first subscript refers to the row and the second subscript to the column.

Matrices will be denoted by capital boldface roman letters (**A**, **B**, **Y**, etc.), or by bracketed subscripted lowercase letters ($[a_{ij}]$, $[b_{ij}]$, $[y_{ij}]$, etc.). Two other notations, also used to denote a matrix, are (a_{ij}) and $\|a_{ij}\|$; we shall consistently use the bracketed notation. Obviously, the general notation, such as **A** or a_{ij}, does not indicate how many rows and columns the matrix has; this information must be established from other sources.

A matrix may have one row and many columns (a row matrix) or one column and many rows (a column matrix). A matrix that has only one column (an $n \times 1$ matrix) or one row (a $1 \times n$ matrix) is often called a *vector*. It could be displayed as the column vector or the row vector*

$$\mathbf{a} = \begin{bmatrix} a_1 \\ a_2 \\ \vdots \\ a_n \end{bmatrix}, \qquad \mathbf{b} = [b_1, b_2, \ldots, b_n] \tag{4.2.2}$$

Note that a lowercase boldface letter will be used to denote a vector quantity.

If $m = n$ in the matrix (4.2.1), it is called a *square matrix*. An example of a square matrix would be the 3×3 matrix

$$\mathbf{A} = \begin{bmatrix} a_{11} & a_{12} & a_{13} \\ a_{21} & a_{22} & a_{23} \\ a_{31} & a_{32} & a_{33} \end{bmatrix} \tag{4.2.3}$$

The number of rows or columns of a square matrix is the *order* of the matrix. The diagonal containing the elements $a_{11}, a_{22}, \ldots, a_{nn}$ of a square matrix of order n is the *main diagonal*. Primary attention will be given to square matrices, since they appear most often in physical applications.

*Note that commas are used to separate the element of a row vector.

It is sometimes useful to form a submatrix of a particular matrix **A**. If we omit all but p rows and q columns of matrix **A**, a $q \times p$ matrix results, called a *submatrix* of **A**. Let us divide a matrix **A** of order four into four submatrices by the dashed lines shown:

$$\mathbf{A} = \left[\begin{array}{ccc:c} a_{11} & a_{12} & a_{13} & a_{14} \\ a_{21} & a_{22} & a_{23} & a_{24} \\ \hdashline a_{31} & a_{32} & a_{33} & a_{34} \\ a_{41} & a_{42} & a_{43} & a_{44} \end{array}\right] \tag{4.2.4}$$

The matrix **A** is said to be *partitioned* into the four submatrices

$$\begin{bmatrix} a_{11} & a_{12} & a_{13} \\ a_{21} & a_{22} & a_{23} \end{bmatrix}, \quad \begin{bmatrix} a_{31} & a_{32} & a_{33} \\ a_{41} & a_{42} & a_{43} \end{bmatrix}, \quad \begin{bmatrix} a_{14} \\ a_{24} \end{bmatrix}, \quad \begin{bmatrix} a_{34} \\ a_{44} \end{bmatrix} \tag{4.2.5}$$

Partitioning is occasionally used in matrix manipulations and may simplify the writing of more complex matrices. Two special cases of partitioning occur when an $m \times n$ matrix **A** is divided into a set of row vectors as

$$\mathbf{A} = \begin{bmatrix} \mathbf{a}_1 \\ \mathbf{a}_2 \\ \vdots \\ \mathbf{a}_m \end{bmatrix}, \quad \begin{array}{l} \mathbf{a}_1 = [a_{11},\ a_{12}, \ldots, a_{1n}] \\ \mathbf{a}_2 = [a_{21},\ a_{22}, \ldots, a_{2n}] \\ \quad \text{etc.} \end{array} \tag{4.2.6}$$

or when an $m \times n$ matrix **A** is divided into a set of column vectors as

$$\mathbf{A} = [\mathbf{a}_1, \mathbf{a}_2, \ldots, \mathbf{a}_n], \quad \mathbf{a}_1 = \begin{bmatrix} a_{11} \\ a_{21} \\ \vdots \\ a_{m1} \end{bmatrix}, \quad \mathbf{a}_2 = \begin{bmatrix} a_{12} \\ a_{22} \\ \vdots \\ a_{m2} \end{bmatrix}, \quad \text{etc.} \tag{4.2.7}$$

Two matrices **A** and **B** are *equal* if each element of **A** is equal to the corresponding element of **B**. This is written as **A** = **B** if $a_{ij} = b_{ij}$. This, of course, demands that the number of rows and columns of **A** be equal, respectively, to the number of rows and columns of **B**. Thus, for the three matrices

$$\mathbf{A} = \begin{bmatrix} 2 & 3 & 0 \\ 1 & 2 & 0 \end{bmatrix}, \quad \mathbf{B} = \begin{bmatrix} 2 & 3 \\ 1 & 2 \end{bmatrix}, \quad \mathbf{C} = \begin{bmatrix} 2 & 3 & 0 \\ 1 & 2 & 0 \end{bmatrix} \tag{4.2.8}$$

we see that

$$\mathbf{A} = \mathbf{C}, \qquad \mathbf{A} \neq \mathbf{B}, \qquad \mathbf{B} \neq \mathbf{C} \tag{4.2.9}$$

A matrix that has all zero elements is called a *zero matrix* or a *null matrix*. Such a matrix need not be square. It will be written as **A** = **0**; this implies that all elements are zero.

4.3 ADDITION OF MATRICES

In the next several sections we shall consider some of the algebraic tasks encountered when performing calculations involving matrices. In this section addition and multiplication by a scalar will be presented. Then we shall consider some special matrices often useful when performing the somewhat difficult task of multiplying matrices, which will be presented in the subsequent sections.

The addition (or subtraction) of two matrices is defined only if both matrices have the same number of rows and the same number of columns. Then the sum **C** of the $(m \times n)$ matrix **A** and the $(m \times n)$ matrix **B** is an $(m \times n)$ matrix whose elements are given by

$$c_{ij} = a_{ij} + b_{ij} \tag{4.3.1}$$

Using boldface notation, this is written as

$$\mathbf{C} = \mathbf{A} + \mathbf{B} \tag{4.3.2}$$

The addition process is displayed as

$$\begin{bmatrix} c_{11} & c_{12} & \cdots & c_{1n} \\ c_{21} & c_{22} & \cdots & c_{2n} \\ \vdots & \vdots & & \vdots \\ c_{m1} & c_{m2} & \cdots & c_{mn} \end{bmatrix} = \begin{bmatrix} a_{11} & a_{12} & \cdots & a_{1n} \\ a_{21} & a_{22} & \cdots & a_{2n} \\ \vdots & \vdots & & \vdots \\ a_{m1} & a_{m2} & \cdots & a_{mn} \end{bmatrix} + \begin{bmatrix} b_{11} & b_{12} & \cdots & b_{1n} \\ b_{21} & b_{22} & \cdots & b_{2n} \\ \vdots & \vdots & & \vdots \\ b_{m1} & b_{m2} & \cdots & b_{mn} \end{bmatrix}$$
$$= \begin{bmatrix} a_{11}+b_{11} & a_{12}+b_{12} & \cdots & a_{1n}+b_{1n} \\ a_{21}+b_{21} & a_{22}+b_{22} & \cdots & a_{2n}+b_{2n} \\ \vdots & \vdots & & \vdots \\ a_{m1}+b_{m1} & a_{m2}+b_{m2} & \cdots & a_{mn}+b_{mn} \end{bmatrix} \tag{4.3.3}$$

From the definition of addition above we note the commutative and associative laws:

$$\begin{aligned} \mathbf{A} + \mathbf{B} &= \mathbf{B} + \mathbf{A} \\ \mathbf{A} + (\mathbf{B} + \mathbf{C}) &= (\mathbf{A} + \mathbf{B}) + \mathbf{C} \end{aligned} \tag{4.3.4}$$

Note that subtraction results if negative signs replace the appropriate positive signs in the equations above.

The sum of k matrices **A** is a matrix whose elements are obtained by multiplying each element of **A** by k; this follows from Eq. 4.3.3. We generalize this as the multiplication of a scalar k with a matrix A, written as

$$\mathbf{A}k = k\mathbf{A} = \begin{bmatrix} ka_{11} & ka_{12} & \cdots & ka_{1n} \\ ka_{21} & ka_{22} & \cdots & ka_{2n} \\ \vdots & \vdots & & \vdots \\ ka_{m1} & ka_{m2} & \cdots & ka_{mn} \end{bmatrix} \tag{4.3.5}$$

The scalar k can be a real number, a complex number, or a function (e.g., x^2). However, it cannot be a matrix quantity; such a product will be defined in the following section. From the equations above we see that

$$\begin{aligned} k(\mathbf{A} + \mathbf{B}) &= k\mathbf{A} + k\mathbf{B} \\ (c + k)\mathbf{A} &= c\mathbf{A} + k\mathbf{A} \end{aligned} \tag{4.3.6}$$

Note that addition or subtraction is not defined if the matrices do not have the same number of rows and the same number of columns. Thus, we could not add the following two matrices:

$$\mathbf{A} = \begin{bmatrix} 1 & 2 & 3 \\ 0 & 1 & 0 \end{bmatrix}, \quad \mathbf{B} = \begin{bmatrix} 1 & 2 \\ 2 & 0 \end{bmatrix} \tag{4.3.7}$$

Example 4.1

Find $\mathbf{A} + \mathbf{B}$, $5\mathbf{A}$, and $\mathbf{B} - 5\mathbf{A}$ for the given matrices:

$$\mathbf{A} = \begin{bmatrix} 0 & 2 & 5 \\ 1 & -2 & 1 \\ 2 & 3 & 1 \end{bmatrix}, \quad \mathbf{B} = \begin{bmatrix} -1 & 2 & 0 \\ 0 & 2 & 1 \\ 6 & -6 & 0 \end{bmatrix}$$

Solution

To find the sum $\mathbf{A} + \mathbf{B}$, we simply add corresponding elements $a_{ij} + b_{ij}$ and obtain

$$\mathbf{A} + \mathbf{B} = \begin{bmatrix} -1 & 4 & 5 \\ 1 & 0 & 2 \\ 8 & -3 & 1 \end{bmatrix}$$

Following Eq. 4.3.5, the product $5\mathbf{A}$ is

$$5\mathbf{A} = \begin{bmatrix} 0 & 10 & 25 \\ 5 & -10 & 5 \\ 10 & 15 & 5 \end{bmatrix}$$

Now we subtract each element of the preceding matrix from the corresponding element of $\mathbf{B}$, that is, $b_{ij} - 5a_{ij}$, and find

$$\mathbf{B} - 5\mathbf{A} = \begin{bmatrix} -1 & -8 & -25 \\ -5 & 12 & -4 \\ -4 & -21 & -5 \end{bmatrix}$$

4.4 THE TRANSPOSE AND SOME SPECIAL MATRICES

Before we consider the multiplication of matrices, the transpose of a matrix and some special matrices, often useful in the multiplication process, will be presented. The addition of matrices is used in this section, hence it logically follows the section on addition.

We shall find it helpful to occasionally interchange the rows with the columns of a matrix. The new matrix that results is called the *transpose* of the original matrix. The transpose $\mathbf{A}^T$ of the matrix displayed by (4.2.1) is displayed as

$$\mathbf{A}^T = \begin{bmatrix} a_{11} & a_{21} & \cdots & a_{m1} \\ a_{12} & a_{22} & \cdots & a_{m2} \\ \vdots & \vdots & & \vdots \\ a_{1n} & a_{2n} & \cdots & a_{mn} \end{bmatrix} \tag{4.4.1}$$

Using index notation we could write

$$a_{ij}^T = a_{ji} \quad \text{or} \quad a_{ji}^T = a_{ij} \tag{4.4.2}$$

Note that if a matrix is square, its transpose is also square; however, if a matrix is $m \times n$, its transpose is $n \times m$. An example of a matrix and its transpose would be

$$\mathbf{A} = \begin{bmatrix} 2 & 0 \\ 3 & -1 \\ 1 & 1 \\ 0 & 0 \end{bmatrix}, \quad \mathbf{A}^T = \begin{bmatrix} 2 & 3 & 1 & 0 \\ 0 & -1 & 1 & 0 \end{bmatrix} \tag{4.4.3}$$

The transpose of a column vector is a row vector, and the transpose of a row vector is a column vector.

If the sum **C** of two matrices **A** and **B** is defined, we would have

$$\mathbf{C}^T = \mathbf{A}^T + \mathbf{B}^T \tag{4.4.4}$$

This is true from the definition of the sum of two matrices.

A *symmetric matrix* is a matrix that is equal to its transpose; that is, **A** is symmetric if

$$\mathbf{A} = \mathbf{A}^T \quad \text{or} \quad a_{ij} = a_{ji} \tag{4.4.5}$$

A symmetric matrix must be a square matrix. An example of a symmetric matrix would be the 4×4 matrix,

$$\mathbf{A} = \begin{bmatrix} 2 & 1 & 3 & 4 \\ 1 & 0 & -2 & 0 \\ 3 & -2 & 1 & -1 \\ 4 & 0 & -1 & 0 \end{bmatrix} \tag{4.4.6}$$

A symmetric matrix is symmetric about its main diagonal, which for the matrix above would contain elements 2, 0, 1, 0.

A *skew-symmetric matrix* is a matrix that is equal to the negative of its transpose; that is, **A** is skew-symmetric if

$$\mathbf{A} = -\mathbf{A}^T \quad \text{or} \quad a_{ij} = -a_{ji} \tag{4.4.7}$$

A skew-symmetric matrix must also be a square matrix. By choosing $i = 1$ and $j = 1$ in Eq. 4.4.7, we see that $a_{11} = 0$; also, $a_{22} = 0$, $a_{33} = 0$, etc. The main diagonal elements of a skew-symmetric matrix are all zero. An example of a skew-symmetric matrix would be the 3×3 matrix

$$\mathbf{A} = \begin{bmatrix} 0 & -1 & 2 \\ 1 & 0 & -3 \\ -2 & 3 & 0 \end{bmatrix} \tag{4.4.8}$$

Any square matrix can be written as the sum of a symmetric matrix and a skew-symmetric matrix. This is done as follows:

$$\mathbf{A} = \left(\frac{\mathbf{A}}{2} + \frac{\mathbf{A}^T}{2}\right) + \left(\frac{\mathbf{A}}{2} - \frac{\mathbf{A}^T}{2}\right) \tag{4.4.9}$$

where the symmetric matrix $\mathbf{A}_s$ and the skew-symmetric matrix* $\mathbf{A}_a$ are given by

$$\mathbf{A}_s = \frac{\mathbf{A}}{2} + \frac{\mathbf{A}^T}{2}, \qquad \mathbf{A}_a = \frac{\mathbf{A}}{2} - \frac{\mathbf{A}^T}{2} \tag{4.4.10}$$

We can use Eq. 4.4.5 to show that $\mathbf{A}_s$ is symmetric, and Eq. 4.4.7 to show that $\mathbf{A}_a$ is skew-symmetric. In doing so, we must observe that

$$(\mathbf{A}^T)^T = \mathbf{A} \tag{4.4.11}$$

There are several special square matrices that are of occasional interest. A square matrix $\mathbf{A}$ is an *upper triangular matrix* if $a_{ij} = 0$ for $i > j$ and is a *lower triangular matrix* if $a_{ij} = 0$ for $i < j$. The matrix

$$\mathbf{A} = \begin{bmatrix} 1 & 0 & 0 \\ -1 & 0 & 0 \\ 0 & 1 & -2 \end{bmatrix} \tag{4.4.12}$$

is an upper triangular matrix; an example of a lower triangular matrix is

$$\mathbf{A} = \begin{bmatrix} 2 & 1 & -1 \\ 0 & 1 & 2 \\ 0 & 0 & 3 \end{bmatrix} \tag{4.4.13}$$

A square matrix that is both upper triangular and lower triangular is called a *diagonal matrix*. The matrix

$$\mathbf{A} = \begin{bmatrix} 2 & 0 & 0 \\ 0 & -1 & 0 \\ 0 & 0 & 3 \end{bmatrix} \tag{4.4.14}$$

is a diagonal matrix.

A *scalar matrix* is a diagonal matrix with all the diagonal elements equal, that is, with $a_{11} = a_{22} = \cdots = a_{nn}$. The matrix

$$\mathbf{A} = \begin{bmatrix} 2 & 0 & 0 & 0 \\ 0 & 2 & 0 & 0 \\ 0 & 0 & 2 & 0 \\ 0 & 0 & 0 & 2 \end{bmatrix} \tag{4.4.15}$$

is a scalar matrix.

Finally, we wish to identify the *unit matrix* (or *identity matrix*) as the diagonal matrix whose diagonal elements are all unity. It is symbolized with $\mathbf{I}$. Thus, the 4×4 unit matrix is

$$\mathbf{I} = \begin{bmatrix} 1 & 0 & 0 & 0 \\ 0 & 1 & 0 & 0 \\ 0 & 0 & 1 & 0 \\ 0 & 0 & 0 & 1 \end{bmatrix} \tag{4.4.16}$$

*We choose the subscript a to represent a skew-symmetric matrix, since the term "asymmetric" is often used in place of "skew-symmetric."

Often, a subscript is used to designate the order of the unit matrix; $\mathbf{I}_2$ would be displayed as

$$\mathbf{I}_2 = \begin{bmatrix} 1 & 0 \\ 0 & 1 \end{bmatrix} \tag{4.4.17}$$

If the order is known, or is not of interest, the subscript is deleted, as in Eq. 4.4.16.

The *Kronecker delta* δ_{ij} is often used to denote the unit matrix; it is defined as

$$\delta_{ij} = \begin{cases} 1, & \text{if } i = j \\ 0, & \text{if } i \neq j \end{cases} \tag{4.4.18}$$

The unit matrix could then be written as

$$\mathbf{I} = [\delta_{ij}] \tag{4.4.19}$$

This notation is occasionally used for the matrix $\mathbf{A}$; that is, $\mathbf{A} = [a_{ij}]$. We shall employ such a notation where useful.

Example 4.2

Express the matrix

$$\mathbf{A} = \begin{bmatrix} 2 & 0 & 3 \\ 2 & 0 & 2 \\ -3 & 4 & 2 \end{bmatrix}$$

as the sum of a symmetric matrix and a skew-symmetric matrix.

Solution

First, let us write the transpose $\mathbf{A}^T$. It is

$$\mathbf{A}^T = \begin{bmatrix} 2 & 2 & -3 \\ 0 & 0 & 4 \\ 3 & 2 & 2 \end{bmatrix}$$

Now, using Eq. 4.4.10, the symmetric part of $\mathbf{A}$ is

$$\mathbf{A}_s = \frac{1}{2}(\mathbf{A} + \mathbf{A}^T) = \frac{1}{2}\begin{bmatrix} 4 & 2 & 0 \\ 2 & 0 & 6 \\ 0 & 6 & 4 \end{bmatrix} = \begin{bmatrix} 2 & 1 & 0 \\ 1 & 0 & 3 \\ 0 & 3 & 2 \end{bmatrix}$$

The skew-symmetric part is given by

$$\mathbf{A}_a = \frac{1}{2}(\mathbf{A} - \mathbf{A}^T) = \frac{1}{2}\begin{bmatrix} 0 & -2 & 6 \\ 2 & 0 & -2 \\ -6 & 2 & 0 \end{bmatrix} = \begin{bmatrix} 0 & -1 & 3 \\ 1 & 0 & -1 \\ -3 & 1 & 0 \end{bmatrix}$$

Obviously the given matrix $\mathbf{A}$ is the sum

$$\mathbf{A} = \mathbf{A}_s + \mathbf{A}_a = \begin{bmatrix} 2 & 1 & 0 \\ 1 & 0 & 3 \\ 0 & 3 & 2 \end{bmatrix} + \begin{bmatrix} 0 & -1 & 3 \\ 1 & 0 & -1 \\ -3 & 1 & 0 \end{bmatrix} = \begin{bmatrix} 2 & 0 & 3 \\ 2 & 0 & 2 \\ -3 & 4 & 2 \end{bmatrix}$$

This provides us with a check on the above manipulations.

4.5 MATRIX MULTIPLICATION—DEFINITION

There are several ways that matrix multiplication could be defined. We shall motivate our definition by considering the simultaneous set of equations

$$\begin{aligned} a_{11}x_1 + a_{12}x_2 + a_{13}x_3 &= r_1 \\ a_{21}x_1 + a_{22}x_2 + a_{23}x_3 &= r_2 \\ a_{31}x_1 + a_{32}x_2 + a_{33}x_3 &= r_3 \end{aligned} \tag{4.5.1}$$

These equations could be written, using the summation symbol, as

$$\sum_{j=1}^{3} a_{ij}x_j = r_i \tag{4.5.2}$$

where the first equation is formed by choosing $i = 1$, the second equation letting $i = 2$, and the third equation with $i = 3$. The quantity a_{ij} contains the nine elements $a_{11}, a_{12}, a_{13}, \ldots, a_{33}$; it is a 3×3 matrix. The quantities x_j and r_i each contain three elements and are treated as column matrices, or vectors. Hence, we write Eqs. 4.5.1 in matrix notation as

$$\mathbf{Ax} = \mathbf{r} \tag{4.5.3}$$

We must define the product of the matrix **A** and the column vector **x** so that Eqs. 4.5.1 result. This, of course, demands that the number of rows in the column matrix **x** equal the number of columns in the matrix **A**. Matrix multiplication is generalized as follows: The matrix product of the matrix **A** and the matrix **B** is the matrix **C** whose elements are computed from

$$c_{ij} = \sum_{k=1}^{r} a_{ik}b_{kj} \tag{4.5.4}$$

For the definition above to be meaningful, the number of columns in **A** must be equal to the number of rows in **B**. If **A** were an $m \times r$ matrix and **B** an $r \times n$ matrix, then **C** would be an $m \times n$ matrix. Note that the matrix multiplication **AB** would not be defined if both **A** and **B** were 2×3 matrices. **AB** would be defined, however, if **A** were 2×3 and **B** were 3×2; the product **AB** would then be a 2×2 matrix and the product **BA** would be a 3×3 matrix. Obviously, matrix multiplication is not, in general, commutative; that is, in general

$$\mathbf{AB} \neq \mathbf{BA} \tag{4.5.5}$$

In fact, the product **BA** may not even be defined, even if **AB** exists.

The multiplication of two matrices **A** and **B** to form the matrix **C** is displayed as

$$\begin{bmatrix} c_{11} & c_{12} & \cdots & c_{1n} \\ c_{21} & c_{22} & \cdots & c_{2n} \\ \vdots & \vdots & & \vdots \\ c_{i1} & \cdots & \boxed{c_{ij}} & c_{in} \\ \vdots & \vdots & & \vdots \\ c_{m1} & c_{m2} & \cdots & c_{mn} \end{bmatrix} = \begin{bmatrix} a_{11} & a_{12} & \cdots & a_{1r} \\ a_{21} & a_{22} & \cdots & a_{2r} \\ \vdots & \vdots & & \vdots \\ \boxed{a_{i1} \quad a_{i2} \quad \cdots \quad a_{ir}} \\ \vdots & \vdots & & \vdots \\ a_{m1} & a_{m2} & \cdots & a_{mr} \end{bmatrix} \begin{bmatrix} b_{11} & b_{12} & \cdots & \boxed{b_{1j}} & \cdots & b_{1n} \\ b_{21} & b_{22} & \cdots & \boxed{b_{2j}} & \cdots & b_{2n} \\ \vdots & \vdots & & \vdots & & \vdots \\ b_{r1} & b_{r2} & \cdots & \boxed{b_{rj}} & \cdots & b_{rn} \end{bmatrix} \tag{4.5.6}$$

Observe that the element c_{ij} depends on the elements in row i of **A** and the elements in column j of **B**. If the elements of row i of **A** and the elements of column j of **B** are considered to be the components of vectors, then the element c_{ij} is simply the scalar (dot) product of the two vectors. Written out we have

$$c_{ij} = a_{i1}b_{1j} + a_{i2}b_{2j} + a_{i3}b_{3j} + \cdots + a_{ir}b_{rj} \tag{4.5.7}$$

This is, of course, the same equation as (4.5.4).

In the matrix product **AB** the matrix **A** may be referred to as the *premultiplier* and the matrix **B** as the *postmultiplier*. The matrix **A** is postmultiplied by **B**, or **B** is premultiplied by **A**.

It is now an easier task to manipulate matrix equations such as Eq. 4.5.3. For example, suppose that the unknown vector **x** were related to another unknown vector **y** by the matrix equation

$$\mathbf{x} = \mathbf{B}\mathbf{y} \tag{4.5.8}$$

where **B** is a known coefficient matrix. We could then substitute Eq. 4.5.8 into Eq. 4.5.3 and obtain

$$\mathbf{A}\mathbf{B}\mathbf{y} = \mathbf{r} \tag{4.5.9}$$

The matrix product **AB** is determined following the multiplication rules outlined above.

Example 4.3

Several examples of the multiplication of two matrices will be given here using the following matrices:

$$\mathbf{A} = \begin{bmatrix} 2 \\ 3 \\ -4 \end{bmatrix}, \quad \mathbf{B} = [2, \ -1, \ 0], \quad \mathbf{C} = \begin{bmatrix} 2 & 3 & -1 \\ 0 & 1 & 4 \end{bmatrix}$$

$$\mathbf{D} = \begin{bmatrix} 3 & 0 & 1 \\ 2 & -2 & 1 \\ 0 & 2 & 0 \end{bmatrix}, \quad \mathbf{E} = \begin{bmatrix} 2 & -1 & 1 \\ 1 & 0 & 0 \\ 2 & 0 & 1 \end{bmatrix}$$

Solution

A is a 3×1 matrix and **B** is a 1×3 matrix. The product matrices are

$$\mathbf{BA} = [2, \ -1, \ 0] \begin{bmatrix} 2 \\ 3 \\ -4 \end{bmatrix} = [2 \times 2 + (-1)(3) + 0(-4)] = [1]$$

$$\mathbf{AB} = \begin{bmatrix} 2 \\ 3 \\ -4 \end{bmatrix} [2, \ -1, \ 0] = \begin{bmatrix} 2 \times 2 & 2 \times -1 & 2 \times 0 \\ 3 \times 2 & 3 \times -1 & 3 \times 0 \\ -4 \times 2 & -4 \times -1 & -4 \times 0 \end{bmatrix} = \begin{bmatrix} 4 & -2 & 0 \\ 6 & -3 & 0 \\ -8 & 4 & 0 \end{bmatrix}$$

From these expressions it is obvious that **AB** ≠ **BA**. In fact, the rows and columns of the product matrix are even different. The first product is often called a *scalar product*, since the product yields only one scalar element.

Now consider the product of a 2 × 3 matrix and a 3 × 1 matrix, **CA**. The product matrix is

$$\mathbf{CA} = \begin{bmatrix} 2 & 3 & -1 \\ 0 & 1 & 4 \end{bmatrix} \begin{bmatrix} 2 \\ 3 \\ -4 \end{bmatrix} = \begin{bmatrix} 2\times 2 + 3\times 3 + -1\times -4 \\ 0\times 2 + 1\times 3 + \ 4\times -4 \end{bmatrix} = \begin{bmatrix} 17 \\ -13 \end{bmatrix}$$

The product **AC** does not exist since matrix multiplication of a 3 × 1 matrix with a 2 × 3 matrix is not defined.

The product of two 3 × 3 matrices will now be displayed. We have

$$\mathbf{DE} = \begin{bmatrix} 3 & 0 & 1 \\ 2 & -2 & 1 \\ 0 & 2 & 0 \end{bmatrix} \begin{bmatrix} 2 & -1 & 1 \\ 1 & 0 & 0 \\ 2 & 0 & 1 \end{bmatrix} = \begin{bmatrix} 8 & -3 & 4 \\ 4 & -2 & 3 \\ 2 & 0 & 0 \end{bmatrix}$$

Check this result using the procedure discussed preceding Eq. 4.5.7. Then verify that

$$\mathbf{ED} = \begin{bmatrix} 2 & -1 & 1 \\ 1 & 0 & 0 \\ 2 & 0 & 1 \end{bmatrix} \begin{bmatrix} 3 & 0 & 1 \\ 2 & -2 & 1 \\ 0 & 2 & 0 \end{bmatrix} = \begin{bmatrix} 4 & 4 & 1 \\ 3 & 0 & 1 \\ 6 & 2 & 2 \end{bmatrix}$$

Notice that even for square matrices **DE** ≠ **ED**.

4.6 MATRIX MULTIPLICATION—ADDITIONAL PROPERTIES

The most useful definition for matrix multiplication was motivated by the solution of linear equations. Other useful properties of matrix multiplication will be presented in this section. We shall make use of the special matrices introduced in Section 4.4.

We shall now list several properties of matrix multiplication. The examples and Problems will serve to verify the equalities. They are

$$\begin{aligned} (k\mathbf{A})\mathbf{B} &= k(\mathbf{AB}) = \mathbf{A}(k\mathbf{B}) \\ (\mathbf{A}+\mathbf{B})\mathbf{C} &= \mathbf{AC} + \mathbf{BC} \\ \mathbf{A}(\mathbf{B}+\mathbf{C}) &= \mathbf{AB} + \mathbf{AC} \\ \mathbf{A}(\mathbf{BC}) &= (\mathbf{AB})\mathbf{C} \end{aligned} \qquad (4.6.1)$$

where k is any scalar quantity, usually a number. Two additional properties will also be noted here. First, if **AB** = **0** it is not necessary for either **A** or **B** to be zero matrices; second, if $\mathbf{AB} = \mathbf{AC}$, it does not necessarily follow that $\mathbf{B} = \mathbf{C}$. We must be careful when manipulating matrices since our experience with the use of scalars may tend to lead us into mistakes.

Using the special matrices of Section 4.4, we observe that the product of a scalar matrix $\mathbf{S}$ and a square matrix $\mathbf{A}$ results in

The scalar matrix can be written as $\mathbf{S} = k\mathbf{I}$.

$$\mathbf{SA} = \mathbf{AS} = k\mathbf{A} \tag{4.6.2}$$

where k is the value of the diagonal elements of $\mathbf{S}$. Letting $k = 1$, we see that

$$\mathbf{IA} = \mathbf{AI} = \mathbf{A} \tag{4.6.3}$$

where $\mathbf{I}$ is the unit matrix; of course, the order of two square matrices must be the same when forming the product of the two matrices. We can, however, form the product of an $m \times n$ matrix $\mathbf{A}$ with $\mathbf{I}_m$ or $\mathbf{I}_n$. Then we would write

$$\mathbf{A} = \mathbf{I}_m\mathbf{A} \quad \text{or} \quad \mathbf{A} = \mathbf{AI}_n \tag{4.6.4}$$

Replacing $\mathbf{A}$ with $\mathbf{I}$ in the above there follows

$$\mathbf{III} = \mathbf{II} = \mathbf{I} \quad \text{or} \quad \mathbf{I}^3 = \mathbf{I}^2 = \mathbf{I} \tag{4.6.5}$$

Similarly, if $\mathbf{A}$ is a square matrix, we may write

$$\mathbf{AA} = \mathbf{A}^2, \quad \mathbf{AAA} = \mathbf{A}^3, \quad \text{etc.} \tag{4.6.6}$$

Obviously, the result of multiplying a square matrix by itself is a square matrix of the same order.

The transpose of the product of two matrices equals the product of the transposes taken in reverse order; that is,

$$(\mathbf{AB})^T = \mathbf{B}^T\mathbf{A}^T \tag{4.6.7}$$

This is most readily verified by writing the equation in index form. Let $\mathbf{C} = \mathbf{AB}$. Then $\mathbf{C}^T = (\mathbf{AB})^T$ and is given by

$$\begin{aligned} c_{ij}^T &= c_{ji} = \sum_{k=1}^{n} a_{jk} b_{ki} \\ &= \sum_{k=1}^{n} a_{kj}^T b_{ik}^T \\ &= \sum_{k=1}^{n} b_{ik}^T a_{kj}^T \end{aligned} \tag{4.6.8}$$

This expression is observed to be the index form of the product $\mathbf{B}^T\mathbf{A}^T$, thereby verifying Eq. 4.6.7. The preceding could also be verified using some particular examples.

Example 4.4

Verify the statement expressed by Eq. 4.6.7 if

$$\mathbf{A} = \begin{bmatrix} 3 \\ 0 \\ -1 \end{bmatrix} \quad \text{and} \quad \mathbf{B} = [2, \quad -1, \quad 1]$$

Solution

The matrix product **AB** is found to be

$$\mathbf{AB} = \begin{bmatrix} 6 & -3 & 3 \\ 0 & 0 & 0 \\ -2 & 1 & -1 \end{bmatrix}$$

The transpose matrices are

$$\mathbf{A}^T[3, \quad 0, \quad -1], \qquad \mathbf{B}^T = \begin{bmatrix} 2 \\ -1 \\ 1 \end{bmatrix}$$

The product $\mathbf{B}^T\mathbf{A}^T$ is found to be

$$\mathbf{B}^T\mathbf{A}^T = \begin{bmatrix} 6 & 0 & -2 \\ -3 & 0 & 1 \\ 3 & 0 & -1 \end{bmatrix}$$

This, obviously, is the transpose of the matrix product **AB**, that is $(\mathbf{AB})^T$.

4.7 DETERMINANTS

A *determinant* is a number computed from the elements of a square matrix according to a specific formula. Determinants should be familiar to the reader since they are introduced in elementary algebra when solving simultaneous, linear equations. We shall review the usual expansions of second- and third-order determinants and then present some rules helpful to the task of evaluating determinants.

Recall that a second-order determinant was introduced to aid in the process of solving the nonhomogeneous* set of two equations

$$\begin{aligned} a_{11}x_1 + a_{12}x_2 &= r_1 \\ a_{21}x_1 + a_{22}x_2 &= r_2 \end{aligned} \tag{4.7.1}$$

*A nonhomogeneous set of equations results whenever there is at least one term in the equations that does not contain one of the unknowns. This term is usually written on the right-hand side. In Eq. 4.7.1 either r_1 or r_2 or both would be nonzero for a nonhomogeneous set of equations.

for x_1 and x_2. The coefficients $a_{11}, a_{12}, a_{21}, a_{22}$ and the numbers on the right-hand side are assumed to be known constants. The solution is written using determinants as

$$x_1 = \frac{\begin{vmatrix} r_1 & a_{12} \\ r_2 & a_{22} \end{vmatrix}}{\begin{vmatrix} a_{11} & a_{12} \\ a_{21} & a_{22} \end{vmatrix}}, \qquad x_2 = \frac{\begin{vmatrix} a_{11} & r_1 \\ a_{21} & r_2 \end{vmatrix}}{\begin{vmatrix} a_{11} & a_{12} \\ a_{21} & a_{22} \end{vmatrix}} \tag{4.7.2}$$

providing the determinant in the denominator, formed from the coefficients of the two unknowns, is nonzero. The value of the second-order determinant is given by

$$\begin{vmatrix} a_{11} & a_{12} \\ a_{21} & a_{22} \end{vmatrix} = a_{11}a_{22} - a_{21}a_{12} \tag{4.7.3}$$

Third-order determinants are useful when solving the three simultaneous, nonhomogeneous, linear equations

$$\begin{aligned} a_{11}x_1 + a_{12}x_2 + a_{13}x_3 &= r_1 \\ a_{21}x_1 + a_{22}x_2 + a_{23}x_3 &= r_2 \\ a_{31}x_1 + a_{32}x_2 + a_{33}x_3 &= r_3 \end{aligned} \tag{4.7.4}$$

In terms of the coefficients and the right-hand sides, the solution can be written as

$$x_1 = \frac{\begin{vmatrix} r_1 & a_{12} & a_{13} \\ r_2 & a_{22} & a_{23} \\ r_3 & a_{32} & a_{33} \end{vmatrix}}{\begin{vmatrix} a_{11} & a_{12} & a_{13} \\ a_{21} & a_{22} & a_{23} \\ a_{31} & a_{32} & a_{33} \end{vmatrix}}, \quad x_2 = \frac{\begin{vmatrix} a_{11} & r_1 & a_{13} \\ a_{21} & r_2 & a_{23} \\ a_{31} & r_3 & a_{33} \end{vmatrix}}{\begin{vmatrix} a_{11} & a_{12} & a_{13} \\ a_{21} & a_{22} & a_{23} \\ a_{31} & a_{32} & a_{33} \end{vmatrix}}, \quad x_3 = \frac{\begin{vmatrix} a_{11} & a_{12} & r_1 \\ a_{21} & a_{22} & r_2 \\ a_{31} & a_{32} & r_3 \end{vmatrix}}{\begin{vmatrix} a_{11} & a_{12} & a_{13} \\ a_{21} & a_{22} & a_{23} \\ a_{31} & a_{32} & a_{33} \end{vmatrix}} \tag{4.7.5}$$

Again, the determinant in the denominator must not be zero; if it is zero, no solution exists. The value of a third-order determinant is found from the formula

$$\begin{aligned} \begin{vmatrix} a_{11} & a_{12} & a_{13} \\ a_{21} & a_{22} & a_{23} \\ a_{31} & a_{32} & a_{33} \end{vmatrix} &= a_{11}a_{22}a_{33} - a_{11}a_{32}a_{23} - a_{12}a_{21}a_{33} + a_{12}a_{31}a_{23} \\ &\quad + a_{13}a_{21}a_{32} - a_{13}a_{31}a_{22} \end{aligned} \tag{4.7.6}$$

This may be found by writing the first two columns after the determinant and then summing the products of the elements of the various diagonals using negative signs with the diagonals sloping upward:

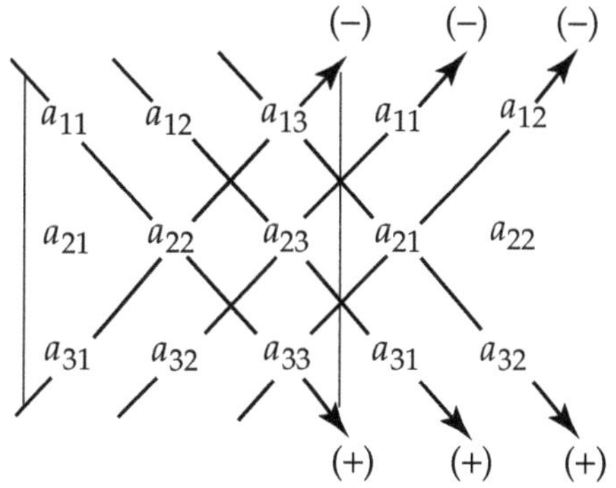

(4.7.7)

In general, the solution to n equations in n unknowns can be written as

$$x_1 = \frac{\begin{vmatrix} r_1 & a_{12} & \cdots & a_{1n} \\ r_2 & a_{22} & \cdots & a_{2n} \\ \vdots & \vdots & & \vdots \\ r_n & a_{n2} & & a_{nn} \end{vmatrix}}{\begin{vmatrix} a_{11} & a_{12} & \cdots & a_{1n} \\ \vdots & & \vdots & \\ a_{n1} & a_{n2} & & a_{nn} \end{vmatrix}}, \quad x_2 = \frac{\begin{vmatrix} a_{11} & r_1 & \cdots & a_{1n} \\ a_{21} & r_2 & \cdots & a_{2n} \\ \vdots & \vdots & & \vdots \\ a_{n1} & r_n & & a_{nn} \end{vmatrix}}{\begin{vmatrix} a_{11} & a_{12} & \cdots & a_{1n} \\ \vdots & & \vdots & \\ a_{n1} & a_{n2} & & a_{nn} \end{vmatrix}}, \ldots$$

$$x_n = \frac{\begin{vmatrix} a_{11} & a_{12} & \cdots & r_1 \\ a_{21} & a_{22} & \cdots & r_2 \\ \vdots & \vdots & & \vdots \\ a_{n1} & a_{n2} & & r_n \end{vmatrix}}{\begin{vmatrix} a_{11} & a_{12} & \cdots & a_{1n} \\ \vdots & & \vdots & \\ a_{n1} & a_{n2} & & a_{nn} \end{vmatrix}} \tag{4.7.8}$$

This application of determinants to solve a set of simultaneous linear algebraic equations is called Cramer's rule.

where the determinant in the denominator must not be zero. However, to evaluate the determinants in Eqs. 4.7.8, the diagonal method used for third-order determinants will not work. Let us now consider determinants of nth order.

Determinants will always be identified with the straight bars on both sides of the elements of a square matrix. We shall use the notation

$$|\mathbf{A}| = |a_{ij}| = \begin{vmatrix} a_{11} & a_{12} & \cdots & a_{1n} \\ a_{21} & a_{22} & \cdots & a_{2n} \\ \vdots & \vdots & & \vdots \\ a_{n1} & a_{n2} & \cdots & a_{nn} \end{vmatrix} \tag{4.7.9}$$

to represent the determinant of the nth-order square matrix $\mathbf{A}$. The nth-order determinant is defined to be the number calculated from

$$|\mathbf{A}| = \sum (-1)^k a_{1i} a_{2j} \cdots a_{ns} \tag{4.7.10}$$

where the summation extends over all possible arrangements of the n second subscripts and k is the total number of inversions* in the sequence of the second subscript. We do not, however, use this definition in actual calculations. Rather, the value of the nth-order determinant is most generally found by expanding in terms of cofactors. The *cofactor* A_{ij} (we do not use straight bars to denote a cofactor) of the element a_{ij} of any determinant is $(-1)^{i+j}$ times the determinant obtained by deleting the ith row and the jth column.† The value of the nth-order determinant is then given by the sum of the products of the elements of any row or column with their respective cofactors; this is written as

$$|\mathbf{A}| = \sum_{j=1}^{n} a_{ij} A_{ij} \quad \text{or} \quad |\mathbf{A}| = \sum_{j=1}^{n} a_{ji} A_{ji} \tag{4.7.11}$$

where i is any value from 1 to n. The first equation represents an expansion by rows and the second an expansion by columns. The expansions of Eqs. 4.7.11 can be shown to yield the same value for the determinant as Eq. 4.7.10 by considering particular examples.

Occasionally, we refer to the "minor" of the element a_{ij}. The *minor* of the element a_{ij} is the determinant formed by removing the elements of the ith row and jth column; hence, a minor is simply the cofactor without the associated sign.

Finally, the expressions above for the expansion of a determinant can be used to verify the following properties of determinants, which are often quite useful when seeking a value for a determinant.

1. If all the elements in any row, or any column, are zero, the value of the determinant is zero.
2. If all the elements in any row, or any column, are multiplied by the same constant k, the value of the determinant is multiplied by k. (Note that in the case of a matrix multiplied by constant k, all elements of the matrix were multiplied by k.)
3. The value of a determinant is not changed if the rows are written as columns and the columns as rows, in the same order. For a square matrix this allows us to write $|\mathbf{A}| = |\mathbf{A}^T|$.
4. The sign of a determinant is changed if any two rows, or any two columns, are interchanged.
5. The value of a determinant is zero if corresponding elements of any two rows, or any two columns, are proportional.
6. The value of a determinant is not changed if the elements of any constant multiple of any row (or column) are added to the corresponding elements of any other row (or column).

---

*The number of inversions is the number of pairs of elements in which a larger number precedes a smaller one; for example, the numbers (1, 5, 2, 4, 3) form the four inversions (5, 2), (5, 4), (5, 3), and (4, 3).

†The cofactor A_{32} in the fourth-order determinant

$$\begin{vmatrix} 2 & 0 & 3 & 1 \\ 1 & 1 & 2 & 1 \\ 0 & 0 & 2 & 0 \\ 1 & 2 & 0 & 1 \end{vmatrix}$$

is formed by omitting the third row and second column. It is the determinant

$$-\begin{vmatrix} 2 & 3 & 1 \\ 1 & 2 & 1 \\ 1 & 0 & 1 \end{vmatrix}$$

There results $A_{32} = -2$.

7. If each element of any row, or any column, of a determinant is expressed as a binomial, the determinant can be written as the sum of two determinants. An example is

$$\begin{vmatrix} a_{11} & a_{12}+k_1 & a_{13} \\ a_{21} & a_{22}+k_2 & a_{23} \\ a_{31} & a_{32}+k_3 & a_{33} \end{vmatrix} = \begin{vmatrix} a_{11} & a_{12} & a_{13} \\ a_{21} & a_{22} & a_{23} \\ a_{31} & a_{32} & a_{33} \end{vmatrix} + \begin{vmatrix} a_{11} & k_1 & a_{13} \\ a_{21} & k_2 & a_{23} \\ a_{31} & k_3 & a_{33} \end{vmatrix} \tag{4.7.12}$$

8. Let **A** and **B** be two square matrices of the same order. If **C** = **AB**, then

$$|\mathbf{C}| = |\mathbf{A}|\ |\mathbf{B}| \tag{4.7.13}$$

The preceding properties of a determinant will be verified either by the following Examples, or by the Problems. We will see that the clever use of one or more of these properties substantially reduces the labor involved in the expansion of a determinant.

Determinants can be easily computed with MATLAB. If a square matrix is entered into a variable `A`, the determinant is obtained with the command `det(A)`. See Section B.3 in the appendix for more information on defining a matrix quantity, and various operations and calculations.

Example 4.5

Using cofactors and the technique of Eq. 4.7.7, find the value of the determinant

$$\begin{vmatrix} 3 & 2 & 1 \\ -1 & 0 & 1 \\ 1 & 2 & 2 \end{vmatrix}$$

by expanding by the first row, and then by the first column.

Solution

Expanding by the first row, we have

$$\begin{vmatrix} 3 & 2 & 1 \\ -1 & 0 & 1 \\ 1 & 2 & 2 \end{vmatrix} = 3\begin{vmatrix} 0 & 1 \\ 2 & 2 \end{vmatrix} - 2\begin{vmatrix} -1 & 1 \\ 1 & 2 \end{vmatrix} + 1\begin{vmatrix} -1 & 0 \\ 1 & 2 \end{vmatrix}$$

$$= 3(-2) - 2(-2-1) + 1(-2) = -2$$

Expanding by the first column, there results

$$\begin{vmatrix} 3 & 2 & 1 \\ -1 & 0 & 1 \\ 1 & 2 & 2 \end{vmatrix} = 3\begin{vmatrix} 0 & 1 \\ 2 & 2 \end{vmatrix} + 1\begin{vmatrix} 2 & 1 \\ 2 & 2 \end{vmatrix} + 1\begin{vmatrix} 2 & 1 \\ 0 & 1 \end{vmatrix}$$

$$= 3(-2) + 1(4-2) + 1(2) = -2$$

Using the diagonal method of Eq. 4.7.7, we obtain

$$\begin{vmatrix} 3 & 2 & 1 \\ -1 & 0 & 1 \\ 1 & 2 & 2 \end{vmatrix}\begin{matrix} 3 & 2 \\ -1 & 0 \\ 1 & 2 \end{matrix} = 0 + 2 - 2 - 0 - 6 + 4 = -2$$

Example 4.6

Evaluate the determinant

$$\mathbf{D} = \begin{vmatrix} 2 & 4 & -2 & 6 \\ 1 & 0 & 3 & 2 \\ 0 & 5 & -1 & 4 \\ -3 & -2 & 1 & 0 \end{vmatrix}$$

Solution

Before we expand using cofactors, let us use some of the properties of a determinant listed in Section 4.7 and simplify the given determinant by attempting to introduce zeros into some row or column. Using property 6, let us subtract from the first row twice the second row. We have

$$\mathbf{D} = \begin{vmatrix} 2 & 4 & -2 & 6 \\ 1 & 0 & 3 & 2 \\ 0 & 5 & -1 & 4 \\ -3 & -2 & 1 & 0 \end{vmatrix} = \begin{vmatrix} 0 & 4 & -8 & 2 \\ 1 & 0 & 3 & 2 \\ 0 & 5 & -1 & 4 \\ -3 & -2 & 1 & 0 \end{vmatrix}$$

Now, add to the fourth row three times the second row and expand by cofactors using the first column:

$$\mathbf{D} = \begin{vmatrix} 0 & 4 & -8 & 2 \\ 1 & 0 & 3 & 2 \\ 0 & 5 & -1 & 4 \\ -3 & -2 & 1 & 0 \end{vmatrix} = \begin{vmatrix} 0 & 4 & -8 & 2 \\ 1 & 0 & 3 & 2 \\ 0 & 5 & -1 & 4 \\ 0 & -2 & 10 & 6 \end{vmatrix} = -\begin{vmatrix} 4 & -8 & 2 \\ 5 & -1 & 4 \\ -2 & 10 & 6 \end{vmatrix}$$

In the third-order determinant, subtract eight times the second row from the first row, and add 10 times the second row to the third row; then expand:

$$\mathbf{D} = -\begin{vmatrix} -36 & 0 & -30 \\ 5 & -1 & 4 \\ 48 & 0 & 46 \end{vmatrix} = -(-1)\begin{vmatrix} -36 & -30 \\ 48 & 46 \end{vmatrix} = -12\begin{vmatrix} 3 & 30 \\ 4 & 46 \end{vmatrix}$$

Finally, the second-order determinant is evaluated and we have

$$D = -12(138 - 120) = -216$$

With similar manipulations, the procedure above can be used to reduce a determinant of any order to a second-order determinant.

This result is easily obtained with MATLAB. The command sequence for defining the matrix and computing its determinant, with output, is as follows.

```
>> A = [2 4 -2 6; 1 0 3 2; 0 5 -1 4; -3 -2 1 0];
>> det(A)
ans =
      -216
```

Example 4.7

Solve the set of equations

$$\begin{aligned} 2x + 3y - z &= 5 \\ x + y + z &= 2 \\ x - y + 2z &= 3 \end{aligned}$$

Solution

The solution for the unknowns x, y, z is given by applying Eqs. 4.7.5. The four determinants are evaluated as follows:

$$\begin{vmatrix} 2 & 3 & -1 \\ 1 & 1 & 1 \\ 1 & -1 & 2 \end{vmatrix} = 5, \quad \begin{vmatrix} 5 & 3 & -1 \\ 2 & 1 & 1 \\ 3 & -1 & 2 \end{vmatrix} = 17, \quad \begin{vmatrix} 2 & 5 & -1 \\ 1 & 2 & 1 \\ 1 & 3 & 2 \end{vmatrix} = -4,$$

$$\begin{vmatrix} 2 & 3 & 5 \\ 1 & 1 & 2 \\ 1 & -1 & 3 \end{vmatrix} = -3$$

The solution is then

$$x = \frac{17}{5}, \quad y = -\frac{4}{5}, \quad z = -\frac{3}{5}$$

4.8 THE ADJOINT AND THE INVERSE MATRICES

We are now in a position to define two additional matrices that are often useful when manipulating linear systems of equations. The *adjoint matrix* $\mathbf{A}^+$ is defined to be the transpose of the matrix obtained from the square matrix $\mathbf{A}$ by replacing each element a_{ij} of $\mathbf{A}$ with its cofactor A_{ij}. It is displayed as

$$\mathbf{A}^+ = \begin{bmatrix} A_{11} & A_{21} & \cdots & A_{n1} \\ A_{12} & A_{22} & \cdots & A_{n2} \\ \vdots & \vdots & & \vdots \\ A_{1n} & A_{2n} & \cdots & A_{nn} \end{bmatrix} \tag{4.8.1}$$

Note that the cofactor A_{ij} occupies the position of a_{ji}, not the position of a_{ij}.

The *inverse matrix* $\mathbf{A}^{-1}$ of the square matrix $\mathbf{A}$ is defined so that*

$$\mathbf{A}\mathbf{A}^{-1} = \mathbf{I} \quad \text{or} \quad \mathbf{A}^{-1}\mathbf{A} = \mathbf{I} \tag{4.8.2}$$

*Note that the inverse of $\mathbf{A}$ is never written as $1/\mathbf{A}$, since we have no rules for dividing by matrices. $\mathbf{A}^{-1}$ is merely a symbol used to denote a matrix, related to the matrix $\mathbf{A}$ such that $\mathbf{A}\mathbf{A}^{-1} = \mathbf{I}$.

This is analogous to the property of a nonmatrix quantity b; that is, $bb^{-1} = b^{-1}b = 1$. The inverse is calculated using the adjoint matrix and the determinant to be

$$\mathbf{A}^{-1} = \frac{\mathbf{A}^{+}}{|\mathbf{A}|} \tag{4.8.3}$$

This relationship can be verified by considering the set of linear equations (see Eq. 4.5.3), written in matrix form,

$$\mathbf{Ax} = \mathbf{r} \tag{4.8.4}$$

where $\mathbf{x}$ is a column vector containing the unknowns x_n in the set of equations. Premultiply both sides of this equation by $\mathbf{A}^{-1}$, and using Eq. 4.8.2, we have

$$\mathbf{Ix} = \mathbf{A}^{-1}\mathbf{r} \tag{4.8.5}$$

Using the property (4.6.3), this is written as

$$\mathbf{x} = \mathbf{A}^{-1}\mathbf{r} \tag{4.8.6}$$

Thus, to find the unknowns, the elements in the column vector $\mathbf{x}$, we simply evaluate the product $\mathbf{A}^{-1}\mathbf{r}$.

To show that Eq. 4.8.6 using $\mathbf{A}^{-1}$ as given by Eq. 4.8.3 gives the correct expression for the unknowns x_n, we know that the solution is given by

$$x_1 = \frac{D_1}{|\mathbf{A}|}, \quad x_2 = \frac{D_2}{|\mathbf{A}|}, \quad \ldots, \quad x_n = \frac{D_n}{|\mathbf{A}|} \tag{4.8.7}$$

where D_i is the determinant obtained from $|\mathbf{A}|$ by replacing the ith column of $|\mathbf{A}|$ with the elements of the column vector $\mathbf{r}$. If we expand D_1 by the first column, D_2 by the second column, etc. (note that each determinant D_i is expanded using the elements of $\mathbf{r}$), we have

$$\begin{aligned} x_1 &= \frac{1}{|\mathbf{A}|}(r_1A_{11} + r_2A_{21} + \cdots + r_nA_{n1}) \\ &\vdots \\ x_n &= \frac{1}{|\mathbf{A}|}(r_1A_{1n} + r_2A_{2n} + \cdots + r_nA_{nn}) \end{aligned} \tag{4.8.8}$$

where A_{ij} is the cofactor of the determinant element a_{ij} of $|\mathbf{A}|$. Now, let us find the first element of the column vector $\mathbf{x}$ using Eq. 4.8.6 with the inverse replaced by Eq. 4.8.3. It is, using Eq. 4.5.7 ($\mathbf{C}$ represents the column vector $\mathbf{x}$ and $\mathbf{B}$ represents the column vector $\mathbf{r}$),

$$x_1 = \frac{1}{|A|}(A_{11}r_1 + A_{21}r_2 + A_{31}r_3 + \cdots + A_{n1}r_n) \tag{4.8.9}$$

where we have used the first row in the matrix $\mathbf{A}^{+}$ (see Eq. 4.8.1). Observe that this is identical to the first equation of Eqs. 4.8.8. Hence, we see that Eq. 4.8.6 gives the correct solution for the unknowns x_n if we calculate the inverse according to Eq. 4.8.3.

An important observation is made regarding the existence of the inverse matrix. The inverse matrix $\mathbf{A}^{-1}$ does not exist if $|\mathbf{A}| = 0$, according to Eq. 4.8.3. If $|\mathbf{A}| \neq 0$, then the inverse does exist. Hence, when working with an inverse matrix, we must make sure that the determinant of the matrix is not zero; that is, the matrix must be nonsingular. A *singular* matrix is a square matrix that possesses a zero determinant.

There are several properties associated with the inverse matrix that are occasionally used when working with matrices. They will be verified with Examples or Problems. They are:

1. The inverse of the inverse is the given square matrix:

$$(\mathbf{A}^{-1})^{-1} = \mathbf{A} \tag{4.8.10}$$

2. The inverse of the product of two square matrices is the product of the inverses in reverse order:

$$(\mathbf{AB})^{-1} = \mathbf{B}^{-1}\mathbf{A}^{-1} \tag{4.8.11}$$

3. The inverse of the transpose is the transpose of the inverse:

$$(\mathbf{A}^{T})^{-1} = (\mathbf{A}^{-1})^{T} \tag{4.8.12}$$

4. The products of two nonsingular matrices cannot be a null matrix.
5. The inverse of a nonsingular diagonal matrix with diagonal elements (k_1, $k_2, \ldots, k_n$) is a diagonal matrix with diagonal elements ($1/k_1, 1/k_2, \ldots, 1/k_n$). This is displayed as

$$\begin{bmatrix} k_1 & 0 & \cdots & 0 \\ 0 & k_2 & \cdots & 0 \\ \vdots & \vdots & & \vdots \\ 0 & 0 & \cdots & k_n \end{bmatrix}^{-1} = \begin{bmatrix} 1/k_1 & 0 & \cdots & 0 \\ 0 & 1/k_2 & \cdots & 0 \\ \vdots & \vdots & & \vdots \\ 0 & 0 & \cdots & 1/k_n \end{bmatrix} \tag{4.8.13}$$

or, more simply,

$$\text{diag}\left[k_1, k_2, \ldots, k_n\right]^{-1} = \text{diag}\left[1/k_1, 1/k_2, \ldots, 1/k_n\right] \tag{4.8.14}$$

Obviously, the inverse of the unit matrix is the unit matrix,

$$\mathbf{I}^{-1} = \mathbf{I} \tag{4.8.15}$$

Matrix inverses can be easily computed with MATLAB. If the elements of a square matrix **A** are entered into a variable `A`, the inverse of **A** is obtained with the command `inv(A)`. The result is a square matrix. That is, `inv(A)` itself can be regarded as a square matrix representing $\mathbf{A}^{-1}$. It is easily confirmed in MATLAB to be the inverse: typing `inv(A)*A` or `A*inv(A)` produces the identity matrix (perhaps with some roundoff error). See Section B.3 in the appendix for more information on defining a matrix quantity, and performing various matrix operations and calculations.

Example 4.8

Find the inverse of the matrix

$$\mathbf{A} = \begin{bmatrix} 3 & 0 & -1 \\ 1 & 2 & 1 \\ 3 & 4 & 0 \end{bmatrix}$$

Solution

Let us first find the adjoint matrix. The cofactors of **A** are $A_{11} = -4$, $A_{12} = 3$, $A_{13} = -2$, $A_{21} = -4$, $A_{22} = 3$, $A_{23} = -12$, $A_{31} = 2$, $A_{32} = -4$, and $A_{33} = 6$. The adjoint matrix $\mathbf{A}^{+}$ is then

$$\mathbf{A}^{+} = \begin{bmatrix} -4 & -4 & 2 \\ 3 & 3 & -4 \\ -2 & -12 & 6 \end{bmatrix}$$

The determinant is evaluated to be

$$|\mathbf{A}| = \begin{vmatrix} 3 & 0 & -1 \\ 1 & 2 & 1 \\ 3 & 4 & 0 \end{vmatrix} = -10$$

Finally, according to Eq. 4.8.3, the inverse matrix $\mathbf{A}^{-1}$ is

$$\mathbf{A}^{-1} = \begin{bmatrix} 0.4 & 0.4 & -0.2 \\ -0.3 & -0.3 & 0.4 \\ 0.2 & 1.2 & -0.6 \end{bmatrix}$$

This result is easily obtained with MATLAB. The command sequence for defining the matrix and computing its inverse, with output, is as follows.

```
>> A = [3 0 -1; 1 2 1; 3 4 0];
>> inv(A)
ans =
       0.4    0.4   -0.2
      -0.3   -0.3    0.4
       0.2    1.2   -0.6
```

Example 4.9

Show that the inverse of the product of two square matrices is the product of the inverses in reverse order, as represented by Eq. 4.8.11.

Solution

Consider the matrix $\mathbf{C} = \mathbf{AB}$. Using Eq. 4.8.2 we have

$$\mathbf{CC}^{-1} = \mathbf{I}$$

or, substituting for $\mathbf{C}$,

$$\mathbf{AB}(\mathbf{AB})^{-1} = \mathbf{I}$$

Premultiply each side by $\mathbf{A}^{-1}$ and use the last property of Eqs. 4.6.1, to obtain

$$\mathbf{B}(\mathbf{AB})^{-1} = \mathbf{A}^{-1}$$

where we have used $\mathbf{A}^{-1}\mathbf{I} = \mathbf{A}^{-1}$ and $\mathbf{IB} = \mathbf{B}$. Now premultiply both sides by $\mathbf{B}^{-1}$:

$$\mathbf{I}(\mathbf{AB})^{-1} = \mathbf{B}^{-1}\mathbf{A}^{-1}$$

Obviously, $\mathbf{IC}^{-1} = \mathbf{C}^{-1}$, and we can write

$$(\mathbf{AB})^{-1} = \mathbf{B}^{-1}\mathbf{A}^{-1}$$

The property stated in the example statement has been shown to be true.

Example 4.10

Find the column vector **x** that represents the solution to the equations

$$\begin{aligned} 4x - y \quad &= 4 \\ 3x + 3y + z &= 2 \\ x + y + z &= 0 \end{aligned}$$

Solution
The set of equations are written as

$$\mathbf{Ax} = \mathbf{r}$$

where

$$\mathbf{A} = \begin{bmatrix} 4 & -1 & 0 \\ 3 & 3 & 1 \\ 1 & 1 & 1 \end{bmatrix}, \quad \mathbf{x} = \begin{bmatrix} x \\ y \\ z \end{bmatrix}, \quad \mathbf{r} = \begin{bmatrix} 4 \\ 2 \\ 0 \end{bmatrix}$$

The solution expressed as the column vector **x** is written as

$$\mathbf{x} = \mathbf{A}^{-1}\mathbf{r}$$

To find $\mathbf{A}^{-1}$ we calculate $\mathbf{A}^{+}$ and $|\mathbf{A}|$ to be

$$\mathbf{A}^{+} = \begin{bmatrix} 2 & 1 & -1 \\ -2 & 4 & -4 \\ 0 & -5 & 15 \end{bmatrix}, \quad |\mathbf{A}| = \begin{vmatrix} 4 & -1 & 0 \\ 3 & 3 & 1 \\ 1 & 1 & 1 \end{vmatrix} = 10$$

Thus,

$$\mathbf{A}^{-1} = \frac{\mathbf{A}^{+}}{|\mathbf{A}|} \begin{bmatrix} 0.2 & 0.1 & -0.1 \\ -0.2 & 0.4 & -0.4 \\ 0 & -0.5 & 1.5 \end{bmatrix}$$

Next, we perform the matrix multiplication to find **x**:

$$\mathbf{x} = \begin{bmatrix} 0.2 & 0.1 & -0.1 \\ -0.2 & 0.4 & -0.4 \\ 0 & -0.5 & 1.5 \end{bmatrix} \begin{bmatrix} 4 \\ 2 \\ 0 \end{bmatrix} = \begin{bmatrix} 1 \\ 0 \\ -1 \end{bmatrix}$$

The solution, represented by the elements of the vector **x**, is

$$x = 1, \qquad y = 0, \qquad z = -1$$

4.9 SOLUTION OF SIMULTANEOUS LINEAR ALGEBRAIC EQUATIONS

The solution of simultaneous linear algebraic equations has arisen as some of the previous topics have been introduced. In this section we consolidate and expound on these solution methods.

As seen previously, a set of linear algebraic equations can be written in matrix form as

$$\mathbf{A}\mathbf{x} = \mathbf{r} \tag{4.9.1}$$

We consider the case where $\mathbf{A}$ is an $n \times n$ square matrix, and the unknowns are the elements of vector $\mathbf{x}$. In such case, Eq. 4.9.1 represents n equations in n unknowns. This set of equations is classified as homogeneous if $\mathbf{r} = \mathbf{0}$, and nonhomogeneous if $\mathbf{r} \neq \mathbf{0}$. We will treat them separately, starting with the nonhomogeneous case.

4.9.1 Nonhomogeneous Sets of Linear Algebraic Equations

We have seen two approaches to the solution of Eq. 4.9.1 in the previous discussions. Firstly, as introduced in Eqs. 4.7.8, we can apply the solution that involves determinants (Cramer's rule), whereby

$$x_i = \frac{D_i}{|\mathbf{A}|} \tag{4.9.2}$$

for $i = 1, \ldots, n$, where D_i is the determinant of the matrix obtained by replacing the ith column of $\mathbf{A}$ with the column vector $\mathbf{r}$.

The other approach, from Eq. 4.8.6, is to use the inverse the matrix $\mathbf{A}$ and write

$$\mathbf{x} = \mathbf{A}^{-1}\,\mathbf{r} \tag{4.9.3}$$

Both of these perspectives are consistent in requiring that $|\mathbf{A}| \neq 0$, i.e., that the matrix $\mathbf{A}$ is nonsingular. In such case, the solution exists and is unique. If the matrix $\mathbf{A}$ is singular, a unique solution to Eq. 4.9.1 does not exist. Either there is no solution at all, or there are infinitely many solutions.

A third method to obtain the solution to Eq. 4.9.1 is to perform Gaussian elimination. In Gaussian elimination, any equation can be multiplied by a scalar constant, and any equation can be replaced by the sum of itself multiplied by a scalar and added to another row multiplied by another scalar. This is facilitated by applying row operations to the augmented matrix form, illustrated here.

As an example of $\mathbf{A}\mathbf{x} = \mathbf{r}$, consider the three linear algebraic equations,

$$\begin{aligned} 2x_1 + 3x_2 &= 2 \\ -x_1 + x_2 + 2x_3 &= 3 \\ x_1 - x_3 &= -1 \end{aligned} \tag{4.9.4}$$

for which $\mathbf{A}$ and $\mathbf{r}$ are

$$\mathbf{A} = \begin{bmatrix} 2 & 3 & 0 \\ -1 & 1 & 2 \\ 1 & 0 & -1 \end{bmatrix} \quad \text{and} \quad \mathbf{r} = \begin{bmatrix} 2 \\ 3 \\ -1 \end{bmatrix} \tag{4.9.5}$$

Here we have $|\mathbf{A}| = 1$, and so a unique solution exists. Equation 4.9.5 can be represented with the augmented matrix $[\mathbf{A}\ \mathbf{r}]$, as

$$\begin{bmatrix} 2 & 3 & 0 & 2 \\ -1 & 1 & 2 & 3 \\ 1 & 0 & -1 & -1 \end{bmatrix} \tag{4.9.6}$$

Next we conduct row operations on the augmented form. Generally, the admissible row operations are to

a. multiply any row by a nonzero constant,
b. add a multiple of one row to another, and
c. interchange any two rows.

For example, multiplying the first row by 1/2 and adding it to the second row, element by element, replaces the second row. Then multiplying the first row by −1/2 and adding it to the third row replaces the third row. The result of these two actions is

$$\begin{bmatrix} 2 & 3 & 0 & 2 \\ 0 & 5/2 & 2 & 4 \\ 0 & -3/2 & -1 & -2 \end{bmatrix} \tag{4.9.7}$$

Multiplying the second and third rows by 2, we have

$$\begin{bmatrix} 2 & 3 & 0 & 2 \\ 0 & 5 & 4 & 8 \\ 0 & -3 & -2 & -4 \end{bmatrix} \tag{4.9.8}$$

Multiplying the second row by 3/5 and adding it to the third row yields

$$\begin{bmatrix} 2 & 3 & 0 & 2 \\ 0 & 5 & 4 & 8 \\ 0 & 0 & 2/5 & 4/5 \end{bmatrix} \tag{4.9.9}$$

Multiplying the third row by 5 we obtain

$$\begin{bmatrix} 2 & 3 & 0 & 2 \\ 0 & 5 & 4 & 8 \\ 0 & 0 & 2 & 4 \end{bmatrix} \tag{4.9.10}$$

Establishing this upper-triangular form is enough to easily obtain the solution to our system (although we could carry the operations further to obtain a more elegant form, namely to diagonalize the three left columns with 1s on the diagonal). This new augmented matrix represents the system of equations reworked as

$$\begin{aligned} 2x_1 + 3x_2 &= 2 \\ 5x_2 + 4x_3 &= 8 \\ 2x_3 &= 4 \end{aligned} \tag{4.9.11}$$

Solving the bottom equation and back substituting into the second and first equations, the solution is $x_1 = 1$, $x_2 = 0$, and $x_3 = 2$.

If instead a matrix $\mathbf{A}$ were such that $|\mathbf{A}| = 0$, then the upper triangular form, instead of Eq. 4.9.10, will have at least one row of zeros for the coefficient part (n left-most elements) of the augmented matrix. If the entire row has elements that are zeros, then there are infinitely many solutions. If an entire row is zero except the right-most element, which is nonzero, then there are no solutions to $\mathbf{Ax} = \mathbf{r}$.

4.9.2 Homogeneous Sets of Linear Algebraic Equations

For the homogeneous case we have $\mathbf{r} = \mathbf{0}$, and the equation to be solved is

$$\mathbf{Ax} = \mathbf{0} \tag{4.9.12}$$

If the matrix $\mathbf{A}$ is nonsingular, then it is invertible. Then applying Eq. 4.9.3 to Eq. 4.9.12 yields $\mathbf{x} = \mathbf{A}^{-1}\mathbf{0} = \mathbf{0}$. That is, if $|\mathbf{A}| \neq 0$, then the solution to Eq. 4.9.12 is $\mathbf{x} = \mathbf{0}$.

Therefore, a necessary condition for obtaining a nonzero solution is that the matrix $\mathbf{A}$ must be singular. That is, for nonzero solutions to Eq. 4.9.12, we must have $|\mathbf{A}| = 0$. But if $\mathbf{A}$ is singular, we cannot apply Eq. 4.9.2 or Eq. 4.9.3 to find a solution to $\mathbf{Ax} = \mathbf{0}$.

An important note is that, if we have found a solution $\mathbf{x}$, then $\alpha\mathbf{x}$ is also a solution, where α is a scalar constant. To see this, consider that $\mathbf{A}\alpha\mathbf{x} = \alpha\mathbf{Ax} = \mathbf{0}$, and hence $\alpha\mathbf{x}$ satisfies Eq. 4.9.12. Therefore, a nonzero solution to Eq. 4.9.12 is nonunique to an arbitrary scalar multiplier, but is unique in shape.

One way to solve Eq. 4.9.12 is to apply Gaussian elimination. We illustrate the Gaussian elimination homogeneous solution for a 3×3 example. Suppose we have

$$\begin{aligned} x_1 + x_2 &= 0 \\ 2x_1 - x_3 &= 0 \\ -x_1 + x_2 + x_3 &= 0 \end{aligned} \tag{4.9.13}$$

for which $\mathbf{A}$ and $\mathbf{r}$ are

$$\mathbf{A} = \begin{bmatrix} 1 & 1 & 0 \\ 2 & 0 & -1 \\ -1 & 1 & 1 \end{bmatrix} \quad \text{and} \quad \mathbf{r} = \begin{bmatrix} 0 \\ 0 \\ 0 \end{bmatrix} \tag{4.9.14}$$

Here we have $|\mathbf{A}| = 0$, and so a nonzero, nonunique solution exists. Equation 4.9.13 can be represented with the augmented matrix $[\mathbf{A}\ \mathbf{r}]$, as

$$\begin{bmatrix} 1 & 1 & 0 & 0 \\ 2 & 0 & -1 & 0 \\ -1 & 1 & 1 & 0 \end{bmatrix} \tag{4.9.15}$$

Next we conduct row operations. Multiplying the first row by –2 and adding it to the second row, element by element, replaces the second row. Then adding the first to the third row replaces the third row. The result of these two actions is

$$\begin{bmatrix} 1 & 1 & 0 & 0 \\ 0 & -2 & -1 & 0 \\ 0 & 2 & 1 & 0 \end{bmatrix} \tag{4.9.16}$$

Adding the second row to the third row replaces the third row, and we have

$$\begin{bmatrix} 1 & 1 & 0 & 0 \\ 0 & -2 & -1 & 0 \\ 0 & 0 & 0 & 0 \end{bmatrix} \tag{4.9.17}$$

In this example we have achieved an upper triangular reduced augmented form with a single row of zeros. The number of rows of zeros in the reduced augmented form of $\mathbf{Ax} = \mathbf{0}$ provides the *nullity* of matrix $\mathbf{A}$. The *rank* of matrix $\mathbf{A}$ is the dimension of $\mathbf{A}$ minus the nullity. In this case, the rank of $\mathbf{A}$ is two. The rank of a matrix is equal to the number of independent rows in the matrix, and is also equal to the number of independent columns.

The new augmented matrix of Eq. 4.9.17 represents the system of equations reworked as

$$\begin{aligned} x_1 + x_2 &= 0 \\ -2x_2 - x_3 &= 0 \\ 0 &= 0 \end{aligned} \qquad (4.9.18)$$

From the top equation, $x_1 = -x_2$, and from the middle equation, $x_3 = -2x_2 = 2x_1$. Hence, a nonunique solution is $x_1 = 1$, $x_2 = -1$, and $x_3 = 2$. Infinitely many solutions can be represented as

$$\mathbf{x} = \alpha \begin{bmatrix} 1 \\ -1 \\ 2 \end{bmatrix} \qquad (4.9.19)$$

where α is an arbitrary scalar multiplier.

Suppose $\mathbf{x}$ is a vector that satisfies $\mathbf{Ax} = \mathbf{0}$. Then $\mathbf{x}$ is called a *null vector* of $\mathbf{A}$. As we see in the example of Eq. 4.9.19, if $\mathbf{x} \neq \mathbf{0}$, then there are infinitely many null vectors via the scalar multiplier α of a single independent vector. The *null space* (also called the *kernel*) of $\mathbf{A}$ is the set of all vectors for which $\mathbf{Ax} = \mathbf{0}$.

The terminology "null vector" can also be used to define a null vector whose elements are zeros, as with the null matrix defined in Section 4.2, which has a different meaning than a null vector of a matrix used here.

It is possible for a matrix to have a nullity other than one. In the example which reduced to the upper-triangular augmented form of Eq. 4.9.10, which had no rows of zeros, the nullity of $\mathbf{A}$ was zero, and the rank of $\mathbf{A}$ was three. In such case, the null space consists trivially of the point at the origin.

If the nullity is larger than one, the null space is defined by more than one independent null vector. For example, consider a matrix

$$\mathbf{A} = \begin{bmatrix} 3 & -3 & -3/2 \\ 2 & -2 & -1 \\ -1 & 1 & 1/2 \end{bmatrix} \qquad (4.9.20)$$

for which row operations on the augmented matrix of $\mathbf{Ax} = \mathbf{0}$ lead to the augmented matrix

$$\begin{bmatrix} 1 & -1 & -1/2 & 0 \\ 0 & 0 & 0 & 0 \\ 0 & 0 & 0 & 0 \end{bmatrix} \qquad (4.9.21)$$

which is upper triangular with two zero rows. The nullity of $\mathbf{A}$ is therefore two. The rank is the dimension of $\mathbf{A}$ minus the nullity. Thus in this example, the rank of $\mathbf{A}$ is one. The first row represents the equation

$$x_1 - x_2 - (1/2)\, x_3 = 0 \qquad (4.9.22)$$

or

$$x_3 = 2x_1 - 2x_2 \qquad (4.9.23)$$

If we choose $x_1 = 1$, and $x_2 = 0$, then $x_3 = 2$. If we choose $x_1 = 0$, and $x_2 = 1$, then $x_3 = -2$. That is, we have found two independent null vectors,

$$\mathbf{x}^1 = \begin{bmatrix} 1 \\ 0 \\ 2 \end{bmatrix} \quad \text{and} \quad \mathbf{x}^2 = \begin{bmatrix} 0 \\ 1 \\ -2 \end{bmatrix} \tag{4.9.24}$$

Any linear combination of these produces a solution to $\mathbf{Ax} = \mathbf{0}$. The set of all linear combinations of $\mathbf{x}^1$ and $\mathbf{x}^2$ (also called the *span* of $\mathbf{x}^1$ and $\mathbf{x}^2$) describes the null space of $\mathbf{A}$.

4.9.3 Solutions to Sets of Linear Equations by MATLAB

For nonhomogeneous equations of Eq. 4.9.1, when $|\mathbf{A}| = 0$, simple MATLAB commands exist for applying Eq. 4.9.2 or Eq. 4.9.3. Given matrix entries stored in variable `A`, and the right hand side stored in variable `r`, the command `inv(A)` produces a matrix that is the inverse of `A`. It is easy to check that `inv(A)*A` and `A*inv(A)` produce the identity matrix. Then the application of Eq. 4.9.3 is through the command `x = inv(A)*r`.

To apply Cramer's rule for solution to x_i, one can first assign the matrix `Ai = A`, and then replace column `i` with `r` by using `Ai(:,i) = r`. The solution to unknown x_i is then obtained from the ratio of determinants, according to the command `x(i) = det(Ai)/det(A)`.

For solving homogeneous sets of equations, namely Eq. 4.9.12, null vectors are the crux. A simple command for obtaining the solution to $\mathbf{Ax} = \mathbf{0}$ in MATLAB is `u = null(A)`, where `A` is the matrix $\mathbf{A}$ specified in the computer code. The command `u = null(A)` will assign variable `u` with independent null vectors, the number of which matches the nullity of the matrix $\mathbf{A}$.

Example 4.11

Write a computer program in MATLAB that will solve the equations

$$\begin{aligned} 2x_1 - 3.5x_2 + 2x_4 &= 0 \\ x_1 + 4x_2 + 2x_3 + x_4 &= 1 \\ -2x_2 - x_3 + x_4 &= -2 \\ 3x_1 + 5x_2 + x_3 - 2x_4 &= 2 \end{aligned}$$

(a) Determine whether there exists a unique solution. (b) Write the whole solution. (c) Write the solution for x_2 only, using the determinant method (Cramer's rule).

Solution

The four equations in four unknowns can be represented by $\mathbf{Ax} = \mathbf{r}$, where

$$\mathbf{A} = \begin{bmatrix} 2 & -3.5 & 0 & 2 \\ 1 & 4 & 2 & 1 \\ 0 & -2 & -1 & 1 \\ 3 & 5 & 1 & -2 \end{bmatrix} \quad \text{and} \quad \mathbf{r} = \begin{bmatrix} 0 \\ 1 \\ -2 \\ 2 \end{bmatrix}$$

The MATLAB script combining parts (a) through (c) is below.

```
% Solve a set of 4 linear algebraic equations of the form Ax = r.

% Enter the elements of the coefficient matrix, A.
A = [2 -3.5 0 2; 1 4 2 1; 0 -2 -1 1; 3 5 1 -2];
% Enter the elements of the right hand side, r.
r = [0; 1; -2; 2];
display('For a set of equations of the form Ax = r, we are given');
A
r

% (a) Check to see if a unique solution exists.
display('(a) The determinant of A is');
D = det(A)
if D ~= 0
    display('The matrix is nonsingular, and so a unique solution exists.');
else
    display('The matrix is singular, and so no unique solution exists.');
end
R = rank(A); dim = min(size(A));
if R == dim
    display('Alternatively, we can confirm that the rank matches the matrix
↙ dimension.');
end

% (b) Solve by inversion, and allow output to be displayed.
display(' ');
display('(b) The solution to the vector of unknowns x is');
x = inv(A)*r

% (c) Seek x_2 = det(A_2)/det(A);
A_2 = A;
A_2(:,2) = r;
% Display solution
display('(c) The solution to x_2, alone, by the determinant method is ');
x_2 = det(A_2)/det(A)
%Check consistency
if x_2 == x(2)
    display('The solutions for x_2 are in agreement.')
else
    display('There is an error in the code.')
end
```

The output of this code is

```
>>Example4_11
For a set of equations of the form Ax = r, we are given

A =

    2.0000   -3.5000         0    2.0000
    1.0000    4.0000    2.0000    1.0000
         0   -2.0000   -1.0000    1.0000
    3.0000    5.0000    1.0000   -2.0000

r =

     0
     1
    -2
     2

(a) The determinant of A is

D =

   47.0000

The matrix is nonsingular, and so a unique solution exists.
Alternatively, we can confirm that the rank matches the matrix
dimension.

(b) The solution to the vector of unknowns x is

x =

    0.1596
   -0.5106
    1.9681
   -1.0532

(c) The solution to x_2, alone, by the determinant method is

x_2 =

   -0.5106

The solutions for x_2 are in agreement.
```

Example 4.12

For the system of equations $\mathbf{Ax} = \mathbf{0}$, where

$$\mathbf{A} = \begin{bmatrix} 1 & 0 & -1 \\ 2 & 1 & 0 \\ 1 & 1 & 1 \end{bmatrix}$$

(a) Determine whether there exists a nonzero solution. If so, (b) use MATLAB to find the solution vector. (c) Apply row operations to the augmented matrix to confirm results in (a) and (b).

Solution

a) The determinant is $|\mathbf{A}| = 0$. Therefore the matrix is singular, and a nonunique solution exists.

b) Using MATLAB interactively,

```
>> A = [1 0 -1; 2 1 0; 1 1 1];
>> det(A)          % Confirming that the solution exists
ans = 0
>> rank(A)         % Rank is less than the dimension of A, also
                   % implying A is singular
ans = 2
>> u = null(A)     % Solution vector x
u =
       0.4082
      -0.8165
       0.4082
```

Any vector x = a*u is a solution to $\mathbf{Ax} = \mathbf{0}$.

c) Continuing by hand, the augmented matrix is

$$\begin{bmatrix} 1 & 0 & -1 & 0 \\ 2 & 1 & 0 & 0 \\ 1 & 1 & 1 & 0 \end{bmatrix}$$

We multiply the first row by –2 and add it to the second row. We then subtract the first row from the third row. The result of these two operations is

$$\begin{bmatrix} 1 & 0 & -1 & 0 \\ 0 & 1 & 2 & 0 \\ 0 & 1 & 2 & 0 \end{bmatrix}$$

Now subtracting the second row from the third row leads to

$$\begin{bmatrix} 1 & 0 & -1 & 0 \\ 0 & 1 & 2 & 0 \\ 0 & 0 & 0 & 0 \end{bmatrix}$$

The reduced augmented system is upper triangular, and the final row is all zeros, indicating that the nullity is 1, and thus the rank is 2. The representative system of equations is

$$\begin{aligned} x_1 - x_3 &= 0 \\ x_2 + 2x_3 &= 0 \\ 0 &= 0 \end{aligned}$$

which can be solved for, say, x_1 and x_2 in terms of x_3. For example, if $x_3 = 1$, then $x_1 = 1$ and $x_2 = -2$. The family of solutions is

$$\mathbf{x} = \alpha \begin{bmatrix} 1 \\ -2 \\ 1 \end{bmatrix}$$

which is consistent with the MATLAB solution in part (b).

Example 4.13

For the system of equations $\mathbf{Ax} = \mathbf{0}$, where

$$\mathbf{A} = \begin{bmatrix} 1 & 0 & 1 & 0 \\ 0 & 1 & 0 & 1 \\ 2 & 1 & 2 & 1 \\ -1 & 1 & -1 & 1 \end{bmatrix}$$

(a) Determine whether there exists a nonzero solution. If so, (b) use MATLAB to find the solution vector. (c) Apply row operations to the augmented matrix to confirm results in (a) and (b).

Solution

a) The determinant is $|\mathbf{A}| = 0$. Therefore the matrix is singular, and a nonunique solution exists.

b) Using MATLAB interactively,

```
>> A = [1 0 1 0; 0 1 0 1; 2 1 2 1; -1 1 -1 1];
>> det(A)        % Confirming that the solution exists
ans = 0
>> rank(A)       % Rank of 2 implies a nullity of 2, and hence we
                 % expect two independent null vectors
ans = 2
>> u = null(A)  % Solution vector(s)
```

```
u =
   -0.7067    0.0251
    0.0251    0.7067
    0.7067   -0.0251
   -0.0251   -0.7067
>> rank(u)        % A rank of 2 confirms that u consists of two
                  % independent columns
ans = 2
```

Thus any vector `x = a*u(:,1) + b*u(:,2)` is a solution to $\mathbf{Ax} = \mathbf{0}$.

c) Continuing by hand, the augmented matrix is

$$\begin{bmatrix} 1 & 0 & 1 & 0 & 0 \\ 0 & 1 & 0 & 1 & 0 \\ 2 & 1 & 2 & 1 & 0 \\ -1 & 1 & -1 & 1 & 0 \end{bmatrix}$$

We multiply the first row by –2 and add it to the third row. We then add the first row to the fourth row. The result of these two operations is

$$\begin{bmatrix} 1 & 0 & 1 & 0 & 0 \\ 0 & 1 & 0 & 1 & 0 \\ 0 & 1 & 0 & 1 & 0 \\ 0 & 1 & 0 & 1 & 0 \end{bmatrix}$$

Next we subtract the second row from both the third and fourth rows to obtain

$$\begin{bmatrix} 1 & 0 & 1 & 0 & 0 \\ 0 & 1 & 0 & 1 & 0 \\ 0 & 0 & 0 & 0 & 0 \\ 0 & 0 & 0 & 0 & 0 \end{bmatrix}$$

The reduced augmented matrix is upper triangular, and the bottom two rows consist of zeros. Therefore the nullity is two, and we expect two independent null vectors. The representative equations are

$$\begin{aligned} x_1 &= -x_3 \\ x_2 &= -x_4 \\ 0 &= 0 \\ 0 &= 0 \end{aligned}$$

If we choose, for example, if $x_3 = 1$ and $x_4 = 0$, then $x_1 = -1$ and $x_2 = 0$. If we choose, for example, if $x_3 = 0$ and $x_4 = 1$, then $x_1 = 0$ and $x_2 = -1$. Thus, two null vectors are

$$\mathbf{x}^1 = \begin{bmatrix} -1 \\ 0 \\ 1 \\ 0 \end{bmatrix} \text{ and } \mathbf{x}^2 = \begin{bmatrix} 0 \\ -1 \\ 0 \\ 1 \end{bmatrix}$$

which are clearly independent (in fact orthogonal) as their inner product is zero. Thus the family of solutions is

$$\mathbf{x} = \alpha\begin{bmatrix}-1\\0\\1\\0\end{bmatrix} + \beta\begin{bmatrix}0\\-1\\0\\1\end{bmatrix}$$

Some ways to check that the solution by row operations is consistent to the solution using MATLAB are as follows. The rank of the matrix `u` from MATLAB is noted to be 2. Similarly the rank of the matrix formed by null vectors obtained from row operations as $\mathbf{U} = [\mathbf{x}^1\ \mathbf{x}^2\ ]$ will be 2, as we know that the null vectors are independent. We can then check the rank of matrix of our MATLAB null vectors and our row-operations null vectors, and verify that it is two, meaning that among the four vectors, there are two linearly independent vectors. In this example, it is easy to see that the null vectors, expressed as the first and second columns `u(:,1)` and `u(:,2)` of matrix `u` obtained from MATLAB, are both linear combinations of $\mathbf{x}^1$ and $\mathbf{x}^2$. Another way to check is to apply the third and fourth elements of `u(:,1)` in the equations representing the upper-triangular reduced augmented matrix, and see that it results in the first and second elements of `u(:,1)`. The same can be done for `u(:,2)`. The choice of linearly independent null vectors is not unique. Different versions of MATLAB, or running MATLAB on different platforms, may generate a different pair of null vectors.

4.10 LEAST-SQUARES FIT AND THE PSEUDO INVERSE

Sometimes we are confronted with a redundant set of equations, that is, more equations than unknowns. A common situation where this happens is in the treatment of data.

Often when working with data, the aim is to get a best curve to represent the trends in the data. One powerful method of doing this is with a least-squares fit. In this section, we introduce the linear least-squares solution in a perspective that involves matrix algebra. The context of our study is that we expect that one of the data variables can be approximately expressed as a function of the other data variables. We discuss the least-squares solution through an example.

Suppose we have data for variables y and z, and we expect a relationship between these variables to be described by the linear equation $z = ay + b$, where a and b are unknown. Suppose our data is tabulated as

y	z
−1	−1
0	1
1	2.2
2	4.2

The question is how to choose a and b to make the "best fit" of $z = f(y) = ay + b$. In a perfect world, if we apply the data to $f(y)$, we hope that the following are true:

$$\begin{aligned} a(-1) + b &= -1 \\ a(0) + b &= 1 \\ a(1) + b &= 2.2 \\ a(2) + b &= 4.2 \end{aligned} \tag{4.10.1}$$

Thus we have four equations and two unknowns, namely a and b. The equations are *overdetermined* and (very probably) have no solution. For example, from the top two equations, we could solve for $b = 1$ and $a = 2$. But then the bottom two expressions do not equate.

A least-squares fit is a choice for a and b that makes the four expressions closest to equating. We choose a and b to minimize the sum of the squares of the errors in $z = f(y) = ay + b$. We define the errors as $r_i = f(y_i) - z_i$ for each data point, i. Namely,

$$\begin{aligned} r_1 &= a(-1) + b - (-1) \\ r_2 &= a(0) + b - 1 \\ r_3 &= a(1) + b - 2.2 \\ r_4 &= a(2) + b - 4.2 \end{aligned} \tag{4.10.2}$$

The errors, r_i, are called *residuals*. We can put the residuals into an array vector as

$$\mathbf{r} = \begin{bmatrix} r_1 \\ r_2 \\ r_3 \\ r_4 \end{bmatrix} \tag{4.10.3}$$

Then we want to choose a and b to minimize $r_1^2 + r_2^2 + r_3^2 + r_4^2 = \mathbf{r}^T\mathbf{r} = e$. We restrict ourselves to the case where the function $f(x)$ is linear in the parameters. In the example, with $f(y) = ay + b$, let the vector of unknowns be listed in a vector

$$\mathbf{x} = \begin{bmatrix} a \\ b \end{bmatrix} \tag{4.10.4}$$

Rewriting Eqs. 4.10.1 in matrix form, we have

$$\begin{bmatrix} -1 & 1 \\ 0 & 1 \\ 1 & 1 \\ 2 & 1 \end{bmatrix} \begin{bmatrix} a \\ b \end{bmatrix} = \begin{bmatrix} -1 \\ 1 \\ 2.2 \\ 4.2 \end{bmatrix} \tag{4.10.5}$$

or

$$\mathbf{Ax} = \mathbf{b} \tag{4.10.6}$$

The errors are given in matrix form by

$$\mathbf{r} = \mathbf{Ax} - \mathbf{b} \tag{4.10.7}$$

If **A** is square ($n \times n$) and nonsingular, then the solution is simply $\mathbf{x} = \mathbf{A}^{-1}\mathbf{b}$. But

1. **A** is not square
2. $\mathbf{A}^{-1}$ does not exist
3. $\mathbf{Ax} = \mathbf{b}$ has no exact solution.

Generally, **A** is ($n \times m$), with $n > m$ (over-determined system), **x** is ($m \times 1$), meaning there are m unknowns, and **b** is ($n \times 1$), i.e., there are n equations.

For a general system, from Eq. 4.10.7, each error, r_i, for $i = 1, ..., n$, is then

$$r_i = a_{i1}x_1 + a_{i2}x_2 + \cdots + a_{im}x_m - b_i \tag{4.10.8}$$

Then $e = \mathbf{r}^T\mathbf{r} = r_1^2 + r_2^2 + \cdots + r_n^2$. We seek to minimize e with respect to our unknown parameters x_j, for $j = 1, ..., m$. Thus we require

$$\frac{\partial e}{\partial x_j} = 2r_1\frac{\partial r_1}{\partial x_j} + 2r_2\frac{\partial r_2}{\partial x_j} + \cdots + 2r_n\frac{\partial r_n}{\partial x_j} = 0 \tag{4.10.9}$$

Using Eq. 4.10.8, we have $\dfrac{\partial r_i}{\partial x_j} = a_{ij}$. Thus we can rewrite Eqs. 4.10.9 as the set of equations

$$\begin{aligned} r_1a_{11} + r_2a_{21} + \cdots + r_na_{n1} &= 0 \\ r_1a_{12} + r_2a_{22} + \cdots + r_na_{n2} &= 0 \\ &\vdots \\ r_1a_1m + r_2a_1m + \cdots + r_na_{nm} &= 0 \end{aligned} \tag{4.10.10}$$

In matrix form, this can be written as $\mathbf{A}^T\mathbf{r} = 0$. Using Eq. 4.10.7, we have $\mathbf{A}^T(\mathbf{Ax} - \mathbf{b}) = \mathbf{0}$, or

$$\mathbf{A}^T\mathbf{Ax} = \mathbf{A}^T\mathbf{b} \tag{4.10.11}$$

In the above equation, $\mathbf{A}^T\mathbf{A}$ is ($m \times m$) and is invertible unless there is redundant data, **x** is ($m \times 1$), and $\mathbf{A}^T\mathbf{b}$ is ($m \times 1$). We can solve Eq. 4.10.11 by the matrix-inverse method as

$$\mathbf{x} = (\mathbf{A}^T\mathbf{A})^{-1}\mathbf{A}^T\mathbf{b} \tag{4.10.12}$$

which is the "best" solution for the parameters.

To help remember this solution, we can reach this result via a simple shortcut as follows. Starting with Eq. 4.10.6, which is our data model, which is linear in parameters listed in **x**, we premultiply both sides of Eq. 4.10.6 by $\mathbf{A}^T$ to produce a square leading matrix, which gets the equation into an invertible (solvable) form. We then invert the leading matrix to solve for **x**. This simple sequence is displayed as

$$\begin{aligned} \mathbf{Ax} &= \mathbf{b} \\ \mathbf{A}^T\mathbf{Ax} &= \mathbf{A}^T\mathbf{b} \\ \mathbf{x} &= (\mathbf{A}^T\mathbf{A})^{-1}\mathbf{A}^T\mathbf{b} \end{aligned} \tag{4.10.13}$$

Then **x** is the least-squares fit for our parameters. The parameter vector **x** is not a solution to Eq. 4.10.6, but can be called a "solution in the least-squares sense".

We have effectively "inverted" the non-square matrix in the over-determined set of equations 4.6.10. This is called a *pseudo inverse*. In this case, the pseudo inverse of the non-square matrix **A** of dimensions ($n \times m$), with $n > m$, is

$$\mathbf{A}^{\dagger} = (\mathbf{A}^T\mathbf{A})^{-1}\mathbf{A}^T \tag{4.10.14}$$

where the superscript † denotes the pseudo inverse.

Example 4.14

Values of the dimensionless fluid lift coefficient $C_L = z$ on an airfoil versus the angle of attack α in degrees are given in the table below (estimated from a graph in Petinrin and Onoja, British Journal of Applied Science and Technology, Vol. 21(3), 2017). Find the least-squares fit of a function $z = f(\alpha) = p_0 + p_1\alpha + p_3\alpha^3$, where the subscripts indicate the power on the associated term. That is, find the values of unknown parameters p_0, p_1, and p_3, that provide the best fit. Plot the data and the least-squares fit of the function.

α	z
–10	–0.6
–6	–0.3
0	0.4
6	1.1
12	1.5
18	1.5

Solution

We write equations of the form $z = f(\alpha)$ using each data pair. In detail

$$\begin{aligned}
p_0 + p_1(-10) + p_3(-10)^3 &= -0.6\\
p_0 + p_1(-6) + p_3(-6)^3 &= -0.3\\
p_0 + p_1(0) + p_3(0)^3 &= 0.4\\
p_0 + p_1(6) + p_3(6)^3 &= 1.1\\
p_0 + p_1(12) + p_3(12)^3 &= 1.5\\
p_0 + p_1(18) + p_3(18)^3 &= 1.5
\end{aligned}$$

which can be written as $\mathbf{Ax} = \mathbf{b}$ with

$$\mathbf{A} = \begin{bmatrix} 1 & -10 & -10^3 \\ 1 & -6 & -6^3 \\ 1 & 0 & 0^3 \\ 1 & 6 & 6^3 \\ 1 & 12 & 12^3 \\ 1 & 18 & 18^3 \end{bmatrix} \quad \mathbf{b} = \begin{bmatrix} -0.6 \\ -0.3 \\ 0.4 \\ 1.1 \\ 1.5 \\ 1.5 \end{bmatrix} \quad \mathbf{x} = \begin{bmatrix} p_0 \\ p_1 \\ p_3 \end{bmatrix}$$

We find the least-squares solution using the pseudo inverse as

$$\hat{\mathbf{x}} = \begin{bmatrix} p_0 \\ p_1 \\ p_3 \end{bmatrix} = (\mathbf{A}^T\mathbf{A})^{-1}\mathbf{A}^T\mathbf{b}$$

Preparing the pseudo inverse, we have

$$\mathbf{A}^T\mathbf{A} = \begin{bmatrix} 1 & 1 & 1 & 1 & 1 & 1 \\ -10 & -6 & 0 & 6 & 12 & 18 \\ -10^3 & -6^3 & 0^3 & 6^3 & 12^3 & 18^3 \end{bmatrix} \begin{bmatrix} 1 & -10 & -10^3 \\ 1 & -6 & -6^3 \\ 1 & 0 & 0^3 \\ 1 & 6 & 6^3 \\ 1 & 12 & 12^3 \\ 1 & 18 & 18^3 \end{bmatrix}$$

The numbers are quite large, and we leave the computation to MATLAB. The code is shown below, also including the plotting commands, and the resulting plot in Fig. 4.1. The least-squares solution from $\hat{\mathbf{x}} = (\mathbf{A}^T\mathbf{A})^{-1}\mathbf{A}^T\mathbf{b}$ is

$$\hat{\mathbf{x}} = \begin{bmatrix} \hat{p}_0 \\ \hat{p}_1 \\ \hat{p}_3 \end{bmatrix} = \begin{bmatrix} 0.40 \\ 0.12 \\ -0.00018 \end{bmatrix}$$

```
%
% Example. Least squares fit of the lift coefficient.
alpha = [-10; -6; 0; 6; 12; 18];
column1 = ones(6,1);
A = [column1 alpha alpha.^3];
b = [-0.6; -0.3; 0.4; 1.1; 1.5; 1.5];
x = inv(A'*A)*A'*b;      % Equivalently, can use built-in function. x = pinv(A)*b;
alphaplot = -10:.1:18;
fofz = x(1) + x(2)*alphaplot + x(3)*alphaplot.^3;
% Plot data and plot best fit
figure(5)
plot(alpha,b,'o'); hold on;
plot(alphaplot,fofz);
set(gca,'FontSize',14);
title('Lift coefficient, C_L','FontSize',18);
xlabel('angle of attack (degrees)','fontsize',18); ylabel('C_L','fontsize',18);
```

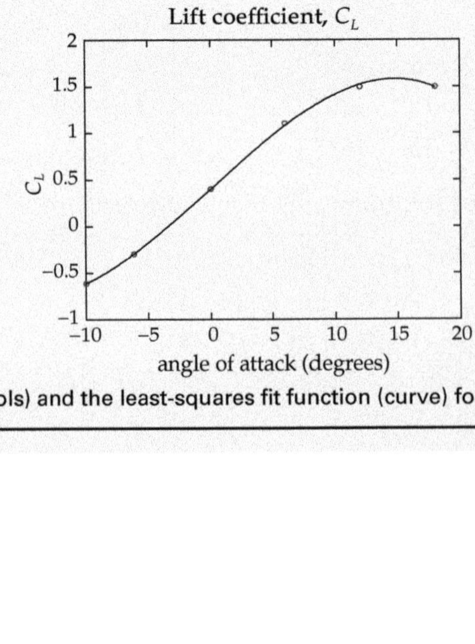

FIGURE 4.1 Data (symbols) and the least-squares fit function (curve) for the lift coefficient, $C_L = z$.

Example 4.15

Given US census data from the years 1900 to 2000, let us fit the population growth to an exponential curve

$$y = p_0 e^{\beta(t-1900)}$$

To make an exponential function linear in the parameters, it is useful to work with the log scale, by noting that

$$z = \ln y = \ln p_0 + \beta(t - 1900) = \alpha + \beta(t - 1900) = \alpha + \beta u$$

where $\alpha = \ln p_0$ and $u = t - 1900$. Working with this form, (a) find the best values of α and β, (b) plot the data with the best fit function, (c) look at the residuals in terms of $\ln y$, (d) estimate the population for October 31, 2017, and predict for the year 2060, and (e) show the MATLAB script for solving this problem. Use the data tabulated below.

year	y (millions)	$u = t - 1900$	$z = \ln y$
1900	77.0	0	4.34
1920	106	20	4.66
1940	132	40	4.88
1960	179	60	5.19
1980	226	80	5.42
2000	281	100	5.63

Solution

a) Based on $z = \alpha + \beta u$ construct $\mathbf{Ax} = \mathbf{b}$ with

$$\mathbf{A} = \begin{bmatrix} 1 & 0 \\ 1 & 20 \\ 1 & 40 \\ 1 & 60 \\ 1 & 80 \\ 1 & 100 \end{bmatrix} \qquad \mathbf{b} = \begin{bmatrix} 4.34 \\ 4.66 \\ 4.88 \\ 5.19 \\ 5.42 \\ 5.63 \end{bmatrix} \qquad \mathbf{x} = \begin{bmatrix} \alpha \\ \beta \end{bmatrix}$$

Find the least-squares solution using the pseudo inverse:

$$\hat{\mathbf{x}} = \begin{bmatrix} \hat{\alpha} \\ \hat{\beta} \end{bmatrix} = (\mathbf{A}^T\mathbf{A})^{-1}\mathbf{A}^T\mathbf{b}$$

Preparing the pseudo inverse, we have

$$\mathbf{A}^T\mathbf{A} = \begin{bmatrix} 1 & 1 & 1 & 1 & 1 & 1 \\ 0 & 20 & 40 & 60 & 80 & 100 \end{bmatrix} \begin{bmatrix} 1 & 0 \\ 1 & 20 \\ 1 & 40 \\ 1 & 60 \\ 1 & 80 \\ 1 & 100 \end{bmatrix} = \begin{bmatrix} 6 & 300 \\ 300 & 22000 \end{bmatrix}$$

and

$$\mathbf{A}^T\mathbf{b} = \begin{bmatrix} 1 & 1 & 1 & 1 & 1 & 1 \\ 0 & 20 & 40 & 60 & 80 & 100 \end{bmatrix} \begin{bmatrix} 4.34 \\ 4.66 \\ 4.88 \\ 5.19 \\ 5.42 \\ 5.63 \end{bmatrix} = \begin{bmatrix} 30.1 \\ 1596 \end{bmatrix}$$

Using the inverse of a 2×2 matrix,

$$(\mathbf{A}^T\mathbf{A})^{-1} = \frac{\begin{bmatrix} 22000 & -300 \\ -300 & 6 \end{bmatrix}}{132000 - 90000} = \begin{bmatrix} 0.5238 & -0.00714 \\ -0.00714 & 0.000143 \end{bmatrix}$$

and so

$$\hat{\mathbf{x}} = \begin{bmatrix} \hat{\alpha} \\ \hat{\beta} \end{bmatrix} = (\mathbf{A}^T\mathbf{A})^{-1}\mathbf{A}^T\mathbf{b} = \begin{bmatrix} 0.5238 & -0.00714 \\ -0.00714 & 0.000143 \end{bmatrix} \begin{bmatrix} 30.1 \\ 1596 \end{bmatrix} = \begin{bmatrix} 4.37 \\ 0.0129 \end{bmatrix}$$

The exponential least-squares fit to the population data was performed using

$$\ln y = \hat{\alpha} + \hat{\beta}(t - 1900)$$

Rewriting in its exponential form, the millions of people approximately follow the model

$$y = e^{\hat{\alpha}+\hat{\beta}(t-1900)} = p_0 e^{\hat{\beta}(t-1900)} = 79.0e^{0.0129(t-1900)}$$

b) A plot of the data, that is $\ln y_i$ versus $(t_i - 1900)$, for $i = 1, \ldots, 6$, is shown in Fig. 4.2 with a plot of the curve of the best fit. Figure 4.3 shows data in y_i versus $(t_i - 1900)$ along with the exponential equation with the parameter values from the least-squares fit. Also included in the plot are data that were not used in the estimation process.

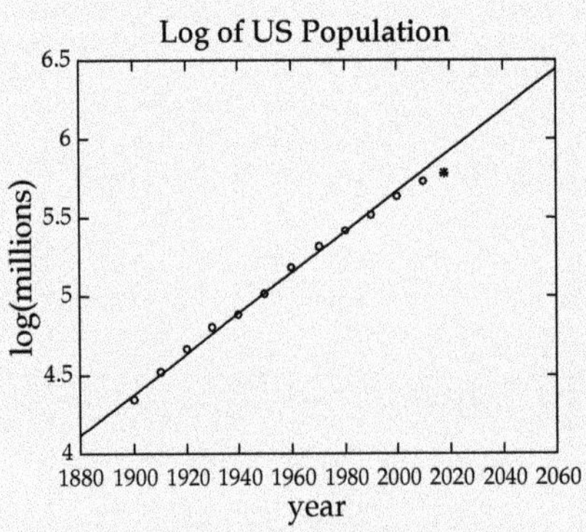

FIGURE 4.2 Log y versus year.

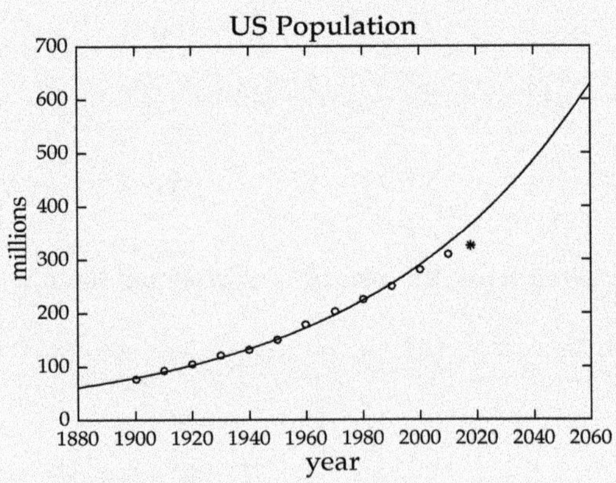

FIGURE 4.3 y versus year.

c) Prof. James V. Beck, who authored many papers and books on parameter estimation, once said, "The residuals tell a story." Here we look at the residuals, defined as

$$r_i = \hat{\alpha} + \hat{\beta}\,(t_i - 1900) - \ln y_i, \qquad i = 1, \ldots, 6$$

In matrix form we have

$$\mathbf{r} = \mathbf{A}\hat{\mathbf{x}} - \mathbf{b}$$

or

$$\mathbf{r} = \begin{bmatrix} 1 & 0 \\ 1 & 20 \\ 1 & 40 \\ 1 & 60 \\ 1 & 80 \\ 1 & 100 \end{bmatrix} \begin{bmatrix} 4.37 \\ 0.0129 \end{bmatrix} - \begin{bmatrix} 4.34 \\ 4.66 \\ 4.88 \\ 5.19 \\ 5.42 \\ 5.63 \end{bmatrix} = \begin{bmatrix} 0.0344 \\ -0.0274 \\ 0.0109 \\ -0.0409 \\ -0.0126 \\ 0.0357 \end{bmatrix}$$

The residuals are plotted in Fig. 4.4 from the data that were used to do the least-squares fit. The residuals "look" random and small, which suggests that the model is reasonably good and the data is good. A "nonrandom" set of residuals would show a trend, which would suggest that something is missing in a model. In our example, with only six data, it may be difficult to realize a trend. Visually observing Figs. 4.2 and 4.3, with added data points, leaves a suspicion that there may be a flattening of the population data relative to our exponential model.

d) The model approximates a population of approximately 370 million on October 31, 2017, which exceeds the 325 million estimated by other means for that day (according to Wikipedia, which perhaps missed all of the trick-or-treaters). The exponential model predicts a population of 629 million people in the US by the year 2060. If an unmodeled "flattening" effect occurs, we may not reach the 629 million predicted by the model by 2060.

e) Sample MATLAB script for the calculations and plots displayed here is displayed below. In the code, the leftward, downward arrow indicates a line continuation.

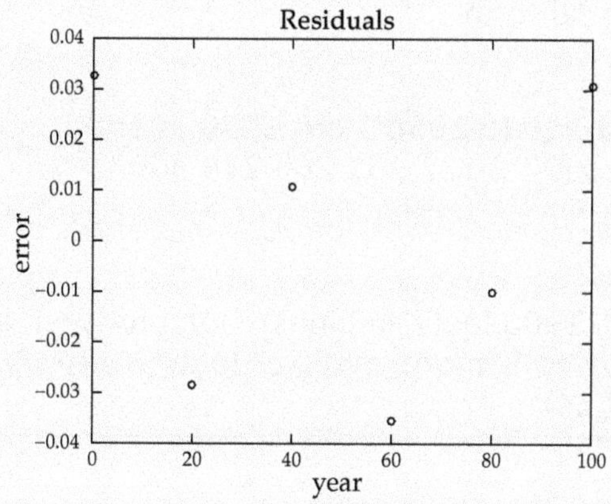

FIGURE 4.4 The residuals are plotted versus year. The error is in ln *y*.

```
% Least squares fit of US population data.
%
% fit y = p0 e^(beta*(t–100)), or ln(y) = ln p0 + beta*(t–100)
% ln(y) = alpha + beta*(t–100)

% data

year = [1900 1920 1940 1960 1980 2000]' – 1900;     % year after 1900
y_millions = [77.0 106 132 179 226 281]';            % population
ln_y = log(y_millions);

% build A x = b.

A = [ones(6,1) year]; % ones(6,1) produces a row of six values of 1. This ↙
column is concatenated with year, which is a column.
b = ln_y;

% LS solution by pseudo inverse

xhat = inv(A'*A)*A'*b;

% Residuals

err = A*xhat - b;

% Extropolate. October 31, 2017, t = 2017+10/12;
t_oct2017 = 2017+10/12;
y_oct2017est = exp(xhat(1))*exp(xhat(2)*(t_oct2017-1900));
y_oct2017 325.246; %wikipedia

% Plots

t = 1880:1:2060;
foft = exp(xhat(1))*exp(xhat(2)*(t-1900));     %f(t) = foft is the estimated ↙
model in exponential form

% additional data
year_added = [1910 1930 1950 1970 1990 2010]';
y_millions_added = [92.2 122 151 203 249 309]';

figure(1);
plot(t,log(foft));
hold on; plot(year+1900,log(y_millions),'bo'); plot(year_added,log ↙
(y_millions_added),'rol'); plot(t_oct2017,log(y_oct2017),'k'); hold off;
set(gca,'FontSize',14);
```

4.11 EIGENVALUES AND EIGENVECTORS

A problem that is frequently encountered while working problems involving physical systems is the *eigenvalue problem* (or *characteristic-value problem*). It is the problem of finding the values of a scalar parameter λ and the components of the vectors $\mathbf{x}$ that satisfy

$$\mathbf{Ax} = \lambda\mathbf{x} \tag{4.11.1}$$

where $\mathbf{A}$ is a known n^{th}-order matrix. The parameter λ is the *eigenvalue* (or *characteristic value)* and the corresponding vector $\mathbf{x}$ is the *eigenvector* (or *characteristic vector*). Clearly, $\mathbf{x} = \mathbf{0}$ satisfies Eq. 4.11.1 for any value λ; this trivial solution is not of interest in the solution to the eigenvalue problem.

Writing $\mathbf{x} = \mathbf{Ix}$, where $\mathbf{I}$ is the unit matrix, Eq. 4.11.1 can be put in the form

$$\mathbf{Ax} = \lambda\mathbf{Ix} \tag{4.11.2}$$

or

$$(\mathbf{A} - \lambda\mathbf{I})\mathbf{x} = \mathbf{0} \tag{4.11.3}$$

This equation can be thought of as $\mathbf{Bx} = \mathbf{0}$, where $(\mathbf{A} - \lambda\mathbf{I}) = \mathbf{B}$. In order for this to have a nontrivial solution, we must demand that $|\mathbf{B}| = 0$; that is,

$$|\mathbf{A} - \lambda\mathbf{I}| = 0 \tag{4.11.4}$$

This determinant can be displayed as

$$\begin{vmatrix} a_{11} - \lambda & a_{12} & \cdots & a_{1n} \\ a_{21} & a_{22} - \lambda & \cdots & a_{2n} \\ \vdots & \vdots & & \vdots \\ a_{n1} & a_{n2} & \cdots & a_{nn} - \lambda \end{vmatrix} = 0 \tag{4.11.5}$$

The expansion of this determinant would include one term that would be the product of the diagonal elements; this term, when expanded, would include the term $(-\lambda)^n$. Other terms containing $(-\lambda)$ to lower powers would also be included. In fact, the expansion of the determinant is a polynomial in $(-\lambda)$ and can be written as

$$|\mathbf{A} - \lambda\mathbf{I}| = k_0 + k_1(-\lambda) + k_2(-\lambda)^2 + \cdots + k_{n-1}(-\lambda)^{n-1} + (-\lambda)^n \tag{4.11.6}$$

where the coefficients $k_0, k_1, \ldots, k_{n-1}$ are determined from the known a_{ij}. Equation 4.11.6 is called the *characteristic equation* (or *secular equation*).

We know that an nth-order equation has at most n roots. The roots of Eq. 4.11.6 are the eigenvalues and will be denoted $\lambda_1, \lambda_2, \ldots, \lambda_n$, or simply λ_i; they may be real or complex. When one of the eigenvalues is substituted back into Eq. 4.11.3, we have

$$(\mathbf{A} - \lambda_i\mathbf{I})\mathbf{x} = 0 \tag{4.11.7}$$

from which the corresponding eigenvector $\mathbf{x}_i$ may be determined. If an eigenvalue is complex, the components of the eigenvector may also be complex. Note that the eigenvector $\mathbf{x}_i$ may be multiplied by any constant and remain a solution to Eq. 4.11.7. Hence, we can make the length of a real eigenvector unity by dividing by the magnitude of $\mathbf{x}_i$; this is often done so that the real eigenvectors are also unit vectors.

The eigenvalues are determined from the expansion of the determinant in Eq. 4.11.4. For a given square matrix **A** this determinant can be formulated. Thus, we often refer to the eigenvalues and eigenvectors of the matrix **A**, without reference to the eigenvalue problem associated with Eq. 4.11.1.

Several properties, which will be verified by Examples or Problems, related to the eigenvalue problem will now be presented. Let a square matrix **A** possess eigenvalues $\lambda_1, \lambda_2, \lambda_3, \ldots, \lambda_n$. Then the following are properties that have occasional use in matrix applications:

1. The eigenvalues of a symmetric matrix are real.
2. The eigenvalues of a skew-symmetric matrix are pure imaginary or zero.
3. The eigenvectors of a symmetric matrix are orthogonal.
4. The transpose $\mathbf{A}^T$ has eigenvalues $\lambda_1, \lambda_2, \ldots, \lambda_n$.
5. The inverse $\mathbf{A}^{-1}$ has eigenvalues $1/\lambda_1, 1/\lambda_2, \ldots, 1/\lambda_n$.
6. The inverse exists if and only if none of the eigenvalues of the matrix **A** is zero.
7. The matrix $\mathbf{A}^m$ has eigenvalues $\lambda_1^m, \lambda_2^m, \ldots, \lambda_n^m$.
8. If matrix **A** is either upper or lower triangular, then $\lambda_1 = a_{11}, \lambda_2 = a_{22}, \ldots, \lambda_n = a_{nn}$.
9. The matrix $k\mathbf{A}$ has eigenvalues $k\lambda_1, k\lambda_2, \ldots, k\lambda_n$.

It should be noted that in any problem where the order of the matrix is greater than 4, the eigenvalues and eigenvectors would usually be found with a numerical method not presented in this book. There are a number of such methods; a book on numerical methods should be consulted if such a problem is encountered.

Example 4.16

Find the eigenvalues and corresponding unit eigenvectors for the matrix $\begin{bmatrix} 4 & 2 \\ -1 & 0 \end{bmatrix}$.

Solution

The characteristic equation is found from

$$\begin{vmatrix} 4-\lambda & 2 \\ -1 & -\lambda \end{vmatrix} = 0$$

It is

$$-\lambda(4-\lambda) + 2 = 0 \quad \text{or} \quad \lambda^2 - 4\lambda + 2 = 0$$

The two roots of this equation, the eigenvalues, are

$$\lambda_1 = 3.414, \qquad \lambda_2 = 0.586$$

We find the eigenvector corresponding to λ_1 from the equations

$$\begin{aligned} (4-\lambda_1)x_1 + 2x_2 &= 0 \\ -x_1 - \lambda_1 x_2 &= 0 \end{aligned} \quad \text{or} \quad \begin{aligned} 0.586x_1 + 2x_2 &= 0 \\ -x_1 - 3.414x_2 &= 0 \end{aligned}$$

If we wish to form unit eigenvectors, we require that

$$x_1^2 + x_2^2 = 1$$

These equations are solved to obtain

$$x_1 = 0.9597, \qquad x_2 = -0.2811$$

The eigenvector $\mathbf{x}_1$ is displayed as

$$\mathbf{x}_1 = \begin{bmatrix} 0.9597 \\ -0.2811 \end{bmatrix}$$

The unit eigenvector corresponding to λ_2 is found from the equations

$$\begin{aligned} (4 - \lambda_2)x_1 + \ 2x_2 &= 0 \\ -x_1 - \lambda_1 x_2 &= 0 \\ x_1^2 + \ x_2^2 &= 1 \end{aligned} \qquad \text{or} \qquad \begin{aligned} 3.414x_1 + \quad 2x_2 &= 0 \\ -x_1 - 0.586x_2 &= 0 \\ x_1^2 + \quad x_2^2 &= 1 \end{aligned}$$

The solution to the equations above is

$$x_1 = 0.5056, \qquad x_2 = -0.8628$$

The eigenvector $\mathbf{x}_2$ is

$$\mathbf{x}_2 = \begin{bmatrix} 0.5056 \\ -0.8628 \end{bmatrix}$$

Note that the eigenvalues of the given matrix of this example are both real, even though the matrix is not symmetric. If the numbers 4 and 2 are interchanged in the matrix, the eigenvalues would be complex.

Example 4.17

Find the eigenvalues and eigenvectors of the symmetric matrix

$$\begin{bmatrix} 4 & 2 & 0 \\ 2 & -6 & 0 \\ 0 & 0 & 8 \end{bmatrix}$$

Because the matrix is symmetric, we know that the eigenvalues will be real.

Solution

The characteristic equation is found from the determinant

$$\begin{vmatrix} 4-\lambda & 2 & 0 \\ 2 & -6-\lambda & 0 \\ 0 & 0 & 8-\lambda \end{vmatrix} = 0$$

It is expanded to give the equation

$$(8-\lambda)[(-6-\lambda)(4-\lambda)-4]=0 \quad \text{or} \quad (8-\lambda)(\lambda^2+2\lambda-28)=0$$

The eigenvalues are found to be

$$\lambda_1=8, \qquad \lambda_2=-6.385, \quad \lambda_3=4.385$$

To find an eigenvector associated with $\lambda_1=8$, we must solve the equations

$$\begin{aligned}(4-\lambda_1)x_1+\qquad 2x_2&=0\\ 2x_1-(6+\lambda_1)x_2&=0\\ (8-\lambda_1)x_3&=0\end{aligned} \qquad \text{or} \qquad \begin{aligned}-4x_1+\ 2x_2&=0\\ 2x_1-14x_2&=0\\ 0\cdot x_3&=0\end{aligned}$$

The solution to these equations is, if we arbitrarily choose $x_3=1$,

$$x_1=0, \quad x_2=0, \quad x_3=1, \quad \text{or} \quad \mathbf{x}_1=\begin{bmatrix}0\\0\\1\end{bmatrix}$$

The eigenvector corresponding to λ_2 is found from

$$\begin{aligned}(4-\lambda_2)x_1+\qquad 2x_2&=0\\ 2x_1-(6+\lambda_2)x_2&=0\\ (8-\lambda_2)x_3&=0\end{aligned} \qquad \text{or} \qquad \begin{aligned}10.385x_1+\qquad 2x_2&=0\\ 2x_1+0.385x_2&=0\\ 14.385x_3&=0\end{aligned}$$

We see from the above that $x_3=0$. If we arbitrarily set $x_1=1$, we have

$$x_1=1, \quad x_2=-5.195, \quad x_3=0, \quad \text{or} \quad \mathbf{x}_2=\begin{bmatrix}1\\-5.195\\0\end{bmatrix}$$

The eigenvector corresponding to λ_3 is found from

$$\begin{aligned}-0.385x_1+\qquad 2x_2&=0\\ 2x_1-10.385x_2&=0\\ 3.615x_3&=0\end{aligned}$$

to be

$$x_1=1, \quad x_2=0.1926, \quad x_3=0, \quad \text{or} \quad \mathbf{x}=\begin{bmatrix}1\\0.1926\\0\end{bmatrix}$$

The eigenvectors of this example should be orthogonal, since the given matrix is symmetric. This is checked by considering the scalar product of each pair of vectors. For vectors $\mathbf{x}_1$ and $\mathbf{x}_2$ we have

$$0\times 1+0\times(-5.195)+1\times 0=0$$

For vectors $\mathbf{x}_1$ and $\mathbf{x}_3$,

$$0 \times 1 + 0 \times 0.1926 + 1 \times 0 = 0$$

Finally, for vectors $\mathbf{x}_2$ and $\mathbf{x}_3$ the scalar product is

$$1 \times 1 + (-5.195) \times 0.1926 + 0 \times 0 = 0$$

All the scalar products are zero, verifying that all the vectors are perpendicular to each other, that is, orthogonal.

Note that we could have arbitrarily made the eigenvectors unit vectors by requiring that $x_1^2 + x_2^2 + x_3^2 = 1$ for each eigenvector. This could easily be done by dividing each component of an eigenvector by the magnitude of that eigenvector, which is $\sqrt{x_1^2 + x_2^2 + x_3^2}$.

The eigensolution can also be obtained by using the following sequence of MATLAB commands:

```
>> A = [4 2 0; 2 -6 0; 0 0 8];
>> [u lam] = eig(A)

u =

   -0.1891   -0.9820         0
    0.9820   -0.1891         0
         0         0    1.0000

lam =

   -6.3852         0         0
         0    4.3852         0
         0         0    8.0000
```

The columns of output variable `u` are the eigenvectors (normalized as unit vectors) and the diagonal elements of `lam` are the associated eigenvalues. The orthogonality of the eigenvectors of the symmetric matrix can be confirmed by typing

```
>> transpose(u)*u

ans =

    1.0000         0         0
         0    1.0000         0
         0         0    1.0000
```

The result is diagonal, because the off-diagonal quantities, from matrix multiplication as illustrated in Eq. 4.5.6, result from inner products between distinct columns of `u`, which means distinct eigenvectors. The diagonal elements all have values 1.0000 because they result from inner products between eigenvectors with themselves, and the eigenvectors produced by this MATLAB command are unit vectors.

Example 4.18

Show that an eigenvalue problem results when solving for the displacements of the two masses shown in Fig. 4.5.

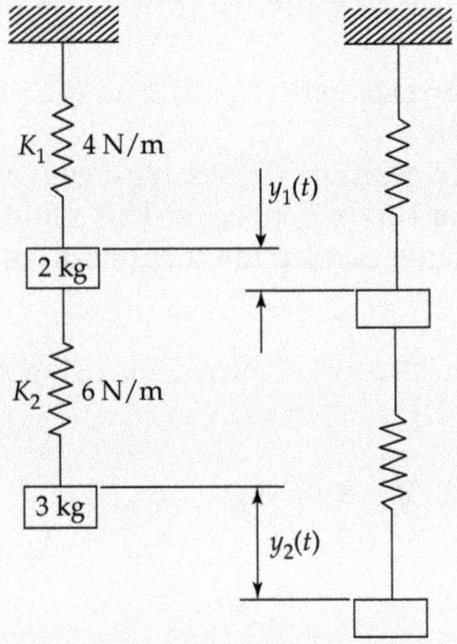

FIGURE 4.5

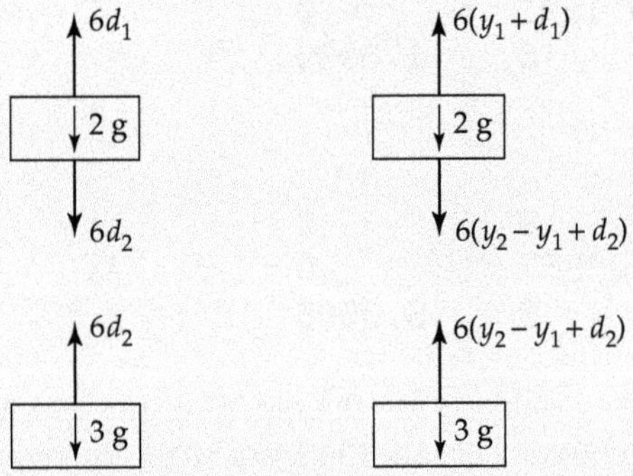

FIGURE 4.6

Solution

We isolate the two masses and show all forces in Fig. 4.6 acting on each. The distances d_1 and d_2 are the amounts the springs are stretched while in static equilibrium. Using Newton's second law we may write:

Static equilibrium	*In motion*
$0 = 6d_2 - 4d_1 + 2g$	$2\ddot{y}_1 = 6(y_2 - y_1 + d_2) + 2_g - 4(y_1 + d_1)$
$0 = 3_g - 6d_2$	$3\ddot{y}_2 = 3_g - 6(y_2 - y_1 + d_2)$

These equations may be simplified to

$$\ddot{y}_1 = -5y_1 + 3y_2$$
$$\ddot{y}_2 = 2y_1 - 2y_2$$

These two equations can be written as the single matrix equation

$$\ddot{\mathbf{y}} = \mathbf{A}\mathbf{y}$$

where

$$\mathbf{y} = \begin{bmatrix} y_1 \\ y_2 \end{bmatrix}, \quad \mathbf{A} = \begin{bmatrix} -5 & 3 \\ 2 & -2 \end{bmatrix}$$

We note that the coefficients of the two independent variables are all constants; hence, as is usual, we assume a solution in the form (see Section 1.6)

$$\mathbf{y} = \mathbf{x}e^{mt}$$

where $\mathbf{x}$ is a constant vector to be determined, and m is a scalar to be determined. Our differential equation is then

$$\mathbf{x}m^2e^{mt} = \mathbf{A}\mathbf{x}e^{mt}$$

or

$$\mathbf{A}\mathbf{x} = \lambda\mathbf{x}$$

where the parameter $\lambda = m^2$. *This is an eigenvalue problem.* The problem would be completed if we found the eigenvalues λ_i and corresponding eigenvectors $\mathbf{x}_i$. The solutions $y_1(t)$ and $y_2(t)$ would then be determined, as in the next example.

Example 4.19

a) Solve the eigenvalue problem, obtained in Example 4.18. b) Find the solutions $y_1(t)$ and $y_2(t)$.

Solution

a) The eigenvalue problem is that of finding the λ_i and the corresponding $\mathbf{x}_i$ of

$$\mathbf{A}\mathbf{x} = \lambda\mathbf{x}$$

The λ_i are found from (see Eq. 4.11.5)

$$\begin{vmatrix} -5-\lambda & 3 \\ 2 & -2-\lambda \end{vmatrix} = 0$$

where the matrix $\mathbf{A}$ is obtained from Example 4.18. The determinant is expanded as

$$\lambda^2 + 7\lambda + 4 = 0$$

The roots to this equation are

$$\lambda_1 = \frac{-7+\sqrt{33}}{2} = -0.6277, \qquad \lambda_2 = \frac{-7-\sqrt{33}}{2} = -6.372$$

The corresponding eigenvectors are found by substituting λ_i into Eq. 4.11.7. We have

$$\text{for } \mathbf{x}_1: \quad \begin{aligned} -4.373x_1 + \quad 3x_2 &= 0 \\ 2x_1 + 4.372x_2 &= 0 \end{aligned}$$

$$\text{for } \mathbf{x}_2: \quad \begin{aligned} 1.372x_1 + \quad 3x_2 &= 0 \\ 2x_1 + 4.372x_2 &= 0 \end{aligned}$$

The two eigenvectors are then

$$\mathbf{x}_1 = \begin{bmatrix} 1 \\ 1.46 \end{bmatrix}, \qquad \mathbf{x}_2 = \begin{bmatrix} 1 \\ -0.457 \end{bmatrix}$$

where the first component was arbitrarily chosen to be unity.

The eigensolution can also be obtained by using the following sequence of MATLAB commands:

```
>> A = [-5 3; 2 -2];
>> [u lam] = eig(A);

u =

   -0.9094   -0.5658
    0.4160   -0.8246

lam =

   -6.3723         0
         0   -0.6277
```

The columns of output variable `u` are the eigenvectors (normalized as unit vectors), and the diagonal elements of `lam` are the associated eigenvalues.

b) The solutions $y_1(t)$ and $y_2(t)$ will now be determined. The constant m is related to the eigenvalues by $m^2 = \lambda$ (see Example 4.18). Thus,

$$m_1^2 = -0.6277 \quad \text{and} \quad m_2^2 = -6.372$$

These give

$$m_1 = \pm 0.7923i, \qquad m_2 = \pm 2.524i$$

We use both positive and negative roots and write the solution as

$$\mathbf{y}(t) = \mathbf{x}_1(c_1e^{0.7923it} + d_1e^{-0.7923it}) + \mathbf{x}_2(c_2e^{2.524it} + d_2e^{-2.524it})$$

where we have superimposed all possible solutions introducing the arbitrary constants c_1, d_1, c_2, and d_2, to obtain the most general solution. The arbitrary constants are then calculated from initial conditions.

The components of the solution vector can be written as, using Eq. 1.6.9,

$$y_1(t) = a_1\cos 0.7923t + b_1\sin 0.7923t + a_2\cos 2.524t + b_2\sin 2.524t$$
$$y_2(t) = 1.46\,[a_1\cos 0.7923t + b_1\sin 0.7923t] - 0.457\,[a_2\cos 2.524t + b_2\sin 2.524t]$$

Note that if we had made the eigenvectors of unit length, the arbitrary constants would simply change accordingly for a particular set of initial conditions.

4.12 EIGENVALUE PROBLEMS IN ENGINEERING

Among the previous examples, we saw that eigenvalue problems arise in the studies of mechanical vibrations. Eigenvalue problems show up in many other applied problems. In this section we will look at some other examples in engineering.

4.12.1 Moments of Inertia

The mass moment of inertia, or second moment of mass, is defined relative to a point and a set of coordinate axes. The illustration in Fig. 4.7 shows a body of mass m in a Cartesian coordinate system centered at point O. We can define the mass moment of inertia about point O through the x, y, and z axes as

$$
\begin{aligned}
I_{xx} &= \int_{Body} r_x^2 dm = \int_{Body} \rho(y^2 + z^2)dV \\
I_{yy} &= \int_{Body} r_y^2 dm = \int_{Body} \rho(x^2 + z^2)dV \qquad (4.12.1) \\
I_{zz} &= \int_{Body} r_z^2 dm = \int_{Body} \rho(x^2 + y^2)dV
\end{aligned}
$$

where $dm = \rho dV$ is a differential mass element, with ρ as the mass density and dV as a differential volume element, and r_x, r_y, and r_z are the distances from the mass element to the x, y, and z axes. I_{xx}, I_{yy}, and I_{zz}, quantify the "inertial resistance" to changes in rotation rate about the x, y, and z axes, analogous to mass quantifying the inertial resistance to rectilinear acceleration.

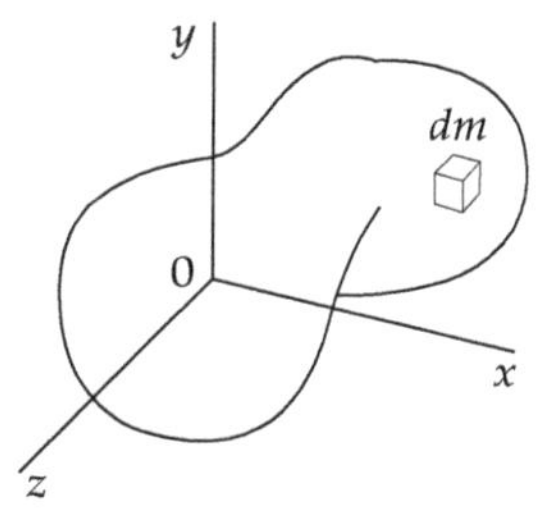

FIGURE 4.7 A rigid body including differential mass element *dm*.

There are also products of inertia:

$$
\begin{aligned}
I_{xy} &= I_{yx} = -\int_{Body} \rho xy dV \\
I_{yz} &= I_{zy} = -\int_{Body} \rho yz dV \qquad (4.12.2) \\
I_{xz} &= I_{zx} = -\int_{Body} \rho xz dV
\end{aligned}
$$

If these definitions are applied with $\rho = 1$, dimensionless, then the I_{ij} are called second moments of area (or "area moments of inertia" and "area products of inertia"), and have units of length to the fourth power.

One can form the moment of inertia matrix as

$$\mathbf{I}_0 = \begin{bmatrix} I_{xx} & I_{xy} & I_{xz} \\ I_{yx} & I_{yy} & I_{yz} \\ I_{zx} & I_{zy} & I_{zz} \end{bmatrix} \tag{4.12.3}$$

Despite the notation, $\mathbf{I}_0$ is not to be confused with the identity matrix, $\mathbf{I}$. Note that $\mathbf{I}_0$ is symmetric.

If we choose a rotated coordinate system, say (X, Y, Z), then the computations of I_{XX}, I_{YY}, I_{ZZ}, I_{XY}, I_{YZ}, and I_{ZX} result in different values. It turns out that the orientation of the coordinate system (X, Y, Z) can be specifically chosen such that $I_{XY} = I_{YX} = I_{YZ} = I_{ZY} = I_{XZ} = I_{ZX} = 0$. This defines the principal coordinate system, and the *principal axes*.

The orientations of the principal axes are of interest because they simplify the governing dynamic equations. They are of physical interest because these are the axes about which a body can spin steadily (although one of the axes is unstable).

Given a calculation of $\mathbf{I}_0$ in the coordinate system (x, y, z), the orientation of the principal coordinate system (X, Y, Z), and the orientation of the principal axes, as well as the associated principal values I_{XX}, I_{YY}, I_{ZZ}, can be determined from an eigenvalue problem. Indeed, if the eigenvalue problem is written as

$$\mathbf{I}_0\mathbf{u} = \lambda\mathbf{u} \tag{4.12.4}$$

then the eigenvalues λ of $\mathbf{I}_0$ indicate the *principal moments of inertia*, and the eigenvectors $\mathbf{u}$ indicate the principal axes.

Example 4.20

The thin slab shown in Fig. 4.8. is shaped as a right triangle with leg lengths L and $2L$, and thickness t. The slab has mass $m = \rho V = \rho t A = \rho t L^2$. (a) Find $\mathbf{I}_0$ based on the right-handed x, y, and z axes shown. (b) Find the principal axes and moments of inertia of the slab.

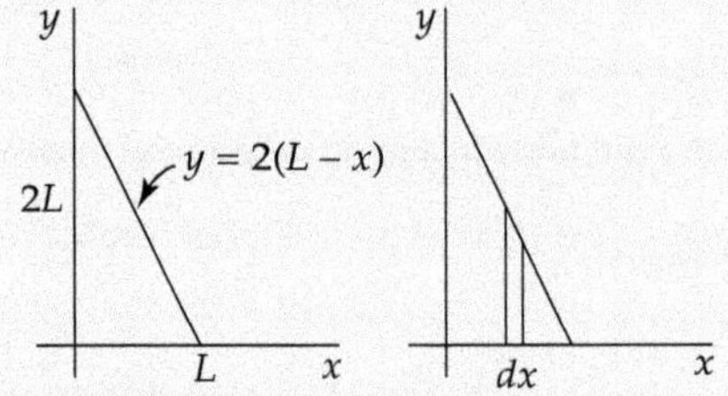

FIGURE 4.8 Left: a triangular slab. Right: an integration scheme, integrating in y first.

Solution

a) To find the matrix of moments of inertia, apply the definitions given above. We see that $\rho t = m/L^2$, and we need $dm = \rho t dA = (m/L^2)dxdy$. We set up an integration scheme where we integrate in y first, and then x, as shown in Fig. 4.8. Then

$$I_{xx} = \int_{Body} \rho 9(y^2 + z^2)dV \cong \int_{x=0}^{L}\int_{y=0}^{2L-2x} \frac{m}{L^2} y^2 dydx = \frac{2}{3}mL^2$$

where the approximation came in letting $z \cong 0$ for a thin slab. Similarly, it can be verified that

$$I_{yy} = \frac{1}{6}mL^2, \quad I_{zz} = \frac{5}{6}mL^2, \quad I_{xy} = I_{yx} = -\frac{1}{6}mL^2, \quad I_{xz} = I_{yz} = 0$$

Thus we have

$$\mathbf{I}_0 = \frac{ml^2}{6}\begin{bmatrix} 4 & -1 & 0 \\ -1 & 1 & 0 \\ 0 & 0 & 5 \end{bmatrix}$$

b) For computation of the principal moments of inertia and principal axes, we let $\mathbf{I}_0 = \frac{mL^2}{6}\mathbf{J}_0$, such that

$$\mathbf{J}_0 = \begin{bmatrix} 4 & -1 & 0 \\ -1 & 1 & 0 \\ 0 & 0 & 5 \end{bmatrix}$$

We aim to solve the eigenvalue problem, $\mathbf{I}_0\mathbf{u} = \lambda\mathbf{u}$ or $\frac{mL^2}{6}\mathbf{J}_0\mathbf{u} = \lambda\mathbf{u}$, which can be rewritten as

$$\mathbf{J}_0\mathbf{u} = \frac{6}{mL^2}\lambda\mathbf{u}$$

Thus the eigenvalue problem

$$\mathbf{J}_0\mathbf{u} = \alpha\mathbf{u}$$

has eigenvalues α such that our desired principal moments of inertia are $\lambda = \frac{mL^2}{6}\alpha$.

We obtain the eigenvalues from

$$|\mathbf{J}_0 - \alpha I| = \begin{vmatrix} 4-\alpha & -1 & 0 \\ -1 & 1-\alpha & 0 \\ 0 & 0 & 5-\alpha \end{vmatrix} = (5-\alpha)\begin{vmatrix} 4-\alpha & -1 \\ -1 & 1-\alpha \end{vmatrix}$$
$$= (5-\alpha)(\alpha^2 - 5\alpha + 3) = 0$$

The roots provide the eigenvalues,

$$\alpha_{1,2} = \frac{5}{2} \mp \frac{\sqrt{13}}{2} \quad \alpha_3 = 5$$

Therefore the principal moments of inertia, from $\lambda = \frac{mL^2}{6}\alpha$, are $\lambda_1 \cong 0.697\frac{mL^2}{6}$, $\lambda_2 \cong 4.303\frac{mL^2}{6}$, and $\lambda_3 = \frac{5mL^2}{6}$.

We obtain the eigenvectors from $(\mathbf{J}_0 - \alpha\mathbf{I})\,\mathbf{u} = 0$. Thus we solve

$$\begin{bmatrix} 4-\alpha_i & -1 & 0 \\ -1 & 1-\alpha_i & 0 \\ 0 & 0 & 5-\alpha_i \end{bmatrix} \begin{bmatrix} u_1 \\ u_2 \\ u_3 \end{bmatrix}^i = \begin{bmatrix} 0 \\ 0 \\ 0 \end{bmatrix}$$

for $i = 1, 2, 3$. The three separate equations are

$$\begin{aligned} (4-\alpha_i)\,u_1 - u_2 + 0 &= 0 \\ -u_1 + (1-\alpha_i)\,u_2 + 0 &= 0 \\ (5-\alpha_i)\,u_3 &= 0 \end{aligned}$$

For the case of α_1 the third equation indicates that $u_3 = 0$. Either of the first two equations produce a relationship

$$u_1 = \left(-\frac{3}{2} + \frac{\sqrt{13}}{2}\right) u_2$$

or, arbitrarily sized, an eigenvector

$$\mathbf{u}^{(1)} = \begin{bmatrix} -\dfrac{3}{2} + \dfrac{\sqrt{13}}{2} \\ 1 \\ 0 \end{bmatrix} \cong \begin{bmatrix} 0.3028 \\ 1 \\ 0 \end{bmatrix}$$

For the case of α_2 the third equation indicates that $u_3 = 0$. Either of the first two equations produce a relationship

$$u_1 = \left(-\frac{3}{2} - \frac{\sqrt{13}}{2}\right) u_2$$

or, arbitrarily sized, an eigenvector

$$\mathbf{u}^{(2)} = \begin{bmatrix} -\dfrac{3}{2} - \dfrac{\sqrt{13}}{2} \\ 1 \\ 0 \end{bmatrix} \cong \begin{bmatrix} -3.3028 \\ 1 \\ 0 \end{bmatrix}$$

For the case of $\alpha_3 = 5$, the first and second equations are only satisfied if $u_1 = u_2 = 0$. Any value u_3 works for the third equation. Thus we choose, for the third eigenvector,

$$\mathbf{u}^{(3)} = \begin{bmatrix} 0 \\ 0 \\ 1 \end{bmatrix}$$

Comments: The eigenvectors and principal axes are shown in Fig. 4.9. If you choose a coordinate system about $\hat{x}$, $\hat{y}$, and $\hat{z}$ axes that are lined up with $\mathbf{u}^{(1)}$, $\mathbf{u}^{(2)}$, and $\mathbf{u}^{(3)}$, you can compute $I_{\hat{x}\hat{x}}, I_{\hat{x}\hat{y}}$, etc., and find that you get a diagonal $\hat{\mathbf{I}}_0$!

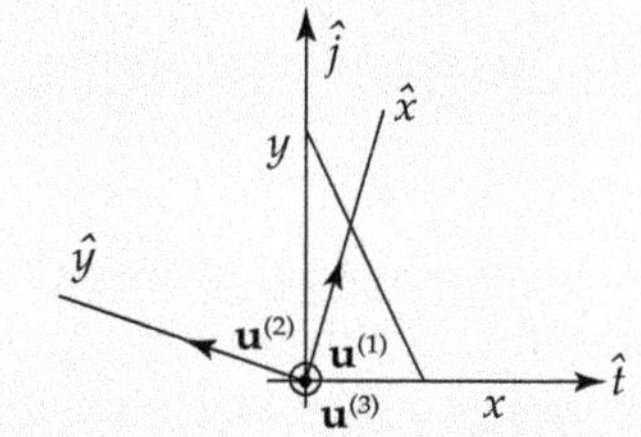

FIGURE 4.9 Plots of the axes of inertia and associated eigenvectors of the thin triangular slab.

4.12.2 Stress

Stress on the face of an element of material is defined as force per unit area of the element face. If the units of force are N (newtons), then stress is in N/m^2 = Pa (pascals). Force is a vector and can have components normal and tangential to the surface. Thus there are stresses that are normal and tangential (shearing) to the element surface. If we consider a square two-dimensional element with x and y axes, there is a normal stress σ_x and a shear stress τ_{xy} on the x-axis face, and a normal stress σ_y and shear stress τ_{yx} on the y-axis face. In order to maintain equilibrium on the square element, we must have $\tau_{xy} = \tau_{yx}$, in the directions drawn in Fig. 4.10. Extending to three dimensions, the stresses on the back faces are equal and opposite to those on the front faces. Two-dimensional stresses are depicted in the figure.

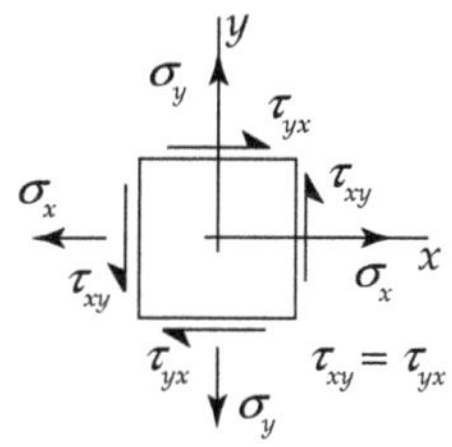

FIGURE 4.10 A square planar element with normal and shearing stresses.

We can put the stresses into a matrix as

$$\boldsymbol{\sigma} = \begin{bmatrix} \sigma_x & \tau_{xy} & \tau_{xz} \\ \tau_{yx} & \sigma_y & \tau_{yz} \\ \tau_{zx} & \tau_{zy} & \sigma_z \end{bmatrix} \tag{4.12.5}$$

where $\tau_{xy} = \tau_{yx}$, $\tau_{xz} = \tau_{zx}$, and $\tau_{yz} = \tau_{zy}$, by equilibrium, meaning $\boldsymbol{\sigma}$ is a symmetric matrix. If the coordinate axes were permutated, for example with the new X axis rotating to the old y axis, a new Y axis becoming the old z axis, and the new Z axis assigned to the old x axis, clearly values in the new $\boldsymbol{\sigma}_{XYZ}$ matrix would be exchanged. For example, in the new matrix, $\sigma_X = \sigma_y$ and $\tau_{YZ} = \tau_{zx}$.

More generally, with an arbitrary rotation of the coordinate axes to a new coordinate system (x', y', z'), the new $\boldsymbol{\sigma}_{x'y'z'}$ matrix would have all new values for its elements. It turns out that we can find a rotated coordinate system (ξ, η, ζ) such that the values of the shear stresses in the $\boldsymbol{\sigma}_{\xi\eta\zeta}$ matrix are zero! Specifically, $\tau_{\xi\eta} = \tau_{\eta\xi} = \tau_{\eta\zeta} = \tau_{\zeta\eta} = \tau_{\zeta\xi} = \tau_{\xi\zeta} = 0$. Then the nonzero normal stresses σ_ξ, σ_η, and σ_ζ are called *principal stresses*, and the orientations of the (ξ, η, ζ) axes are the *principal axes*. Starting with the original (x, y, z) frame and expressing $\boldsymbol{\sigma}$, the principal stresses and principal directions are the eigenvalues and eigenvectors of $\boldsymbol{\sigma}$. That is, the eigenvalue problem is

$$\boldsymbol{\sigma}\mathbf{n} = \lambda\mathbf{n} \tag{4.12.6}$$

Thus the eigenvalues λ_i of matrix $\boldsymbol{\sigma}$ are the principal stresses, and the eigenvectors $\mathbf{n}_i$ are the principle stress orientations, i.e., the orientations in which the stress state has normal stresses only, and no shearing stresses. When normalized, the eigenvectors $\mathbf{n}_i$ are unit normal vectors on the element cube oriented in the principle directions.

Example 4.21

Suppose a stress state at a point in a loaded material is given by

$$\boldsymbol{\sigma} = \begin{bmatrix} \sigma_x & \tau_{xy} & \tau_{xz} \\ \tau_{yx} & \sigma_y & \tau_{yz} \\ \tau_{zx} & \tau_{zy} & \sigma_z \end{bmatrix} = \begin{bmatrix} 10 & 5 & 0 \\ 5 & 20 & 0 \\ 0 & 0 & 8 \end{bmatrix} \text{MPa}$$

Determine the principal stresses and their orientations.

Solution

We obtain the principal stresses and principal directions from the eigenvalue problem $\boldsymbol{\sigma}\mathbf{n} = \lambda\mathbf{n}$, or $(\boldsymbol{\sigma} - \lambda\mathbf{I})\mathbf{n} = \mathbf{0}$. The eigenvalues are computed from

$$\begin{vmatrix} 10-\lambda & 5 & 0 \\ 5 & 20-\lambda & 0 \\ 0 & 0 & 8-\lambda \end{vmatrix} = (8-\lambda)\left(\lambda^2 - 30\lambda + 175\right) = 0$$

The eigenvalues are the roots, $\lambda_{1,2} = 15 \mp 5\sqrt{2}$, and $\lambda_3 = 8$, all in MPa; they are the principal stresses.

The eigenvectors are obtained from

$$\begin{bmatrix} 10-\lambda_i & 5 & 0 \\ 5 & 20-\lambda_i & 0 \\ 0 & 0 & 8-\lambda_i \end{bmatrix} \begin{bmatrix} n_1 \\ n_2 \\ n_3 \end{bmatrix} = \begin{bmatrix} 0 \\ 0 \\ 0 \end{bmatrix}$$

Written out, this represents three equations of the form

$$\begin{aligned} (10-\lambda_i)\, n_1 + 5n_2 + 0 &= 0 \\ 5n_1 + (20-\lambda_i)\, n_2 + 0 &= 0 \\ (8-\lambda_i)\, n_3 &= 0 \end{aligned}$$

For the case of λ_1 the third equation indicates that $n_3 = 0$. Either of the first two equations produce a relationship

$$n_2 = (1-\sqrt{2})n_1$$

or, arbitrarily sized, an eigenvector

$$\mathbf{n}^{(1)} = \begin{bmatrix} 1 \\ 1-\sqrt{2} \\ 0 \end{bmatrix}$$

For the case of λ_2 the third equation indicates that $n_3 = 0$. Either of the first two equations produce a relationship

$$n_2 = (1+\sqrt{2})n_1$$

or, arbitrarily sized, an eigenvector

$$\mathbf{n}^{(2)} = \begin{bmatrix} 1 \\ 1+\sqrt{2} \\ 0 \end{bmatrix}$$

For the case of $\lambda_3 = 8$, the first and second equations are only satisfied if that $u_1 = u_2 = 0$. Any value u_3 works for the third equation. Thus we choose, for the third eigenvector,

$$\mathbf{n}^{(3)} = \begin{bmatrix} 0 \\ 0 \\ 1 \end{bmatrix}$$

If we normalize the eigenvectors as unit vectors, we have

$$\mathbf{n}^{(1)} \cong \begin{bmatrix} 0.924 \\ -0.383 \\ 0 \end{bmatrix} \qquad \mathbf{n}^{(2)} \cong \begin{bmatrix} 0.383 \\ 0.924 \\ 0 \end{bmatrix} \qquad \mathbf{n}^{(3)} = \begin{bmatrix} 0 \\ 0 \\ 1 \end{bmatrix}$$

as the principal stress directions. The z axis is a principal direction, and the stress state is shown in Fig. 4,11, oriented along the x, y axes in the left-hand image, and oriented in the principal directions on the right. The angle of orientation is obtained from either $\mathbf{n}^{(1)}$ or $\mathbf{n}^{(2)}$, as $\theta = \tan^{-1} 0.383/0.924 = 0.3929$ radians or 22.5 degrees.

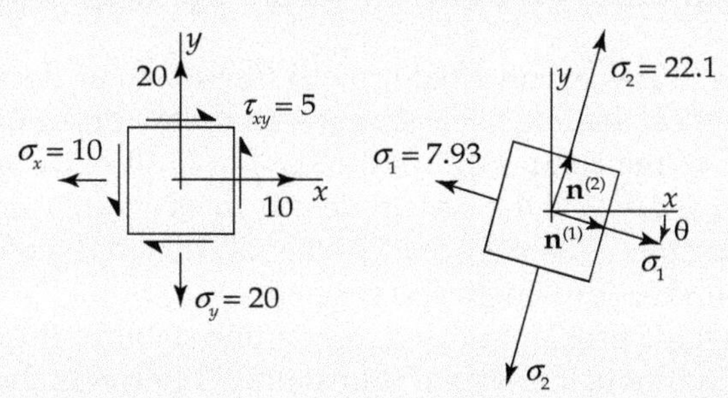

FIGURE 4.11 The stress states in the *x*, *y* frame, and in the principal orientation. The stress in the out-of-plane direction is not depicted.

4.12.3 Linear Dynamic Systems and Stability

Linear dynamic systems are often modeled with systems of linear ordinary differential equations. Suppose we have a system of n linear homogeneous differential equations in the form

$$\frac{dx_i}{dt} = a_{i1}x_1 + a_{i2}x_2 + \cdots + a_{in}x_n \qquad i = 1,2,\cdots,n \tag{4.12.7}$$

with initial conditions $x_i(0) = x_{0i}$. This set of equations can be written in matrix form as

$$\frac{d\mathbf{x}}{dt} = \mathbf{Ax} \tag{4.12.8}$$

with initial conditions $\mathbf{x}(0) = \mathbf{x}_0$, where $\mathbf{x}$ is an $n \times 1$ array of elements x_i and $\mathbf{A}$ is an $n \times n$ matrix of elements a_{ij}. We seek an exponential solution of the form $\mathbf{x} = \mathbf{u}\mathrm{e}^{\alpha t}$ and insert it into Eq. 4.12.8 to obtain

$$\mathbf{u}\alpha \mathrm{e}^{\alpha t} = \mathbf{Au}\mathrm{e}^{\alpha t} \tag{4.12.9}$$

or

$$\mathbf{Au} = \alpha\mathbf{u} \tag{4.12.10}$$

which is an eigenvalue problem with eigenvalues α_i and eigenvectors $\mathbf{u}_i$, for $i = 1, \ldots, n$, which can be real or complex conjugates. The general solution is then

$$\mathbf{x}(t) = C_1\mathbf{u}_1 e^{\alpha_1 t} + C_2\mathbf{u}_2 e^{\alpha_2 t} + \cdots + C_n\mathbf{u}_n e^{\alpha_n t} \tag{4.12.11}$$

for the case when the eigenvectors are distinct. The coefficients C_i are constants of integration and are determined by the initial conditions. (If there are repeated eigenvalues for which only one eigenvector can be found, then the above expression does not hold.)

The solution represents a combination of exponential (could be complex) components of motion about an equilibrium at the origin, and the eigenvalues indicate the characteristics of these exponential components. Generally, if the eigenvalues are complex, and we write them as $\alpha_j = a_j + b_j i$, where $i = \sqrt{-1}$, and $j = 1, \ldots, n$, then a component of the response is

$$\mathbf{x}_j = C_j\mathbf{u}_j e^{a_j t} = C_j\mathbf{u}_j e^{(a_j + b_j i)t} = C_j\mathbf{u}_j e^{a_j t}(\cos b_j t + i\sin b_j t) \tag{4.12.12}$$

This shows that a_j indicates exponential growth (for $a_j > 0$) or decay (for $a_j < 0$), and b_j indicates frequency of oscillation. By adding the complex conjugate, such as $\mathbf{x}_j + \mathbf{x}_j^*$, we obtain the real form of the component of oscillation. If all of the $a_j < 0$, then all of the terms of the solution decay, and the system decays to its equilibrium (the origin). However, if $a_j > 0$ for at least one j, there is a term in the solution that grows exponentially and the solution is unbounded (provided C_j is nonzero).

Hence the eigenvalue problem is used to examine stability. If the real part of at least one eigenvalue α_j is positive, the system is unstable. Otherwise, the system is stable.

Example 4.22

Suppose a dynamic system is modeled by the set of first order differential equations in matrix form below:

$$\begin{bmatrix} \dot{x}_1 \\ \dot{x}_2 \end{bmatrix} = \begin{bmatrix} 0 & 1 \\ -\sigma & -3+\sigma \end{bmatrix} \begin{bmatrix} x_1 \\ x_2 \end{bmatrix}$$

(a) Find the eigenvalues and eigenvectors. (b) Find the specific solution for the case of $\sigma = 0$, when the initial conditions are $x_1(0) = 0$, $x_2(0) = 1$. (c) For general values of σ, under what conditions is the system stable?

Solution

Seeking a solution of the form $\mathbf{x} = \mathbf{u}e^{\alpha t}$, or

$$\begin{bmatrix} x_1 \\ x_2 \end{bmatrix} = \begin{bmatrix} u_1 \\ u_2 \end{bmatrix} e^{\alpha t}$$

inserting into the differential equations, and canceling out the exponential function, leads to the eigenvalue problem

$$\begin{bmatrix} 0 & 1 \\ -\sigma & -3+\sigma \end{bmatrix} \begin{bmatrix} u_1 \\ u_2 \end{bmatrix} = \alpha \begin{bmatrix} u_1 \\ u_2 \end{bmatrix}$$

The eigenvalues are obtained from

$$\begin{vmatrix} -\alpha & 1 \\ -\sigma & -3+\sigma-\alpha \end{vmatrix} = -\alpha(-3+\sigma-\alpha)+\sigma = \alpha^2 + (3-\sigma)\alpha + \sigma = 0$$

The eigenvalues are

$$\alpha_{1,2} = \frac{\sigma - 3 \pm \sqrt{(\sigma-3)^2 - 4\sigma}}{2}$$

Then the first equation of the eigenvalue problem leads to

$$u_2 = \alpha_i u_1$$

or

$$\mathbf{u}^{(i)} = \begin{bmatrix} 1 \\ \alpha_i \end{bmatrix}$$

as a representation of the eigenvalues associated with $\alpha_{1,2}$.

(a) For the case of $\sigma = 0$, we have distinct eigenvalues, $\alpha_1 = 0$ and $\alpha_2 = -3$. The associated eigenvectors are

$$\mathbf{u}^{(1)} = \begin{bmatrix} 1 \\ 0 \end{bmatrix} \quad \mathbf{u}^{(2)} = \begin{bmatrix} 1 \\ -3 \end{bmatrix}$$

In such case, the general solution is

$$\begin{bmatrix} x_1(t) \\ x_2(t) \end{bmatrix} = C_1 \begin{bmatrix} 1 \\ 0 \end{bmatrix} e^{0t} + C_2 \begin{bmatrix} 1 \\ -3 \end{bmatrix} e^{-3t}$$

Applying the initial conditions,

$$\begin{bmatrix} x_1(0) \\ x_2(0) \end{bmatrix} = C_1 \begin{bmatrix} 1 \\ 0 \end{bmatrix} + C_2 \begin{bmatrix} 1 \\ -3 \end{bmatrix} = \begin{bmatrix} 0 \\ 1 \end{bmatrix}$$

we determine that $C_1 = 1/3$ and $C_2 = -1/3$. Thus the specific solution is

$$\begin{bmatrix} x_1(t) \\ x_2(t) \end{bmatrix} = \frac{1}{3} \begin{bmatrix} 1 \\ 0 \end{bmatrix} - \frac{1}{3} \begin{bmatrix} 1 \\ -3 \end{bmatrix} e^{-3t}$$

(b) The solution is stable if the real parts of both eigenvalues are less than or equal to zero. If the eigenvalues are real, we require that the largest eigenvalue be nonpositive. If the eigenvalues are complex, we require that the real parts are nonpositive. We treat these two cases separately, and then combine the conditions into the final stability requirements.

The eigenvalues are real if $(\sigma - 3)^2 - 4\sigma \geq 0$, or if $(\sigma - 1)(\sigma - 9) \geq 0$. Thus, if $\sigma \geq 9$ or if $\sigma \leq 1$ the eigenvalues are real. The largest eigenvalue (plus sign in front of the square root) is nonpositive if $\sigma - 3 + \sqrt{(\sigma - 3)^2 - 4\sigma} \leq 0$. This is true if $0 \leq \sigma \leq 3$. Under the condition for real eigenvalues, the condition for nonpositive eigenvalues is $0 \leq \sigma \leq 1$.

The eigenvalues are complex if $1 \leq \sigma \leq 9$. Then the real part must be nonpositive for stability, and hence $\sigma \leq 3$. Under the condition that the eigenvalues are complex, the condition for nonpositive real parts is $1 \leq \sigma \leq 3$.

Assemb ling the stability conditions for the real and imaginary cases, the system is stable if $0 \leq \sigma \leq 3$.

PROBLEMS

4.1 Identify which of the following groups of numbers are matrices.

(a) $\begin{bmatrix} 0 & 2 \end{bmatrix}$ **(b)** $\begin{bmatrix} 0 \\ 2 \end{bmatrix}$

(c) $\begin{bmatrix} & 0 & \\ 1 & & 2 \end{bmatrix}$ **(d)** $\begin{bmatrix} 0 & 1 \\ 2 & 3 \end{bmatrix}$

(e) $\begin{bmatrix} & 0 & \\ 1 & & 2 \\ & 3 & \end{bmatrix}$ **(f)** $\begin{bmatrix} 1 & 2 \\ 0 & \end{bmatrix}$

(g) $\begin{bmatrix} 2x & x^2 \\ 2 & 0 \end{bmatrix}$ **(h)** $\begin{bmatrix} 2x & x^2 \end{bmatrix}$

(i) $\begin{bmatrix} x \\ xy \\ z^2 \end{bmatrix}$ **(j)** $\begin{bmatrix} x & y & 2x \\ 1 & 0 & 1 \\ x^2 & z^2 & y^2 \end{bmatrix}$

(k) $\begin{bmatrix} 2-i & i \\ 2 & 3+i \end{bmatrix}$

4.2 For Problem 4.1 identify all a) vectors, b) square matrices, c) row matrices, and d) column matrices.

4.3 **(a)** Partition the matrix **A**,

$$\mathbf{A} = \begin{bmatrix} 5 & 2 & 0 & -3 \\ 4 & -2 & 7 & 0 \\ 1 & 0 & 6 & 8 \\ -2 & 4 & 0 & 9 \end{bmatrix}$$

to include a 3×3 matrix such that the elements of the principal diagonal total 13; display all submatrices.

(b) Display the three row vectors contained in the 3×3 matrix of part (a).

4.4 Identify the following elements from the matrix of Problem 4.3.

(a) a_{22} **(b)** a_{32}
(c) a_{23} **(d)** a_{11}
(e) a_{14}

4.5 **(a)** The matrix of Problem 4.3 represents 4 column vectors. Display them.

(b) Display the 4 row vectors of the matrix of Problem 4.3.

4.6 For the three matrices

$$\mathbf{A} = \begin{bmatrix} 2 & 1 & 0 \\ 1 & -1 & -2 \\ 4 & 2 & 0 \end{bmatrix}, \quad \mathbf{B} = \begin{bmatrix} 1 & 1 & 1 \\ 0 & 0 & 0 \\ 2 & 1 & -3 \end{bmatrix},$$

$$\mathbf{C} = \begin{bmatrix} 2 & 3 & 1 \\ 0 & 2 & 0 \\ -1 & 2 & -1 \end{bmatrix}$$

determine:

(a) $\mathbf{A} + \mathbf{B}$, $\mathbf{B} + \mathbf{A}$ **(b)** $\mathbf{A} - \mathbf{B}$, $\mathbf{B} - \mathbf{A}$

(c) $\mathbf{A} + (\mathbf{B} - \mathbf{C})$, $(\mathbf{A} + \mathbf{B}) - \mathbf{C}$ **(d)** $4\mathbf{A} + 4\mathbf{B}$, $4(\mathbf{A} + \mathbf{B})$

(e) $2\mathbf{A} - 4\mathbf{C}$, $2(\mathbf{A} - 2\mathbf{C})$

4.7 Write the transpose of each of the matrices in Problems 4.6 and 4.3.

4.8 For the matrices of Problem 4.6, show that (a) $(\mathbf{A} + \mathbf{B})^T = \mathbf{A}^T + \mathbf{B}^T$, (b) $\mathbf{C} = (\mathbf{C}^T)^T$, (c) $\mathbf{A} + \mathbf{A}^T$ is symmetric, and (d) $\mathbf{A} - \mathbf{A}^T$ is skew-symmetric.

4.9 Using index notation, show that $\mathbf{A}_s$ is symmetric and $\mathbf{A}_a$ is skew-symmetric. Refer to Eq. 4.4.10.

4.10 Using the two matrices

$$\mathbf{A} = \begin{bmatrix} 2 & 4 & 6 \\ 0 & 4 & -2 \\ -4 & 2 & 2 \end{bmatrix}, \quad \mathbf{B} = \begin{bmatrix} 0 & -8 & 6 \\ -2 & 0 & 2 \\ 2 & -4 & 4 \end{bmatrix}$$

find: (a) $\mathbf{A}_s$ and $\mathbf{A}_a$, (b) $\mathbf{B}_s$ and $\mathbf{B}_a$, and (c) $(\mathbf{A} + \mathbf{B})_s$ and $(\mathbf{A} - \mathbf{B})_s$.

4.11 Write the matrix

$$\mathbf{A} = \begin{bmatrix} 2 & -2 & 4 \\ 2 & 2 & 2 \\ -4 & -2 & 2 \end{bmatrix}$$

as the sum of a symmetric and a skew-symmetric matrix. Can the result be written in terms of the unit matrix, at least in part?

4.12 Let

$$\mathbf{A} = \begin{bmatrix} 1 \\ -1 \\ 2 \end{bmatrix}, \quad \mathbf{B} = [2, \ 4, \ -1],$$

$$\mathbf{C} = \begin{bmatrix} 3 & 2 & 1 \\ -2 & 0 & -1 \\ 1 & 0 & 1 \end{bmatrix}, \quad \mathbf{D} = \begin{bmatrix} -1 & 0 & 2 \\ 1 & 2 & 1 \\ 2 & -1 & -1 \end{bmatrix}$$

Find:

(a) **AB** (b) **BA** (c) (**AB**)**C**
(d) **BC** (e) **A**(**BC**) (f) **CA**
(g) **CD** (h) **BD** (i) **DA**

4.13 Let

$$\mathbf{x} = \begin{bmatrix} x \\ y \\ z \end{bmatrix}, \quad \mathbf{x}' = \begin{bmatrix} x' \\ y' \\ z' \end{bmatrix},$$

$$\mathbf{A} = \begin{bmatrix} \cos\theta & \sin\theta & 0 \\ \sin\theta & -\cos\theta & 0 \\ 0 & 0 & 1 \end{bmatrix}$$

Write out the equations represented by $\mathbf{x}' = \mathbf{A}\mathbf{x}$. What do these equations represent?

4.14 A set of linear equations is given as

$$\begin{aligned} 2x_1 + 3x_2 - x_3 &= r_1 \\ x_1 - x_2 + x_3 &= r_2 \\ x_1 - x_3 &= r_3 \end{aligned}$$

The vector $\mathbf{x}$ is related to the vector $\mathbf{y}$ by the equations

$$\begin{aligned} x_1 &= 2y_1 + y_2 - y_3 \\ x_2 &= y_1 - y_2 \\ x_3 &= y_3 \end{aligned}$$

Find the set of equations that relate the vector $\mathbf{y}$ to the vector $\mathbf{r}$.

4.15 Let

$$\mathbf{A} = \begin{bmatrix} 0 & 3 & 1 \\ -1 & 2 & 0 \\ 0 & 0 & 1 \end{bmatrix}, \quad \mathbf{B} = \begin{bmatrix} 1 & 0 & 0 \\ -1 & 2 & 1 \\ 3 & 1 & 0 \end{bmatrix},$$

$$\mathbf{C} = \begin{bmatrix} 2 \\ 0 \\ -1 \end{bmatrix}, \quad \mathbf{D} = [1, \ 2, \ 0]$$

Determine the following products (identify those that are not defined).

(a) (**A** + **B**)**C** and **AC** + **BC**
(b) **D**(**A** + **B**) and **DA** + **DB**
(c) **A**(**BC**) and (**AB**)**C**
(d) **A**(**BD**) and (**AB**)**D**
(e) $(\mathbf{AB})^T$ and $\mathbf{B}^T\mathbf{A}^T$
(f) $\mathbf{A}^T\mathbf{A}$
(g) $(\mathbf{ABC})^T$ and $\mathbf{C}^T\mathbf{B}^T\mathbf{A}^T$
(h) $\mathbf{C}^T\mathbf{C}$
(i) $\mathbf{C}\mathbf{D}^T$
(j) $\mathbf{A}^2$ and $\mathbf{A}^3$
(k) $\mathbf{C}^2$
(l) **A** + **C**
(m) $\mathbf{A}^2 - 2\mathbf{B} + 3\mathbf{I}$
(n) $2\mathbf{AC} + \mathbf{DB} - 4\mathbf{I}$

4.16 Use $k = 3$ and show that $\mathbf{SA} = \mathbf{AS} = k\mathbf{A}$, where **S** is a scalar matrix, using

$$\mathbf{A} = \begin{bmatrix} 2 & -1 & 3 \\ 0 & 1 & 2 \\ -2 & 0 & 0 \end{bmatrix}$$

4.17 Multiply a 3×3 diagonal matrix **A** with diagonal elements a_{11}, a_{22}, a_{33}, and a general 3×3 matrix **B**. Display the resulting matrix **AB**.

4.18 A diagonal matrix **A** is given by

$$\mathbf{A} = \begin{bmatrix} 2 & 0 & 0 \\ 0 & -1 & 0 \\ 0 & 0 & 3 \end{bmatrix}$$

Find the products **AB** and **AC** if

$$\mathbf{B} = \begin{bmatrix} 2 & 1 & 3 \\ 1 & -1 & 2 \\ 1 & 3 & 2 \end{bmatrix} \quad \text{and} \quad \mathbf{C} = \begin{bmatrix} 2 \\ 1 \\ -1 \end{bmatrix}$$

4.19 Using the products of the elements of diagonals, evaluate each of the following determinants.

(a) $\begin{vmatrix} 2 & 0 \\ -1 & 3 \end{vmatrix}$ (b) $\begin{vmatrix} 1 & 2 \\ 1 & 3 \end{vmatrix}$

(c) $\begin{vmatrix} 2 & -2 \\ -1 & 1 \end{vmatrix}$ (d) $\begin{vmatrix} 3 & 1 & 0 \\ 1 & 3 & -1 \\ 2 & -1 & 0 \end{vmatrix}$

(e) $\begin{vmatrix} 4 & -1 & 3 \\ 2 & 2 & 2 \\ 1 & -2 & 4 \end{vmatrix}$

4.20 Using cofactors, evaluate

$$\begin{vmatrix} 3 & 2 & -1 \\ 3 & 0 & 3 \\ -1 & 2 & 1 \end{vmatrix}$$

(a) Expand by the first row.
(b) Expand by the second row.
(c) Expand by the first column.
(d) Expand by the second column.

4.21 Solve each of the following systems of linear, algebraic equations.

(a) $x - y = 6$, $x + y = 0$

(b) $x - 2y = 4$, $2x + y = 3$

(c) $3x + 4y = 7$, $2x - 5y = 2$

(d) $3x + 2y - 6z = 0$, $x - y + z = 4$, $y + z = 3$

(e) $x - 3y + z = -2$, $x - 3y - z = 0$, $-3y + z = 0$

(f) $x_1 + x_2 + x_3 = 4$, $x_1 - x_2 - x_3 = 2$, $x_1 - 2x_2 = 0$

4.22 Determine the value of

$$\begin{vmatrix} 2 & 0 & 8 & 6 \\ -1 & 4 & 2 & 0 \\ 0 & -1 & 3 & 0 \\ 3 & 5 & 7 & 3 \end{vmatrix}$$

(a) Expand by the first row.
(b) Expand by the third row.
(c) Expand by the first column.
(d) Expand by the fourth column.

4.23 Show the following, by computation.

(a) $\begin{vmatrix} 3 & 2 & -1 \\ 6 & 3 & 0 \\ 3 & 1 & 2 \end{vmatrix} = 3\begin{vmatrix} 1 & 2 & -1 \\ 2 & 3 & 0 \\ 1 & 1 & 2 \end{vmatrix}$

(b) $\begin{vmatrix} 3 & 2 & -1 \\ 6 & 3 & 0 \\ 3 & 1 & 2 \end{vmatrix} = -\begin{vmatrix} 2 & 3 & -1 \\ 3 & 6 & 0 \\ 1 & 3 & 2 \end{vmatrix}$

(c) $\begin{vmatrix} 3 & 2 & -1 \\ 6 & 3 & 0 \\ 3 & 1 & 2 \end{vmatrix} = \begin{vmatrix} 3+2 & 2 & -1 \\ 6+3 & 3 & 0 \\ 3+1 & 1 & 2 \end{vmatrix}$

(d) $\begin{vmatrix} 3 & 2 & -1 \\ 6 & 3 & 0 \\ 3 & 1 & 2 \end{vmatrix} = \begin{vmatrix} 3+10 & 2 & -1 \\ 6+15 & 3 & 0 \\ 3+5 & 1 & 2 \end{vmatrix}$

(e) $\begin{vmatrix} 3 & 2 & -1 \\ 6 & 3 & 0 \\ 3 & 1 & 2 \end{vmatrix} = \begin{vmatrix} 3+3 & 2+1 & -1+2 \\ 6 & 3 & 0 \\ 3 & 1 & 2 \end{vmatrix}$

(f) $\begin{vmatrix} 3 & -3 & -1 \\ 6 & -6 & 0 \\ 3 & -3 & 2 \end{vmatrix} = 0$

4.24 Evaluate the following determinants by reducing a row or column to only one nonzero element.

(a) $\begin{vmatrix} 3 & 1 & 3 \\ 2 & 0 & 4 \\ -1 & 2 & -2 \end{vmatrix}$ **(b)** $\begin{vmatrix} 2 & 3 & 4 \\ -1 & 0 & 3 \\ 1 & 2 & 3 \end{vmatrix}$

(c) $\begin{vmatrix} 1 & 1 & 1 \\ -1 & -2 & 2 \\ 1 & 2 & 3 \end{vmatrix}$ **(d)** $\begin{vmatrix} 2 & 1 & 3 \\ 4 & 2 & 6 \\ -3 & 1 & 0 \end{vmatrix}$

(e) $\begin{vmatrix} 4 & 3 & 1 & 4 \\ 3 & 0 & 0 & 3 \\ 1 & 2 & 2 & 1 \\ 0 & -1 & 3 & 2 \end{vmatrix}$ **(f)** $\begin{vmatrix} 3 & 1 & -1 & 0 \\ 2 & 2 & 2 & 1 \\ -1 & 3 & 0 & 4 \\ 8 & 6 & -2 & 2 \end{vmatrix}$

(g) $\begin{vmatrix} 1 & 1 & 1 & 1 \\ -2 & 3 & 1 & 0 \\ 4 & 3 & 8 & 1 \\ 7 & 5 & -2 & 0 \end{vmatrix}$ **(h)** $\begin{vmatrix} 2 & -1 & 6 & 3 \\ -2 & 4 & 5 & -1 \\ 3 & 4 & 3 & 2 \\ 1 & -1 & 2 & 3 \end{vmatrix}$

4.25 Verify by computation the eighth property listed for a determinant, using

$$\mathbf{A} = \begin{bmatrix} 2 & 3 & 1 \\ -1 & 0 & 2 \\ 3 & 4 & 1 \end{bmatrix}, \quad \mathbf{B} = \begin{bmatrix} 2 & -1 & 3 \\ 6 & 7 & -1 \\ 3 & 4 & 2 \end{bmatrix}$$

First, using matrix multiplications, find $\mathbf{C} = \mathbf{AB}$. Then show that $|\mathbf{C}| = |\mathbf{A}||\mathbf{B}|$.

4.26 Find the value of the determinant of an nth-order diagonal matrix with diagonal elements $(k_1, k_2, \ldots, k_n)$.

4.27 Find the adjoint matrix $\mathbf{A}^+$ and the inverse matrix $\mathbf{A}^{-1}$ (if one exists) if $\mathbf{A}$ is given by:

(a) $\begin{bmatrix} 1 & -1 \\ 1 & 1 \end{bmatrix}$ **(b)** $\begin{bmatrix} 2 & 6 \\ 1 & 3 \end{bmatrix}$

(c) $\begin{bmatrix} 2 & 0 \\ 0 & 1 \end{bmatrix}$ **(d)** $\begin{bmatrix} 1 & 2 \\ 0 & 0 \end{bmatrix}$

(e) $\begin{bmatrix} -3 & 5 \\ -2 & 6 \end{bmatrix}$ **(f)** $\begin{bmatrix} 1 & 0 & 2 \\ 2 & 1 & 1 \\ 1 & 1 & 1 \end{bmatrix}$

(g) $\begin{bmatrix} 1 & 2 & 2 \\ 1 & 1 & 2 \\ 1 & -2 & 2 \end{bmatrix}$ **(h)** $\begin{bmatrix} 3 & 1 & 2 \\ -1 & 2 & 1 \\ 0 & 1 & 1 \end{bmatrix}$

(i) $\begin{bmatrix} 0 & 1 & 1 & 0 \\ 0 & 0 & 1 & 1 \\ 1 & 0 & 1 & 1 \\ 1 & 1 & 1 & 1 \end{bmatrix}$

4.28 Determine the inverse of each of the following diagonal matrices.

(a) $\begin{bmatrix} 2 & 0 \\ 0 & -1 \end{bmatrix}$ **(b)** $\begin{bmatrix} 1 & 0 \\ 0 & 1 \end{bmatrix}$

(c) $\begin{bmatrix} 2 & 0 & 0 \\ 0 & 2 & 0 \\ 0 & 0 & 1 \end{bmatrix}$ **(d)** $\begin{bmatrix} 1 & 0 & 0 \\ 0 & 1 & 0 \\ 0 & 0 & 0 \end{bmatrix}$

4.29 Find the inverse of each of the following symmetric matrices and conclude that the inverse of a symmetric matrix is also symmetric.

(a) $\begin{bmatrix} 2 & 1 \\ 1 & 1 \end{bmatrix}$ **(b)** $\begin{bmatrix} 3 & 1 & 2 \\ 1 & 0 & 1 \\ 2 & 1 & -1 \end{bmatrix}$

(c) $\begin{bmatrix} 0 & 2 & 3 \\ 2 & 0 & 2 \\ 3 & 2 & 0 \end{bmatrix}$ **(d)** $\begin{bmatrix} 2 & 1 & 1 & 1 \\ 1 & 2 & 0 & 0 \\ 1 & 0 & 0 & 1 \\ 1 & 0 & 1 & 2 \end{bmatrix}$

4.30 Using matrix notation, show that (see Example 4.9):

(a) The inverse of the inverse is the given square matrix.
(b) The inverse of the transpose is the transpose of the inverse.
(c) The product of two nonsingular matrices cannot be a null matrix. (*Hint:* Assume that the product is a null matrix.)

4.31 Find the column vector representing the solution to each of the following sets of algebraic equations. See Eq. 4.8.6.

(a) $x - y = 2$, $x + y = 0$ **(b)** $x + z = 4$, $2x + 3z = 8$

(c)
$$x + 2y + z = -2$$
$$x + y = 3$$
$$x + z = 4$$

(d)
$$x_1 + x_2 + x_3 = 5$$
$$2x_1 - 4x_2 + 3x_3 = 0$$
$$x_1 - 6x_2 + 2x_3 = 3$$

4.32 For $\mathbf{Ax} = \mathbf{r}$, where $\mathbf{A}$ and $\mathbf{r}$ are given below, if a unique solution exists, find the solution for x_2 using Cramer's rule (by hand or using MATLAB).

(a) $\mathbf{A} = \begin{bmatrix} 2 & 3 \\ 1 & -1 \end{bmatrix}$ $\mathbf{r} = \begin{bmatrix} -5 \\ 5 \end{bmatrix}$

(b) $\mathbf{A} = \begin{bmatrix} 1 & 0 & -1 \\ 2 & 1 & 1 \\ 0 & 1 & -1 \end{bmatrix}$ $\mathbf{r} = \begin{bmatrix} 3 \\ 1 \\ 3 \end{bmatrix}$

(c) $\mathbf{A} = \begin{bmatrix} 1 & 3 & -1 \\ 0 & 2 & 2 \\ -2 & 0 & 1 \end{bmatrix}$ $\mathbf{r} = \begin{bmatrix} 4 \\ -2 \\ 0 \end{bmatrix}$

(d) $\mathbf{A} = \begin{bmatrix} 1 & 0 & -1 & 0 \\ 0 & 1 & 1 & 0 \\ 2 & 0 & 0 & 1 \\ 0 & 1 & 0 & 1 \end{bmatrix}$ $\mathbf{r} = \begin{bmatrix} 2 \\ -1 \\ 0 \\ 2 \end{bmatrix}$

4.33 For $\mathbf{Ax} = \mathbf{r}$, where $\mathbf{A}$ and $\mathbf{r}$ are given below, if a unique solution exists, find the solution for all variables using the inverse method (by hand or using MATLAB).

(a) $\mathbf{A} = \begin{bmatrix} 2 & 3 \\ 1 & -1 \end{bmatrix}$ $\mathbf{r} = \begin{bmatrix} -5 \\ 5 \end{bmatrix}$

(b) $\mathbf{A} = \begin{bmatrix} 1 & 0 & -1 \\ 2 & 1 & 1 \\ 0 & 1 & -1 \end{bmatrix}$ $\mathbf{r} = \begin{bmatrix} 3 \\ 1 \\ 3 \end{bmatrix}$

(c) $\mathbf{A} = \begin{bmatrix} 1 & 3 & -1 \\ 0 & 2 & 2 \\ -2 & 0 & 1 \end{bmatrix}$ $\mathbf{r} = \begin{bmatrix} 4 \\ -2 \\ 0 \end{bmatrix}$

(d) $\mathbf{A} = \begin{bmatrix} 1 & 0 & -1 & 0 \\ 0 & 1 & 1 & 0 \\ 2 & 0 & 0 & 1 \\ 0 & 1 & 0 & 1 \end{bmatrix}$ $\mathbf{r} = \begin{bmatrix} 2 \\ -1 \\ 0 \\ 2 \end{bmatrix}$

4.34 For $\mathbf{Ax} = \mathbf{r}$, where $\mathbf{A}$ and $\mathbf{r}$ are given below, if a unique solution exists, find the solution for all variables using row operations.

(a) $\mathbf{A} = \begin{bmatrix} 2 & 3 \\ 1 & -1 \end{bmatrix}$ $\mathbf{r} = \begin{bmatrix} -5 \\ 5 \end{bmatrix}$

(b) $\mathbf{A} = \begin{bmatrix} 1 & 0 & -1 \\ 2 & 1 & 1 \\ 0 & 1 & -1 \end{bmatrix}$ $\mathbf{r} = \begin{bmatrix} 3 \\ 1 \\ 3 \end{bmatrix}$

(c) $\mathbf{A} = \begin{bmatrix} 1 & 3 & -1 \\ 0 & 2 & 2 \\ -2 & 0 & 1 \end{bmatrix}$ $\mathbf{r} = \begin{bmatrix} 4 \\ -2 \\ 0 \end{bmatrix}$

(d) $\mathbf{A} = \begin{bmatrix} 1 & 0 & -1 & 0 \\ 0 & 1 & 1 & 0 \\ 2 & 0 & 0 & 1 \\ 0 & 1 & 0 & 1 \end{bmatrix}$ $\mathbf{r} = \begin{bmatrix} 2 \\ -1 \\ 0 \\ 2 \end{bmatrix}$

4.35 Use (a) MATLAB, and (b) row operations, to express the solution to $\mathbf{Ax} = \mathbf{0}$, where $\mathbf{A}$ is given below:

$$\mathbf{A} = \begin{bmatrix} -1 & 3 \\ -2 & -1 \end{bmatrix}$$

4.36 Use (a) MATLAB, and (b) row operations, to express the solution to $\mathbf{Ax} = \mathbf{0}$, where $\mathbf{A}$ is given below:

$$\mathbf{A} = \begin{bmatrix} 1 & 2 & 3 \\ 4 & 5 & 6 \\ 7 & 8 & 9 \end{bmatrix}$$

4.37 Use (a) MATLAB, and (b) row operations, to express the solution to $\mathbf{Ax} = \mathbf{0}$, where $\mathbf{A}$ is:

$$\begin{bmatrix} 1 & 2 & 3 & 4 \\ 3 & 4 & 5 & 6 \\ 5 & 6 & 7 & 8 \\ 7 & 8 & 9 & 10 \end{bmatrix}$$

4.38 The "quarterback rating" is a rather mysterious football statistic, and is dependent on these quantities: passing attempts A, completions C, passing yards Y, touchdowns T, and interceptions I. Assume that the rating r is based on the formula

$$r = a\,(C/A) + b(Y/A) + c(T/A) + d(I/A)$$

Several players' statistics are given below, along with their quarterback ratings r, as reported on the date of November 3, 2017.

(a) Use the data to identify the parameters in the formula.

(b) Look at the residuals. Do you think there is anything missing from the formula above?

You may wish to use a software package such as MATLAB.

Player	C	A	Y	T	I	*Rating, r*
Tom Brady	209	309	2541	16	2	106.5
Alex Smith	179	259	2181	16	0	115.4
Carson Wentz	161	264	2063	19	5	101.6
Russell Wilson	164	258	2008	15	4	100.4
Kirk Cousins	161	237	1900	13	4	103.3
Matt Stafford	163	270	1851	12	4	89.6
Cam Newton	166	263	1841	10	11	79.1
Aaron Rodgers	128	193	1385	13	3	103.2
Jamies Winston	152	246	1853	10	6	88.3
Joe Flacco	153	239	1290	6	8	72.3
DeShone Kizer	111	213	1144	3	11	51.1
Eli Manning	167	260	1600	10	5	86.1
Carson Palmer	164	267	1978	9	7	84.4

4.39 A forced mass-spring oscillator on a linear bearing is excited by a harmonically moving base, and is modeled to have both linear and Coulomb damping. The damping coefficients are very difficult to obtain from first principles, and so we turn to experimental estimation of these parameters. An analysis of the model leads to a relationship between the base-excitation amplitude Y and steady-state vibration amplitude X in the form

$$\pi Y = 4x_k + 2\pi\zeta X$$

when the excitation is at the resonant frequency, where x_k is the scaled Coulomb coefficient (units in mm), and ζ is the nondimensional damping factor (based on Liang and Feeny, 2006, Journal of Sound and Vibration **295**). Two sets of steady-state vibration experiments were conducted, and the input and steady-state output amplitudes are tabulated together below.

(a) Perform a least-squares fit to determine the estimated values of x_k and ζ.

(b) Plot the data (Y vs. X) using unconnected symbols, and overlay a line plot of the least-squares fit of the above equation to visualize quality.

(c) Examine the residuals and comment on the quality of the model.

You may use a computational package such as MATLAB.

Data	
Y(mm)	X(mm)
4.50	4.40
4.95	6.1
5.22	7.2
5.49	8.9
5.87	11.2
4.45	4.50
4.85	6.20
5.20	6.80
5.58	9.75
5.96	12.01

4.40 Determine the eigenvalues and unit eigenvectors of the following matrices.

(a) $\begin{bmatrix} 0 & 4 \\ 1 & 0 \end{bmatrix}$ **(b)** $\begin{bmatrix} 2 & 0 \\ 0 & -1 \end{bmatrix}$

(c) $\begin{bmatrix} 0 & 3 \\ 3 & 8 \end{bmatrix}$ **(d)** $\begin{bmatrix} 2 & 2 \\ -1 & -1 \end{bmatrix}$

(e) $\begin{bmatrix} 5 & 4 \\ 4 & -1 \end{bmatrix}$ **(f)** $\begin{bmatrix} 2 & -2 \\ 2 & 2 \end{bmatrix}$

(g) $\begin{bmatrix} 2 & 0 & 0 \\ 0 & -1 & 0 \\ 0 & 0 & 2 \end{bmatrix}$ **(h)** $\begin{bmatrix} 10 & 8 & 0 \\ 8 & -2 & 0 \\ 0 & 0 & 4 \end{bmatrix}$

4.41 The eigenvalues are λ_1 and λ_2 for the matrix

$$\mathbf{A} = \begin{bmatrix} 3 & 1 \\ 5 & -1 \end{bmatrix}$$

(a) Find λ_1 and λ_2.
(b) Find the eigenvalues of $\mathbf{A}^T$.
(c) Show that the eigenvalues of $\mathbf{A}^{-1}$ are $1/\lambda_1$ and $1/\lambda_2$.
(d) Find the eigenvalues of $\mathbf{A}^3$.

4.42 Repeat Problem 4.41 using the matrix of Problem 4.40(f).

4.43 The eigenvalues for the matrix

$$\mathbf{A} = \begin{bmatrix} -1 & 3 & 0 \\ 3 & 7 & 0 \\ 0 & 0 & 6 \end{bmatrix}$$

are λ_1, λ_2, and λ_3.

(a) Find λ_1, λ_2, and λ_3.
(b) Find the eigenvalues of $\mathbf{A}^T$.
(c) Show that the eigenvalues of $\mathbf{A}^{-1}$ are $1/\lambda_1$, $1/\lambda_2$, and $1/\lambda_3$.
(d) Find the eigenvalues of $\mathbf{A}^2$.

4.44 Repeat Problem 4.43 using the matrix

$$\mathbf{A} = \begin{bmatrix} 3 & 0 & 1 \\ 0 & 2 & 0 \\ 5 & 0 & -1 \end{bmatrix}.$$

4.45 Consider the electrical circuit shown.

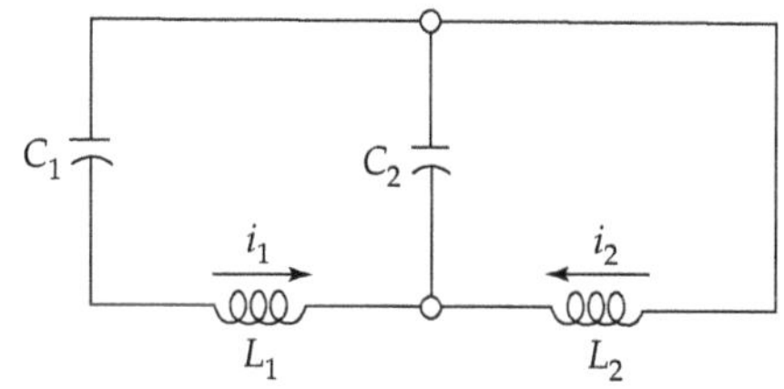

(a) Derive the differential equations that describe the currents $i_1(t)$ and $i_2(t)$.
(b) Write the differential equations in the matrix form $\mathbf{i}'' = \mathbf{A}\mathbf{i}$ and identify the elements in the coefficient matrix $\mathbf{A}$.
(c) Let $\mathbf{i} = \mathbf{x}e^{mt}$ and show that an eigenvalue problem results.
(d) Find the eigenvalues and unit eigenvectors if $L_1 = 1$, $L_2 = 2$, $C_1 = 0.02$, and $C_2 = 0.01$.
(e) Determine the general form of the solutions $i_1(t)$ and $i_2(t)$.
(f) Find the specific solutions for $i_1(t)$ and $i_2(t)$ if $i_1(0) = 1$, $i_2(0) = 0$, $i_1'(0) = 0$, and $i_2'(0) = 0$.

4.46 In Example 4.18, without assigning the spring constants and masses any specific values, determine the elements in the coefficient matrix $\mathbf{A}$.

Using $M_1 = 2$, $M_2 = 2$, $K_1 = 12$, and $K_2 = 8$,

(a) What are the eigenvalues and unit eigenvectors?
(b) What is the most general solution?
(c) Find the specific solutions for $y_1(t)$ and $y_2(t)$ if $y_1(0) = 0$, $y_2(0) = 0$, $\dot{y}_1(0) = 0$, and $\dot{y}_2(0) = 10$.

4.47 The following apply to either Problem 4.45 or Problem 4.46. Find the specific solutions for each.

(a) $L_1 = \dfrac{10}{3}$, $L_2 = 5$, $C_1 = \dfrac{1}{200}$, $C_2 = \dfrac{1}{300}$, $i_1(0) = 0$, $i_2(0) = 0$, $i_1'(0) = 50$, $i_2'(0) = 0$.

(b) $M_1 = 1$, $M_2 = 4$, $K_1 = 20$, $K_2 = 40$, $y_1(0) = 2$, $y_2(0) = 0$, $\dot{y}_1(0) = 0$, $\dot{y}_2(0) = 0$.

4.48 A uniform slender rod of length $2L$ and mass m is fixed in the x, y plane as shown, at 45° to the x axis. The mass moment of inertia about the point O, in the x, y, z coordinate frame, is given below. Determine the principal moments of inertia and the principal axes of inertia.

$$\mathbf{I}_0 = \frac{mL^2}{3\sqrt{2}}\begin{bmatrix} 1 & -1 & 0 \\ -1 & 1 & 0 \\ 0 & 0 & \sqrt{2} \end{bmatrix}$$

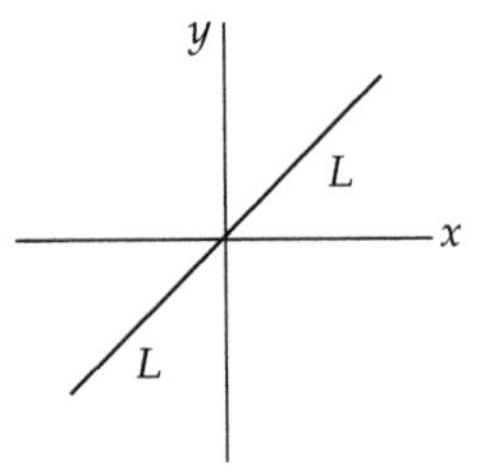

4.49 Find the principal axes of inertia about the point O of the rectangular plate below, for which the mass moment of inertia matrix about the point O is given, for the case of h = 3 and w = 5 meters. Plot the principal axes on the sketch of the plate.

$$\mathbf{I}_0 = \begin{bmatrix} \frac{h^2}{3} & \frac{wh}{4} & 0 \\ \frac{wh}{4} & \frac{w^2}{3} & 0 \\ 0 & 0 & \frac{h^2+w^2}{3} \end{bmatrix}$$

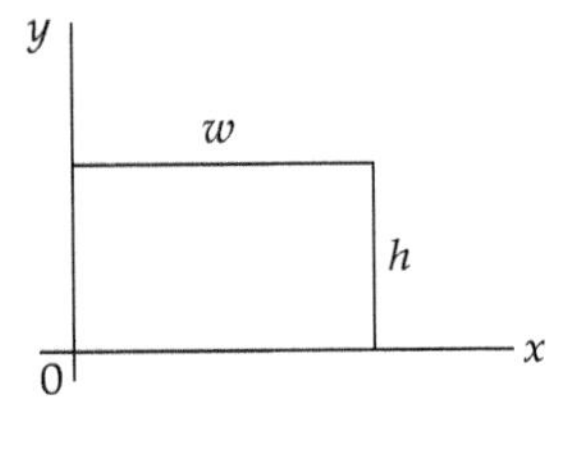

4.50 A structure of connected thin rods of total mass $6m$ is shown below. Its mass moment of inertia matrix in the x, y, z frame is given. Determine the principal moments of inertia and the principal axes of inertia.

$$\mathbf{I}_0 = \frac{mL^2}{3}\begin{bmatrix} 28 & 0 & -27 \\ 0 & 71 & 0 \\ -27 & 0 & 45 \end{bmatrix}$$

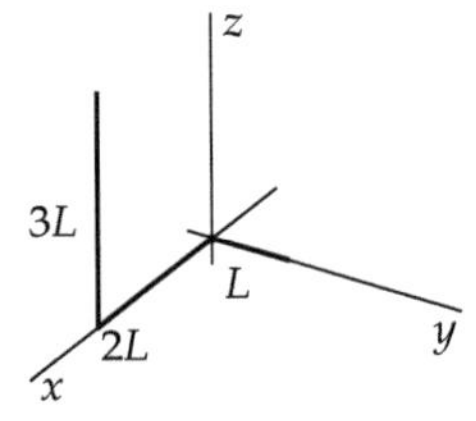

4.51 A material has a state of stress that depends on the loading conditions, manifested in the parameter a in the stress matrix σ. Find the value of a such that two of the principal stresses have the value 40.

$$\sigma = \begin{bmatrix} 100 & 40a & 0 \\ 40a & 0 & 0 \\ 0 & 0 & 40 \end{bmatrix}$$

4.52 A material has a state of stress quantified in an (x, y, z) coordinate system, as given in the matrix σ below. Find the principal stresses and their directions.

$$\sigma = \begin{bmatrix} \sigma_{xx} & \tau_{xy} & \tau_{xz} \\ \tau_{yx} & \sigma_{yy} & \tau_{yz} \\ \tau_{zx} & \tau_{zy} & \sigma_{zz} \end{bmatrix} = \begin{bmatrix} -10 & -10 & -10 \\ -10 & 10 & -30 \\ -10 & -30 & 10 \end{bmatrix}$$

4.53 A linearized oscillator in state-variable form is modeled by $\dot{\mathbf{x}} = \mathbf{A}\mathbf{x}$, where

$$\mathbf{A} = \begin{bmatrix} 0 & 1 \\ -c & -k \end{bmatrix}$$

where $c > 0$, and k can be positive or negative. Determine the values of k for which the system is stable.

4.54 A dynamic system is modeled by $\dot{\mathbf{x}} = \mathbf{A}\mathbf{x}$, where

$$\mathbf{A} = \begin{bmatrix} 2 & -4 \\ \sigma & -1 \end{bmatrix}$$

Determine the values of σ for which the system is stable.

4.55 A dynamic system is modeled by $\dot{\mathbf{x}} = \mathbf{A}\mathbf{x}$, where

$$\mathbf{A} = \begin{bmatrix} 3 & -5 \\ -1 & \sigma \end{bmatrix}$$

Determine the values of σ for which the system is stable.

5 Vector Analysis

Outline

5.1 Introduction
5.2 Vector Algebra
5.2.1 Definitions
5.2.2 Addition and Subtraction
5.2.3 Components of a Vector
5.2.4 Multiplication
5.3 Vector Differentiation
5.3.1 Ordinary Differentiation
5.3.2 Partial Differentiation
5.4 The Gradient
5.5 Cylindrical and Spherical Coordinates
5.5.1 Cylindrical Coordinates
5.5.2 Spherical Coordinates
5.6 Integral Theorems
5.6.1 The Divergence Theorem
5.6.2 Stokes's Theorem
Problems

5.1 INTRODUCTION

One of the major changes in undergraduate science curricula, and in engineering curricula in particular, brought about by the modern space age has been the introduction of vector analysis into several courses. The use of vector analysis comes rather naturally, since many of the quantities encountered in the modeling of physical phenomena are vector quantities; examples of such quantities are velocity, acceleration, force, electric and magnetic fields, and heat flux. It is not absolutely necessary that we use vector analysis when working with these quantities; we could use the components of the vectors and continue to manipulate scalars. Vector analysis though, simplifies many of the operations demanded in the solution to problems or in the derivation of mathematical models; thus, it has been introduced in most undergraduate science curricula.

Vector analysis is often introduced in an introductory mathematics course. Vector algebra, which includes addition and subtraction; multiplication including the scalar (dot) and vector (cross) products; and scalar differentiation of vectors, is usually presented. These operations will be reviewed here. In addition, vector calculus will be presented; this includes vector differentiation using a vector operator, and the use of vector integral theorems.

5.2 VECTOR ALGEBRA

5.2.1 Definitions

A quantity that is completely defined by both magnitude and direction is a *vector*. Force is such a quantity. We must be careful, though, since not all quantities that have

M. C. Potter and B. F. Feeny, *Mathematical Methods for Engineering and Science*,
https://doi.org/10.1007/978-3-031-26151-0_5

both magnitude and direction are vectors. Stress is such a quantity; it is not completely characterized by a magnitude and a direction and thus it is not a vector. (It is a second-order tensor, a quantity that we will not study in this text.) A *scalar* is a quantity that is completely characterized by only a magnitude. Temperature and density are two such quantities.

A vector will be denoted by placing an arrow* over the top of an italic letter, e.g., $\vec{A}$. The magnitude of the vector will be denoted by the letter without the arrow, e.g., A or with bars, e.g., $|\vec{A}|$. A vector $\vec{A}$ is graphically represented by an arrow that points in the direction of $\vec{A}$ and whose length is proportional to the magnitude of $\vec{A}$, as in Fig. 5.1. Two vectors are equal if they have the same magnitude and direction; they need not act at the same location. A vector with the same magnitude as $\vec{A}$ but acting in the opposite direction will be denoted $-\vec{A}$.

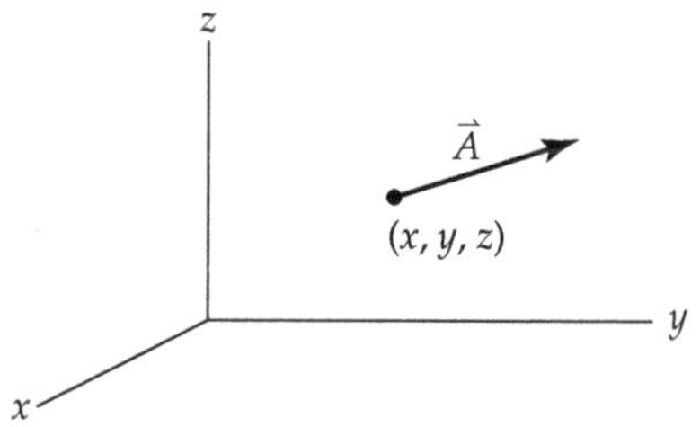

FIGURE 5.1 Graphical representation of a vector.

A *unit vector* is a vector having a magnitude of 1. If we divide a vector $\vec{A}$ by its magnitude A we obtain a unit vector in the direction of $\vec{A}$. Such a unit vector will be denoted $\hat{i}_A$, where the caret over the top signifies a unit vector. It is given by

$$\hat{i}_A = \frac{\vec{A}}{A} \tag{5.2.1}$$

Any vector can be represented by its unit vector times its magnitude.

Three unit vectors which are used extensively are the unit vectors in the coordinate directions of the rectangular, Cartesian reference frame (this reference frame will be used primarily in this chapter). No subscripts will be used to denote these unit vectors. They are $\hat{i}, \hat{j}$, and $\hat{k}$ acting in the x, y, and z directions, respectively, and are shown in Fig. 5.2.

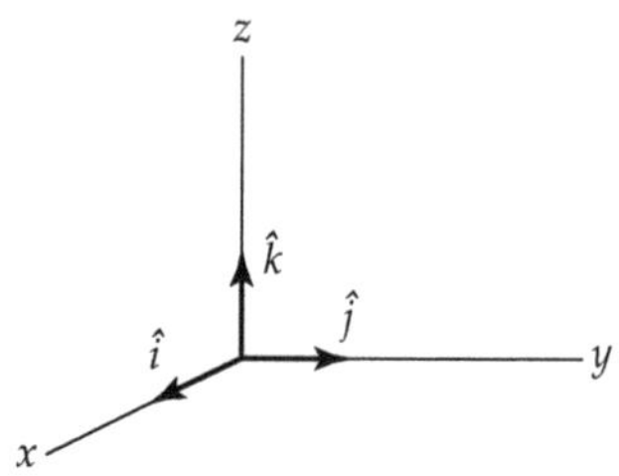

FIGURE 5.2 The three unit vectors, $\hat{i}, \hat{j}$, and $\hat{k}$.

*A vector is often denoted using a bold letter, but when writing a vector it is quite difficult to write bold letters, hence, the arrow is selected.

5.2.2 Addition and Subtraction

Two vectors $\vec{A}$ and $\vec{B}$ are added by placing the beginning of one at the tip of the other, as shown in Fig. 5.3. The sum $\vec{A}+\vec{B}$ is the vector obtained by connecting the point of beginning of the first vector with the tip of the second vector, as shown. If the two

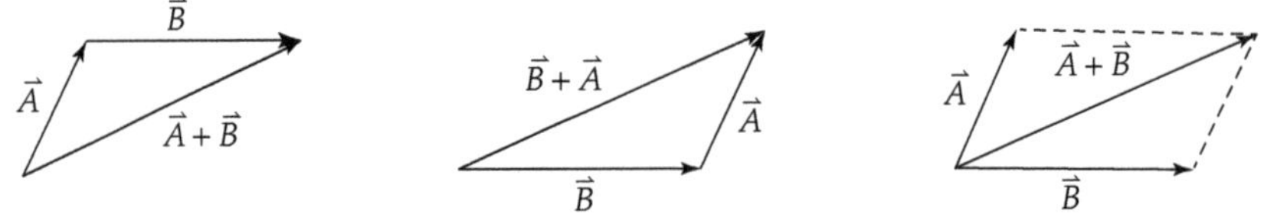

FIGURE 5.3 Vector addition.

vectors to be added are considered to be the sides of a parallelogram, their sum is given by the diagonal shown in the figure. It is clear from the geometry that vector addition is commutative, written as

$$\vec{A}+\vec{B}=\vec{B}+\vec{A} \tag{5.2.2}$$

Subtraction of vectors may be taken as a special case of addition; that is,

$$\vec{A}-\vec{B}=\vec{A}+(-\vec{B}) \tag{5.2.3}$$

Subtraction is illustrated in Fig. 5.4. Note that $\vec{A}-\vec{B}$ is the other diagonal (see Fig. 5.3) of the parallelogram whose sides are $\vec{A}$ and $\vec{B}$.

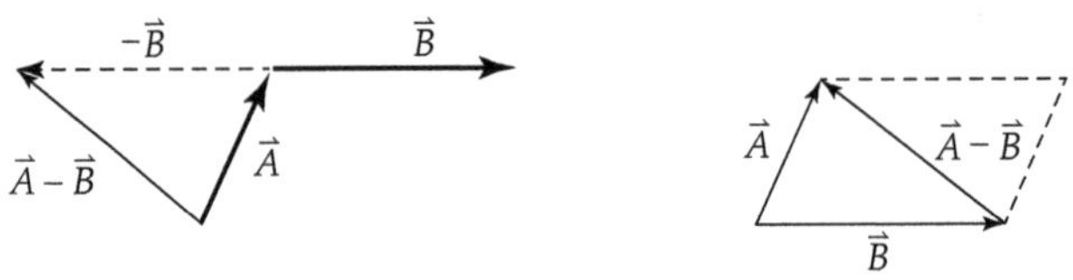

FIGURE 5.4 Vector subtraction.

Finally, we show in Fig. 5.5 a graphical illustration to demonstrate that vector addition is associative; that is,

$$\vec{A}+(\vec{B}+\vec{C})=(\vec{A}+\vec{B})+\vec{C} \tag{5.2.4}$$

Note that the resultant vector is simply written as $\vec{A}+\vec{B}+\vec{C}$.

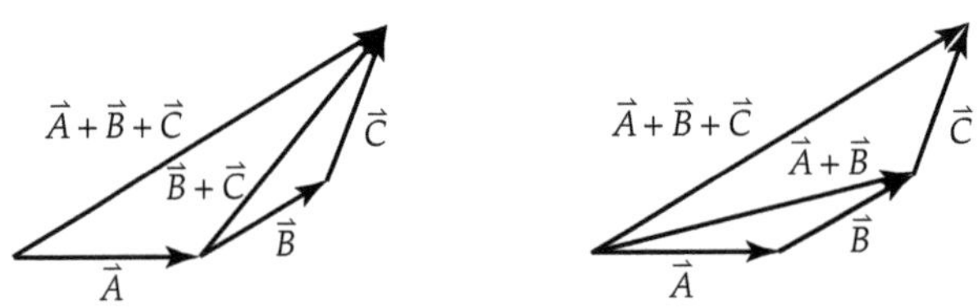

FIGURE 5.5 The associative property of vector addition.

5.2.3 Components of a Vector

So far we have not actually written a vector with magnitude and direction. To do so, we must choose a coordinate system and express the vector in terms of its *components*, which are the projections of the vector along the three coordinate directions. We shall illustrate using a rectangular, Cartesian coordinate system. Consider the beginning of the vector to be at the origin, as shown in Fig. 5.6. The projections on the x, y, and z axes are A_x, A_y, and A_z, respectively. Using the unit vectors defined earlier, we can then write the vector $\vec{A}$ as

$$\vec{A} = A_x\hat{i} + A_y\hat{j} + A_z\hat{k} \tag{5.2.5}$$

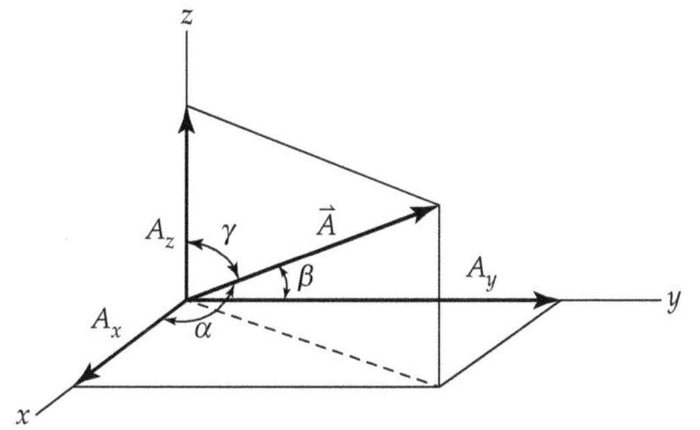

FIGURE 5.6 The components of a vector.

The vector $\vec{A}$ makes angles α, β, and γ with the x, y, and z axes, respectively. Thus, we have

$$A_x = A\cos\alpha, \qquad A_y = A\cos\beta, \qquad A_z = A\cos\gamma \tag{5.2.6}$$

where A, the magnitude of $\vec{A}$, is related geometrically to the components by

$$A = \sqrt{A_x^2 + A_y^2 + A_z^2} \tag{5.2.7}$$

Substituting Eqs. 5.2.6 into Eq. 5.2.7, we have the familiar result,

$$\cos^2\alpha + \cos^2\beta + \cos^2\gamma = 1 \tag{5.2.8}$$

Also, if we use Eqs. 5.2.6 in Eq. 5.2.5, we see that

$$\vec{A} = A(\cos\alpha\,\hat{i} + \cos\beta\,\hat{j} + \cos\gamma\,\hat{k}) \tag{5.2.9}$$

Comparing this with Eq. 5.2.1, we can express the unit vector $\hat{i}_A$ as

$$\hat{i}_A = \cos\alpha\,\hat{i} + \cos\beta\,\hat{j} + \cos\gamma\,\hat{k} \tag{5.2.10}$$

The quantities $\cos\alpha$, $\cos\beta$, and $\cos\gamma$ are often denoted ℓ, m, and n, respectively, and called the *direction cosines* of $\vec{A}$.

Two other coordinate systems are naturally encountered in physical situations, a cylindrical coordinate system and a spherical coordinate system. These coordinate systems will be presented in a subsequent section.

5.2.4 Multiplication

There are three distinct operations when considering multiplication involving vectors. First, a vector may be multiplied by a scalar. Consider the vector $\vec{A}$ multiplied by the scalar ϕ. The scalar ϕ simply multiplies each component of the vector and there results

$$\phi\vec{A} = \phi A_x\hat{i} + \phi A_y\hat{j} + \phi A_z\hat{k} \tag{5.2.11}$$

in rectangular coordinates. The resultant vector acts in the same direction as the vector $\vec{A}$, unless ϕ is negative, in which case $\phi\vec{A}$ acts in the opposite direction of $\vec{A}$.

The second multiplication is the *scalar product,* also known as the *dot product.* It involves the multiplication of two vectors so that a scalar quantity results. The scalar product is defined to be the product of the magnitudes of the two vectors and the cosine of the angle between the two vectors. This is written as

$$\vec{A}\cdot\vec{B} = AB\cos\theta \tag{5.2.12}$$

where θ is the angle shown in Fig. 5.7. Note that the scalar product $\vec{A}\cdot\vec{B}$ is equal to the length of $\vec{B}$ multiplied by the projection of $\vec{A}$ on $\vec{B}$, or the length of $\vec{A}$ multiplied by the projection of $\vec{B}$ on $\vec{A}$. We recall that the scalar quantity *work* was defined in much the same way, that is, force multiplied by the distance the force moved in the direction of the force, or

$$W = \vec{F}\cdot\vec{d} \tag{5.2.13}$$

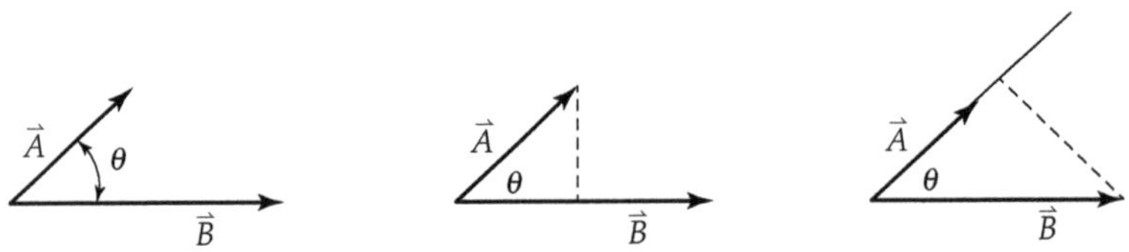

FIGURE 5.7 The dot product.

If the two vectors $\vec{A}$ and $\vec{B}$ are perpendicular, so that $\theta = 90°$, then $\vec{A}\cdot\vec{B} = 0$. This particular property of the dot product will be used quite often in applications. From the definition of the dot product we note that the dot product is commutative, so that

$$\vec{A}\cdot\vec{B} = \vec{B}\cdot\vec{A} \tag{5.2.14}$$

and the dot product is distributive, that is,

$$\vec{A}\cdot(\vec{B}+\vec{C}) = \vec{A}\cdot\vec{B} + \vec{A}\cdot\vec{C} \tag{5.2.15}$$

Using the unit vectors $\hat{i}$, $\hat{j}$, and $\hat{k}$, the definition of the dot product yields

$$\hat{i}\cdot\hat{i} = 1, \quad \hat{j}\cdot\hat{j} = 1, \quad \hat{k}\cdot\hat{k} = 1 \tag{5.2.16}$$

and

$$\hat{i}\cdot\hat{j} = 0, \quad \hat{i}\cdot\hat{k} = 0, \quad \hat{j}\cdot\hat{k} = 0 \tag{5.2.17}$$

These allow us to express the dot product in rectangular coordinates as

$$\vec{A}\cdot\vec{B} = (A_x\hat{i} + A_y\hat{j} + A_z\hat{k})\cdot(B_x\hat{i} + B_y\hat{j} + B_z\hat{k})$$
$$= A_xB_x + A_yB_y + A_zB_z \tag{5.2.18}$$

The dot product of a vector $\vec{A}$ with itself can be written as

$$\vec{A}\cdot\vec{A} = A^2 = A_x^2 + A_y^2 + A_z^2 \tag{5.2.19}$$

Finally, in our discussion of the dot product we note that the component of $\vec{A}$ in the x direction is found by taking the dot product of $\vec{A}$ with $\hat{i}$; that is, using Eq. 5.2.17,

$$\vec{A}\cdot\hat{i} = (A_x\hat{i} + A_y\hat{j} + A_z\hat{k})\cdot\hat{i} = A_x \tag{5.2.20}$$

Similarly,

$$\vec{A}\cdot\hat{j} = (A_x\hat{i} + A_y\hat{j} + A_z\hat{k})\cdot\hat{j} = A_y \tag{5.2.21}$$

Considering equations 5.2.6, these dot products can be related to the direction cosines. For example, $A\cos\alpha = \vec{A}\cdot\hat{i}$.

In general, the component of a vector in any direction is given by the dot product of the vector with a unit vector in the desired direction.

The third multiplication operation is the *vector* product, also called the *cross product*. It is a product between two vectors that yields a third vector, the magnitude of which is defined to be the product of the magnitudes of the two vectors and the sine of the angle between them. The third vector acts in a direction perpendicular to the plane of the two vectors so that the three vectors form a right-handed set of vectors. We write the cross product as

$$\vec{C} = \vec{A}\times\vec{B} \tag{5.2.22}$$

The magnitude of $\vec{C}$ is given by

$$C = AB\sin\theta \tag{5.2.23}$$

The vectors are shown in Fig. 5.8. We see that the magnitude of $\vec{C}$ is equal to the area of the parallelogram with sides $\vec{A}$ and $\vec{B}$.

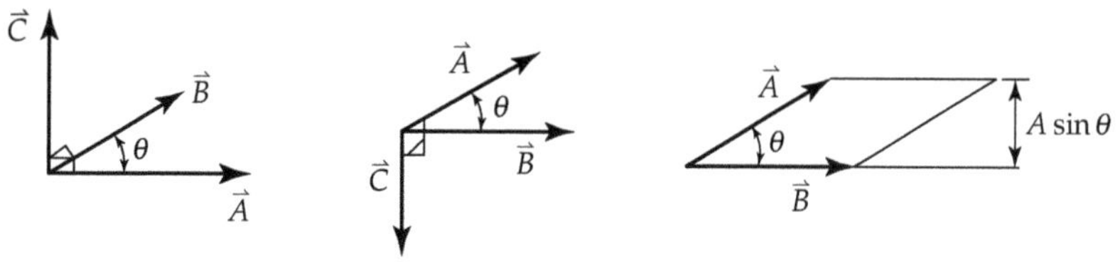

FIGURE 5.8 The cross product.

There are two common techniques for determining the sense of the vector $\vec{C}$. First $\vec{C}$ acts in the direction of the advance of a right-handed screw as it is turned from $\vec{A}$ to $\vec{B}$. Second, if the fingers curl $\vec{A}$ into $\vec{B}$, the thumb will point in the direction of $\vec{C}$.

From the definition we see that the cross product is not commutative, since

$$\vec{A}\times\vec{B} = -\vec{B}\times\vec{A} \tag{5.2.24}$$

However, it is true that

$$\vec{A} \times (\vec{B} + \vec{C}) = \vec{A} \times \vec{B} + \vec{A} \times \vec{C} \tag{5.2.25}$$

If two vectors act in the same direction, the angle θ is zero degrees and the cross product vanishes. It follows that

$$\vec{A} \times \vec{A} = 0 \tag{5.2.26}$$

The unit vectors $\hat{i}$, $\hat{j}$, and $\hat{k}$ form the cross products

$$\hat{i} \times \hat{i} = 0, \quad \hat{j} \times \hat{j} = 0, \quad \hat{k} \times \hat{k} = 0 \tag{5.2.27}$$

and

$$\begin{aligned} \hat{i} \times \hat{j} &= \hat{k}, & \hat{j} \times \hat{k} &= \hat{i}, & \hat{k} \times \hat{i} &= \hat{j}, \\ \hat{j} \times \hat{i} &= -\hat{k}, & \hat{k} \times \hat{j} &= -\hat{i}, & \hat{i} \times \hat{k} &= -\hat{j} \end{aligned} \tag{5.2.28}$$

These relationships are easily remembered by visualizing a display of unit vectors. The cross product of a unit vector into its neighbor is the following vector when going clockwise, and is the negative of the following vector when going counterclockwise.

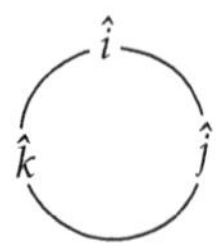

Using the relationship above we can express the cross product of $\vec{A}$ and $\vec{B}$ in rectangular coordinates as

$$\begin{aligned} \vec{A} \times \vec{B} &= (A_x\hat{i} + A_y\hat{j} + A_z\hat{k}) \times (B_x\hat{i} + B_y\hat{j} + B_z\hat{k}) \\ &= (A_yB_z + A_zB_y)\hat{i} + (A_zB_x - A_xB_z)\hat{j} + (A_xB_y - A_yB_x)\hat{k} \end{aligned} \tag{5.2.29}$$

A convenient way to recall this expansion of the cross product is to utilize a determinant formed by the unit vectors, the components of $\vec{A}$, and the components of $\vec{B}$. The cross product of $\vec{A} \times \vec{B}$ is then related to the determinant by

$$\vec{A} \times \vec{B} = \begin{vmatrix} \hat{i} & \hat{j} & \hat{k} \\ A_x & A_y & A_z \\ B_x & B_y & B_z \end{vmatrix} \tag{5.2.30}$$

Two applications of the cross product are the torque $\vec{T}$ produced about a point by a force $\vec{F}$ acting at the distance $\vec{r}$ from the point, and the velocity $\vec{V}$ induced by an angular velocity $\vec{\omega}$ at a point $\vec{r}$ from the axis of rotation. The magnitude of the torque is given by the magnitude of $\vec{F}$ multiplied by the perpendicular distance from the point to the line of action of the force, that is,

$$T = Fd \tag{5.2.31}$$

where d is $r \sin \theta$; see Fig. 5.9a. It can be represented by a vector normal to the plane of $\vec{F}$ and $\vec{r}$, given by

$$\vec{T} = \vec{r} \times \vec{F} \tag{5.2.32}$$

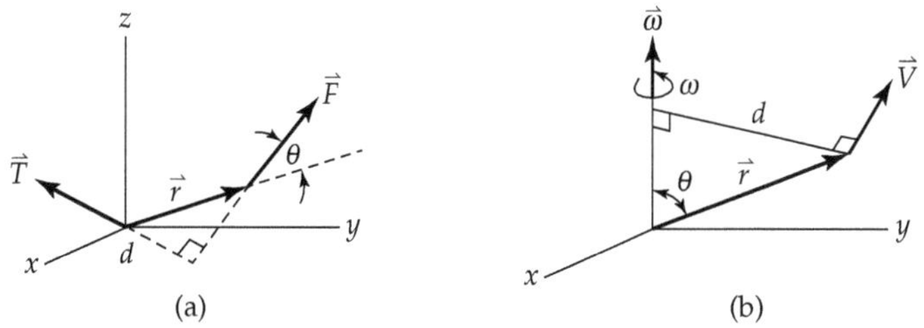

FIGURE 5.9 Examples of the cross product.

We are often interested in the torque produced about an axis, for example the z axis. It would be the vector $\vec{T}$ dotted with the unit vector in the direction of the axis. About the z axis it would be

$$T_z = \vec{T} \cdot \hat{k} \tag{5.2.33}$$

The magnitude of the velocity induced by an angular velocity $\vec{\omega}$ is the magnitude ω of the angular velocity multiplied by the perpendicular distance d from the axis to the point where the velocity is desired, as shown in Fig. 5.9b. If $\vec{r}$ is the position vector from the origin of a coordinate system to the point where the velocity is desired, then $r \sin \theta$ is the distance d, where θ is the angle between $\vec{\omega}$ and $\vec{r}$. The velocity $\vec{V}$ is then given by

$$\vec{V} = \vec{\omega} \times \vec{r} \tag{5.2.34}$$

where in the figure we have let the axis of rotation be the z axis. Note that the vector $\vec{V}$ is perpendicular to the plane of $\vec{\omega}$ and $\vec{r}$.

In concluding this section on multiplication, we shall present the *scalar triple product.* Consider three vectors $\vec{A}, \vec{B}$, and $\vec{C}$, shown in Fig. 5.10. The scalar triple product is the dot product of one of the vectors with the cross product of the remaining two. For example, the product $(\vec{A} \times \vec{B}) \cdot \vec{C}$ is a scalar quantity given by

$$(\vec{A} \times \vec{B}) \cdot \vec{C} = ABC \sin \alpha \cos \beta \tag{5.2.35}$$

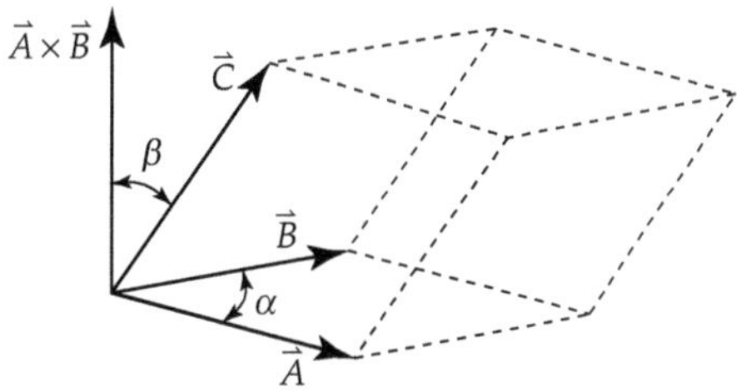

FIGURE 5.10 The scalar triple product.

where α is the angle between $\vec{A}$ and $\vec{B}$, and β is the angle between $\vec{A} \times \vec{B}$ and $\vec{C}$. The quantity $AB \sin \alpha$ is the area of the parallelogram with sides $\vec{A}$ and $\vec{B}$. The quantity $C \cos \beta$ is the component of $\vec{C}$ in a direction perpendicular to the parallelogram with sides $\vec{A}$ and $\vec{B}$. Thus, the scalar triple product represents the volume of the parallelepiped with sides $\vec{A}$, $\vec{B}$, and $\vec{C}$. Since the volume is the same regardless of how we form the product, we see that

$$(\vec{A} \times \vec{B}) \cdot \vec{C} = \vec{A} \cdot (\vec{B} \times \vec{C}) = (\vec{C} \times \vec{A}) \cdot \vec{B} \tag{5.2.36}$$

Also, the parentheses in Eq. 5.2.36 are usually omitted since the cross product must be performed first. If the dot product were performed first, the quantity would be meaningless since the cross product requires two vectors.

Using rectangular coordinates the scalar triple product is

$$\vec{A}\times\vec{B}\cdot\vec{C} = C_x(A_yB_z - A_zB_y) + C_y(A_zB_x - A_xB_z) + C_z(A_xB_y - A_yB_x)$$
$$= \begin{vmatrix} A_x & A_y & A_z \\ B_x & B_y & B_z \\ C_x & C_y & C_z \end{vmatrix} \tag{5.2.37}$$

The vector triple product will be presented in an Example.

Example 5.1

Prove that the diagonals of a parallelogram bisect each other, as shown in Fig. 5.11.

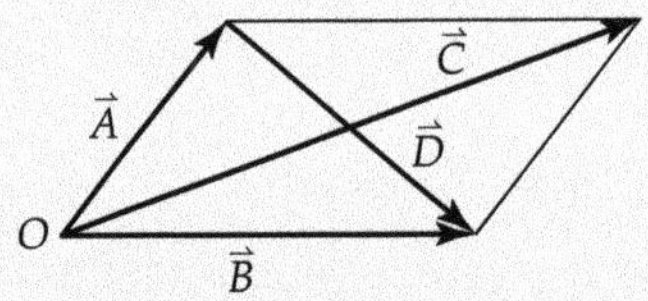

FIGURE 5.11

Solution

From the parallelogram shown, we observe that

$$\vec{A}+\vec{B}=\vec{C}, \qquad \vec{B}-\vec{A}=\vec{D}$$

The vector from point O to the intersection of the two diagonals is some fraction of $\vec{C}$, say $m\vec{C}$; and the vector representing part of the shorter diagonal is assumed to be $n\vec{D}$, displayed in Fig. 5.12. If m and n are both $\frac{1}{2}$, then the diagonals bisect each other.

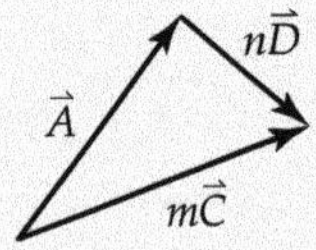

FIGURE 5.12

Now, using the triangle formed with the vectors shown, we can write

$$\vec{A}+n\vec{D}=m\vec{C}$$

Substituting for the vectors $\vec{C}$ and $\vec{D}$, the equation above becomes

$$\vec{A}+n(\vec{B}-\vec{A})=m(\vec{A}+\vec{B})$$

Rearranging, we have

$$(1-n-m)\vec{A}=(m-n)\vec{B}$$

The quantity on the left is a vector in the direction of $\vec{A}$, and the quantity on the right is a vector in the direction of $\vec{B}$. Since the direction of $\vec{A}$ is different from the direction of $\vec{B}$, we must demand that the coefficients be zero. Thus,

$$1-n-m=0$$
$$m-n=0$$

The solution to this set of equations is

$$n=m=\frac{1}{2}$$

Hence, the diagonals bisect each other.

Example 5.2

For the vectors $\vec{A}=3\hat{i}-2\hat{j}+\hat{k}$ and $\vec{B}=2\hat{i}-\hat{k}$, determine (a) $\vec{A}-\vec{B}$, (b) $\vec{A}\cdot\vec{B}$, and (c) $\vec{A}\times\vec{B}\cdot\vec{A}$.

Solution

a) To find the difference of two vectors, we simply find the difference of the respective components. We have

$$\begin{aligned}\vec{A}-\vec{B}&=(3\hat{i}-2\hat{j}+\hat{k})-(2\hat{i}-\hat{k})\\&=(3-2)\hat{i}-2\hat{j}+(1+1)\hat{k}=\hat{i}-2\hat{j}+2\hat{k}\end{aligned}$$

b) The dot product is given by

$$\begin{aligned}\vec{A}\cdot\vec{B}&=(3\hat{i}-2\hat{j}+\hat{k})\cdot(2\hat{i}-\hat{k})\\&=6\hat{i}\cdot\hat{i}-3\hat{i}\cdot\hat{k}-4\hat{j}\cdot\hat{i}+2\hat{j}\cdot\hat{k}+2\hat{k}\cdot\hat{i}-\hat{k}\cdot\hat{k}\\&=6-1=\underline{5}\end{aligned}$$

c) To perform the indicated product we must first find $\vec{A}\times\vec{B}$. It is

$$\begin{aligned}\vec{A}\times\vec{B}&=(A_yB_z-A_zB_y)\hat{i}+(A_zB_x-A_xB_z)\hat{j}+(A_xB_y-A_yB_x)\hat{k}\\&=[(-2)(-1)-1\cdot 0]\hat{i}+[1\cdot 2-3(-1)]\hat{j}+[3\cdot 0-(-2)(2)]\hat{k}\\&=2\hat{i}+5\hat{j}+4\hat{k}\end{aligned}$$

We then dot this vector with $\vec{A}$ and obtain

$$\vec{A} \times \vec{B} \cdot \vec{A} = (2\hat{i} + 5\hat{j} + 4\hat{k}) \cdot (3\hat{i} - 2\hat{j} + \hat{k})$$
$$= 6 - 10 + 4 = \underline{0}$$

We are not surprised that we get zero, since the vector $\vec{A} \times \vec{B}$ is perpendicular to $\vec{A}$, and the dot product of two perpendicular vectors is always zero, since the cosine of the angle between the two vectors is zero.

Example 5.3

Find a unit vector in the direction of $\vec{A} = 2\hat{i} + 3\hat{j} + 6\hat{k}$.

Solution

The magnitude of the vector $\vec{A}$ is

$$A = \sqrt{A_x^2 + A_y^2 + A_z^2}$$
$$= \sqrt{2^2 + 3^2 + 6^2} = 7$$

The unit vector is then

$$\hat{i}_A = \frac{\vec{A}}{A}$$
$$= \frac{2\hat{i} + 3\hat{j} + 6\hat{k}}{7} = \frac{2}{7}\hat{i} + \frac{3}{7}\hat{j} + \frac{6}{7}\hat{k}$$

Example 5.4

Using the definition of the dot product of the two vectors displayed in Fig. 5.13, show that $\cos(\alpha - \beta) = \cos\alpha \cos\beta + \sin\alpha \sin\beta$.

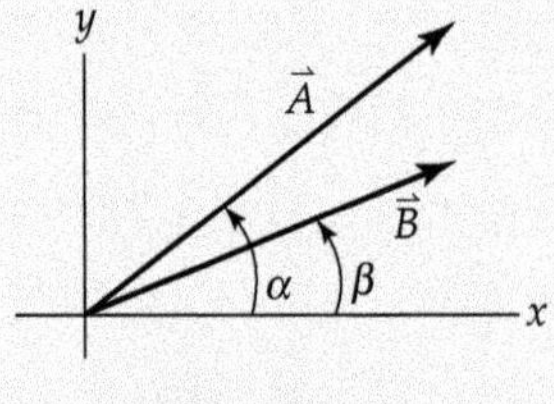

FIGURE 5.13

Solution

The dot product of the two vectors $\vec{A}$ and $\vec{B}$ is

$$\vec{A} \cdot \vec{B} = AB\cos\theta$$

where θ is the angle between the two vectors, that is,

$$\theta = \alpha - \beta$$

We know that

$$A = \sqrt{A_x^2 + A_y^2}, \qquad B = \sqrt{B_x^2 + B_y^2}$$

and

$$\vec{A} \cdot \vec{B} = A_x B_x + A_y B_y$$

Thus,

$$\cos\theta = \frac{\vec{A} \cdot \vec{B}}{AB} = \frac{A_x B_x + A_y B_y}{\sqrt{A_x^2 + A_y^2}\sqrt{B_x^2 + B_y^2}}$$

This can be written as

$$\cos(\alpha - \beta) = \frac{A_x}{\sqrt{A_x^2 + A_y^2}} \frac{B_x}{\sqrt{B_x^2 + B_y^2}} + \frac{A_y}{\sqrt{A_x^2 + A_y^2}} \frac{B_y}{\sqrt{B_x^2 + B_y^2}} = \cos\alpha\cos\beta + \sin\alpha\sin\beta$$

and the trigonometric identity is verified.

Example 5.5

Find the projection of $\vec{A}$ on $\vec{B}$, if $\vec{A} = 12\hat{i} - 3\hat{j} + 6\hat{k}$ and $\vec{B} = 2\hat{i} + 4\hat{j} + 4\hat{k}$.

Solution

Let us first find a unit vector $\hat{i}_B$ in the direction of $\vec{B}$. Then the projection of $\vec{A}$ on $\vec{B}$ will be $\vec{A} \cdot \hat{i}_B$. We have

$$\hat{i}_B = \frac{\vec{B}}{B} = \frac{2\hat{i} + 4\hat{j} + 4\hat{k}}{\sqrt{2^2 + 4^2 + 4^2}} = \frac{2\hat{i} + 4\hat{j} + 4\hat{k}}{6}$$

The projection of $\vec{A}$ on $\vec{B}$ is then

$$\vec{A} \cdot \hat{i}_B = (12\hat{i} - 3\hat{j} + 6\hat{k}) \cdot \left(\frac{1}{3}\hat{i} + \frac{2}{3}\hat{j} + \frac{2}{3}\hat{k}\right) = 4 - 2 + 4 = \underline{6}$$

Example 5.6

Find a unit vector $\hat{i}_C$ perpendicular to the plane of $\vec{A}$ and $\vec{B}$, if $\vec{A} = 2\hat{i} + 3\hat{j}$ and $\vec{B} = \hat{i} - \hat{j} + 2\hat{k}$.

Solution

Let $\vec{C} = \vec{A} \times \vec{B}$. Then $\vec{C}$ is perpendicular to the plane of $\vec{A}$ and $\vec{B}$. It is given by

$$\vec{C} = \vec{A} \times \vec{B}$$
$$= \begin{vmatrix} \hat{i} & \hat{j} & \hat{k} \\ 2 & 3 & 0 \\ 1 & -1 & 2 \end{vmatrix} = 6\hat{i} - 4\hat{j} - 5\hat{k}$$

The unit vector is then

$$\hat{i}_C = \frac{\vec{C}}{C}$$
$$= \frac{6\hat{i} - 4\hat{j} - 5\hat{k}}{\sqrt{6^2 + 4^2 + 5^2}} = 0.684\hat{i} - 0.456\hat{j} - 0.570\hat{k}$$

Example 5.7

Find an equivalent vector expression for the vector triple product $(\vec{A} \times \vec{B}) \times \vec{C}$.

Solution

We will expand the triple product in rectangular coordinates. First, the cross product $\vec{A} \times \vec{B}$ is

$$\vec{A} \times \vec{B} = (A_yB_z - A_zB_y)\hat{i} + (A_zB_x - A_xB_z)\hat{j} + (A_xB_y - A_yB_x)\hat{k}$$

Now, write the cross product of the vector above with $\vec{C}$. It is

$$(\vec{A} \times \vec{B}) \times \vec{C} = [(A_zB_x - A_xB_z)C_z - (A_xB_y - A_yB_x)C_y]\hat{i}$$
$$+ [(A_xB_y - A_yB_x)C_x - (A_yB_z - A_zB_y)C_z]\hat{j}$$
$$+ [(A_yB_z - A_zB_y)C_y - (A_zB_x - A_xB_z)C_x]\hat{k}$$

The above equation can be rearranged in the form

$$\begin{aligned}(\vec{A}\times\vec{B})\times\vec{C} = {} & (A_zC_z + A_yC_y + A_xC_x)B_x\hat{i} \\ & \qquad - (B_zC_z + B_yC_y + B_xC_x)A_x\hat{i} \\ & + (A_xC_x + A_zC_z + A_yC_y)B_y\hat{j} \\ & \qquad - (B_xC_x + B_zC_z + B_yC_y)A_y\hat{j} \\ & + (A_yC_y + A_xC_x + A_zC_z)B_z\hat{k} \\ & \qquad - (B_yC_y + B_xC_x + B_zC_z)A_z\hat{k}\end{aligned}$$

where the last terms in the parentheses have been inserted so that they cancel each other but help to form a dot product. Now we recognize that the equation above can be written as

$$\begin{aligned}(\vec{A}\times\vec{B})\times\vec{C} = {} & (\vec{A}\cdot\vec{C})B_x\hat{i} + (\vec{A}\cdot\vec{C})B_y\hat{j} + (\vec{A}\cdot\vec{C})B_z\hat{k} \\ & -(\vec{B}\cdot\vec{C})A_x\hat{i} - (\vec{B}\cdot\vec{C})A_y\hat{j} - (\vec{B}\cdot\vec{C})A_z\hat{k}\end{aligned}$$

or, finally,

$$(\vec{A}\times\vec{B})\times\vec{C} = (\vec{A}\cdot\vec{C})\vec{B} - (\vec{B}\cdot\vec{C})\vec{A}$$

Similarly, we could show that

$$\vec{A}\times(\vec{B}\times\vec{C}) = (\vec{A}\cdot\vec{C})\vec{B} - (\vec{B}\cdot\vec{A})\vec{C}$$

Note that

$$(\vec{A}\times\vec{B})\times\vec{C} \neq \vec{A}\times(\vec{B}\times\vec{C})$$

5.3 VECTOR DIFFERENTIATION

5.3.1 Ordinary Differentiation

A vector function may depend on one independent variable, for example $\vec{u} = \vec{u}(t)$. The derivative of the vector $\vec{u}(t)$ with respect to t is defined, as usual, to be

$$\frac{d\vec{u}}{dt} = \lim_{\Delta t\to 0}\frac{\vec{u}(t+\Delta t) - \vec{u}(t)}{\Delta t} \tag{5.3.1}$$

where

$$\vec{u}(t+\Delta t) - \vec{u}(t) = \Delta\vec{u} \tag{5.3.2}$$

This is illustrated in Fig. 5.14. Note that the direction of $\Delta\vec{u}$ is, in general, unrelated to the direction of $\vec{u}(t)$.

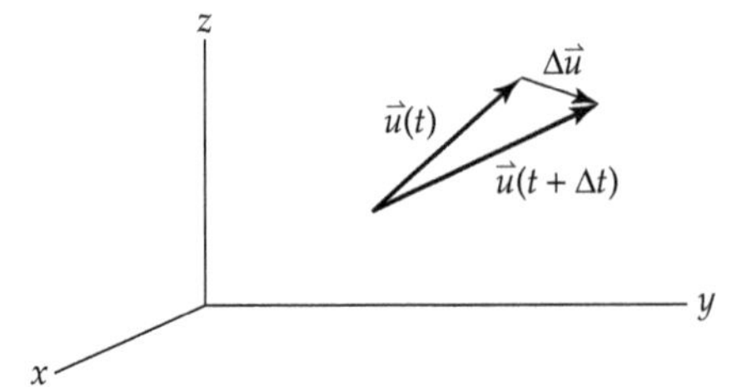

FIGURE 5.14 Vectors used in the definition of the derivative $d\vec{u}/dt$.

From this definition it follows that the sums and products involving vector quantities can be differentiated as in ordinary calculus; that is,

$$\begin{aligned}\frac{d}{dt}(\phi\,\vec{u}) &= \phi\frac{d\vec{u}}{dt} + \vec{u}\frac{d\phi}{dt}\\ \frac{d}{dt}(\vec{u}\cdot\vec{v}) &= \vec{u}\cdot\frac{d\vec{v}}{dt} + \vec{v}\cdot\frac{d\vec{u}}{dt}\\ \frac{d}{dt}(\vec{u}\times\vec{v}) &= \vec{u}\times\frac{d\vec{v}}{dt} + \vec{v}\times\frac{d\vec{u}}{dt}\end{aligned} \tag{5.3.3}$$

If we express the vector $\vec{u}(t)$ in rectangular coordinates, as

$$\vec{u}(t) = u_x\hat{i} + u_y\hat{j} + u_z\hat{k} \tag{5.3.4}$$

it can be differentiated term by term to yield

$$\frac{d\vec{u}}{dt} = \frac{dv_x}{dt}\hat{i} + \frac{du_y}{dt}\hat{j} + \frac{du_z}{dt}\hat{k} \tag{5.3.5}$$

The noninertial reference frame can translate and rotate relative to the inertial frame, which is fixed.

provided that the unit vectors $\hat{i}, \hat{j}$, and $\hat{k}$ are independent of t. If t represents time, such a reference frame is referred to as an *inertial reference frame.*

We shall illustrate differentiation by considering the motion of a particle in a noninertial reference frame. Let us calculate the velocity and acceleration of such a particle. The particle occupies the position (x, y, z) measured in the noninertial xyz reference frame which is rotating with an angular velocity $\vec{\omega}$, as shown in Fig. 5.15. The xyz

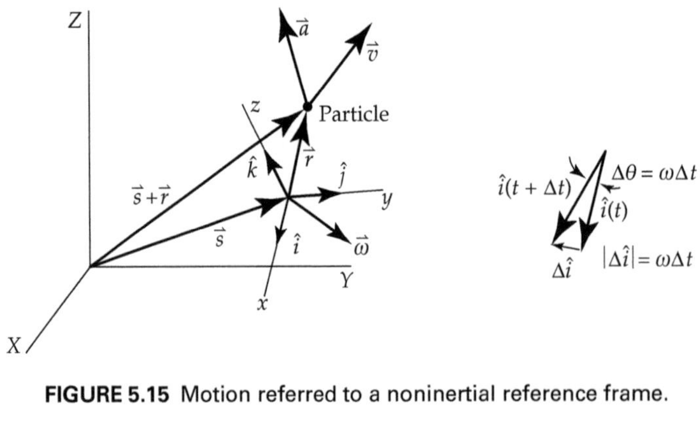

FIGURE 5.15 Motion referred to a noninertial reference frame.

reference frame is located by the vector $\vec{s}$ relative to the inertial XYZ reference frame.* The velocity $\vec{V}$ referred to the XYZ frame is

$$\vec{V} = \frac{d}{dt}(\vec{s} + \vec{r}) = \frac{d\vec{s}}{dt} + \frac{d\vec{r}}{dt} \tag{5.3.6}$$

The quantity $d\vec{s}/dt$ is the velocity of the xyz reference frame and is denoted $\vec{V}_{\text{ref}}$. The vector $d\vec{r}/dt$ is, using $\vec{r} = x\hat{i} + y\hat{j} + z\hat{k}$,

The non-inertial reference frame translates and rotates, and its unit vectors $\hat{i}, \hat{j}$, and $\hat{k}$ rotate (only) with it.

$$\frac{d\vec{r}}{dt} = \frac{dx}{dt}\hat{i} + \frac{dy}{dt}\hat{j} + \frac{dz}{dt}\hat{k} + x\frac{d\hat{i}}{dt} + y\frac{d\hat{j}}{dt} + z\frac{d\hat{k}}{dt} \tag{5.3.7}$$

To determine an expression for the time derivatives of the unit vectors, which are due to the angular velocity $\vec{\omega}$ of the xyz frame, consider the unit vector $\hat{i}$ to rotate through a small angle during the time Δt, illustrated in Fig. 5.15. Using the definition of a derivative, there results

$$\begin{aligned}\frac{d\hat{i}}{dt} &= \lim_{\Delta t \to 0} \frac{\hat{i}(t+\Delta t) - \hat{i}(t)}{\Delta t} \\ &= \lim_{\Delta t \to 0} \frac{\Delta \hat{i}}{\Delta t} = \lim_{\Delta t \to 0} \frac{\omega \Delta t \left(\dfrac{\vec{\omega} \times \hat{i}}{\omega} \right)}{\Delta t} = \vec{\omega} \times \hat{i}\end{aligned} \tag{5.3.8}$$

where the quantity $\vec{\omega} \times \hat{i}/\omega$ is a unit vector perpendicular to $\hat{i}$ in the direction of $\Delta\hat{i}$. Similarly,

$$\frac{d\hat{j}}{dt} = \vec{\omega} \times \hat{j}, \qquad \frac{d\hat{k}}{dt} = \vec{\omega} \times \hat{k} \tag{5.3.9}$$

Substituting these and Eq. 5.3.7 in to Eq. 5.3.6, we have

$$\vec{V} = \vec{V}_{\text{ref}} + \frac{dx}{dt}\hat{i} + \frac{dy}{dt}\hat{j} + \frac{dz}{dt}\hat{k} + x\vec{\omega} \times \hat{i} + y\vec{\omega} \times \hat{j} + z\vec{\omega} \times \hat{k} \tag{5.3.10}$$

The velocity $\vec{v}$ of the particle relative to the xyz frame is

$$\vec{v} = \frac{dx}{dt}\hat{i} + \frac{dy}{dt}\hat{j} + \frac{dz}{dt}\hat{k} \tag{5.3.11}$$

Hence, we can write the expression for the absolute velocity as

$$\vec{V} = \vec{V}_{\text{ref}} + \vec{v} + \vec{\omega} \times \vec{r} \tag{5.3.12}$$

The absolute acceleration $\vec{A}$ is obtained by differentiating $\vec{V}$ with respect to time to obtain

$$\vec{A} = \frac{d\vec{V}}{dt} = \frac{d\vec{V}_{\text{ref}}}{dt} + \frac{d\vec{v}}{dt} + \frac{d\vec{\omega}}{dt} \times \vec{r} + \vec{\omega} \times \frac{d\vec{r}}{dt} \tag{5.3.13}$$

*This xyz reference frame would be attached to the ground for the case of a projectile or a rotating device such as a dishwasher arm, or it would be attached to the sun when describing the motion of satellites.

In this equation

$$\frac{d\vec{V}_{\text{ref}}}{dt} = \vec{A}_{\text{ref}} \tag{5.3.14}$$

$$\begin{aligned}\frac{d\vec{v}}{dt} &= \frac{d}{dt}(v_x\hat{i} + v_y\hat{j} + v_z\hat{k}) \\ &= \frac{dv_x}{dt}\hat{i} + \frac{dv_y}{dt}\hat{j} + \frac{dv_z}{dt}\hat{k} + v_x\frac{d\hat{i}}{dt} + v_y\frac{d\hat{j}}{dt} + v_z\frac{d\hat{k}}{dt} \\ &= \vec{a} + \vec{\omega} + \vec{v}\end{aligned} \tag{5.3.15}$$

$$\frac{d\vec{r}}{dt} = \vec{v} + \vec{\omega}\times\vec{r} \tag{5.3.16}$$

where $d\vec{v}/dt$ comes from Eq. 5.3.7 and $\vec{a}$ is the acceleration of the particle observed in the *xyz* frame. The absolute acceleration is then

$$\vec{A} = \vec{A}_{\text{ref}} + \vec{a} + \vec{\omega} + \vec{v} + \frac{d\vec{\omega}}{dt}\times\vec{r} + \vec{\omega}\times(\vec{v} + \vec{\omega}\times\vec{r}) \tag{5.3.17}$$

This is reorganized in the form

$$\vec{A} = \vec{A}_{\text{ref}} + \vec{a} + 2\vec{\omega}\times\vec{v} + \vec{\omega}\times(\vec{\omega}\times\vec{r}) + \frac{d\vec{\omega}}{dt}\times\vec{r} \tag{5.3.18}$$

The quantity $2\vec{\omega}\times\vec{v}$ is often referred to as the *Coriolis acceleration*, and $d\vec{\omega}/dt$ is the angular acceleration of the *xyz* frame. The $\vec{\omega}\times(\vec{\omega}\times\vec{r})$ term is the centripetal acceleration. For a rigid body $\vec{a}$ and $\vec{v}$ would be zero.

Example 5.8

Using the definition of a derivative, show that

$$\frac{d}{du}(\vec{u}\cdot\vec{v}) = \vec{u}\cdot\frac{d\vec{v}}{dt} + \vec{v}\cdot\frac{d\vec{u}}{dt}$$

Solution

The definition of a derivative allows us to write

$$\frac{d}{dt}(\vec{u}\cdot\vec{v}) = \lim_{\Delta t\to 0}\frac{\vec{u}(t+\Delta t)\cdot\vec{v}(t+\Delta t) - \vec{u}(t)\cdot\vec{v}(t)}{\Delta t}$$

But we know that (see Fig. 5.14)

$$\vec{u}(t+\Delta t) - \vec{u}(t) = \Delta\vec{u}$$

$$\vec{v}(t+\Delta t) - \vec{v}(t) = \Delta\vec{v}$$

Substituting for $\vec{u}(t+\Delta t)$ and $\vec{v}(t+\Delta t)$, there results

$$\frac{d}{dt}(\vec{u}\cdot\vec{v}) = \lim_{\Delta t\to 0}\frac{[\Delta\vec{u} + \vec{u}(t)]\cdot[\Delta\vec{v} + \vec{v}(t)] - \vec{u}(t)\cdot\vec{v}(t)}{\Delta t}$$

This product is expanded to yield

$$\frac{d}{dt}(\vec{u}\cdot\vec{v}) = \lim_{\Delta t\to 0}\frac{\Delta\vec{u}\cdot\Delta\vec{v} + \vec{u}\cdot\Delta\vec{v} + \vec{v}\cdot\Delta\vec{u} + \cancel{\vec{u}\cdot\vec{v}} - \cancel{\vec{u}\cdot\vec{v}}}{\Delta t}$$

In the limit as $\Delta t \to 0$, both $\Delta\vec{u} \to 0$ and $\Delta\vec{v} \to 0$. Hence,

$$\lim_{\Delta t\to 0}\frac{\Delta\vec{u}\cdot\Delta\vec{v}}{\Delta t} \to 0$$

We are left with

$$\begin{aligned}\frac{d}{dt}(\vec{u}\cdot\vec{v}) &= \lim_{\Delta t\to 0}\left(\vec{u}\cdot\frac{\Delta\vec{v}}{\Delta t} + \vec{v}\cdot\frac{\Delta\vec{u}}{\Delta t}\right)\\ &= \vec{u}\cdot\frac{d\vec{v}}{dt} + \vec{v}\cdot\frac{d\vec{u}}{dt}\end{aligned}$$

and the given relationship is shown to be true.

Example 5.9

The position of a particle is given by $\vec{r} = t^2\hat{i} + 2\hat{j} + 5(t-1)\hat{k}$ meters, measured in a reference frame that has no translational velocity but that has an angular velocity of 20 rad/s about the z axis. Determine the absolute velocity at $t = 2$ s.

Solution

Given that $\vec{V}_{\text{ref}} = 0$ the absolute velocity is

$$\vec{V} = \vec{v} + \vec{\omega}\times\vec{r}$$

The velocity, as viewed from the rotating reference frame, is

$$\begin{aligned}\vec{v} &= \frac{d\vec{r}}{dt} = \frac{d}{dt}[t^2\hat{i} + 2\hat{j} + 5(t-1)\hat{k}]\\ &= 2t\hat{i} + 5\hat{k}\end{aligned}$$

The contribution due to the angular velocity is

$$\begin{aligned}\vec{\omega}\times\vec{r} &= 20\hat{k}\times[t^2\hat{i} + 2\hat{j} + 5(t-1)\hat{k}]\\ &= 20t^2\hat{j} - 40\hat{i}\end{aligned}$$

Thus, the absolute velocity is

$$\vec{V} = 2t\hat{i} + 5\hat{k} + 20t^2\hat{j} - 40\hat{i}$$
$$= (2t - 40)\hat{i} + 20t^2\hat{j} + 5\hat{k}$$

At $t = 2$ s this becomes

$$\vec{V} = -36\hat{i} + 80\hat{j} + 5\hat{k} \quad \text{m/s}$$

Example 5.10

A person is walking toward the center of a merry-go-round along a radial line at a constant rate of 6 m/s. The angular velocity of the merry-go-round is 1.2 rad/s. Calculate the absolute acceleration when the person reaches a position 3 m from the axis of rotation.

Solution

The acceleration $\vec{A}_{\text{ref}}$ is assumed to be zero, as is the angular acceleration $d\vec{\omega}/dt$ of the merry-go-round. Also, the acceleration $\vec{a}$ of the person relative to the merry-go-round is zero. Thus, the absolute acceleration is

$$\vec{A} = 2\vec{\omega} \times \vec{v} + \vec{\omega} \times (\vec{\omega} \times \vec{r})$$

Attach the xyz reference frame to the merry-go-round with the z axis vertical and the person walking along the x axis toward the origin. Then

$$\vec{\omega} = 1.2\hat{k}, \qquad \vec{r} = 3\hat{i}, \qquad \vec{v} = -6\hat{i}$$

The absolute acceleration is then

$$\vec{A} = 2[1.2\hat{k} \times (-6\hat{i})] + 1.2\hat{k} \times (1.2\hat{k} \times 3\hat{i})$$
$$= -4.32\hat{i} - 14.4\hat{j} \quad \text{m/s}^2$$

Note the y component of acceleration that is normal to the direction of motion, which makes the person sense a tugging in that direction.

5.3.2 Partial Differentiation

Many phenomena require that a quantity be defined at all points in a region of interest. The quantity may also vary with time. Such quantities are often referred to as *field quantities:* electric fields, magnetic fields, velocity fields, and pressure fields are examples. Partial derivatives are necessary when describing fields. Consider a vector function $\vec{u}(x, y, z, t)$. The partial derivative of $\vec{u}$ with respect to x is defined to be

$$\frac{\partial \vec{u}}{\partial x} = \lim_{\Delta x \to 0} \frac{\vec{u}(x + \Delta x, y, z, t) - \vec{u}(x, y, z, t)}{\Delta x} \tag{5.3.19}$$

In terms of the components we would have

$$\frac{\partial \vec{u}}{\partial x} = \frac{\partial u_x}{\partial x}\hat{i} + \frac{\partial u_y}{\partial x}\hat{j} + \frac{\partial u_z}{\partial x}\hat{k} \tag{5.3.20}$$

where each component could be a function of x, y, z, and/or t.

The incremental quantity $\Delta \vec{u}$ between the two points (x, y, z) and $(x + \Delta x, y + \Delta y, z + \Delta z)$ at the same instant in time would be

$$\Delta \vec{u} = \frac{\partial \vec{u}}{\partial x}\Delta x + \frac{\partial \vec{u}}{\partial y}\Delta y + \frac{\partial \vec{u}}{\partial z}\Delta z \tag{5.3.21}$$

At a fixed point in space $\Delta \vec{u}$ would be given by

$$\Delta \vec{u} = \frac{\partial \vec{u}}{\partial t}\Delta t \tag{5.3.22}$$

If we are interested in the acceleration of a particular particle in a region fully occupied by particles, a *continuum*, we would write the incremental velocity $\Delta \vec{v}$ between two points, shown in Fig. 5.16, as

$$\Delta \vec{v} = \frac{\partial \vec{v}}{\partial x}\Delta x + \frac{\partial \vec{v}}{\partial y}\Delta y + \frac{\partial \vec{v}}{\partial z}\Delta z + \frac{\partial \vec{v}}{\partial t}\Delta t \tag{5.3.23}$$

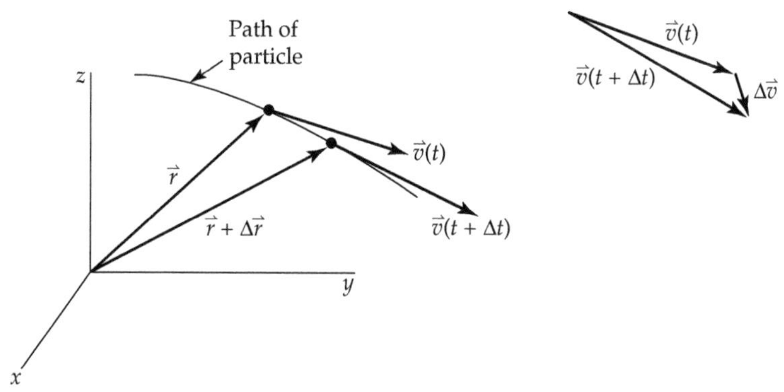

FIGURE 5.16 Motion of a particle.

where we recognize that not only is the position of the particle changing but so is time increasing. Acceleration is defined to be

$$\vec{a} = \frac{d\vec{v}}{dt} = \lim_{\Delta t \to 0} \frac{\vec{v}(t + \Delta t) - \vec{v}(t)}{\Delta t} = \lim_{\Delta t \to 0} \frac{\Delta \vec{v}}{\Delta t} \tag{5.3.24}$$

Using the expression from Eq. 5.3.23 we have

$$\frac{d\vec{v}}{dt} = \lim_{\Delta t \to 0} \left[\frac{\partial \vec{v}}{\partial x}\frac{\Delta x}{\Delta t} + \frac{\partial \vec{v}}{\partial y}\frac{\Delta y}{\Delta t} + \frac{\partial \vec{v}}{\partial z}\frac{\Delta z}{\Delta t} \right] + \frac{\partial \vec{v}}{\partial t} \tag{5.3.25}$$

Realizing that we are following a particular particle,

$$\lim_{\Delta t \to 0} \frac{\Delta x}{\Delta t} = v_x, \qquad \lim_{\Delta t \to 0} \frac{\Delta y}{\Delta t} = v_y, \qquad \lim_{\Delta t \to 0} \frac{\Delta z}{\Delta t} = v_z \tag{5.3.26}$$

Then there follows

$$\vec{a} = \frac{D\vec{v}}{Dt} = v_x \frac{\partial \vec{v}}{\partial x} + v_y \frac{\partial \vec{v}}{\partial y} + v_z \frac{\partial \vec{v}}{\partial z} + \frac{\partial \vec{v}}{\partial t} \tag{5.3.27}$$

where we have adopted the popular convention to use D/Dt to emphasize that we have followed a material particle. It is called the *material* or *substantial derivative,* and from Eq. 5.3.27 is observed to be

$$\frac{D}{Dt} = v_x \frac{\partial}{\partial x} + v_y \frac{\partial}{\partial y} + v_z \frac{\partial}{\partial z} + \frac{\partial}{\partial t} \tag{5.3.28}$$

$\frac{D}{Dt}$ is used to emphasize the material derivative in continuum mechanics, but mathematically is equivalent to the total derivative, $\frac{d}{dt}$.

We could form derivatives as above for any quantity of interest. For example, the rate of change of temperature of a particle would be given by

$$\frac{DT}{Dt} = v_x \frac{\partial T}{\partial x} + v_y \frac{\partial T}{\partial y} + v_z \frac{\partial T}{\partial z} + \frac{\partial T}{\partial t} \tag{5.3.29}$$

Example 5.11

A velocity field is given by $\vec{v} = x^2\hat{i} + xy\hat{j} + 2t^2\hat{k}$ m/s. Determine the acceleration at the point (2, 1, 0) meters and $t = 2$ s.

Solution

The acceleration is given by

$$\begin{aligned}\vec{a} = \frac{D\vec{v}}{Dt} &= \left[v_x \frac{\partial}{\partial x} + v_y \frac{\partial}{\partial y} + v_z \frac{\partial}{\partial z} + \frac{\partial}{\partial t} \right] \vec{v} \\ &= \left[x^2 \frac{\partial}{\partial x} + xy \frac{\partial}{\partial y} + 2t^2 \frac{\partial}{\partial z} + \frac{\partial}{\partial t} \right] (x^2\hat{i} + xy\hat{j} + 2t^2\hat{k}) \\ &= x^2(2x\hat{i} + y\hat{j}) + xy(x\hat{j}) + 2t^2 \cdot 0 + 4t\hat{k} \\ &= 2x^3\hat{i} + 2x^2y\hat{j} + 4t\hat{k}\end{aligned}$$

At the point (2, 1, 0) and at $t = 2$ s, there results

$$\vec{a} = 16\hat{i} + 8\hat{j} + 8\hat{k} \quad \text{m/s}^2$$

5.4 THE GRADIENT

When studying phenomena that occur in a region of interest certain variables often change from point to point, and this change must usually be accounted for. Consider a scalar variable represented at the point (x, y, z) by the function $\phi(x, y, z)$.* This could be the temperature, for example. The incremental change in ϕ, as we move to a neighboring point $(x + \Delta x, y + \Delta y, z + \Delta z)$, is given by

$$\Delta\phi = \frac{\partial\phi}{\partial x}\Delta x + \frac{\partial\phi}{\partial y}\Delta y + \frac{\partial\phi}{\partial z}\Delta z \tag{5.4.1}$$

where $\partial\phi/\partial x$, $\partial\phi/\partial y$, and $\partial\phi/\partial z$ represent the rate of change of ϕ in the x, y, and z directions, respectively. If we divide by the incremental distance between the two points, shown in Fig. 5.17, we have

$$\frac{\Delta\phi}{\Delta r} = \frac{\partial\phi}{\partial x}\frac{\Delta x}{\Delta r} + \frac{\partial\phi}{\partial y}\frac{\Delta y}{\Delta r} + \frac{\partial\phi}{\partial z}\frac{\Delta z}{\Delta r} \tag{5.4.2}$$

Now we can let Δx, Δy, and Δz approach zero and we arrive at the derivative of ϕ in the direction of $\Delta\vec{r}$,

$$\frac{d\phi}{dr} = \frac{\partial\phi}{\partial x}\frac{dx}{dr} + \frac{\partial\phi}{\partial y}\frac{dy}{dr} + \frac{\partial\phi}{\partial z}\frac{dz}{dr} \tag{5.4.3}$$

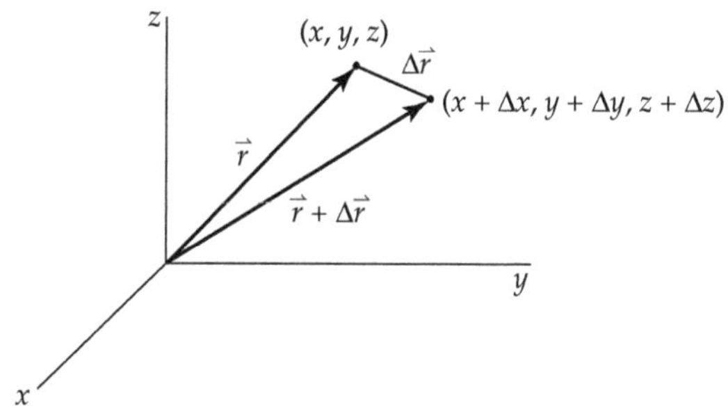

FIGURE 5.17 Change in the position vector.

The form of this result suggests that it be written as the dot product of two vectors; that is,

$$\frac{d\phi}{dr} = \left(\frac{\partial\phi}{\partial x}\hat{i} + \frac{\partial\phi}{\partial y}\hat{j} + \frac{\partial\phi}{\partial z}\hat{k}\right)\cdot\left(\frac{dx}{dr}\hat{i} + \frac{dy}{dr}\hat{j} + \frac{dz}{dr}\hat{k}\right) \tag{5.4.4}$$

Recognizing that

$$d\vec{r} = dx\hat{i} + dy\hat{j} + dz\hat{k} \tag{5.4.5}$$

we can write Eq. 5.4.4 as

$$\frac{d\phi}{dr} = \left(\frac{\partial\phi}{\partial x}\hat{i} + \frac{\partial\phi}{\partial y}\hat{j} + \frac{\partial\phi}{\partial z}\hat{k}\right)\cdot\frac{d\vec{r}}{dr} \tag{5.4.6}$$

*We shall use rectangular coordinates in this section. Cylindrical and spherical coordinates will be presented in Section 5.5.

The vector in parentheses is called the *gradient* of ϕ and is usually written

$$\vec{\nabla}\phi = \text{grad}\,\phi = \frac{\partial\phi}{\partial x}\hat{i} + \frac{\partial\phi}{\partial y}\hat{j} + \frac{\partial\phi}{\partial z}\hat{k} \tag{5.4.7}$$

The symbol $\vec{\nabla}$ is called *del* and is the vector differential operator

$$\vec{\nabla} = \frac{\partial}{\partial x}\hat{i} + \frac{\partial}{\partial y}\hat{j} + \frac{\partial}{\partial z}\hat{k} \tag{5.4.8}$$

The quantity $d\vec{r}/dr$ is obviously a unit vector in the direction of $d\vec{r}$. Thus, returning to Eq. 5.4.6 we observe that the rate of change of ϕ in a particular direction is given by $\vec{\nabla}\phi$ dotted with a unit vector in that direction; that is,

$$\frac{d\phi}{dn} = \vec{\nabla}\phi \cdot \hat{i}_n \tag{5.4.9}$$

where $\hat{i}_n$ is a unit vector in the n direction.

Another important property of $\vec{\nabla}\phi$ is that $\vec{\nabla}\phi$ is normal to a constant ϕ surface. To show this, consider the constant ϕ surface and the differential displacement vector $d\vec{r}$, illustrated in Fig. 5.18. If $\vec{\nabla}\phi$ is normal to a constant ϕ surface, then $\vec{\nabla}\phi \cdot d\vec{r}$ should be zero since $d\vec{r}$ is a vector that lies in the surface. The quantity $\vec{\nabla}\phi \cdot d\vec{r}$ is given by (see Eqs. 5.4.5 and 5.4.7)

$$\vec{\nabla}\phi \cdot d\vec{r} = \frac{\partial\phi}{\partial x}dx + \frac{\partial\phi}{\partial y}dy + \frac{\partial\phi}{\partial z}dz \tag{5.4.10}$$

We recognize that this expression is simply $d\phi$. But $d\phi = 0$ along a constant ϕ surface; thus, $\vec{\nabla}\phi \cdot d\vec{r} = 0$ and $\vec{\nabla}\phi$ is normal to a constant ϕ surface.

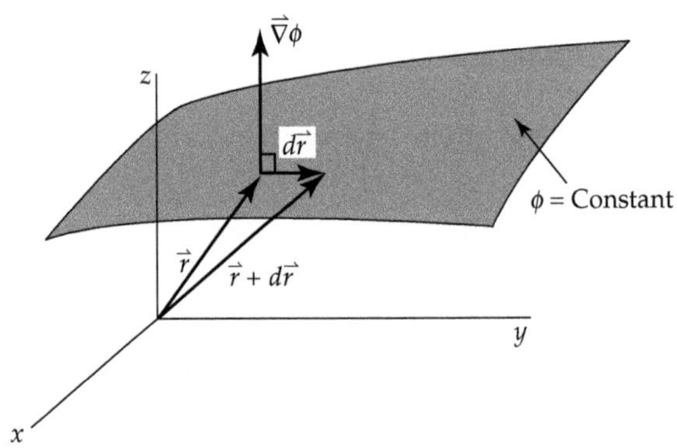

FIGURE 5.18 Constant ϕ surface.

We also note that $\vec{\nabla}\phi$ points in a direction in which the derivative of ϕ is numerically the greatest since Eq. 5.4.9 shows that $d\phi/dn$ is maximum when $\hat{i}_n$ is in the direction of $\vec{\nabla}\phi$. Because of this, $\vec{\nabla}\phi$ may be referred to as the *maximum directional derivative.*

The vector character of the del operator suggests that we form the dot and cross products with $\vec{\nabla}$ and a vector function. Consider a general vector function $\vec{u}(x,y,z)$

in which each component is a function of x, y, and z. The dot product of the $\vec{\nabla}$ operator with $\vec{u}(x,y,z)$ is written in rectangular coordinates* as

$$\vec{\nabla}\cdot\vec{u} = \left(\frac{\partial}{\partial x}\hat{i} + \frac{\partial}{\partial y}\hat{j} + \frac{\partial}{\partial z}\hat{k}\right)\cdot(u_x\hat{i} + u_y\hat{j} + u_z\hat{k})$$

$$= \frac{\partial u_x}{\partial x} + \frac{\partial u_y}{\partial y} + \frac{\partial u_z}{\partial z} \qquad (5.4.11)$$

It is known as the *divergence* of the vector field $\vec{u}$.

The cross product in rectangular coordinates is

$$\vec{\nabla}\times\vec{u} = \left(\frac{\partial}{\partial x}\hat{i} + \frac{\partial}{\partial y}\hat{j} + \frac{\partial}{\partial z}\hat{k}\right)\times(u_x\hat{i} + u_y\hat{j} + u_z\hat{k})$$

$$= \left(\frac{\partial u_z}{\partial y} - \frac{\partial u_y}{\partial z}\right)\hat{i} + \left(\frac{\partial u_x}{\partial z} - \frac{\partial u_z}{\partial x}\right)\hat{j} + \left(\frac{\partial u_y}{\partial x} - \frac{\partial u_x}{\partial y}\right)\hat{k} \qquad (5.4.12)$$

and is known as the *curl* of the vector field $\vec{u}$. Using a determinant, the curl is

$$\vec{\nabla}\times\vec{u} = \begin{vmatrix} \hat{i} & \hat{j} & \hat{k} \\ \frac{\partial}{\partial x} & \frac{\partial}{\partial y} & \frac{\partial}{\partial z} \\ u_x & u_y & u_z \end{vmatrix} \qquad (5.4.13)$$

The divergence and the curl of a vector function appear quite often when deriving the mathematical models for various physical phenomena. For example, let us determine the rate at which material is leaving the incremental volume shown in Fig. 5.19. The volume of material crossing a face in a time period Δt is indicated as the component

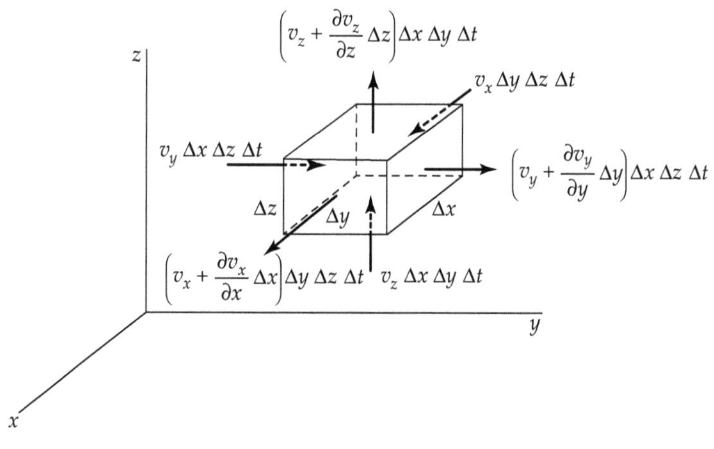

FIGURE 5.19 Flow from an incremental volume.

*Expressions in cylindrical and spherical coordinates will be given in Section 5.5.

of velocity normal to a face multiplied by the area of the face and the time Δt. If we account for all the material leaving the element, we have

$$\begin{aligned}\text{net loss} &= \left(v_x + \frac{\partial v}{\partial x}\Delta x\right)\Delta y\,\Delta z\,\Delta t - v_x \Delta y\,\Delta z\,\Delta t + \left(v_y + \frac{\partial v_y}{\partial y}\Delta y\right)\Delta x\,\Delta z\,\Delta t \\ &\quad - v_y \Delta x\,\Delta z\,\Delta t + \left(v_z + \frac{\partial v_z}{\partial z}\Delta z\right)\Delta x\,\Delta y\,\Delta t - v_z \Delta x\,\Delta y\,\Delta t \\ &= \left(\frac{\partial v_x}{\partial x} + \frac{\partial v_y}{\partial y} + \frac{\partial v_z}{\partial z}\right)\Delta x\,\Delta y\,\Delta z\,\Delta t \end{aligned} \tag{5.4.14}$$

If we divide by the elemental volume $\Delta x\,\Delta y\,\Delta z$ and the time increment Δt, there results

$$\text{rate of loss per unit volume} = \frac{\partial v_x}{\partial x} + \frac{\partial v_y}{\partial y} + \frac{\partial v_z}{\partial z} = \vec{\nabla} \cdot \vec{v} \tag{5.4.15}$$

For a constant density material, the amount of material in a volume remains constant; thus, the rate of loss must be zero; that is,

$$\vec{\nabla} \cdot \vec{v} = 0 \tag{5.4.16}$$

It is the *continuity equation*. The same equation applies to a static electric field, in which case $\vec{v}$ represents the current density.

As we let Δx, Δy, and Δz shrink to zero, we note that the volume element approaches a point. If we consider material or electric current to occupy all points in a region of interest, then the divergence is valid at a point, and it represents the flux (quantity per second) emanating per unit volume.

For a physical interpretation of the curl, let us consider a rectangle undergoing motion while a material is deforming, displayed in Fig. 5.20. The velocity components at P are v_x and v_y, at Q they are $[v_x + (\partial v_x/\partial x)\,\Delta x]$ and $[v_y + (\partial v_y/\partial x)\,\Delta x]$, and at R they are $[v_x + (\partial v_x/\partial y)\Delta y]$ and $[v_y + (\partial v_y/\partial y)\,\Delta y]$. Point P will move to P' a distance $v_y\,\Delta t$ above P and a distance $v_x\,\Delta t$ to the right of P; Q will move to Q' a distance $[v_y + (\partial v_y/\partial x)\,\Delta x]\,\Delta t$ above Q; and R' will move a distance $[v_x + (\partial v_x/\partial y)\Delta y]\Delta t$ to the right of R. The quantity $(d/dt)[(\alpha + \beta)/2)]$, approximated by $(\Delta\alpha + \Delta\beta)/(2\,\Delta t)$ (the angles $\Delta\alpha$ and $\Delta\beta$ are shown),

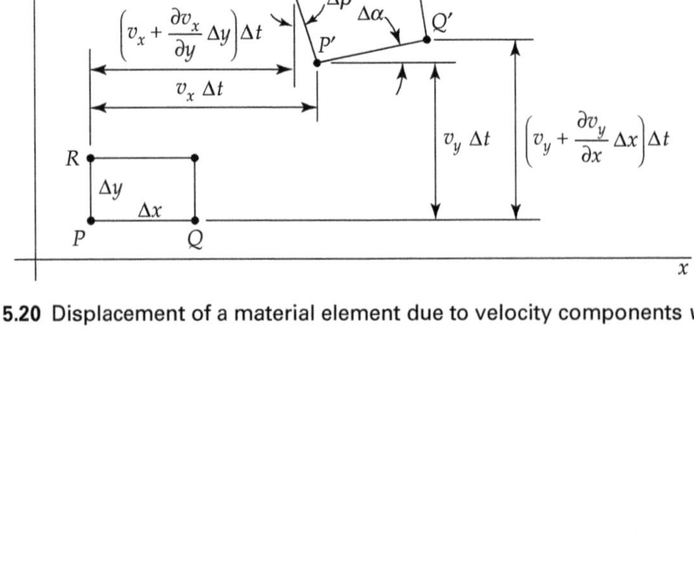

FIGURE 5.20 Displacement of a material element due to velocity components v_x and v_y.

would represent the rate at which the element is rotating. In terms of the velocity components, referring to the figure, we have

$$\frac{d}{dt}\left(\frac{\alpha+\beta}{2}\right)$$

$$\cong \frac{\Delta\alpha+\Delta\beta}{2\Delta t}$$

$$= \frac{\left[\left(v_y + \frac{\partial v_y}{\partial x}\Delta x\right)\Delta t - v_y\Delta t\right]\Big/\Delta x + \left[v_x\Delta t - \left(v_x + \frac{\partial v_x}{\partial y}\Delta y\right)\Delta t\right]\Big/\Delta y}{2\Delta t}$$

$$= \frac{1}{2}\left(\frac{\partial v_y}{\partial x} - \frac{\partial v_x}{\partial y}\right) \tag{5.4.17}$$

where we have used $\Delta\alpha \cong \tan\Delta\alpha$ since $\Delta\alpha$ is small. Thus, we see that the z component of $\vec{\nabla}\times\vec{v}$, which is $[(\partial v_y/\partial x) - (\partial v_x/\partial y)]$, represents twice the rate of rotation of a material element about the z axis. Likewise, the x and y components of $\vec{\nabla}\times\vec{V}$ would represent twice the rate of rotation about the x and y axes, respectively. If we let $\vec{\Omega}$ be the angular velocity (rate of rotation), then

$$\vec{\Omega} = \frac{1}{2}\vec{\nabla}\times\vec{v} \tag{5.4.18}$$

As we let Δx and Δy again approach zero, we note that the element again approaches a point. Thus, the curl of a vector function is valid at a point and it represents twice the rate at which a material element occupying the point is rotating. In electric and magnetic fields, the curl does not possess this physical meaning; it does, however, appear quite often and for a static field the electric current density $\vec{J}$ is given by the curl of the magnetic field intensity $\vec{H}$; that is,

$$\vec{J} = \vec{\nabla}\times\vec{H} \tag{5.4.19}$$

There are several combinations of vector operations involving the $\vec{\nabla}$ operator which are encountered in applications. A very common one is the divergence of the gradient of a scalar function, written as $\vec{\nabla}\cdot\vec{\nabla}\phi$. In rectangular coordinates it is

$$\vec{\nabla}\cdot\vec{\nabla}\phi = \left(\frac{\partial}{\partial x}\hat{i} + \frac{\partial}{\partial y}\hat{j} + \frac{\partial}{\partial z}\hat{k}\right)\cdot\left(\frac{\partial\phi}{\partial x}\hat{i} + \frac{\partial\phi}{\partial y}\hat{j} + \frac{\partial\phi}{\partial z}\hat{k}\right)$$

$$= \frac{\partial^2\phi}{\partial x^2} + \frac{\partial^2\phi}{\partial y^2} + \frac{\partial^2\phi}{\partial z^2} \tag{5.4.20}$$

It is usually written $\nabla^2\phi$ and is called the *Laplacian* of ϕ. If it is zero, that is,

$$\nabla^2\phi = 0 \tag{5.4.21}$$

It is referred to as *Laplace's equation.*

The divergence of the curl of a vector function, and the curl of the gradient of a scalar function are also quantities of interest, but they can be shown to be zero by expanding in rectangular coordinates. Written out, they are

$$\begin{aligned}\vec{\nabla}\cdot\vec{\nabla}\times\vec{u} &= 0\\ \vec{\nabla}\times\vec{\nabla}\phi &= 0\end{aligned} \tag{5.4.22}$$

Two special kinds of vector fields exist. One is a *solenoidal vector field,* in which the divergence is zero, that is,

$$\vec{\nabla} \cdot \vec{u} = 0 \tag{5.4.23}$$

and the other is an *irrotational* (or *conservative*) *vector field,* in which the curl is zero, that is,

$$\vec{\nabla} \times \vec{u} = 0 \tag{5.4.24}$$

If the vector field $\vec{u}$ is given by the gradient of a scalar function ϕ, that is,

$$\vec{u} = \vec{\nabla}\phi \tag{5.4.25}$$

then, according to Eq. 5.4.22, the curl of $\vec{u}$ is zero and $\vec{u}$ is irrotational. The function ϕ is referred to as the *scalar potential function* of the vector field $\vec{u}$.

Several vector identities are often useful, and these are presented in Table 5.1. They can be verified by expanding in a particular coordinate system.

TABLE 5.1 Some Vector Identities

$\vec{\nabla} \times \vec{\nabla}\phi = 0$
$\vec{\nabla} \cdot \vec{\nabla} \times \vec{u} = 0$
$\vec{\nabla} \cdot (\phi \vec{u}) = \vec{\nabla}\phi \cdot \vec{u} + \phi \vec{\nabla} \cdot \vec{u}$
$\vec{\nabla} \times (\phi \vec{u}) = \vec{\nabla}\phi \times \vec{u} + \phi \vec{\nabla} \times \vec{u}$
$\vec{\nabla} \times (\vec{\nabla} \times \vec{u}) = \vec{\nabla}(\vec{\nabla} \cdot \vec{u}) - \nabla^2 \vec{u}$
$\vec{u} \times (\vec{\nabla} \times \vec{u}) = \frac{1}{2}\vec{\nabla}u^2 - (\vec{u} \cdot \vec{\nabla})\vec{u}$
$\vec{\nabla} \cdot (\vec{u} \times \vec{v}) = (\vec{\nabla} \times \vec{u}) \cdot \vec{v} - \vec{u} \cdot (\vec{\nabla} \times \vec{v})$
$\vec{\nabla} \cdot (\vec{u} \times \vec{v}) = \vec{u}(\vec{\nabla} \cdot \vec{v}) - \vec{v}(\vec{\nabla} \cdot \vec{u}) + (\vec{v} \cdot \vec{\nabla})\vec{u} - (\vec{u} \cdot \vec{\nabla})\vec{v}$
$\vec{\nabla}(\vec{u} \cdot \vec{v}) = (\vec{u} \cdot \vec{\nabla})\vec{v} + (\vec{v} \cdot \vec{\nabla})\vec{u} + \vec{u} \times (\vec{\nabla} \times \vec{v}) + \vec{v} \times (\vec{\nabla} \times \vec{u})$

Example 5.12

Find the derivative of the function $\phi = x^2 - 2xy + z^2$ at the point (2, –1, 1) in the direction of the vector $\vec{A} = 2\hat{i} - 4\hat{j} + 4\hat{k}$.

Solution

To find the derivative of a function in a particular direction we use Eq. 5.4.9. We must first find the gradient of the function. It is

$$\begin{aligned}\vec{\nabla}\phi &= \left(\frac{\partial}{\partial x}\hat{i} + \frac{\partial}{\partial y}\hat{j} + \frac{\partial}{\partial z}\hat{k}\right)(x^2 - 2xy + z^2) \\ &= (2x - 2y)\hat{i} - 2x\hat{j} + 2z\hat{k}\end{aligned}$$

At the point (2, –1, 1), it is

$$\vec{\nabla}\phi = 6\hat{i} - 4\hat{j} + 2\hat{k}$$

The unit vector in the desired direction is

$$\hat{i}_n = \frac{\vec{A}}{A} = \frac{2\hat{i} - 4\hat{j} + 4\hat{k}}{6} = \frac{1}{3}\hat{i} - \frac{2}{3}\hat{j} + \frac{2}{3}\hat{k}$$

Finally, the derivative in the direction of $\vec{A}$ is

$$\begin{aligned}\frac{d\phi}{dn} &= \vec{\nabla}\phi \cdot \hat{i}_n \\ &= (6\hat{i} - 4\hat{j} + 2\hat{k}) \cdot \left(\frac{1}{3}\hat{i} - \frac{2}{3}\hat{j} + \frac{2}{3}\hat{k}\right) \\ &= 2 + \frac{8}{3} + \frac{4}{3} = \underline{6}\end{aligned}$$

Example 5.13

Find a unit vector $\hat{i}_n$ normal to the surface represented by the equation $x^2 - 8y^2 + z^2 = 0$ at the point (8, 1, 4).

Solution

We know that the gradient $\vec{\nabla}\phi$ is normal to a constant ϕ surface. So, with

$$\phi = x^2 - 8y^2 + z^2$$

we can form the gradient, to get

$$\begin{aligned}\vec{\nabla}\phi &= \frac{\partial\phi}{\partial x}\hat{i} + \frac{\partial\phi}{\partial y}\hat{j} + \frac{\partial\phi}{\partial z}\hat{k} \\ &= 2x\hat{i} - 16y\hat{j} + 2z\hat{k}\end{aligned}$$

At the point (8, 1, 4) we have

$$\vec{\nabla}\phi = 16\hat{i} - 16\hat{j} + 8\hat{k}$$

This vector is normal to the surface at (8, 1, 4). To find the unit vector, we simply divide the vector by its own magnitude, obtaining

$$\begin{aligned}\hat{i}_n &= \frac{\vec{\nabla}\phi}{\left|\vec{\nabla}\phi\right|} \\ &= \frac{16\hat{i} - 16\hat{j} + 8\hat{k}}{\sqrt{256 + 256 + 64}} = \frac{2}{3}\hat{i} - \frac{2}{3}\hat{j} + \frac{1}{3}\hat{k}\end{aligned}$$

Example 5.14

Find the equation for the plane which is tangent to the surface $x^2 + y^2 - z^2 = 4$ at the point (1, 2, –1).

Solution

The gradient of ϕ is normal to a constant ϕ surface. Hence, with $\phi = x^2 + y^2 - z^2$, the vector

$$\begin{aligned}\vec{\nabla}\phi &= \frac{\partial\phi}{\partial x}\hat{i} + \frac{\partial\phi}{\partial y}\hat{j} + \frac{\partial\phi}{\partial z}\hat{k} \\ &= 2x\hat{i} + 2y\hat{j} - 2z\hat{k}\end{aligned}$$

is normal to the given surface. At the point (1, 2, –1) the normal vector is

$$\vec{\nabla}\phi = 2\hat{i} + 4\hat{j} - 2\hat{k}$$

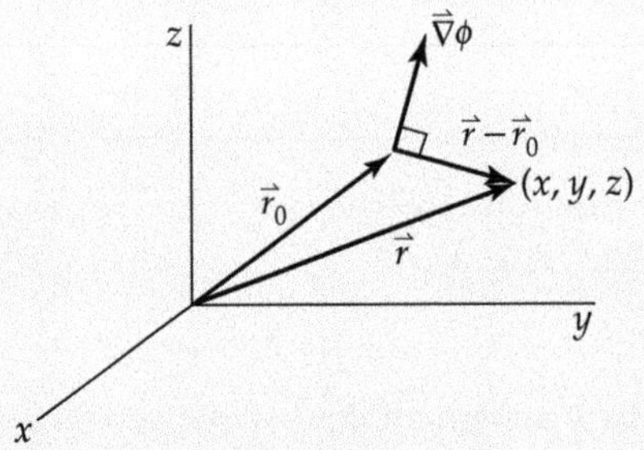

FIGURE 5.21

Consider the sketch shown in Fig. 5.21. The vector to the given point $\vec{r}_0 = \hat{i} + 2\hat{j} - \hat{k}$ subtracted from the vector to the general point $\vec{r} = x\hat{i} + y\hat{j} + z\hat{k}$ is a vector in the desired plane. It is

$$\begin{aligned}\vec{r} - \vec{r}_0 &= (x\hat{i} + y\hat{j} + z\hat{k}) - (\hat{i} + 2\hat{j} - \hat{k}) \\ &= (x-1)\hat{i} + (y-2)\hat{j} + (z+1)\hat{k}\end{aligned}$$

This vector, when dotted with a vector normal to it, namely $\vec{\nabla}\phi$, must yield zero; that is,

$$\begin{aligned}\vec{\nabla}\phi\cdot(\vec{r} - \vec{r}_0) &= (2\hat{i} + 4\hat{j} + 2\hat{k})\cdot[(x-1)\hat{i} + (y-2)\hat{j} + (z+1)\hat{k}] \\ &= 2(x-1) + 4(y-2) + 2(z+1) \\ &= 0\end{aligned}$$

Thus, the tangent plane is given by

$$x + 2y + z = 4$$

Example 5.15

A vector field is given by $\vec{u} = y^2\hat{i} + 2xy\hat{j} - z^2\hat{k}$. Determine the divergence of $\vec{u}$ and curl of $\vec{u}$ at the point (1, 2, 1). Also, determine if the vector field is solenoidal or irrotational.

Solution

The divergence of $\vec{u}$ is given by Eq. 5.4.11. It is

$$\begin{aligned}\vec{\nabla} \cdot \vec{u} &= \frac{\partial u_x}{\partial x} + \frac{\partial u_y}{\partial y} + \frac{\partial u_z}{\partial z} \\ &= 0 + 2x - 2z\end{aligned}$$

At the point (1, 2, 1) this scalar function has the value

$$\vec{\nabla} \cdot \vec{u} = 2 - 2 = \underline{0}$$

The curl of $\vec{u}$ is given by Eq. 5.4.12. It is

$$\begin{aligned}\vec{\nabla} \times \vec{u} &= \left(\frac{\partial u_z}{\partial y} - \frac{\partial u_y}{\partial z}\right)\hat{i} + \left(\frac{\partial u_x}{\partial z} - \frac{\partial u_z}{\partial x}\right)\hat{j} + \left(\frac{\partial u_y}{\partial x} - \frac{\partial u_x}{\partial y}\right)\hat{k} \\ &= 0\hat{i} + 0\hat{j} + (2y - 2y)\hat{k} \\ &= \underline{0}\end{aligned}$$

The curl of $\vec{u}$ is zero at all points in the field; hence, it is an irrotational vector field. However, $\vec{\nabla} \cdot \vec{u}$ is not zero at *all* points in the field; thus, $\vec{u}$ is not solenoidal.

Example 5.16

For the vector field $\vec{u} = y^2\hat{i} + 2xy\hat{j} - z^2\hat{k}$, find the associated scalar potential function $\phi(x, y, z)$, providing that one exists.

Solution

The scalar potential function $\phi(x, y, z)$ is related to the vector field by

$$\vec{\nabla}\phi = \vec{u}$$

providing that the curl of $\vec{u}$ is zero. The curl of $\vec{u}$ was shown to be zero in Example 5.15; hence, a potential function ϕ does exist. Writing the above using rectangular components, we have

$$\frac{\partial \phi}{\partial x}\hat{i} + \frac{\partial \phi}{\partial y}\hat{j} + \frac{\partial \phi}{\partial z}\hat{k} = y^2\hat{i} + 2xy\hat{j} - z^2\hat{k}$$

This vector equation contains three scalar equations which result from equating the x component, the y component, and the z component, respectively, from each side of the equation. This gives

$$\frac{\partial\phi}{\partial x} = y^2$$
$$\frac{\partial\phi}{\partial y} = 2xy$$
$$\frac{\partial\phi}{\partial z} = -z^2$$

The first of these is integrated to give the solution

$$\phi(x, y, z) = xy^2 + f(y, z)$$

Note that in solving partial differential equations the "constant of integration" is a function. In the first equation we are differentiating with respect to x, holding y and z fixed; thus, this "constant of integration" may be a function of y and z, namely $f(y, z)$. Now, substitute the solution above into the second equation and obtain

$$2xy + \frac{\partial f}{\partial y} = 2xy$$

This results in $\partial f/\partial y = 0$, which means that f does not depend on y. Thus, f must be at most a function of z. So substitute the solution into the third equation, and there results

$$\frac{df}{dz} = -z^2$$

where we have used an ordinary derivative since $f = f(z)$. This equation is integrated to give

$$f(z) = -\frac{z^3}{3} + C$$

where C is a constant of integration. Finally, the scalar potential function is

$$\phi(x, y, z) = xy^2 - \frac{z^3}{3} + C$$

To check that we have made no mistakes, let us find the gradient of this function. It is

$$\begin{aligned}\vec{\nabla}\phi &= \frac{\partial\phi}{\partial x}\hat{i} + \frac{\partial\phi}{\partial y}\hat{j} + \frac{\partial\phi}{\partial z}\hat{k}\\ &= y^2\hat{i} + 2xy\hat{j} - z^2\hat{k}\end{aligned}$$

This is equal to the given vector function $\vec{u}$, as it must be since $\vec{\nabla}\phi = \vec{u}$.

5.5 CYLINDRICAL AND SPHERICAL COORDINATES

There are several coordinate systems that are convenient to use with various geometries encountered in physical applications. The most common is the rectangular coordinate system (often referred to as simply the Cartesian coordinate system), used primarily in this text. There are situations, however, when solutions to problems are much simpler if a more natural coordinate system is chosen. Two other coordinate systems that attract much attention are the cylindrical coordinate system and the spherical coordinate system. We shall relate the rectangular coordinates to both the cylindrical and spherical coordinates, and express the various vector quantities of previous sections in cylindrical and spherical coordinates.

The cylindrical coordinates* (r, θ, z), with respective orthogonal unit vectors $\hat{i}_r, \hat{i}_\theta$, and $\hat{i}_z$, and the spherical coordinates (r, θ, ϕ), with respective orthogonal unit vectors $\hat{i}_r, \hat{i}_\theta$, and $\hat{i}_\phi$, are shown in Fig. 5.22. A vector is expressed in cylindrical coordinates as

$$\vec{A} = A_r\hat{i}_r + A_\theta\hat{i}_\theta + A_z\hat{i}_z \tag{5.5.1}$$

where the components A_r, A_θ, and A_z are functions of r, θ, and z. In spherical coordinates a vector would be expressed as

$$\vec{A} = A_r\hat{i}_r + A_\theta\hat{i}_\theta + A_\phi\hat{i}_\phi \tag{5.5.2}$$

where A_r, A_θ, and A_ϕ are functions of r, θ, and ϕ.

Cylindrical coordinates are useful in applications such as mechanics of circular shafts, flow about cylinders, and fields about a straight wire. Spherical coordinates are applicable, for example, in planetary systems and electrical fields around charged particles.

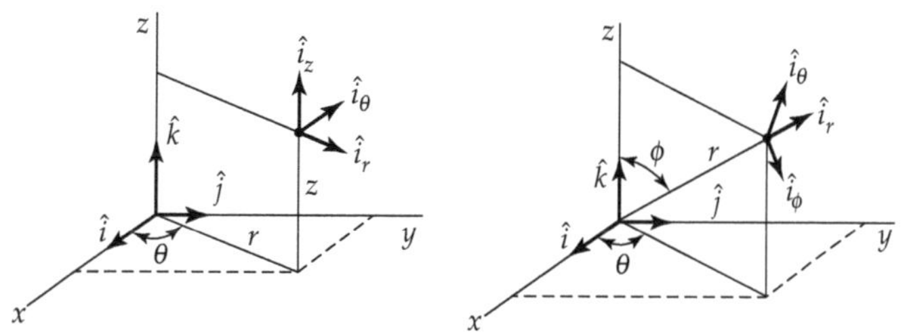

FIGURE 5.22 The cylindrical and spherical coordinate systems.

We have, in previous sections, expressed all vector quantities in rectangular coordinates. Let us transform some of the more important quantities to cylindrical and spherical coordinates. We will do this first for cylindrical coordinates.

5.5.1 Cylindrical Coordinates

The cylindrical coordinates are related to rectangular coordinates by (refer to Fig. 5.22)

$$x = r\cos\theta, \qquad y = r\sin\theta, \qquad z = z \tag{5.5.3}$$

*Note that in cylindrical coordinates it is conventional to use r as the distance from the z axis to the point of interest. Do not confuse it with the distance from the origin $|\vec{r}|$.

where we are careful* to note that $r \neq |\vec{r}|, \vec{r}$ being the position vector measured from the origin. From the geometry of Fig. 5.22 we can write

$$\begin{aligned} \hat{i}_r &= \cos\theta\,\hat{i} + \sin\theta\,\hat{j} \\ \hat{i}_\theta &= -\sin\theta\,\hat{i} + \cos\theta\,\hat{j} \\ \hat{i}_z &= \hat{k} \end{aligned} \tag{5.5.4}$$

These three equations can be solved simultaneously to give

$$\begin{aligned} \hat{i} &= \cos\theta\,\hat{i}_r - \sin\theta\,\hat{i}_\theta \\ \hat{j} &= \sin\theta\,\hat{i}_r + \cos\theta\,\hat{i}_\theta \\ \hat{k} &= \hat{i}_z \end{aligned} \tag{5.5.5}$$

We have thus related the unit vectors in the cylindrical and rectangular coordinate systems. They are collected in Table 5.2.

TABLE 5.2 Relationship of Cylindrical and Spherical Coordinates to Rectangular Coordinates

Cylindrical		Spherical	
$x = r\cos\theta$	$r = \sqrt{x^2 + y^2}$	$x = r\sin\phi\cos\theta$	$r = \sqrt{x^2 + y^2 + z^2}$
$y = r\sin\theta$	$\theta = \tan^{-1} y/x$	$y = r\sin\phi\sin\theta$	$\theta = \tan^{-1} y/x$
$z = z$	$z = z$	$z = r\cos\phi$	$\phi = \tan^{-1}\sqrt{x^2 + y^2}/z$
$\hat{i}_r = \cos\theta\,\hat{i} + \sin\theta\,\hat{j}$		$\hat{i}_r = \sin\phi\cos\theta\,\hat{i} + \sin\phi\sin\theta\,\hat{j} + \cos\phi\,\hat{k}$	
$\hat{i}_\theta = -\sin\theta\,\hat{i} + \cos\theta\,\hat{j}$		$\hat{i}_\theta = -\sin\theta\,\hat{i} + \cos\theta\,\hat{j}$	
$\hat{i}_z = \hat{k}$		$\hat{i}_\phi = \cos\phi\cos\theta\,\hat{i} + \cos\phi\sin\theta\,\hat{j} - \sin\phi\,\hat{k}$	
$\hat{i} = \cos\theta\,\hat{i}_r - \sin\theta\,\hat{i}_\theta$		$\hat{i} = \sin\phi\cos\theta\,\hat{i}_r - \sin\theta\,\hat{i}_\theta + \cos\phi\cos\theta\,\hat{i}_\phi$	
$\hat{j} = \sin\theta\,\hat{i}_r + \cos\theta\,\hat{i}_\theta$		$\hat{j} = \sin\phi\sin\theta\,\hat{i}_r + \cos\theta\,\hat{i}_\theta + \cos\phi\sin\theta\,\hat{i}_\phi$	
$\hat{k} = \hat{i}_z$		$\hat{k} = \cos\phi\,\hat{i}_r - \sin\phi\,\hat{i}_\phi$	

To express the gradient of the scalar function ϕ in cylindrical coordinates, we observe from Fig. 5.23 that

$$d\vec{r} = dr\,\hat{i}_r + r\,d\theta\,\hat{i}_\theta + dz\,\hat{i}_z \tag{5.5.6}$$

The quantity $d\phi$ is, by the chain rule,

$$d\phi = \frac{\partial\phi}{\partial r}dr + \frac{\partial\phi}{\partial\theta}d\theta + \frac{\partial\phi}{\partial z}dz \tag{5.5.7}$$

Let the gradient of ϕ be the vector

$$\vec{\nabla}\phi = \lambda_r\hat{i}_r + \lambda_\theta\hat{i}_\theta + \lambda_z\hat{i}_z \tag{5.5.8}$$

*This is a rather unfortunate choice, but it is the most conventional. Occasionally, ρ (rho) is used in place of r, which helps avoid confusion.

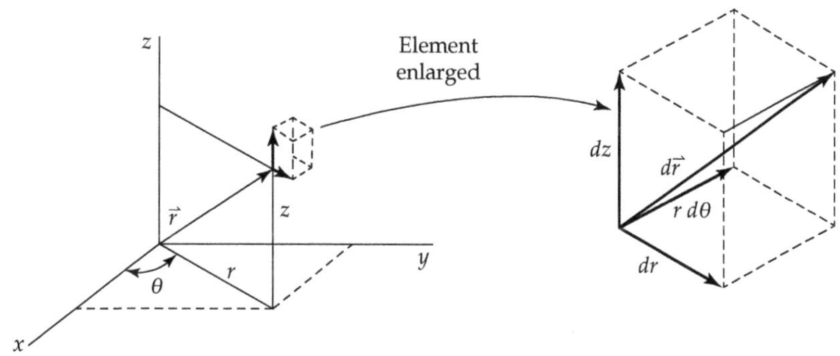

FIGURE 5.23 Differential changes in cylindrical coordinates.

where λ_r, λ_θ, and λ_z are the components of $\vec{\nabla}\phi$ that we wish to determine. We refer to Eq. 5.4.10 and recognize that

$$d\phi = \vec{\nabla}\phi \cdot d\vec{r} \tag{5.5.9}$$

Substituting the preceding expressions for $d\phi$, $\vec{\nabla}\phi$, and $d\vec{r}$ into this equation results in

$$\begin{aligned} \frac{\partial\phi}{\partial r}dr + \frac{d\phi}{\partial\theta}d\theta + \frac{\partial\phi}{\partial z}dz &= (\lambda_r\hat{i}_r + \lambda_\theta\hat{i}_\theta + \lambda_z\hat{i}_z)\cdot(dr\hat{i}_r + r\,d\theta\,\hat{i}_\theta + dz\hat{i}_z) \\ &= \lambda_r dr + \lambda_\theta r\,d\theta + \lambda_z dz \end{aligned} \tag{5.5.10}$$

Since r, θ, and z are independent quantities, the coefficients of the differential quantities allow us to write

$$\lambda_r = \frac{\partial\phi}{\partial r}, \qquad r\lambda_\theta = \frac{\partial\phi}{\partial\theta}, \qquad \lambda_z = \frac{\partial\phi}{\partial z} \tag{5.5.11}$$

Hence, the gradient of ϕ, in cylindrical coordinates, becomes

$$\vec{\nabla}\phi = \frac{\partial\phi}{\partial r}\hat{i}_r + \frac{1}{r}\frac{\partial\phi}{\partial\theta}\hat{i}_\theta + \frac{\partial\phi}{\partial z}\hat{i}_z \tag{5.5.12}$$

The gradient operator $\vec{\nabla}$ is, from the equation above,

$$\vec{\nabla} = \frac{\partial}{\partial r}\hat{i}_r + \frac{1}{r}\frac{\partial}{\partial\theta}\hat{i}_\theta + \frac{\partial}{\partial z}\hat{i}_z \tag{5.5.13}$$

Now we wish to find an expression for the divergence $\vec{\nabla}\cdot\vec{u}$. In cylindrical coordinates, it is

$$\vec{\nabla}\cdot\vec{u} = \left(\frac{\partial}{\partial r}\hat{i}_r + \frac{1}{r}\frac{\partial}{\partial\theta}\hat{i}_\theta + \frac{\partial}{\partial z}\hat{i}_z\right)\cdot(u_r\hat{i}_r + u_\theta\hat{i}_\theta + u_z\hat{i}_z) \tag{5.5.14}$$

When we perform the dot products above, we must be sure to account for the changes in $\hat{i}_r$ and $\hat{i}_\theta$ as the angle θ changes; that is, the quantities $\partial\hat{i}_r/\partial\theta$ and $\partial\hat{i}_\theta/\partial\theta$ are not zero. For example, consider the term $[(1/r)(\partial/\partial\theta)\hat{i}_\theta]\cdot(u_r\hat{i}_r)$. It yields

$$\left(\frac{1}{r}\frac{\partial}{\partial\theta}\hat{i}_\theta\right)\cdot(u_r\hat{i}_r) = \frac{1}{r}\frac{\partial u_r}{\partial\theta}\overset{0}{\hat{i}_\theta\cdot\hat{i}_r} + \frac{u_r}{r}\hat{i}_\theta\cdot\frac{\partial\hat{i}_r}{\partial\theta} \tag{5.5.15}$$

The product $\hat{i}_\theta \cdot \hat{i}_r = 0$ since $\hat{i}_\theta$ is normal to $\hat{i}_r$. The other term, however, is not zero. By referring to Fig. 5.24, we see that

$$\begin{aligned} \frac{\partial \hat{i}_r}{\partial \theta} &= \lim_{\Delta\theta \to 0} = \frac{\Delta \hat{i}_r}{\Delta\theta} = \lim_{\Delta\theta\to 0} \frac{\Delta\theta \hat{i}_\theta}{\Delta\theta} = \hat{i}_\theta \\ \frac{\partial \hat{i}_\theta}{\partial \theta} &= \lim_{\Delta\theta \to 0} \frac{\Delta \hat{i}_\theta}{\Delta\theta} = \lim_{\Delta\theta\to 0} \frac{-\Delta\theta \hat{i}_r}{\Delta\theta} = -\hat{i}_r \end{aligned} \tag{5.5.16}$$

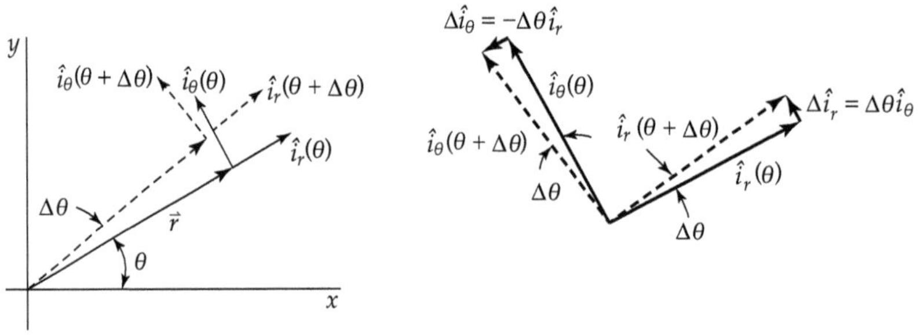

FIGURE 5.24 Change in unit vectors with the angle θ.

Since $\hat{i}_z$ never changes direction, $\partial\hat{i}_z/\partial\theta = 0$. Recalling that

$$\hat{i}_r \cdot \hat{i}_r = \hat{i}_\theta \cdot \hat{i}_\theta = \hat{i}_z \cdot \hat{i}_z = 1, \qquad \hat{i}_r \cdot \hat{i}_\theta = \hat{i}_r \cdot \hat{i}_z = \hat{i}_\theta \cdot \hat{i}_z = 0 \tag{5.5.17}$$

the divergence is then, referring to Eqs. 5.5.14 and 5.5.16,

$$\begin{aligned} \vec{\nabla} \cdot \vec{u} &= \frac{\partial u_r}{\partial r} + \frac{1}{r}\frac{\partial u_\theta}{\partial \theta} + \frac{\partial u_z}{\partial z} + \frac{u_r}{r}\hat{i}_\theta \cdot \frac{\partial \hat{i}_r}{\partial\theta} + \frac{u_\theta}{r}\hat{i}_\theta \cdot \overbrace{\frac{\partial \hat{i}_\theta}{\partial\theta}}^{0} \\ &= \frac{\partial u_r}{\partial r} + \frac{1}{r}\frac{\partial u_\theta}{\partial \theta} + \frac{\partial u_z}{\partial z} + \frac{u_r}{r} \end{aligned} \tag{5.5.18}$$

This can be rewritten in the more conventional form

$$\vec{\nabla} \cdot \vec{u} = \frac{1}{r}\frac{\partial}{\partial r}(r u_r) + \frac{1}{r}\frac{\partial u_\theta}{\partial\theta} + \frac{\partial u_z}{\partial z} \tag{5.5.19}$$

Now, the curl $\vec{\nabla} \times \vec{u}$ will be expressed in cylindrical coordinates. It is

$$\vec{\nabla} \times \vec{u} = \left(\frac{\partial}{\partial r}\hat{i}_r + \frac{1}{r}\frac{\partial}{\partial\theta}\hat{i}_\theta + \frac{\partial}{\partial z}\hat{i}_z\right) \times (u_r\hat{i}_r + u_\theta\hat{i}_\theta + u_z\hat{i}_z) \tag{5.5.20}$$

Carrying out the cross products term by term, we have

$$\begin{aligned} \vec{\nabla} \times \vec{u} = &\left(\frac{1}{r}\frac{\partial u_z}{\partial\theta} - \frac{\partial u_\theta}{\partial z}\right)\hat{i}_r + \left(\frac{\partial u_r}{\partial z} - \frac{\partial u_r}{\partial r}\right)\hat{i}_\theta + \left(\frac{\partial u_\theta}{\partial r} - \frac{1}{r}\frac{\partial u_r}{\partial\theta}\right)\hat{i}_z \\ &+ \frac{u_r}{r}\hat{i}_\theta \times \overbrace{\frac{\partial \hat{i}_r}{\partial\theta}}^{0} + \frac{u_\theta}{r}\hat{i}_\theta \times \frac{\partial\hat{i}_\theta}{\partial\theta} \end{aligned} \tag{5.5.21}$$

where we have used

$$\hat{i}_r \times \hat{i}_r = \hat{i}_\theta \times \hat{i}_\theta = \hat{i}_z \times \hat{i}_z = 0, \qquad \hat{i}_r \times \hat{i}_\theta = \hat{i}_z,\ \hat{i}_\theta \times \hat{i}_z = \hat{i}_r,\ \hat{i}_z \times \hat{i}_r = \hat{i}_\theta \tag{5.5.22}$$

Using Eqs. 5.5.16, there results, writing $(\partial u_\theta/\partial r) + (u_\theta/r) = (1/r)(\partial/\partial r)(ru_\theta)$,

$$\vec{\nabla} \times \vec{u} = \left[\frac{1}{r}\frac{\partial u_z}{\partial \theta} - \frac{\partial u_\theta}{\partial z}\right]\hat{i}_r + \left[\frac{\partial u_r}{\partial z} - \frac{\partial u_z}{\partial r}\right]\hat{i}_\theta + \left[\frac{1}{r}\frac{\partial}{\partial r}(ru_\theta) - \frac{1}{r}\frac{\partial u_r}{\partial \theta}\right]\hat{i}_z \quad (5.5.23)$$

Finally, the Laplacian of a scalar function ϕ, in cylindrical coordinates, is

$$\begin{aligned}\vec{\nabla} \cdot \vec{\nabla}\phi = \nabla^2\phi &= \left(\frac{\partial}{\partial r}\hat{i}_r + \frac{1}{r}\frac{\partial}{\partial \theta}\hat{i}_\theta + \frac{\partial}{\partial z}\hat{i}_z\right) \cdot \left(\frac{\partial \phi}{\partial r}\hat{i}_r + \frac{1}{r}\frac{\partial \phi}{\partial \theta}\hat{i}_\theta + \frac{\partial \phi}{\partial z}\hat{i}_z\right) \\ &= \frac{\partial^2 \phi}{\partial r^2} + \frac{1}{r^2}\frac{\partial^2 \phi}{\partial \theta^2} + \frac{\partial^2 \phi}{\partial z^2} + \frac{1}{r}\frac{\partial \phi}{\partial r}\hat{i}_\theta \cdot \frac{\partial \hat{i}_r}{\partial \theta} \\ &= \frac{1}{r}\frac{\partial}{\partial r}\left(r\frac{\partial \phi}{\partial r}\right) + \frac{1}{r^2}\frac{\partial^2 \phi}{\partial \theta^2} + \frac{\partial^2 \phi}{\partial z^2} \end{aligned} \quad (5.5.24)$$

5.5.2 Spherical Coordinates

If we follow the same procedure using spherical coordinates we would find that the coordinates are related by

$$x = r\sin\phi\cos\theta, \quad y = r\sin\phi\sin\theta, \quad z = r\cos\phi \quad (5.5.25)$$

The unit vectors are related by the following equations:

$$\begin{aligned}\hat{i}_r &= \sin\phi\cos\theta\,\hat{i} + \sin\phi\sin\theta\,\hat{j} + \cos\phi\,\hat{k} \\ \hat{i}_\theta &= -\sin\phi\,\hat{i} + \cos\theta\,\hat{j} \\ \hat{i}_\phi &= \cos\phi\cos\theta\,\hat{i} + \cos\phi\sin\theta\,\hat{j} - \sin\theta\,\hat{k}\end{aligned} \quad (5.5.26)$$

$$\begin{aligned}\hat{i} &= \sin\phi\cos\theta\,\hat{i}_r - \sin\theta\,\hat{i}_\theta\,\cos\phi\cos\theta\,\hat{i}_\theta \\ \hat{j} &= \sin\phi\sin\theta\,\hat{i}_r + \cos\theta\,\hat{i}_\theta + \cos\phi\sin\theta\,\hat{i}_\phi \\ \hat{k} &= \cos\phi\,\hat{i}_r - \sin\phi\,\hat{i}_\phi\end{aligned} \quad (5.5.27)$$

Using Fig. 5.25, the gradient of the scalar function ϕ is found to be, following the steps of Section 5.5.1

$$\vec{\nabla}\phi = \frac{\partial \phi}{\partial r}\hat{i}_r + \frac{1}{r\sin\phi}\frac{\partial \phi}{\partial \theta}\hat{i}_\theta + \frac{1}{r}\frac{\partial \phi}{\partial \phi}\hat{i}_\phi \quad (5.5.28)$$

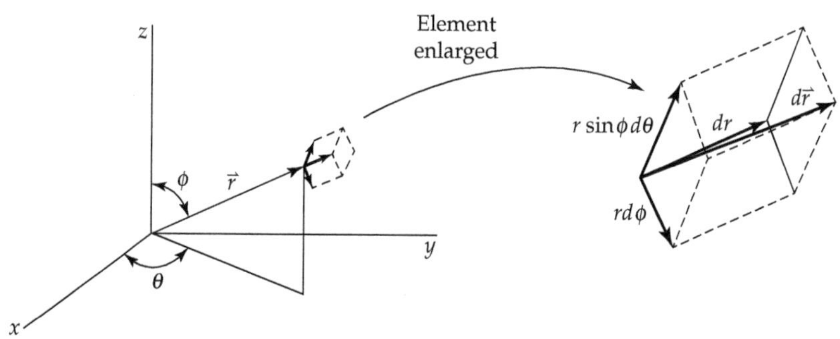

FIGURE 5.25 Differential changes in spherical coordinates.

allowing us to write

$$\vec{\nabla} = \frac{\partial}{\partial r}\hat{i}_r + \frac{1}{r\sin\phi}\frac{\partial}{\partial\theta}\hat{i}_\theta \frac{1}{r}\frac{\partial}{\partial\theta}\hat{i}_\phi \tag{5.5.29}$$

The divergence of a vector field is

$$\vec{\nabla}\cdot\vec{u} = \frac{1}{r^2}\frac{\partial}{\partial r}(r^2u_r) + \frac{1}{r\sin\phi}\frac{\partial u_\theta}{\partial\theta} + \frac{1}{r\sin\phi}\frac{\partial}{\partial\phi}(u_\phi\sin\phi) \tag{5.5.30}$$

and the curl is

$$\vec{\nabla}\times\vec{u} = \frac{1}{r\sin\phi}\left[\frac{\partial}{\partial\phi}(u_\theta\sin\phi) - \frac{\partial u_\phi}{\partial\theta}\right]\hat{i}_r + \frac{1}{r}\left[\frac{\partial}{\partial r}(ru_\phi) - \frac{\partial u_r}{\partial\phi}\right]\hat{i}_\theta + \frac{1}{r}\left[\frac{1}{\sin\phi}\frac{\partial u_r}{\partial\theta} - \frac{\partial}{\partial r}(ru_\theta)\right]\hat{i}_\phi \tag{5.5.31}$$

The Laplacian of a scalar function ϕ is

$$\nabla^2\phi = \frac{1}{r^2}\frac{\partial}{\partial r}\left(r^2\frac{\partial\phi}{\partial r}\right) + \frac{1}{r^2\sin^2\phi}\frac{\partial^2\phi}{\partial\theta^2} + \frac{1}{r^2\sin^2\phi}\frac{\partial}{\partial\phi}\left(\sin\phi\frac{\partial\phi}{\partial\phi}\right) \tag{5.5.32}$$

The relationships involving the gradient operator $\vec{\nabla}$ are collected in Table 5.3.

Example 5.17

Express the vector $\vec{u} = 2x\hat{i} - z\hat{j} + y\hat{k}$ in a) cylindrical coordinates, and b) spherical coordinates.

Solution

a) The coordinates are related by Eqs. 5.5.3 and the unit vectors by Eqs. 5.5.5. Thus, we can write the vector as

$$\vec{u} = 2r\cos\theta(\cos\theta\hat{i}_r - \sin\theta\hat{i}_\theta) - z(\sin\theta\hat{i}_r + \cos\theta\hat{i}_\theta) + r\sin\phi\hat{i}_z$$

This is rearranged in the conventional form,

$$\vec{u} = (2r\cos^2\theta - z\sin\theta)\hat{i}_r - (2r\cos\theta\sin\theta + z\cos\theta)\hat{i}_\theta + r\sin\theta\,\hat{i}_z$$

b) For spherical coordinates use Eqs. 5.5.25 and Eqs. 5.5.27. There results

$$\begin{aligned}\vec{u} &= 2r\sin\phi\cos\theta(\sin\phi\cos\theta\hat{i}_r - \sin\theta\hat{i}_\theta + \cos\phi\cos\theta\hat{i}_\phi)\\ &\quad -r\cos\phi(\sin\phi\sin\theta\hat{i}_r + \cos\theta\hat{i}_\theta + \cos\phi\sin\theta\hat{i}_\phi)\\ &\quad +r\sin\phi\sin\theta(\cos\phi\hat{i}_r - \sin\phi\hat{i}_\phi)\\ &= 2r\sin^2\phi\cos^2\theta\hat{i}_r + r\cos\theta(2\sin\phi\sin\theta - \cos\phi)\hat{i}_\theta\\ &\quad + r(2\sin\phi\cos\phi\cos^2\theta - \sin\theta)\hat{i}_\phi\end{aligned}$$

Note the relatively complex forms that the vector takes when expressed in cylindrical and spherical coordinates. This, however, is not always the case; the vector $\vec{u} = x\hat{i} + y\hat{j} + z\hat{k}$ becomes simply $\vec{u} = r\hat{i}_r$ in spherical coordinates. We shall obviously choose the particular coordinate system that simplifies the analysis.

TABLE 5.3 Relationships Involving $\vec{\nabla}$ in Rectangular, Cylindrical, and Spherical Coordinates

Rectangular

$$\vec{\nabla}\phi = \frac{\partial\phi}{\partial x}\hat{i} + \frac{\partial\phi}{\partial y}\hat{j} + \frac{\partial\phi}{\partial z}\hat{k}$$

$$\vec{\nabla}\cdot\vec{u} = \frac{\partial u_x}{\partial x} + \frac{\partial u_y}{\partial y} + \frac{\partial u_z}{\partial z}$$

$$\vec{\nabla}\times\vec{u} = \left(\frac{\partial u_z}{\partial y} - \frac{\partial u_y}{\partial z}\right)\hat{i} + \left(\frac{\partial u_x}{\partial z} - \frac{\partial u_z}{\partial x}\right)\hat{j} + \left(\frac{\partial u_y}{\partial x} - \frac{\partial u_x}{\partial y}\right)\hat{k}$$

$$\nabla^2\phi = \frac{\partial^2\phi}{\partial x^2} + \frac{\partial^2\phi}{\partial y^2} + \frac{\partial^2\phi}{\partial z^2}$$

$$\nabla^2\vec{u} = \nabla^2 u_x\hat{i} + \nabla^2 u_y\hat{j} + \nabla^2 u_z\hat{k}$$

Cylindrical

$$\vec{\nabla}\phi = \frac{\partial\phi}{\partial r}\hat{i}_r + \frac{1}{r}\frac{\partial\phi}{\partial\theta}\hat{i}_\theta + \frac{\partial\phi}{\partial\theta}\hat{i}_z$$

$$\vec{\nabla}\cdot\vec{u} = \frac{1}{r^2}\frac{\partial}{\partial r}(ru_r) + \frac{1}{r}\frac{\partial u_\theta}{\partial\theta} + \frac{\partial u_z}{\partial z}$$

$$\vec{\nabla}\times\vec{u} = \left[\frac{1}{r}\frac{\partial u_z}{\partial\theta} - \frac{\partial u_\theta}{\partial z}\right]\hat{i}_r + \left[\frac{\partial u_r}{\partial z} - \frac{\partial u_z}{\partial r}\right]\hat{i}_\theta + \left[\frac{1}{r}\frac{\partial}{\partial r}(ru_\theta) - \frac{1}{r}\frac{\partial u_r}{\partial\theta}\right]\hat{i}_z$$

$$\nabla^2\phi = \frac{1}{r}\frac{\partial}{\partial r}\left(r\frac{\partial\phi}{\partial r}\right) + \frac{1}{r^2}\frac{\partial^2\phi}{\partial\theta^2} + \frac{\partial^2\phi}{\partial z^2}$$

$$\nabla^2\vec{u} = \left(\nabla^2 u_r - \frac{u_r}{r^2} - \frac{2}{r^2}\frac{\partial u_\theta}{\partial\theta}\right)\hat{i}_r + \left(\nabla^2 u_\theta - \frac{u_\theta}{r^2} + \frac{2}{r^2}\frac{\partial u_r}{\partial\theta}\right)\hat{i}_\theta + \nabla^2 u_z\hat{i}_z$$

Spherical

$$\vec{\nabla}\phi = \frac{\partial\phi}{\partial r}\hat{i}_r + \frac{1}{r\sin\phi}\frac{\partial\phi}{\partial\theta}\hat{i}_\theta + \frac{1}{r}\frac{\partial\phi}{\partial\phi}\hat{i}_\phi$$

$$\vec{\nabla}\cdot\vec{u} = \frac{1}{r^2}\frac{\partial}{\partial r}(r^2u_r) + \frac{1}{r\sin\phi}\frac{\partial u_\theta}{\partial\theta} + \frac{1}{r\sin\phi}\frac{\partial}{\partial\phi}(u_\phi\sin\phi)$$

$$\vec{\nabla}\times\vec{u} = \frac{1}{r\sin\phi}\left[\frac{\partial}{\partial\phi}(u\phi\sin\phi) - \frac{\partial u_\theta}{\partial\theta}\right]\hat{i}_r + \frac{1}{r}\left[\frac{\partial}{\partial r}(ru_\phi) - \frac{\partial u_r}{\partial\phi}\right]\hat{i}_\theta + \frac{1}{r}\left[\frac{1}{\sin\phi}\frac{\partial u_r}{\partial\theta} - \frac{\partial}{\partial r}(ru_\theta)\right]\hat{i}_\phi$$

$$\nabla^2\phi = \frac{1}{r^2}\frac{\partial}{\partial r}\left(r^2\frac{\partial\phi}{\partial r}\right) + \frac{1}{r^2\sin^2\phi}\frac{\partial^2\phi}{\partial\theta^2} + \frac{1}{r^2\sin\phi}\frac{\partial}{\partial\phi}\left(\sin\phi\frac{\partial\phi}{\partial\phi}\right)$$

$$\begin{aligned}\nabla^2\vec{u} = &\left[\nabla^2 u_r - \frac{2u_r}{r^2} - \frac{2}{r^2\sin\phi}\frac{\partial u_\theta}{\partial\theta} - \frac{2}{r^2\sin\phi}\frac{\partial}{\partial\phi}(u_\phi\sin\phi)\right]\hat{i}_r \\ &+\left[\nabla^2 u_\theta - \frac{u_\theta}{r^2\sin^2\phi} + \frac{2\cos\phi}{r^2\sin^2\phi}\frac{\partial u_\phi}{\partial\theta} + \frac{2}{r^2\sin\phi}\frac{\partial u_r}{\partial\theta}\right]\hat{i}_\theta \\ &+\left[\nabla^2 u_\phi - \frac{2\cos\phi}{r^2\sin^2\phi}\frac{\partial u_\theta}{\partial\theta} - \frac{u_\phi}{r^2\sin^2\phi} + \frac{2}{r^2}\frac{\partial u_r}{\partial\phi}\right]\hat{i}_\phi\end{aligned}$$

Example 5.18

A particle moves in three-dimensional space. Determine an expression for its acceleration in cylindrical coordinates.

Solution

Remember r is the distance from the z axis and $\vec{r}$ is measured from the origin.

The particle is positioned by the vector

$$\vec{r} = r\hat{i}_r + z\hat{i}_z$$

The velocity is found by differentiating with respect to time; that is,

$$\vec{v} = \frac{d\vec{r}}{dt} = \frac{dr}{dt}\hat{i}_r + r\frac{d\hat{i}_r}{dt} + \frac{dz}{dt}\hat{i}_z$$

We find an expression for $d\hat{i}_r/dt$ by using Eq. 5.5.4 to get

$$\begin{aligned}\frac{d\hat{i}_r}{dt} &= -\sin\theta\frac{d\theta}{dt}\hat{i} + \cos\theta\frac{d\theta}{dt}\hat{j} \\ &= \frac{d\theta}{dt}(-\sin\theta\,\hat{i} + \cos\theta\,\hat{j}) \\ &= \theta\hat{i}_\theta\end{aligned}$$

Thus,

$$\vec{v} = \dot{r}\hat{i}_r + r\dot{\theta}\hat{i}_\theta + \dot{z}\hat{i}_z$$

where $\dot{\theta} = d\theta/dt$.

Differentiate again with respect to time. We have

$$\vec{a} = \ddot{r}\hat{i}_r + \dot{r}\frac{d\hat{i}_r}{dt} + \dot{r}\dot{\theta}\hat{i}_\theta + r\ddot{\theta}\hat{i}_\theta + r\dot{\theta}\frac{d\hat{i}_\theta}{dt} + \ddot{z}\hat{i}_z$$

The quantity $d\hat{i}_\theta/dt$ is (see Eq. 5.5.4)

$$\begin{aligned}\frac{d\hat{i}_\theta}{dt} &= (-\cos\theta\,\hat{i} - \sin\theta\,\hat{j})\frac{d\theta}{dt} \\ &= -\hat{i}_r\dot{\theta}\end{aligned}$$

The acceleration is then

$$\begin{aligned}\vec{a} &= \ddot{r}\hat{i}_r + \dot{r}\dot{\theta}\hat{i}_\theta + \dot{r}\dot{\theta}\hat{i}_\theta + r\ddot{\theta}\hat{i}_\theta - r\dot{\theta}^2\hat{i}_r + \ddot{z}\hat{i}_z \\ &= (\ddot{r} - r\dot{\theta}^2)\hat{i}_r + (2\dot{r}\dot{\theta} + r\ddot{\theta})\hat{i}_\theta + \ddot{z}\hat{i}_z\end{aligned}$$

5.6 INTEGRAL THEOREMS

Many of the derivations of the mathematical models used to describe physical phenomena make use of integral theorems, theorems that enable us to transform surface integrals to volume integrals or line integrals to surface integrals. In this section we shall present the more commonly used integral theorems, with emphasis on the divergence theorem and Stokes's theorem, the two most important ones.

5.6.1 The Divergence Theorem

The *divergence theorem* (also referred to as *Gauss's theorem*) states that if a volume V is completely enclosed by the surface S, then for the vector function $\vec{u}(x, y, z)$, which is continuous with continuous derivatives,

$$\iiint_V \vec{\nabla} \cdot \vec{u}\, dV = \oiint_S \vec{u} \cdot \hat{n}\, dS \tag{5.6.1}$$

where $\hat{n}$ is an outward pointing unit vector normal to the elemental area dS. In rectangular component form the divergence theorem is

$$\iiint_V \left(\frac{\partial u_x}{\partial x} + \frac{\partial u_y}{\partial y} + \frac{\partial u_z}{dz} \right) dx\, dy\, dz = \oiint_S (u_x \hat{i} + u_y \hat{j} + u_z \hat{k}) \cdot \hat{n}\, dS \tag{5.6.2}$$

Recall that $\vec{u} = u_x \hat{i} + u_y \hat{j} + u_z \hat{k}$.

To show that this equation is true, consider the volume V of Fig. 5.26. Let S be a special surface which has the property that any line drawn parallel to a coordinate axis intersects S in at most two points. Let the equation of the lower surface S_1 be given by $f(x, y)$ and of the upper surface S_2 by $g(x, y)$, and the projection of the surface on the xy plane be denoted R. Then the third term of the volume integral of Eq. 5.6.2 can be written as

$$\iiint_V \frac{\partial u_z}{\partial z} dx\, dy\, dz = \iint_R \left[\int_{f(x,y)}^{g(x,y)} \frac{\partial u_z}{\partial z} dz \right] dx\, dy \tag{5.6.3}$$

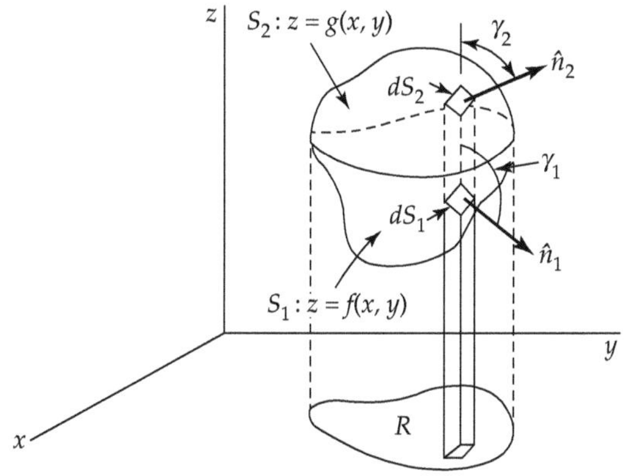

FIGURE 5.26 Volume used in proof of the divergence theorem.

The integral in the brackets is integrated to give (we hold x and y fixed in this integration)

$$\int_{f(x,y)}^{g(x,y)} \frac{\partial u_z}{\partial z} dz = \int_{f(x,y)}^{g(x,y)} du_z = u_z(x, y, g) - u_z(x, y, f) \tag{5.6.4}$$

Our integral then becomes

$$\iiint_V \frac{\partial u_z}{\partial z} dx\,dy\,dz = \iint_R [u_z(x, y, g) - u_z(x, y, f)]\,dx\,dy \tag{5.6.5}$$

The unit vector $\hat{n}$ is related to the direction cosines by $\hat{n} = \cos\alpha\hat{i} + \cos\beta\hat{j} + \cos\gamma\hat{k}$. Hence, for the upper surface S_2, we have

$$\begin{aligned} \cos\gamma_2\, dS_2 &= \hat{n}_2 \cdot \hat{k}\, dS_2 \\ &= dx\,dy \end{aligned} \tag{5.6.6}$$

For the lower surface S_1, realizing that γ_1 is an obtuse angle so that cos γ_1 is negative, there results

$$\begin{aligned} \cos\gamma_1\, dS_1 &= \hat{n}_1 \cdot \hat{k}\, dS_1 \\ &= -dx\,dy \end{aligned} \tag{5.6.7}$$

Now, with the results above substituted for $dx\,dy$, we can write Eq. 5.6.5 as

$$\begin{aligned} \iiint_V \frac{\partial u_z}{\partial z} dx\,dy\,dz &= \iint_{S_2} u_z(x, y, g)\hat{n}_2 \cdot \hat{k}\, dS_2 + \iint_{S_1} u_z(x, y, f)\hat{n}_1 \cdot \hat{k}\, dS_1 \\ &= \iint_{S_2} u_z\hat{n} \cdot \hat{k}\, dS_2 + \iint_{S_1} u_z\hat{n} \cdot \hat{k}\, dS_1 = \oiint_S u_z\hat{n} \cdot \hat{k}\, dS \end{aligned} \tag{5.6.8}$$

where the complete surface S is equal to $S_1 + S_2$.

Likewise, by taking volume stripes parallel to the x axis and the y axis, we can show that

$$\begin{aligned} \iiint_V \frac{\partial u_x}{\partial x} dx\,dy\,dz &= \oiint_S u_x\hat{n} \cdot \hat{i}\, dS \\ \iiint_V \frac{\partial u_y}{\partial z} dx\,dy\,dz &= \oiint_S u_y\hat{n} \cdot \hat{j}\, dS \end{aligned} \tag{5.6.9}$$

Summing Eqs. 5.6.8 and 5.6.9, we have

$$\begin{aligned} \iiint_V \left[\frac{\partial u_x}{\partial x} + \frac{\partial u_y}{\partial y} + \frac{\partial u_z}{\partial z}\right] dx\,dy\,dz &= \oiint_S [u_x\hat{n} \cdot \hat{i} + u_y\hat{n} \cdot \hat{j} + u_z\hat{n} \cdot \hat{k}]dS \\ &= \oiint_S [u_x\hat{i} + u_y\hat{j} + u_z\hat{k}] \cdot \hat{n}\, dS \end{aligned} \tag{5.6.10}$$

This is identical to Eq. 5.6.2, and the divergence theorem is shown to be valid. If the surface is not the special surface of Fig. 5.26, divide the volume into subvolumes, each of which satisfies the special condition. Then argue that the divergence theorem is valid for the original region.

The divergence theorem is often used to define the divergence, rather than Eq. 5.4.11, which utilizes rectangular coordinates. If we let the volume V shrink to the incremental volume ΔV, the divergence theorem takes the form

$$\overline{\vec{\nabla}\cdot\vec{u}}\,\Delta V = \oiint_{\Delta S} \vec{u}\cdot\hat{n}\,dS \tag{5.6.11}$$

where ΔS is the incremental area surrounding ΔV, and $\overline{\vec{\nabla}\cdot\vec{u}}$ is the average value of $\vec{\nabla}\cdot\vec{u}$ in ΔV. If we then allow ΔV to shrink to a point, $\overline{\vec{\nabla}\cdot\vec{u}}$ becomes $\vec{\nabla}\cdot\vec{u}$ at the point, and there results

$$\vec{\nabla}\cdot\vec{u} = \lim_{\Delta V\to 0} \frac{\oiint_{\Delta S} \vec{u}\cdot\hat{n}\,dS}{\Delta V} \tag{5.6.12}$$

This definition of the divergence is obviously independent of a particular coordinate system.

By letting the vector function $\vec{u}$ of the divergence theorem take on various forms, such as $\vec{\nabla}\phi$, $\phi\hat{i}$, and $\psi\vec{\nabla}\phi$, we can derive other useful integral formulas. These will be included in the Examples and Problems. Also, Table 5.4 tabulates these formulas.

TABLE 5.4 Integral Formulas

$$\iiint_V \vec{\nabla}\cdot\vec{u}\,dV = \oiint_S \vec{u}\cdot\hat{n}\,dS$$

$$\iiint_V \nabla^2\phi\,dV = \oiint_S \frac{\partial\phi}{\partial n}dS$$

$$\iiint_V \vec{\nabla}\phi\,dV = \oiint_S \phi\hat{n}\,dS$$

$$\iiint_V \vec{\nabla}\times\vec{u}\,dV = \oiint_S \hat{n}\times\vec{u}\,dS$$

$$\iiint_V (\psi\nabla^2\phi - \phi\nabla^2\psi)dV = \oiint_S \left(\psi\frac{\partial\phi}{\partial n} - \phi\frac{\partial\psi}{\partial n}\right)dS$$

$$\iiint_V (\psi\nabla^2\phi + \vec{\nabla}\phi\cdot\vec{\nabla}\psi)dV = \oiint_S \psi\frac{\partial\phi}{\partial n}dS$$

$$\iiint_S (\vec{\nabla}\times\vec{u})\cdot\hat{n}\,dS = \oint_C \vec{u}\cdot d\vec{l}$$

$$\iint_S (\hat{n}\times\vec{\nabla})\times\vec{u}\,dS = -\oint_C \vec{u}\times d\vec{l}$$

5.6.2 Stokes's Theorem

Let a surface S be surrounded by a simple curve C as shown in Fig. 5.27. For the vector function $\vec{u}(x, y, z)$, *Stokes's theorem* states that

$$\oint_C \vec{u} \cdot d\vec{l} = \iint_S (\vec{\nabla} \times \vec{u}) \cdot \hat{n}\, dS \tag{5.6.13}$$

where $d\vec{l}$ is a directed line element of C and $\hat{n}$ is a unit vector normal to dS. Using rectangular coordinates, this can be written as

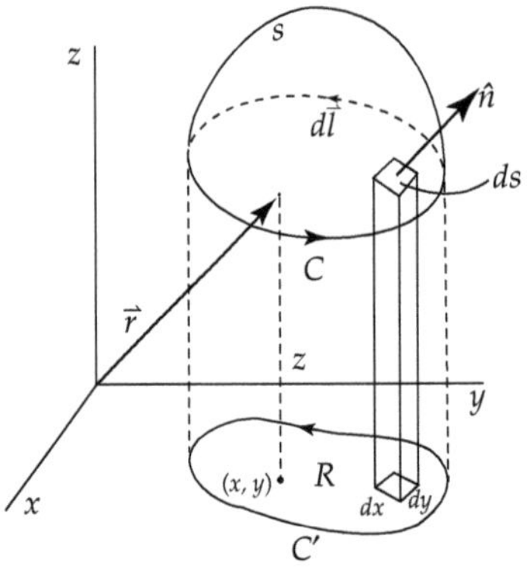

FIGURE 5.27 Surface used in the proof of Stokes's theorem.

$$\oint_C u_x dx + u_y dy + u_z dz = \iint_S \left[\left(\frac{\partial u_z}{\partial y} - \frac{\partial u_y}{\partial z} \right) \hat{i} \cdot \hat{n} + \left(\frac{\partial u_x}{\partial z} - \frac{\partial u_z}{\partial x} \right) \hat{j} \cdot \hat{n} \right.$$
$$\left. + \left(\frac{\partial u_y}{\partial x} - \frac{\partial u_x}{\partial y} \right) \hat{k} \cdot \hat{n} \right] dS \tag{5.6.14}$$

We will show that the terms involving u_x equate; that is,

$$\oint_C u_x dx = \iint_S \left[\frac{\partial u_x}{\partial z} \hat{j} \cdot \hat{n} - \frac{\partial u_x}{\partial y} \hat{k} \cdot \hat{n} \right] dS \tag{5.6.15}$$

To show that Eq. 5.6.15 is true, assume that the projection of S on the xy plane forms a simple curve C', which is intersected by lines parallel to the y axis at most twice. Let the surface S be located by the function $z = f(x, y)$; then a position vector to a point on S is

$$\begin{aligned} \vec{r} &= x\hat{i} + y\hat{j} + z\hat{k} \\ &= x\hat{i} + y\hat{j} + f(x, y)\hat{k} \end{aligned} \tag{5.6.16}$$

If we increment y an amount Δy the position vector locating a neighboring point on S becomes $\vec{r} + \Delta\vec{r} = x\hat{i} + (y + \Delta y)\hat{j} + (f + \Delta f)\hat{k}$, so that $\Delta\vec{r}$ is a vector approximately tangent to the surface S. In the limit as $\Delta y \to 0$, it is tangent. Hence, the vector

$$\frac{\partial \vec{r}}{\partial y} = \lim_{\Delta y \to 0} \frac{\Delta \vec{r}}{\Delta y} = \hat{j} + \frac{\partial f}{\partial y}\hat{k} \tag{5.6.17}$$

is tangent to S and thus normal to $\hat{n}$. We can then write

$$\hat{n} \cdot \frac{\partial \vec{r}}{\partial y} = \hat{n} \cdot \hat{j} + \frac{\partial f}{\partial y}\hat{n} \cdot \hat{k} = 0 \tag{5.6.18}$$

so

$$\hat{n} \cdot \hat{j} = -\frac{\partial f}{\partial y}\hat{n} \cdot \hat{k} \tag{5.6.19}$$

Substitute this back into Eq. 5.6.15 and obtain

$$\oint_C u_x\, dx = -\iint_S \left[\frac{\partial u_x}{\partial z}\frac{\partial f}{\partial y} + \frac{\partial u_x}{\partial y}\right]\hat{n} \cdot \hat{k}\, dS \tag{5.6.20}$$

Now, on the surface S,

$$u_x = u_x[x, y, f(x, y)] = g(x, y) \tag{5.6.21}$$

Using the chain rule from calculus we have, using $z = f(x, y)$,

$$\frac{\partial g}{\partial y} = \frac{\partial u_x}{\partial y} + \frac{\partial u_x}{\partial z}\frac{\partial f}{\partial y} \tag{5.6.22}$$

Equation 5.6.20 can then be written in the form (see Eq. 5.6.6)

$$\oint_C u_x\, dx = -\iint_R \frac{\partial g}{\partial y}\, dx\, dy \tag{5.6.23}$$

The area integral above can be written as* (see Fig. 5.28)

$$\begin{aligned}\iint_R \frac{\partial g}{\partial y}\, dx\, dy &= \int_{x_1}^{x_2}\left[\int_{h_1(x)}^{h_2(x)} \frac{\partial g}{\partial y}\, dy\right] dx = \int_{x_1}^{x_2} [g(x, h_2) - g(x, h_1)]\, dx \\ &= -\int_{C_2'} g\, dx - \int_{C_1'} g\, dx\end{aligned} \tag{5.6.24}$$

where $h_1(x)$ and $h_2(x)$ are defined in Fig. 5.28.

*This can also be accomplished by using Green's theorem in the plane, derived in Example 5.21, by letting $\phi = 0$ in that theorem.

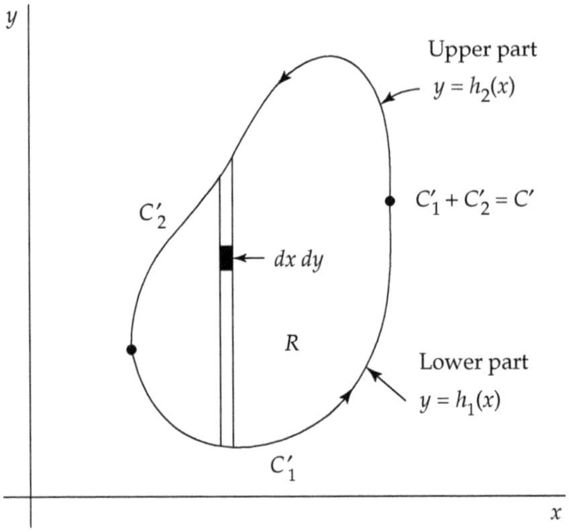

FIGURE 5.28 Plane surface *R* from Fig. 5.27.

where the negative sign on the C_2' integral is necessary to account for changing the direction of integration. Since $C_1' + C_2' = C'$, we see that

$$\oint_C u_x \, dx = \oint_{C'} g \, dx \tag{5.6.25}$$

From Eq. 5.6.21 we see that g on C' is the same as u_x on C. Thus, our proof of Eq. 5.6.15 is complete.

Likewise, by projections on the other coordinate planes, we can verify that

$$\begin{aligned}\oint_C u_y \, dy &= \iint_S \left[\frac{\partial u_y}{\partial x} \hat{k} \cdot \hat{n} - \frac{\partial u_y}{\partial z} \hat{i} \cdot \hat{n} \right] dS \\ \oint_C u_z \, dz &= \iint_S \left[\frac{\partial u_z}{\partial y} \hat{i} \cdot \hat{n} - \frac{\partial u_z}{\partial x} \hat{j} \cdot \hat{n} \right] dS\end{aligned} \tag{5.6.26}$$

If we add Eq. 5.6.15 to the two equations above, Eq. 5.6.14 results and our proof of Stokes's theorem is accomplished, a rather difficult task!

The scalar quantity resulting from the integration in Stokes's theorem is called the *circulation* of the vector $\vec{u}$ around the curve C. It is usually designated Γ and is

$$\Gamma = \oint_C \vec{u} \cdot d\vec{l} \tag{5.6.27}$$

It is of particular interest in aerodynamics since the quantity $\rho \Gamma U$ (ρ is the density of air and U is the speed) gives the magnitude of the lift on an airfoil. Note that for an irrotational vector field Eq. 5.6.13 implies that the circulation is zero.

Example 5.19

In the divergence theorem let the vector function $\vec{u} = \phi\hat{i}$, where ϕ is a scalar function. Derive the resulting integral theorem.

Solution

The divergence theorem is given by Eq. 5.6.1. Let $\vec{u} = \phi\hat{i}$ and there results

$$\iiint_V \vec{\nabla}\cdot(\phi\hat{i})dV = \oiint_S \phi\hat{i}\cdot\hat{n}\,dS$$

The unit vector $\hat{i}$ is constant and thus $\vec{\nabla}\cdot(\phi\hat{i}) = \hat{i}\cdot\vec{\nabla}\phi$ (see Table 5.1). Removing the constant $\hat{i}$ from the integrals yields

$$\hat{i}\cdot\iiint_V \vec{\nabla}\phi\,dV = \hat{i}\cdot\oiint_S \phi\hat{n}\,dS$$

This can be rewritten as

$$\hat{i}\cdot\left[\iiint_V \vec{\nabla}\phi\,dV - \oiint_S \phi\hat{n}\,dS\right] = 0$$

Since $\hat{i}$ is never zero and the quantity in brackets is not, in general, perpendicular to $\hat{i}$, we must demand that the quantity in brackets be zero. Consequently,

$$\iiint_V \vec{\nabla}\phi\,dV = \oiint_S \phi\hat{n}\,dS$$

This is another useful form of the divergence theorem.

Example 5.20

Let $\vec{u} = \psi\vec{\nabla}\phi$ in the divergence theorem, and then let $\vec{u} = \phi\vec{\nabla}\psi$. Subtract the resulting equations, thereby deriving Green's theorem.

Solution

Substituting $\vec{u} = \psi\vec{\nabla}\phi$ into the divergence theorem given by Eq. 5.6.1, we have

$$\iiint_V \vec{\nabla}\cdot(\psi\vec{\nabla}\phi)dV = \oiint_S \psi\vec{\nabla}\phi\cdot\hat{n}\,dS$$

Using Table 5.1, we can write

$$\vec{\nabla}\cdot(\psi\vec{\nabla}\phi) = \vec{\nabla}\psi\cdot\vec{\nabla}\phi + \psi\vec{\nabla}\cdot\vec{\nabla}\phi$$
$$= \vec{\nabla}\psi\cdot\vec{\nabla}\phi + \psi\nabla^2\phi$$

The divergence theorem takes the form, using Eq. 5.4.9,

$$\iiint_V [\psi\nabla^2\phi + \vec{\nabla}\psi\cdot\vec{\nabla}\phi]dV = \oint\!\!\oint_S \psi\frac{\partial\phi}{\partial n}dS$$

Now, with $\vec{u} = \phi\vec{\nabla}\psi$, we find that

$$\iiint_V [\phi\nabla^2\psi + \vec{\nabla}\psi\cdot\vec{\nabla}\phi]\,dV = \oint\!\!\oint_S \phi\frac{\partial\psi}{\partial n}\,dS$$

Subtract the two equations above and obtain

$$\iiint_V [\psi\nabla^2\phi - \phi\nabla^2\psi]dV = \oint\!\!\oint_S \left[\psi\frac{\partial\phi}{\partial n} - \phi\frac{\partial\psi}{\partial n}\right]dS$$

This is known as *Green's theorem,* or alternately the *second form of Green's theorem.*

Example 5.21

Let the volume V be the simply connected region shown in Fig. 5.29. The top and bottom are parallel to the xy plane.

Derive the resulting form of the divergence theorem if

$$\vec{u} = \vec{u}(x, y) = \phi\hat{i} + \psi\hat{j}$$

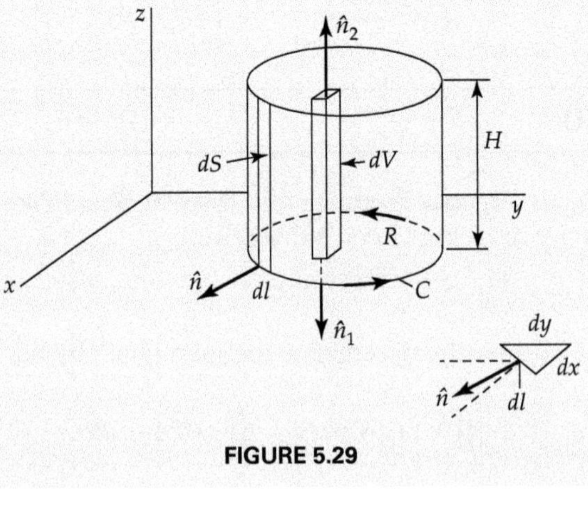

FIGURE 5.29

Solution

The divergence theorem, Eq. 5.6.1, takes the form

$$\iint_R \left[\int_0^H \vec{\nabla} \cdot \vec{u} \, dz \right] dx\,dy = \oiint_S \vec{u} \cdot \hat{n} \, dS$$

The quantity $\vec{\nabla} \cdot \vec{u}$ is independent of z, since $\vec{u} = \vec{u}(x, y)$. Thus,

$$\int_0^H \vec{\nabla} \cdot \vec{u} \, dz = H \vec{\nabla} \cdot \vec{u}$$

Also, on the top surface $\vec{u} \cdot \hat{n}_2 = \vec{u} \cdot \hat{k} = 0$, since $\vec{u}$ has no z component. Likewise, on the bottom surface $\vec{u} \cdot \hat{n}_1 = 0$. Consequently, only the side surface contributes to the surface integral. For this side surface we can write $dS = H\,dl$ and perform the integration around the closed curve C. The divergence theorem then takes the form

$$H \iint_R \vec{\nabla} \cdot \vec{u} \, dx\,dy = \oint_C \vec{u} \cdot \hat{n} H \, dl$$

Since

$$\vec{u} = \phi \hat{i} + \psi \hat{j} \quad \text{and} \quad \hat{n} = \frac{dy}{dl}\hat{i} - \frac{dx}{dl}\hat{j}$$

(see the small sketch in Fig. 5.29), there results

$$\vec{\nabla} \cdot \vec{u} = \frac{\partial \phi}{\partial x} + \frac{\partial \psi}{\partial y}$$

$$\vec{u} \cdot \hat{n} = \phi \frac{dy}{dl} - \psi \frac{dx}{dl}$$

Finally, we have the useful result

$$\iint_R \left(\frac{\partial \phi}{\partial x} + \frac{\partial \psi}{\partial y} \right) dy\,dx = \oint_C (\phi\,dy - \psi\,dx)$$

It is known as *Green's theorem in the plane.*

Example 5.22

Determine the circulation of the vector function $\vec{u} = 2y\hat{i} + x\hat{j}$ around the triangle of Fig. 5.30 by a) direct integration, and b) Stokes's theorem.

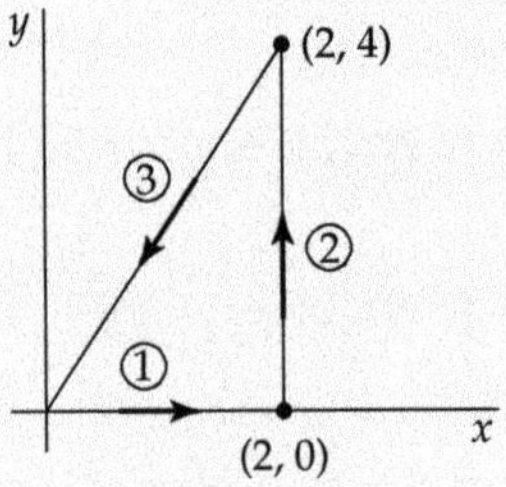

FIGURE 5.30

Solution

a) The circulation is given by

$$\Gamma = \oint_C \vec{u} \cdot d\vec{l} = \oint_C u_x dx + u_y dy + u_z dz$$

For the three lines of the curve, we have

$$\Gamma = \int_{①} u_x dx + u_y \overset{0}{\cancel{dy}} + u_z \overset{0}{\cancel{dz}} + \int_{②} u_x \overset{0}{\cancel{dx}} + u_y dy + u_z \overset{0}{\cancel{dz}}$$

$$+\int_{③} u_x dx + u_y dy + u_z \overset{0}{\cancel{dz}}$$

Along line 1, $u_x = 2y = 0$; along line 2, $u_y = x = 2$; and along line 3, $2x = y$, so $2dx = dy$. Thus, we have

$$\Gamma = \int_0^4 2dy + \int_2^0 (2 \cdot 2x\,dx + x\,2dx)$$

$$= 2 \cdot 4 + 3x^2\Big|_2^0 = \underline{-4}$$

b) Using Stokes's theorem, we have

$$\Gamma = \iint_S (\vec{\nabla} \times \vec{u}) \cdot \hat{n}\,dS$$

$$= \iint_S \left[\left(\frac{\partial u_z}{\partial y} - \overset{0}{\cancel{\frac{\partial u_y}{\partial z}}}\right)\hat{i} + \left(\frac{\partial u_x}{\partial z} - \overset{0}{\cancel{\frac{\partial u_z}{\partial x}}}\right)\hat{j} + \left(\frac{\partial u_y}{\partial x} - \frac{\partial u_x}{\partial y}\right)\hat{k}\right] \cdot \hat{k}\,dS$$

$$= \iint_S (1-2)\hat{k} \cdot \hat{k}\,dS = -\iint dS = \underline{-4}$$

PROBLEMS

5.1 State which of the following quantities are vectors.

(a) Volume
(b) Position of a particle
(c) Force
(d) Energy
(e) Momentum
(f) Color
(g) Pressure
(h) Frequency
(i) Magnetic field intensity
(j) Centrifugal acceleration
(k) Voltage

5.2 Two vectors $\vec{A}$ and $\vec{B}$ act as shown in Fig. 5.31. Find $\vec{A}+\vec{B}$ and $\vec{A}-\vec{B}$ for each by determining both magnitude and direction.

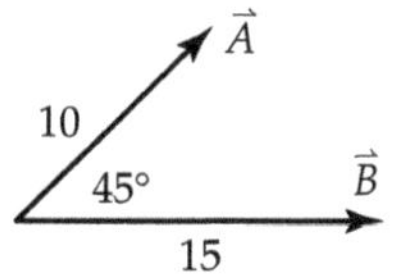

(a)

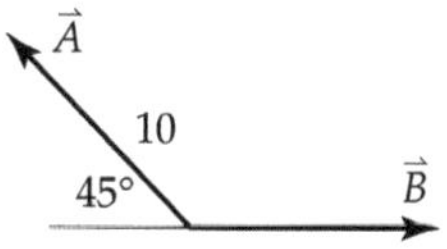

(b)

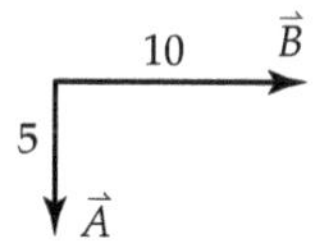

(c)

FIGURE 5.31

5.3 Express each of the vectors of Fig. 5.32 in component form, and write an expression for the unit vector that acts in the direction of each.

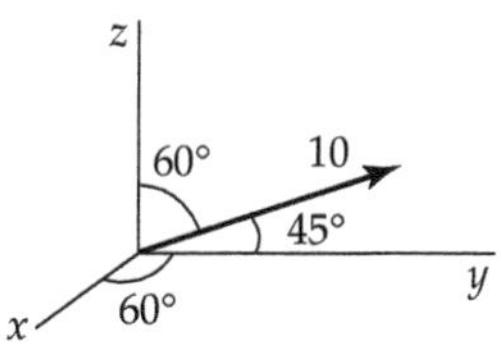

(a)

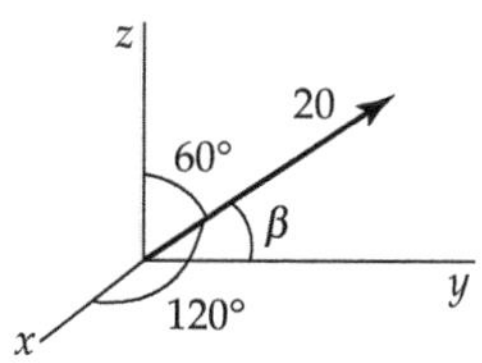

(b)

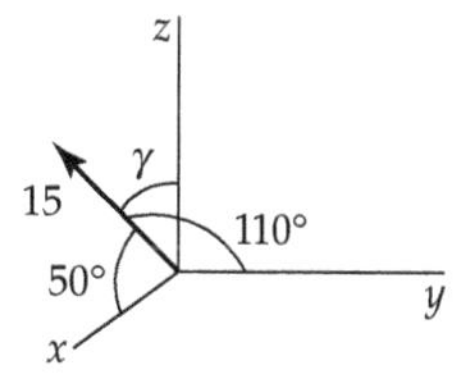

(c)

FIGURE 5.32

5.4 Given the vectors $\vec{A} = 2\hat{i} - 4\hat{j} - 4\hat{k}$, $\vec{B} = 4\hat{i} + 7\hat{j} - 4\hat{k}$, and $\vec{C} = 3\hat{i} - 4\hat{k}$. Find

(a) $\vec{A}+\vec{B}$	**(b)** $\vec{A}-\vec{C}$
(c) $\vec{A}+\vec{B}-\vec{C}$	**(d)** $\lvert\vec{A}-\vec{B}\rvert$
(e) $\vec{A}\cdot\vec{B}$	**(f)** $\vec{A}\times\vec{C}$
(g) $\lvert\vec{A}\times\vec{B}\rvert$	**(h)** $\vec{A}\times\vec{B}\cdot\vec{C}$
(i) $\vec{A}\cdot\vec{A}\times\vec{B}$	**(j)** $(\vec{A}\times\vec{B})\times\vec{C}$
(k) $\vec{A}\times(\vec{B}\times\vec{C})$	**(l)** $\lvert\vec{A}\times(\vec{B}\times\vec{C})\rvert$

5.5 Two vector fields are given by $\vec{A} = x\hat{i} - y\hat{j} + 2t\hat{k}$ and $\vec{B} = (x^2 - z^2)\hat{i} - y^2\hat{j}$. Find the following quantities at the point (0, 2, 2) at $t = 2$.

(a) $\vec{A}\cdot\vec{B}$	**(b)** $\vec{A}\times\vec{B}$
(c) $\lvert(\vec{A}\times\vec{B})\times\vec{A}\rvert$	**(d)** $\vec{A}\cdot\vec{A}\times\vec{B}$

5.6 Show that the diagonals of a rhombus (a parallelogram with equal sides) are perpendicular.

5.7 Verify each of following trigonometric identities.

(a) $\cos(\alpha+\beta) = \cos\alpha\cos\beta - \sin\alpha\sin\beta$
(b) $\sin(\alpha-\beta) = \sin\alpha\cos\beta - \sin\beta\cos\alpha$
(c) $\sin(\alpha+\beta) = \sin\alpha\cos\beta + \sin\beta\cos\alpha$

5.8 Find the projection of $\vec{A}$ on $\vec{B}$ if:

(a) $\vec{A} = 3\hat{i} - 6\hat{j} + 2\hat{k}$, $\vec{B} = 7\hat{i} - 4\hat{j} + 4\hat{k}$
(b) $\vec{A} = 3\hat{i} - 6\hat{j} + 9\hat{k}$, $\vec{B} = 4\hat{i} - 4\hat{j} + 2\hat{k}$
(c) $\vec{A} = 4\hat{i} - 3\hat{j} + 7\hat{k}$, $\vec{B} = 2\hat{i} - 5\hat{j} + 7\hat{k}$

5.9 Determine a unit vector $\hat{i}_c$ perpendicular to the plane of $\vec{A} = 3\hat{i} + 6\hat{j} - \hat{k}$ and $\vec{B} = 2\hat{i} - 3\hat{j} + 4\hat{k}$.

5.10 Find a unit vector $\hat{i}_c$ perpendicular to both $\vec{A} = 3\hat{i} - 2\hat{j}$ and $\vec{B} = \hat{i} - 2\hat{j} + \hat{k}$.

5.11 Determine m such that $\vec{A} = 2\hat{i} - m\hat{j} + \hat{k}$ is perpendicular to $\vec{B} = 3\hat{i} - 2\hat{j}$.

5.12 The direction cosines of a vector $\vec{A}$ of length 15 are $\frac{1}{3}, \frac{2}{3}, -\frac{2}{3}$. Find the component of $\vec{A}$ along the line passing through the points (1, 3, 2) and (3, –2, 6).

5.13 Find the equation of the plane perpendicular to $\vec{B} = 3\hat{i} + 2\hat{j} - 4\hat{k}$. (Let $\vec{r} = x\hat{i} + y\hat{j} + z\hat{k}$ be a point on the plane.)

5.14 An object is moved from the point (3, 2, –4) to the point (5, 0, 6), where the distance is measured in meters. If the force acting on the object is $\vec{F} = 3\hat{i} - 10\hat{j}$ newtons, determine the work done.

5.15 An object weighs 10 newtons and falls 10 m while a force of $3\hat{i} - 5\hat{j}$ newtons acts on the object. Find the work done if the z axis is positive upward. Include the work done by the weight.

5.16 A rigid device is rotating with a speed of 45 rad/s about an axis oriented by the direction cosines $\frac{7}{9}, -\frac{4}{9}$, and $\frac{4}{9}$. Determine the velocity of a point on the device located by the position vector $\vec{r} = 2\hat{i} + 3\hat{j} - \hat{k}$ meters.

5.17 The velocity at the point (–4, 2, –3), distances measured in meters, due to an angular velocity $\vec{\omega}$ is measured to be $\vec{V} = 10\hat{i} + 20\hat{j}$ m/s. What is $\vec{\omega}$ if $\omega_x = 2$ rad/s?

5.18 A force of 50 N acts at a point located by the position vector $\vec{r} = 4\hat{i} - 2\hat{j} + 4\hat{k}$ meters. The line of action of the force is oriented by the unit vector $\hat{i}_F = \frac{2}{3}\hat{i} - \frac{2}{3}\hat{j} + \frac{1}{3}\hat{k}$. Determine the moment of the force about the a) x axis, and b) a line oriented by the unit vector $\hat{i}_L = -\frac{2}{3}\hat{i} + \frac{2}{3}\hat{j} - \frac{1}{3}\hat{k}$.

5.19 By using the definition of the derivative show that

$$\frac{d}{dt}(\phi \vec{u}) = \phi\frac{d\vec{u}}{dt} + \vec{u}\frac{d\phi}{dt}$$

5.20 Given the two vectors $\vec{u} = 2t\hat{i} + t^2\hat{k}$ and $\vec{v} = \cos 5t\hat{i} + \sin 5t\hat{j} - 10\hat{k}$. At $t = 2$, evaluate the following:

(a) $\dfrac{d\vec{u}}{dt}$ **(b)** $\dfrac{d\vec{v}}{dt}$

(c) $\dfrac{d}{dt}(\vec{u} \cdot \vec{v})$ **(d)** $\dfrac{d}{dt}(\vec{u} \times \vec{v})$

(e) $\dfrac{d^2}{dt^2}(\vec{u} \cdot \vec{v})$

5.21 Find a unit vector in the direction of $d\vec{u}/dt$ if $\vec{u} = 2t^2\hat{i} - 3t\hat{j}$, at $t = 1$.

5.22 The velocity of a particle of water moving down the dishwasher arm of Fig. 5.33 at a distance of 0.2 m is 10 m/s. It is decellerating at a rate of 30 m/s^2. The arm is rotating at 30 rad/s. Determine the absolute acceleration of the particle.

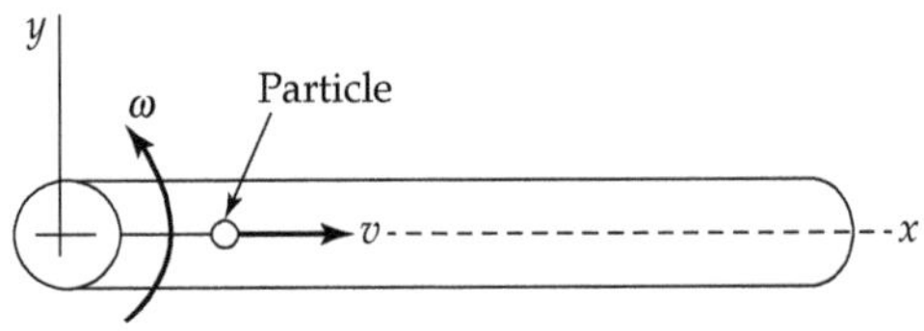

FIGURE 5.33

5.23 Show why it is usually acceptable to consider a reference frame attached to the earth as an inertial reference frame. The average radius of the earth is 6400 m.

5.24 The wind is blowing straight south at 90 km/hour. At a latitude of 45°, calculate the magnitudes of the Coriolis acceleration and the $\vec{\omega} \times (\vec{\omega} \times \vec{r})$ component of the acceleration of an air particle.

5.25 A velocity field is given by $\vec{v} = x^2\hat{i} - 2xy\hat{j} + 4t\hat{k}$ m/s. Determine the acceleration at the point (2, 1, –4) meters.

5.26 A temperature field is calculated to be $T(x, y, z, t) = e^{-0.1t}\sin 5x$. Determine the rate at which the temperature of a particle is changing if $\vec{v} = 10\hat{i} - 5\hat{j}$ m/s. Evaluate DT/Dt at $x = 2$ m and $t = 10$ s.

5.27 Find the gradient of each scalar function ($\vec{r} = x\hat{i} + y\hat{j} + z\hat{k}$).

(a) $\phi = x^2 + y^2$ (b) $\phi = 2xy$

(c) $\phi = r^2$ (d) $\phi = e^x \sin 2y$

(e) $\phi = x^2 + 2xy - z^2$ (f) $\phi = \ln r$

(g) $\phi = 1/r$ (h) $\phi = \tan^{-1} y/x$

(i) $\phi = r^n$

5.28 Find a unit vector $\hat{i}_n$ normal to each of the following surfaces at the point indicated.

(a) $x^2 + y^2 = 5$, (2, 1, 0)
(b) $r = 5$, (4, 0, 3)
(c) $2x^2 - y^2 = 7$, (2, 1, –1)
(d) $x^2 + yz = 3$, (2, –1, 1)
(e) $x + y^2 - 2z^2 = 6$, (4, 2, 1)
(f) $x^2y + yz = 6$, (2, 3, –2)

5.29 Determine the equation of the plane tangent to the given surface at the point indicated.

(a) $x^2 + y^2 + z^2 = 25$, (3, 4, 0)
(b) $r = 6$, (2, 4, 4)
(c) $x^2 - 2xy = 0$, (2, 2, 1)
(d) $xy^2 - zx + y^2 = 0$, (1, –1, 2)

5.30 The temperature in a region of interest is determined to be given by the function $T = x^2 + xy + yz$. At the point (2, 1, 4), answer the following questions. What is the unit vector that points in the direction of maximum change of temperature? What is the value of the derivative of the temperature

(a) in the x direction?
(b) in the direction of the vector $\hat{i} - 2\hat{j} + 2\hat{k}$?
(c) in the direction of $\hat{i} + \hat{j} + \hat{k}$?

5.31 Find the divergence of each of the following vector fields at the point (2, 1, –1).

(a) $\vec{u} = x^2\hat{i} + yz\hat{j} + y^2\hat{k}$

(b) $\vec{u} = y\hat{i} + xz\hat{j} + xy\hat{k}$

(c) $\vec{u} = x\hat{i} + y\hat{j} + z\hat{k}$

(d) $\vec{u} = xy\hat{i} + y^2\hat{j} + z^2\hat{k}$

(e) $\vec{u} = \vec{r}/r$

(f) $\vec{u} = \vec{r}/r^3$

5.32 Show that $\vec{\nabla} \cdot (\phi \vec{u}) = \phi \vec{\nabla} \cdot \vec{u} + \vec{u} \cdot \vec{\nabla} \phi$ by expanding in rectangular coordinates.

5.33 One person claims that the velocity field in a certain water flow is $\vec{v} = x^2\hat{i} - y^2\hat{j} + 2\hat{k}$, and another claims that it is $\vec{v} = y^2\hat{i} - x^2\hat{j} + 2\hat{k}$. Which one is obviously wrong and why?

5.34 It is known that the x component of velocity in a certain plane water flow (no z component of velocity) is given by x^2. Determine the velocity vector if $v_y = 0$ along the x axis.

5.35 Find the curl of each of the following vector fields at the point (–2, 4, 1).

(a) $\vec{u} = x^2\hat{i} - y^2\hat{j} + z^2\hat{k}$

(b) $\vec{u} = y^2\hat{i} + 2xy\hat{j} + z^2\hat{k}$

(c) $\vec{u} = xy\hat{i} + y^2\hat{j} + xz\hat{k}$

(d) $\vec{u} = \sin y\hat{i} + x\cos y\,\hat{j}$

(e) $\vec{u} = e^x \sin y\,\hat{i} + e^x \cos y\, j + e^x\hat{k}$

(f) $\vec{u} = \vec{r}/r^3$

5.36 Using the vector functions $\vec{u} = xy\hat{i} + y^2\hat{j} + z\hat{k}$ and $\vec{v} = x^2\hat{i} + xy\hat{j} + yzk$, evaluate each of the following at the point (–1, 2, 2).

(a) $\vec{\nabla} \cdot \vec{u}$ (b) $\vec{\nabla} \cdot \vec{v}$

(c) $\vec{\nabla} \times \vec{u}$ (d) $\vec{\nabla} \times \vec{v}$

(e) $\vec{\nabla} \cdot \vec{u} \times \vec{v}$ (f) $(\vec{\nabla} \times \vec{u}) \times \vec{v}$

(g) $\vec{\nabla} \times (\vec{u} \times \vec{v})$ (h) $\vec{u} \times (\vec{\nabla} \times \vec{v})$

(i) $(\vec{u} \times \vec{\nabla}) \times \vec{v}$ (j) $(\vec{u} \cdot \vec{\nabla})\vec{v}$

(k) $\vec{\nabla}(\vec{u} \cdot \vec{v})$ (l) $(\vec{v} \cdot \vec{\nabla})\vec{v}$

5.37 Determine if the following vector fields are solenoidal and/or irrotational.

(a) $x\hat{i} + y\hat{j} + z\hat{k}$

(b) $x\hat{i} - 2y\hat{j} + z\hat{k}$

(c) $y\hat{i} + x\hat{j}$

(d) $x^2\hat{i} \times y^2\hat{j} + z^2\hat{k}$

(e) $y^2\hat{i} + 2xy\hat{j} + z^2\hat{k}$

(f) $yz\hat{i} + xz\hat{j} + xy\hat{k}$

(g) $\sin y\,\hat{i} + \sin x\,\hat{j} + e^3\hat{k}$

(h) $x^2y\hat{i} + y^2x\hat{j} + z^2\hat{k}$

(i) $\vec{r}/r^3$

5.38 Verify each of the following vector identities by expanding in rectangular coordinates,

(a) $\vec{\nabla} \times \vec{\nabla}\phi = 0$

(b) $\vec{\nabla} \cdot \vec{\nabla} \times \vec{u} = 0$

(c) $\vec{\nabla} \cdot (\phi\vec{u}) = \vec{\nabla}\phi \cdot \vec{u} + \phi\vec{\nabla} \cdot \vec{u}$

(d) $\vec{\nabla} \times (\phi\vec{u}) = \vec{\nabla}\phi \times \vec{u} + \phi\vec{\nabla} \times \vec{u}$

(e) $\vec{\nabla} \times (\vec{u} \times \vec{v}) = \vec{u}(\vec{\nabla} \cdot \vec{v}) - \vec{v}(\vec{\nabla} \cdot \vec{u}) + (\vec{v} \cdot \vec{\nabla})\vec{u} - (\vec{u} \cdot \vec{\nabla})\vec{v}$

(f) $\vec{\nabla} \times (\vec{\nabla} \times \vec{u}) = \vec{\nabla}(\vec{\nabla} \cdot \vec{u}) - \nabla^2 \vec{u}$

(g) $\vec{\nabla} \cdot (\vec{u} \times \vec{v}) = \vec{\nabla} \times \vec{u} \cdot \vec{v} - \vec{u} \cdot \vec{\nabla} \times \vec{v}$

(h) $\vec{u} \times (\vec{\nabla} \times \vec{u}) = \frac{1}{2}\vec{\nabla}(\vec{u} \cdot \vec{u}) - (\vec{u} \cdot \vec{\nabla})\vec{u}$

5.39 Determine the scalar potential function ϕ, provided that one exists, associated with each of the following vector fields.

(a) $\vec{u} = x\hat{i} + y\hat{j} + z\hat{k}$

(b) $\vec{u} = x^2\hat{i} + y^2\hat{j} + z^2\hat{k}$

(c) $\vec{u} = y^2\hat{i} + 2xy\hat{j} + z\hat{k}$

(d) $\vec{u} = e^x \sin y\,\hat{i} + e^x \cos y\,\hat{j}$

(e) $\vec{u} = 2x\sin y\,\hat{i} + x^2 \cos y\,\hat{j} + z^2\,\hat{k}$

(f) $\vec{u} = 2xz\,\hat{i} + y^2\,\hat{j} + x^2\hat{k}$

5.40 Show that the unit vectors are orthogonal in a) the cylindrical coordinate system, and b) the spherical coordinate system. See Table 5.2.

5.41 By Using Eqs. 5.5.4, show that $d\hat{i}_r/dt = \dot{\theta}\hat{i}_\theta$ and $d\hat{i}_\theta/dt = -\dot{\theta}\hat{i}_r$. Sketch the unit vectors at two neighboring points and graphically display $\Delta\hat{i}_r$ and $\Delta\hat{i}_\theta$.

5.42 Find an expression for each of the following at the same point. The subscript c identifies cylindrical coordinates and subscript s spherical coordinates.

(a) $\hat{i} \cdot \hat{i}_{rc}$ (b) $\hat{i} \cdot \hat{i}_{rs}$

(c) $\hat{j} \cdot \hat{i}_{rs}$ (d) $\hat{i}_{rc} \cdot \hat{i}_{rs}$

(e) $\hat{i}_{\theta c} \cdot \hat{i}_\phi$ (f) $\hat{i}_{\theta c} \cdot \hat{i}_{\theta s}$

(g) $\hat{i}_z \cdot \hat{i}_{rs}$ (h) $\hat{i}_{rs} \cdot \hat{i}_{\theta s}$

(i) $\hat{i}_\phi \cdot \hat{i}_{rc}$

5.43 Relate the cylindrical coordinates at a point to the spherical coordinates at the same point.

5.44 A point is established in three-dimensional space by the intersection of three surfaces; for example, in rectangular coordinates they are three planes. What are the surfaces in a) cylindrical coordinates, and b) spherical coordinates? Sketch the intersecting surfaces for all three coordinate systems.

5.45 Express each of the following vectors as indicated:

(a) $\vec{u} = 2r\hat{i}_r + r\sin\phi\hat{i}_\theta + r^2\sin\phi\hat{i}_\phi$ in rectangular coordinates.

(b) $\vec{u} = r\hat{i}_\theta$ in rectangular coordinates.

(c) $\vec{u} = 2z\hat{i} + x\hat{j} + y\hat{k}$ in spherical coordinates.

(d) $\vec{u} = 2z\hat{i} + x\hat{j} + y\hat{k}$ in cylindrical coordinates.

5.46 Express the square of the differential arc length, that is, ds^2, in all three coordinate systems.

5.47 Following the procedure of Section 5.5, derive the expression for the gradient of scalar function ϕ in spherical coordinates, Eq. 5.5.28.

5.48 Determine the scalar potential function provided that one exists, associated with each of the following.

(a) $\vec{u} = r\hat{i}_r + \hat{i}_z$

(b) $\vec{u} = \left(A - \frac{B}{r^2}\right)\cos\theta\hat{i}_r - \left(A + \frac{B}{r^2}\right)\hat{i}_\theta$

(cylindrical coordinates)

(c) $\vec{u} = \left(A - \frac{B}{r^3}\right)\cos\phi\hat{i}_r - \left(A + \frac{B}{2r^3}\right)\sin\phi\hat{i}_\phi$

5.49 By using the divergence theorem, evaluate $\oint\!\!\oint_S \vec{u} \cdot \hat{n}\, dS$, where:

(a) $\vec{u} = x\hat{i} + y\hat{j} + z\hat{k}$ and S is the sphere $x^2 + y^2 + z^2 = 9$.

(b) $\vec{u} = xy\hat{i} + xz\hat{j} + (1 - z)\hat{k}$ and S is the unit cube bounded by $x = 0, y = 0, z = 0, x = 1, y = 1$, and $z = 1$.

(c) $\vec{u} = x\hat{i} + x\hat{j} + z^2\hat{k}$ and S is the cylinder $x^2 + y^2 = 4$ bounded by $z = 0$ and $z = 8$.

5.50 Recognizing that $\hat{i} \cdot \hat{n}\, dS = dy\, dz$, $\hat{j} \cdot \hat{n}\, dS = dx\, dz$, and $\hat{k} \cdot \hat{n}\, dS = dx\, dy$, evaluate the following using the divergence theorem.

(a) $\iint_S (x\, dy\, dz + 2y\, dx\, dz + y^2 dx\, dy)$, where S is the sphere $x^2 + y^2 + z^2 = 4$.

(b) $\iint_S (x^2\, dy\, dz + 2xy\, dx\, dz + xy\, dx\, dy)$, where S is the cube of Problem 5.49(b).

(c) $\iint_S xz\, dx\, dy$, where S is the cylinder of Problem 5.49(c).

5.51 Let $\vec{u} = \vec{\nabla}\phi$, and derive one of the forms of the divergence theorem given in Table 5.4.

5.52 With $\vec{u} = \vec{v} \times \hat{i}$, derive one of the forms of the divergence theorem given in Table 5.4.

5.53 Assume that ϕ is a harmonic function, that is, ϕ satisfies Laplace's equation $\nabla^2\phi = 0$. Show that:

(a) $\oiint_S \dfrac{\partial\phi}{\partial n} dS = 0$

(b) $\iiint_V \vec{\nabla}\phi \cdot \vec{\nabla}\phi\, dV = \oiint_S \phi \dfrac{\partial\phi}{\partial n} dS$

5.54 If no fluid is being introduced into a volume V, that is, there are no sources or sinks, the conservation of mass is written in integral form as

$$-\iiint_V \frac{\partial\rho}{\partial t} dV = \oiint_S \rho\vec{v} \cdot \hat{n}\, dS$$

where the surface S surrounds V, $\rho(x, y, z, t)$ is the density (mass per unit volume), and $\vec{v}(x, y, z, t)$ is the velocity. Convert the area integral to a volume integral and combine the two volume integrals. Then, since the equation is valid for any arbitrary volume, extract the differential form of the conservation of mass.

5.55 The integral form of the energy equation of a stationary material equates the rate of change of energy contained by the material to the rate at which heat enters the material by conduction; that is,

$$\iiint_V \rho \frac{\partial e}{\partial t} dV = -\oiint_S \vec{q} \cdot \hat{n}\, dS$$

where ρ is the density, e is the internal energy, and $\vec{q}$ is the heat flux. Empirical evidence allows us to write

$$\Delta e = C\Delta T \quad \text{and} \quad \vec{q} = -K\vec{\nabla}T$$

where C is the specific heat and K is the conductivity. If this is true for any arbitrary volume, derive the differential heat equation if the coefficients are assumed constant.

5.56 Derive Green's theorem in the plane by letting $\vec{u} = \phi\hat{i} + \psi\hat{j}$ and S be the xy plane in Stokes's theorem.

5.57 Calculate the circulation of the vector $\vec{u} = y^2\hat{i} + xy\hat{j} + z^2\hat{k}$ around a triangle with vertices at the origin, (2, 2, 0) and (0, 2, 0), by a) direct integration, and b) using Stokes's theorem.

5.58 Calculate the circulation of $\vec{u} = y\hat{i} - x\hat{j} + z\hat{k}$ around a unit circle in the xy plane with center at the origin by a) direct integration, and b) using Stokes's theorem.

5.59 Evaluate the circulation of the following vector functions around the curves specified. Use either direct integration or Stokes's theorem.

(a) $\vec{u} = 2z\hat{i} + y\hat{j} + x\hat{k}$; the triangle with vertices at the origin, (1, 0, 0) and (0, 0, 4).

(b) $\vec{u} = 2xy\hat{i} + y^2z\hat{j} + xy\hat{k}$; the rectangle with corners at (0, 0, 0), (0, 4, 0), (6, 4, 0), (6, 0, 0).

(c) $\vec{u} = 2x^2\hat{i} + y^2\hat{j} + z^2\hat{k}$; the unit circle in the xy plane with center at the origin.

6 Partial Differential Equations

Outline

6.1 Introduction
6.2 Wave Motion
  6.2.1 Vibration of a Stretched, Flexible String
  6.2.2 The Vibrating Membrane
  6.2.3 Longitudinal Vibrations of an Elastic Bar
  6.2.4 Transmission-Line Equations
6.3 The D'Alembert Solution of the Wave Equation
6.4 Separation of Variables
6.5 Diffusion
6.6 Solution of the Diffusion Equation
  6.6.1 A Long, Insulated Rod with Ends at Fixed Temperatures
  6.6.2 A Long, Totally Insulated Rod
  6.6.3 Two-Dimensional Heat Conduction in a Long, Rectangular Bar
6.7 Electric Potential About a Spherical Surface
6.8 Heat Transfer in a Cylindrical Body
6.9 Gravitational Potential

Problems

6.1 INTRODUCTION

The physical systems studied thus far have been described primarily by ordinary differential equations. We are now interested in studying phenomena that require partial derivatives in the describing equations as they are formed in modeling the particular phenomena. Partial differential equations arise where the dependent variable depends on two or more independent variables. The assumption of lumped parameters in a physical problem usually leads to ordinary differential equations, whereas the assumption of a continuously distributed quantity, a field, generally leads to a partial differential equation. A field approach is quite common now in such undergraduate courses as deformable solids, electromagnetics, and fluid mechanics; hence, the study of partial differential equations is often included in undergraduate programs. Many applications (fluid flow, heat transfer, wave motion) involve second-order equations; thus, they will be emphasized.

The order of the highest derivative is again the order of the equation. The questions of linearity and homogeneity are answered as before in ordinary differential equations. Solutions are superposable as long as the equation is linear. In general, the number of solutions of a partial differential equation is very large. The unique solution corresponding to a particular physical problem is obtained by use of additional information

M. C. Potter and B. F. Feeny, *Mathematical Methods for Engineering and Science*,
https://doi.org/10.1007/978-3-031-26151-0_6

arising from the physical situation. If this information is given on the boundary as *boundary conditions*, a *boundary-value problem* results. If the information is given at one instant as *initial conditions*, an *initial-value problem* results. A *well-posed problem* has just the right number of these conditions specified to give the solution. We shall not delve into the mathematical theory of making a well-posed problem. We shall, instead, rely on our physical understanding to determine problems that are well posed. We caution the reader that:

1. A problem that has too many boundary and/or initial conditions specified is not well posed and is an overspecified problem.
2. A problem that has too few boundary and/or initial conditions does not possess a unique solution.

In general, a partial differential equation with independent variables x and t which is second order on each of the variables requires two bits of information (this could be dependent on time t) at some x location (or x locations) and two bits of information at some time t, usually $t = 0$.

We are presenting a mathematical tool by way of physical motivation. We shall derive the describing equations of some common phenomena to illustrate the modeling process; other phenomena could have been chosen such as those encountered in magnetic fields, elasticity, fluid flows, aerodynamics, diffusion of pollutants, and so on. An analytical solution technique will then be introduced in this chapter. In a later chapter a numerical technique will be reviewed so that solutions may be obtained to problems that cannot be solved analytically.

We shall be particularly concerned with second-order partial differential equations involving two independent variables, because of the many phenomena that they model. The general form is written as

$$A\frac{\partial^2 u}{\partial x^2} + B\frac{\partial^2 u}{\partial x\,\partial y} + C\frac{\partial^2 u}{\partial y^2} + D\frac{\partial u}{\partial x} + E\frac{\partial u}{\partial y} + Fu = G \tag{6.1.1}$$

where the coefficients may depend on x and y but are most often constants. The equations are classed depending on the coefficients A, B, and C. They are said to be

$$\begin{aligned} &1.\ \text{Elliptic} && \text{if } B^2 - 4AC < 0 \\ &2.\ \text{Parabolic} && \text{if } B^2 - 4AC = 0 \\ &3.\ \text{Hyperbolic} && \text{if } B^2 - 4AC > 0 \end{aligned} \tag{6.1.2}$$

We shall derive equations of each class and illustrate the different types of solutions for each. The boundary conditions are specified depending on the class of the partial differential equation. That is, for an elliptic equation the function $u(x, y)$ (or its derivative) will be specified around the entire boundary enclosing a region of interest, whereas for the hyperbolic and parabolic equations the function cannot be specified around an entire boundary. It is also possible to have an elliptic equation in part of a region of interest and a hyperbolic equation in the remaining part. A discontinuity would separate the two parts of the region; a shock wave would be an example of such a discontinuity.

In the following three sections we shall derive the mathematical equations that describe several phenomena of general interest. The remaining sections will be devoted to the solutions of the equations.

6.2 WAVE MOTION

One of the first phenomena to be modeled with a partial differential equation was that of wave motion. Wave motion occurs in a variety of physical situations; these include vibrating strings, vibrating membranes (drum heads), waves traveling through a solid bar, waves traveling through a solid media (earthquakes), acoustic waves, water waves, compression waves (shock waves), electromagnetic radiation, vibrating beams, and oscillating shafts, to mention a few. We shall illustrate wave motion with several examples.

6.2.1 Vibration of a Stretched, Flexible String

The motion of a tightly stretched, flexible string was first modeled with a partial differential equation approximately 250 years ago. It still serves as an excellent introductory example for wave motion. We shall derive the equation that describes the motion and then in later sections present methods of solution.

Suppose that we wish to describe the position for all time of the string shown in Fig. 6.1. In fact, we shall seek a describing equation for the deflection u of the string for any position x and for any time t. The initial and boundary conditions will be considered in detail when the solution is presented.

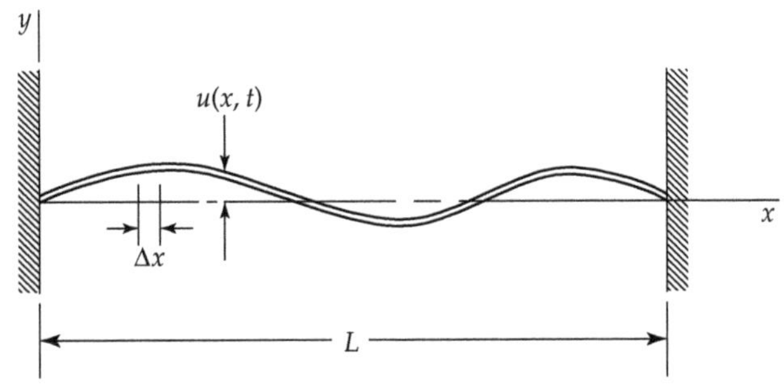

FIGURE 6.1 Deformed, flexible string at an instant t.

Consider an element of the string at a particular instant enlarged in Fig. 6.2. We shall make the following assumptions:

1. The string offers no resistance to bending so that no shearing force exists on a surface normal to the string.
2. The tension P is so large that the weight of the string is negligible.
3. Every element of the string moves normal to the x axis.

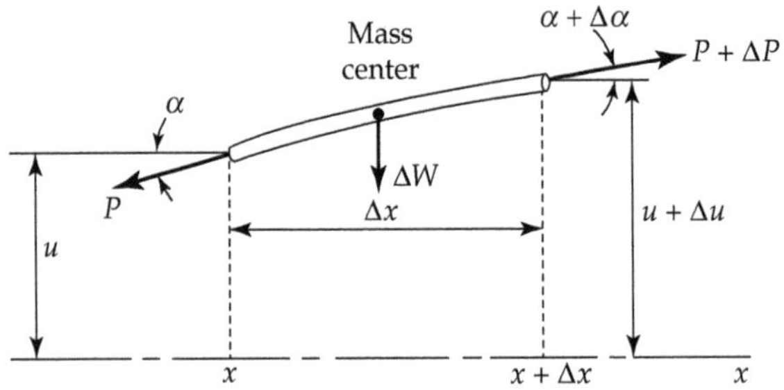

FIGURE 6.2 Small element of the vibrating string.

4. The slope of the deflection curve is small.
5. The mass m per unit length of the string is constant.
6. The effects of friction are negligible.

Newton's second law states that the net force acting on a body of constant mass equals the mass M of the body multiplied by the acceleration $\vec{a}$ of the center of mass of the body. This is expressed as

$$\sum \vec{F} = M\vec{a} \tag{6.2.1}$$

Consider the forces acting in the x direction on the element of the string. Using assumption 3 there is no acceleration of the element in the x direction; hence,

$$\sum F_x = 0 \tag{6.2.2}$$

or, referring to Fig. 6.2,

$$(P + \Delta P)\cos(\alpha + \Delta\alpha) - P\cos\alpha = 0 \tag{6.2.3}$$

Using assumption 4 we have

$$\cos\alpha \cong \cos(\alpha + \Delta\alpha) \cong 1 \tag{6.2.4}$$

Equation 6.2.3 then gives us

$$\Delta P = 0 \tag{6.2.5}$$

showing us that the tension is constant along the string.

For the y direction we have, neglecting friction and the weight of the string,

$$P\sin(\alpha + \Delta\alpha) - P\sin\alpha = m\Delta x \frac{\partial^2}{\partial t^2}\left(u + \frac{\Delta u}{2}\right) \tag{6.2.6}$$

where $m\Delta x$ is the mass of the element and $\partial^2/\partial t^2(u + \Delta u/2)$ is the acceleration of the mass center. Again, using assumption 4 we have

$$\begin{aligned} \sin(\alpha + \Delta\alpha) &\cong \tan(\alpha + \Delta\alpha) = \frac{\partial u}{\partial x}(x + \Delta x, t) \\ \sin\alpha &\cong \tan\alpha = \frac{\partial u}{\partial x}(x, t) \end{aligned} \tag{6.2.7}$$

Equation 6.2.6 can then be written as

$$P\left[\frac{\partial u}{\partial x}(x + \Delta x, t) - \frac{\partial u}{\partial x}(x, t)\right] = m\Delta x \frac{\partial^2}{\partial t^2}\left(u + \frac{\Delta u}{2}\right) \tag{6.2.8}$$

or, equivalently,

$$P\frac{\dfrac{\partial u}{\partial x}(x + \Delta x, t) - \dfrac{\partial u}{\partial x}(x, t)}{\Delta x} = m\frac{\partial^2}{\partial t^2}\left(u + \frac{\Delta u}{2}\right) \tag{6.2.9}$$

Now, we let $\Delta x \to 0$, which also implies that $\Delta u \to 0$. Then, using the definition,

$$\lim_{\Delta x \to 0} \frac{\frac{\partial u}{\partial x}(x + \Delta x, t) - \frac{\partial u}{\partial x}(x, t)}{\Delta x} = \frac{\partial^2 u}{\partial x^2}, \tag{6.2.10}$$

our describing equation becomes

$$P\frac{\partial^2 u}{\partial x^2} = m\frac{\partial^2 u}{\partial t^2} \tag{6.2.11}$$

This is usually written in the form

$$\frac{\partial^2 u}{\partial t^2} = a^2\frac{\partial^2 u}{\partial x^2} \tag{6.2.12}$$

where we have set

$$a = \sqrt{\frac{P}{m}} \tag{6.2.13}$$

Equation 6.2.12 is the *one-dimensional wave equation* and a is the *wave speed.* It is a transverse wave; that is, it moves normal to the string. This hyperbolic equation will be solved in a subsequent section.

6.2.2 The Vibrating Membrane

A stretched vibrating membrane, such as a drumhead, is simply an extension into another dimension of the vibrating-string problem. We shall derive a partial differential equation that describes the deflection u of the membrane for any position (x, y) and for any time t. The simplest equation results if the following assumptions are made:

1. The membrane offers no resistance to bending, so shearing stresses are absent.
2. The tension τ per unit length is so large that the weight of the membrane is negligible.
3. Every element of the membrane moves normal to the xy plane.
4. The slope of the deflection surface is small.
5. The mass m of the membrane per unit area is constant.
6. Frictional effects are neglected.

With these assumptions we can now apply Newton's second law to an element of the membrane as shown in Fig. 6.3. Assumption 3 leads to the conclusion that τ is constant throughout the membrane, since there are no accelerations of the element in the x and y directions. This is shown on the element. In the z direction we have

$$\sum F_z = Ma_z \tag{6.2.14}$$

For our element this becomes

$$\tau\,\Delta x \sin(\alpha + \Delta\alpha) - \tau\,\Delta x \sin\alpha + \tau\,\Delta y \sin(\beta + \Delta\beta) - \tau\,\Delta y \sin\beta = m\,\Delta x\,\Delta y\frac{\partial^2 u}{\partial t^2} \tag{6.2.15}$$

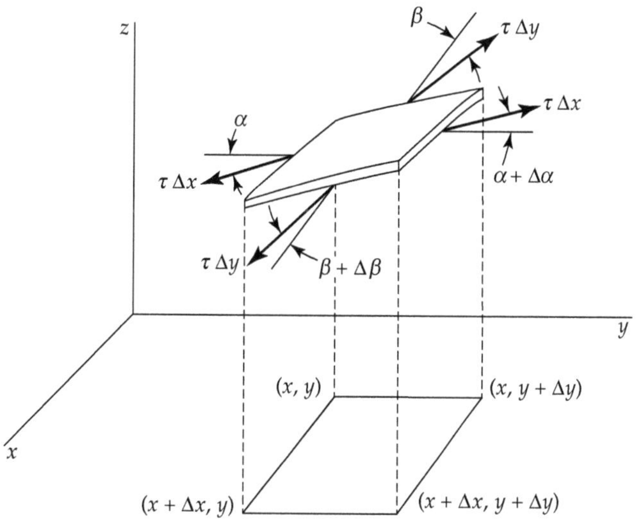

FIGURE 6.3 Element from a stretched, flexible membrane.

where the mass of the element is $m\,\Delta x\,\Delta y$ and the acceleration a_z is $\partial^2 u/\partial t^2$. Recognizing that for small angles

$$\begin{aligned}
\sin(\alpha + \Delta\alpha) &\cong \tan(\alpha + \Delta\alpha) = \frac{\partial u}{\partial y}\left(x + \frac{\Delta x}{2},\, y + \Delta y,\, t\right) \\
\sin\alpha &\cong \tan\alpha = \frac{\partial u}{\partial y}\left(x + \frac{\Delta x}{2},\, y,\, t\right) \\
\sin(\beta + \Delta\beta) &\cong \tan(\beta + \Delta\beta) = \frac{\partial u}{\partial x}\left(x + \Delta x,\, y + \frac{\Delta y}{2},\, t\right) \\
\sin\beta &\cong \tan\beta = \frac{\partial u}{\partial x}\left(x,\, y + \frac{\Delta y}{2},\, t\right)
\end{aligned} \tag{6.2.16}$$

we can then write Eq. 6.2.15 as

$$\begin{aligned}
&\tau\,\Delta x\left[\frac{\partial u}{\partial y}\left(x + \frac{\Delta x}{2}, y + \Delta y, t\right) - \frac{\partial u}{\partial y}\left(x + \frac{\Delta x}{2}, y, t\right)\right] \\
&\quad + \tau\,\Delta y\left[\frac{\partial u}{\partial x}\left(x + \Delta x,\, y + \frac{\Delta y}{2}, t\right) - \frac{\partial u}{\partial x}\left(x, y + \frac{\Delta y}{2}, t\right)\right] = m\,\Delta x\,\Delta y\frac{\partial^2 u}{\partial t^2}
\end{aligned} \tag{6.2.17}$$

or, by dividing by $\Delta x\,\Delta y$,

$$\tau\left[\frac{\dfrac{\partial u}{\partial y}\left(x + \dfrac{\Delta x}{2},\, y + \Delta y,\, t\right) - \dfrac{\partial u}{\partial y}\left(x + \dfrac{\Delta x}{2},\, y,\, t\right)}{\Delta y} + \frac{\dfrac{\partial u}{\partial x}\left(x + \Delta x, y + \dfrac{\Delta y}{2},\, t\right) - \dfrac{\partial u}{\partial x}\left(x,\, y + \dfrac{\Delta y}{2},\, t\right)}{\Delta x}\right] = m\frac{\partial^2 u}{\partial t^2} \tag{6.2.18}$$

Taking the limit as $\Delta x \to 0$ and $\Delta y \to 0$, we arrive at

$$\frac{\partial^2 u}{\partial t^2} = a^2\left(\frac{\partial^2 u}{\partial x^2} + \frac{\partial^2 u}{\partial y^2}\right) \tag{6.2.19}$$

where

This is an example where the dependent variable, u, depends on three independent variables, x, y, and t.

$$a = \sqrt{\frac{\tau}{m}} \tag{6.2.20}$$

Equation 6.2.19 is the *two-dimensional wave equation* and a is the wave speed.

6.2.3 Longitudinal Vibrations of an Elastic Bar

For another example of wave motion, let us determine the equation describing the motion of an elastic bar (steel, for example) that is subjected to an initial displacement or velocity. An example would be striking the bar on the end with a hammer, Fig. 6.4. We make the following assumptions:

1. The bar has a constant cross-sectional area A in the unstrained state.
2. All cross-sectional planes remain plane.
3. The density ρ remains constant throughout the bar.
4. Hooke's law may be used to relate stress and strain.

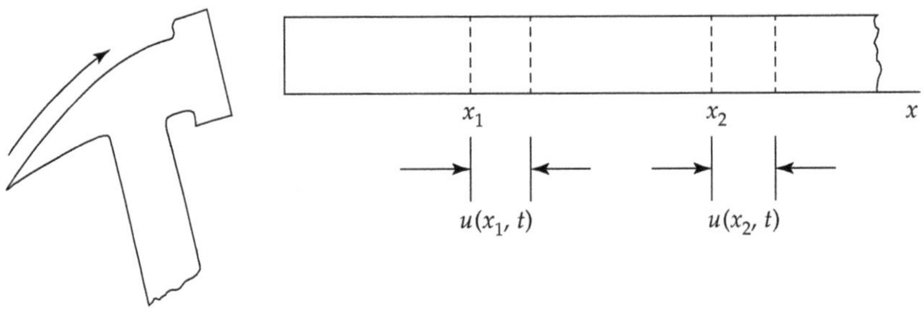

FIGURE 6.4 Wave motion in an elastic bar.

We let $u(x, t)$ denote the displacement of the plane of particles that were at x at $t = 0$. Consider the element of the bar between x_1 and x_2, shown in Fig. 6.5. We assume that the bar has mass ρ per unit volume. The force exerted on the element at x_1 is, by Hooke's law,

$$F_x = \text{area} \times \text{stress} = \text{area} \times E \times \text{strain}, \tag{6.2.21}$$

where E is the modulus of elasticity. The strain ϵ at x_1 is given by

$$\epsilon = \frac{\text{elongation}}{\text{unstrained length}} \tag{6.2.22}$$

Thus, for Δx_1 small, we have the strain at x_1 as

$$\epsilon = \frac{u(x_1 + \Delta x_1, t) - u(x_1, t)}{\Delta x} \tag{6.2.23}$$

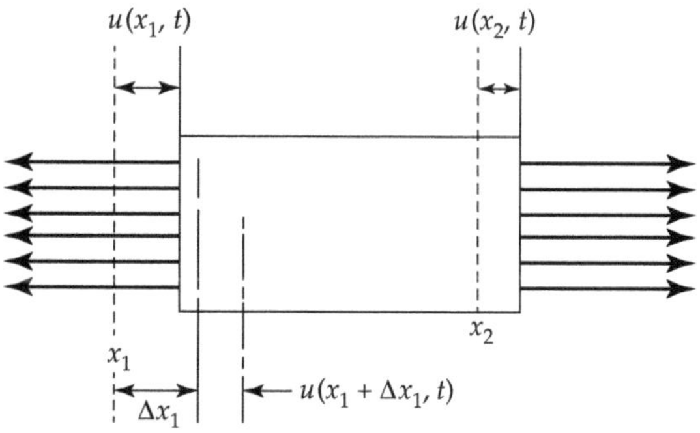

FIGURE 6.5 Element of an elastic bar.

Letting $\Delta x_1 \to 0$, we find that

$$\epsilon = \frac{\partial u}{\partial x} \tag{6.2.24}$$

Returning to the element, the force acting in the x direction is

$$F_x = AE\left[\frac{\partial u}{\partial x}(x_2, t) - \frac{\partial u}{\partial x}(x_1, t)\right] \tag{6.2.25}$$

Newton's second law states that

$$\begin{aligned} F_x &= ma \\ &= \rho A(x_2 - x_1)\frac{\partial^2 u}{\partial t^2} \end{aligned} \tag{6.2.26}$$

Hence, Eqs. 6.2.25 and 6.2.26 give

$$\rho A(x_2 - x_1)\frac{\partial^2 u}{\partial t^2} = AE\left[\frac{\partial u}{\partial x}(x_2, t) - \frac{\partial u}{\partial x}(x_1, t)\right] \tag{6.2.27}$$

We divide Eq. 6.2.27 by $(x_2 - x_1)$ and let $x_1 \to x_2$, to give

$$\frac{\partial^2 u}{\partial t^2} = a^2 \frac{\partial^2 u}{\partial x^2} \tag{6.2.28}$$

where the longitudinal wave speed a is given by

$$a = \sqrt{\frac{E}{\rho}} \tag{6.2.29}$$

Therefore, longitudinal displacements in an elastic bar may be described by the one-dimensional wave equation with wave speed $\sqrt{E/\rho}$.

6.2.4 Transmission-Line Equations

As a final example of wave motion, we shall derive the transmission-line equations. Electricity flows in the transmission line shown in Fig. 6.6, resulting in a current flow between conductors due to the capacitance and conductance between the conductors.

The cable also possesses both resistance and inductance resulting in voltage drops along the line. We shall choose the following symbols in our analysis:

$$v(x, t) = \text{voltage at any point along the line}$$
$$i(x, t) = \text{current at any point along the line}$$
$$R = \text{resistance per meter}$$
$$L = \text{self-inductance per meter}$$
$$C = \text{capacitance per meter}$$
$$G = \text{conductance per meter}$$

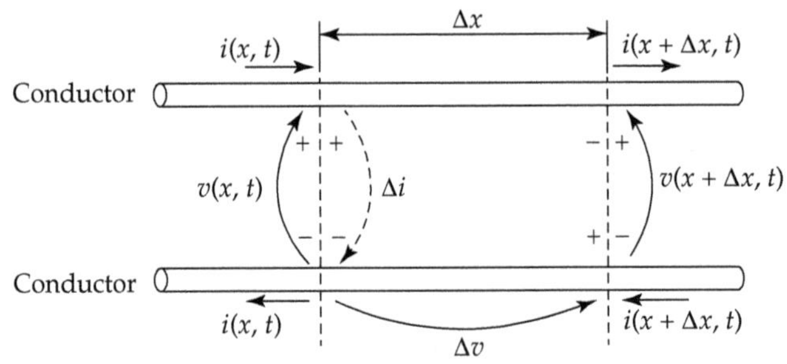

(a) Actual element, composed of the two conductors

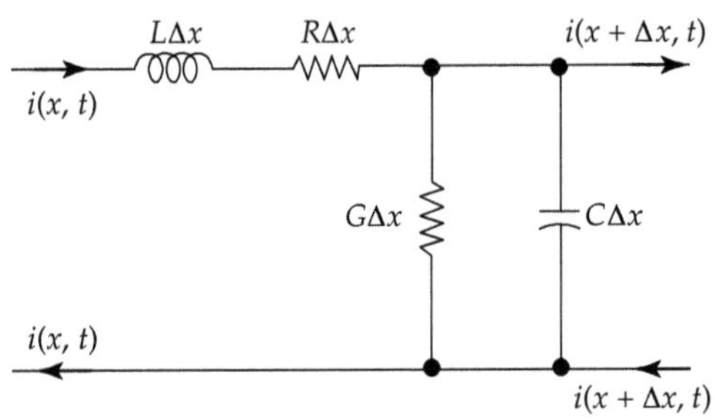

(b) Equivalent circuit

FIGURE 6.6 Element from a transmission line.

The voltage drop over the incremental length Δx at a particular instant (see Eqs. 1.4.3) is

$$\Delta v = v(x + \Delta x, t) - v(x, t) = -iR\Delta x - L\Delta x \frac{\partial i}{\partial t} \tag{6.2.30}$$

Dividing by Δx and taking the limit as $\Delta x \to 0$ yields the partial differential equation relating $v(x, t)$ and $i(x, t)$,

$$\frac{\partial v}{\partial x} + iR + L\frac{\partial i}{\partial t} = 0 \tag{6.2.31}$$

Now, let us find an expression for the change in the current over the length Δx. The current change is

$$\Delta i = i(x + \Delta x, t) - i(x, t) = -G\Delta x\, v - C\Delta x \frac{\partial v}{\partial t} \tag{6.2.32}$$

Again, dividing by Δx and taking the limit as $\Delta x \to 0$ gives a second equation,

$$\frac{\partial i}{\partial x} + vG + C\frac{\partial v}{\partial t} = 0 \tag{6.2.33}$$

Take the partial derivative of Eq. 6.2.31 with respect to x and of Eq. 6.2.33 with respect to t. Multiplying the second equation by L and subtracting the resulting two equations, using $\partial^2 i/\partial x\,\partial t = \partial^2 i/\partial t\,\partial x$, presents us with

$$\frac{\partial^2 v}{\partial x^2} + R\frac{\partial i}{\partial x} = LG\frac{\partial v}{\partial t} + LC\frac{\partial^2 v}{\partial t^2} \tag{6.2.34}$$

Then, substituting for $\partial i/\partial x$ from Eq. 6.2.33 results in an equation for $v(x, t)$ only. It is

$$\frac{\partial^2 v}{\partial x^2} = LC\frac{\partial^2 v}{\partial t^2} + (LG + RC)\frac{\partial v}{\partial t} + RGv \tag{6.2.35}$$

Take the partial derivative of Eq. 6.2.31 with respect to t and multiply by C; take the partial derivative of Eq. 6.2.33 with respect to x, subtract the resulting two equations and substitute for $\partial v/\partial x$ from Eq. 6.2.31; there results

$$\frac{\partial^2 i}{\partial x^2} = LC\frac{\partial^2 i}{\partial t^2} + (LG + RC)\frac{\partial i}{\partial t} + RGi \tag{6.2.36}$$

The two equations above are difficult to solve in the general form presented; two special cases are of interest. First, there are conditions under which the self-inductance, and leakage due to the conductance between conductors, are negligible; that is, $L \cong 0$, and $G \cong 0$. Then our equations become

$$\begin{aligned} \frac{\partial^2 v}{\partial x^2} &= RC\frac{\partial v}{\partial t} \\ \frac{\partial^2 i}{\partial x^2} &= RC\frac{\partial i}{\partial t} \end{aligned} \tag{6.2.37}$$

Second, for a condition of high frequency a time derivative increases the magnitude of a term*; that is, $\partial^2 i/\partial t^2 \gg \partial i/\partial t \gg i$. Thus, our general equations can be approximated by

$$\begin{aligned} \frac{\partial^2 v}{\partial t^2} &= \frac{1}{LC}\frac{\partial^2 v}{\partial x^2} \\ \frac{\partial^2 i}{\partial t^2} &= \frac{1}{LC}\frac{\partial^2 i}{\partial x^2} \end{aligned} \tag{6.2.38}$$

---

*As an example, consider the term $\sin(\omega t + x/L)$ where $\omega \gg 1$. Then

$$\frac{\partial}{\partial t}\left[\sin\left(\omega t + \frac{x}{L}\right)\right] = \omega\cos\left(\omega t + \frac{x}{L}\right)$$

We see that

$$\left|\omega\cos\left(\omega t + \frac{x}{L}\right)\right| \gg \left|\sin\left(\omega t + \frac{x}{L}\right)\right|$$

These latter two equations are wave equations with the units on $\sqrt{1/LC}$ of meters/second.

Although we shall not discuss any other wave phenomenon, it is well for the reader to be aware that sound waves, light waves, water waves, quantum-mechanical systems, and many other physical systems are described, at least in part, by a wave equation.

6.3 THE D'ALEMBERT SOLUTION OF THE WAVE EQUATION

It is possible to solve all the partial differential equations that we have derived in this chapter by a general method, the separation of variables. The wave equation can, however, be solved by a more special technique that will be presented in this section. It gives a quick look at the motion of a wave. We will obtain a general solution to the wave equation

$$\frac{\partial^2 u}{\partial t^2} = a^2 \frac{\partial^2 u}{\partial x^2} \tag{6.3.1}$$

The Greek letter xi is pronounced kasi (the second syllable sounds like an "I").

by a proper transformation of variables. Introduce the new independent variables ξ(xi) and η(eta):

$$\xi = x - at, \qquad \eta = x + at \tag{6.3.2}$$

Then, using the chain rule we find that

$$\begin{aligned} \frac{\partial u}{\partial x} &= \frac{\partial u}{\partial \xi}\frac{\partial \xi}{\partial x} + \frac{\partial u}{\partial \eta}\frac{\partial \eta}{\partial x} = \frac{\partial u}{\partial \xi} + \frac{\partial u}{\partial \eta} \\ \frac{\partial u}{\partial t} &= \frac{\partial u}{\partial \xi}\frac{\partial \xi}{\partial t} + \frac{\partial u}{\partial \eta}\frac{\partial \eta}{\partial t} = -a\frac{\partial u}{\partial \xi} + a\frac{\partial u}{\partial \eta} \end{aligned} \tag{6.3.3}$$

and

$$\begin{aligned} \frac{\partial^2 u}{\partial x^2} &= \frac{\partial\left(\frac{\partial u}{\partial x}\right)}{\partial \xi}\frac{\partial \xi}{\partial x} + \frac{\partial\left(\frac{\partial u}{\partial x}\right)}{\partial \eta}\frac{\partial \eta}{\partial x} = \frac{\partial^2 u}{\partial \xi^2} + 2\frac{\partial^2 u}{\partial \xi\,\partial \eta} + \frac{\partial^2 u}{\partial \eta^2} \\ \frac{\partial^2 u}{\partial t^2} &= \frac{\partial\left(\frac{\partial u}{\partial t}\right)}{\partial \xi}\frac{\partial \xi}{\partial t} + \frac{\partial\left(\frac{\partial u}{\partial t}\right)}{\partial \eta}\frac{\partial \eta}{\partial t} = a^2\frac{\partial^2 u}{\partial \xi^2} - 2a^2\frac{\partial^2 u}{\partial \xi\,\partial \eta} + a^2\frac{\partial^2 u}{\partial \eta^2} \end{aligned} \tag{6.3.4}$$

Substitute the expressions above into the wave equation to obtain

$$a^2\left[\frac{\partial^2 u}{\partial \xi^2} - 2\frac{\partial^2 u}{\partial \xi\,\partial \eta} + \frac{\partial^2 u}{\partial \eta^2}\right] = a^2\left[\frac{\partial^2 u}{\partial \xi^2} + 2\frac{\partial^2 u}{\partial \xi\,\partial \eta} + \frac{\partial^2 u}{\partial \eta^2}\right] \tag{6.3.5}$$

and there results

$$\frac{\partial^2 u}{\partial \xi\,\partial \eta} = 0 \tag{6.3.6}$$

Integration with respect to ξ gives

$$\frac{\partial u}{\partial \eta} = h(\eta) \tag{6.3.7}$$

where $h(\eta)$ is an arbitrary function of η (for an ordinary differential equation, this would be a constant). A second integration yields

$$u(\xi, \eta) = \int h(\eta) d\eta + g(\xi) \tag{6.3.8}$$

The integral is a function of η only and is replaced by $f(\eta)$, so the solution is

$$u(\xi, \eta) = g(\xi) + f(\eta) \tag{6.3.9}$$

or, equivalently,

The D'Alembert solution shows that a differentiable wave shape keeps a fixed shape as it travels at speed a.

$$u(x,t) = g(x - at) + f(x + at) \tag{6.3.10}$$

This is the D'Alembert solution of the wave equation.

Inspection of the equation above shows the wave nature of the solution. Consider an infinite string, stretched from $-\infty$ to $+\infty$, with an initial displacement $u(x, 0) = g(x) + f(x)$, as shown in Fig. 6.7. At some later time $t = t_1$ the curves $g(x)$ and $f(x)$ will simply be displaced to the right and left, respectively, a distance at_1. The original deflection curves move without distortion at the speed of propagation a.

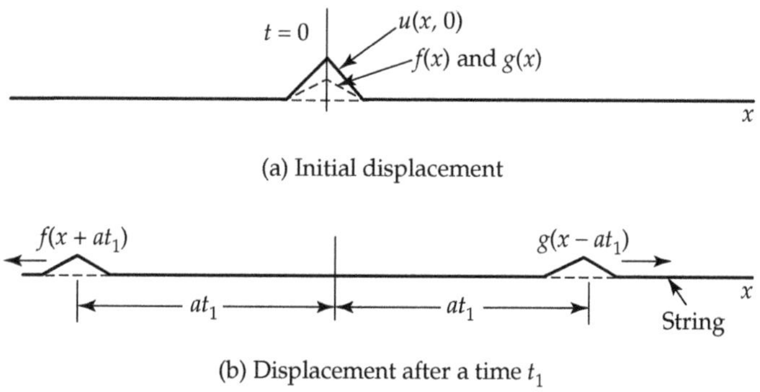

FIGURE 6.7 Traveling wave in a string.

To determine the form of the functions $g(x)$ and $f(x)$ when $u(x, 0)$ is given, we use the initial conditions. The term $\partial^2 u/\partial t^2$ in the wave equation demands that two bits of information be given at $t = 0$. Let us assume, for example, that the initial velocity is zero and that the initial displacement is given by

$$u(x,0) = f(x) + g(x) = \phi(x) \tag{6.3.11}$$

The velocity is

$$\frac{\partial u}{\partial t} = \frac{dg}{d\xi}\frac{\partial \xi}{\partial t} + \frac{df}{d\eta}\frac{\partial \eta}{\partial t} \tag{6.3.12}$$

At $t = 0$ this becomes (see Eqs. 6.3.2 and 6.3.10)

$$\frac{\partial u}{\partial t} = \frac{dg}{dx}(-a) + \frac{df}{dx}(a) = 0 \tag{6.3.13}$$

Hence, we have the requirement that

$$\frac{dg}{dx} = \frac{df}{dx} \tag{6.3.14}$$

which is integrated to provide us with

$$g = f + C \tag{6.3.15}$$

Inserting this in Eq. 6.3.11 gives

$$f(x) = \frac{\phi(x)}{2} - \frac{C}{2} \tag{6.3.16}$$

so that

$$g(x) = \frac{\phi(x)}{2} + \frac{C}{2} \tag{6.3.17}$$

Finally, replacing x in $f(x)$ with $x + at$ and x in $g(x)$ with $x - at$, there results the specific solution for the prescribed initial conditions,

$$u(x, t) = \frac{1}{2}\phi(x - at) + \frac{1}{2}\phi(x + at) \tag{6.3.18}$$

Our result shows that, for the infinite string, two initial conditions are necessary to determine the solution. A finite string will be discussed in the following section.

Example 6.1

Consider that the string in this article is given an initial velocity $\theta(x)$ and zero initial displacement. Determine the form of the solution.

Solution

The velocity is given by Eq. 6.3.12:

$$\frac{\partial u}{\partial t} = \frac{dg}{d\xi}\frac{\partial \xi}{\partial t} + \frac{df}{d\eta}\frac{\partial \eta}{\partial t}$$

At $t = 0$ this takes the form (using Eq. 6.3.13)

$$\theta(x) = a\frac{df}{dx} - a\frac{dg}{dx}$$

This is integrated to yield

$$f - g = \frac{1}{a}\int_0^x \theta(s)\,ds + C$$

where s is a dummy variable of integration. The initial displacement is zero, giving

$$u(x,0) = f(x) + g(x) = 0$$

or,

$$f(x) = -g(x)$$

The constant of integration C is thus evaluated to be

$$C = 2f(0) = -2g(0)$$

Combining this with the relation above results in

$$f(x) = \frac{1}{2a}\int_0^x \theta(s)\,ds + f(0)$$

$$g(x) = -\frac{1}{2a}\int_0^x \theta(s)\,ds + g(0)$$

Returning to Eq. 6.3.10, we can obtain the solution $u(x, t)$ using the forms above for $f(x)$ and $g(x)$ simply by replacing x by the appropriate quantity. We then have the solution

$$\begin{aligned} u(x,t) &= \frac{1}{2a}\left[\int_0^{x+at} \theta(s)ds - \int_0^{x-at} \theta(s)ds\right] \\ &= \frac{1}{2a}\left[\int_0^{x+at} \theta(s)\,ds + \int_{x-at}^{0} \theta(s)\,ds\right] \\ &= \frac{1}{2a}\int_{x-at}^{x+at} \theta(s)\,ds \end{aligned}$$

For a given $\theta(x)$ this expression would provide us with the solution.

Example 6.2

An infinite string is subjected to the initial displacement

$$\phi(x) = \frac{0.02}{1+9x^2}$$

Find an expression for the subsequent motion of the string if it is released from rest. The tension is 20 N and the mass per unit length is 5×10^{-4} kg/m. Also, sketch the solution for $t = 0$, $t = 0.002$ s, and $t = 0.01$ s.

Solution

The motion is given by the solution of this section. Equation 6.3.18 gives it as

$$u(x,t) = \frac{1}{2}\frac{0.02}{1+9(x-at)^2} + \frac{1}{2}\frac{0.02}{1+9(x+at)^2}$$

The wave speed a is given by

$$a = \sqrt{\frac{P}{m}}$$

$$= \sqrt{\frac{20}{5 \times 10^{-4}}} = 200 \text{ m/s}$$

The solution is then

$$u(x, t) = \frac{0.01}{1 + 9(x - 200t)^2} + \frac{0.01}{1 + 9(x + 200t)^2}$$

The sketches are presented in Fig. 6.8.

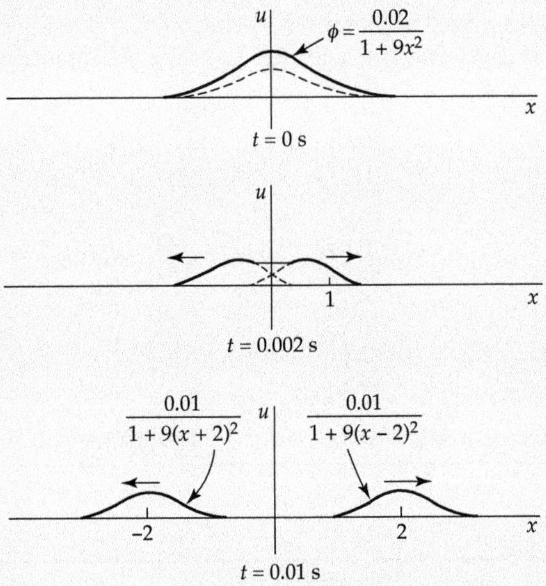

FIGURE 6.8

6.4 SEPARATION OF VARIABLES

We shall now present a powerful technique used to solve many of the partial differential equations encountered in physical applications in which the domains of interest are finite. It is the method of *separation of variables.* Even though it has limitations, it is a widely used technique. It involves the idea of reducing a more difficult problem to several simpler problems; here, we shall reduce a partial differential equation to several ordinary differential equations for which we already have methods of solution. Then, hopefully, by satisfying the initial and boundary conditions, a solution to the partial differential equation can be found.

To illustrate the details of the method, let us use the mathematical description of a finite string of length L that is fixed at both ends and is released from rest with an initial displacement. The motion of the string is described by the wave equation

$$\frac{\partial^2 u}{\partial t^2} = a^2 \frac{\partial^2 u}{\partial x^2} \tag{6.4.1}$$

We shall, as usual, consider the wave speed a to be a constant. The boundary conditions of fixed ends may be written as

$$u(0, t) = 0 \tag{6.4.2}$$

and

$$u(L, t) = 0 \tag{6.4.3}$$

Since the string is released from rest, the initial velocity is zero; hence,

$$\frac{\partial u}{\partial t}(x, 0) = 0 \tag{6.4.4}$$

The initial displacement will be denoted by $f(x)$; we then have

$$u(x, 0) = f(x) \tag{6.4.5}$$

Eqs. 6.4.2 and 6.4.3 make up the boundary conditions, while Eqs. 6.4.4 and 6.4.5 comprise the initial conditions.

We assume that the solution of our problem can be written in the separated form

$$u(x, t) = X(x)T(t) \tag{6.4.6}$$

$X(x)$ represents the spatial shape of the solution and $T(t)$ represents the temporal modulation of this shape.

that is, the x variable separates from the t variable. Substitution of this relationship into Eq. 6.4.1 yields

$$X(x)T''(t) = a^2 X''(x)T(t) \tag{6.4.7}$$

where the primes denote differentiation with respect to the associated independent variable. Rewriting Eq. 6.4.7 results in

$$\frac{T''}{a^2 T} = \frac{X''}{X} \tag{6.4.8}$$

The left side of this equation is a function of t only and the right side is a function of x only. Thus, as we vary t holding x fixed, the right side cannot change; this means that $T''(t)/a^2T(t)$ must be the same for all t. As we vary x holding t fixed the left side must not change. Thus, the quantity $X''(x)/X(x)$ must be the same for all x. Therefore, both sides must equal the same *constant* value μ (mu) sometimes called the *separation* constant. Equation 6.4.8 may then be written as two equations:

$$T'' - \mu a^2 T = 0 \tag{6.4.9}$$

$$X'' - \mu X = 0 \tag{6.4.10}$$

We note at this point that we have *separated* the *variables* and reduced a *partial differential equation* to two *ordinary differential equations*. If the boundary conditions can be satisfied, then we have succeeded with our separation of variables. We shall assume that we need to consider μ only as a real number. Thus, we are left with the three cases;

$$\begin{aligned} \mu &> 0 \\ \mu &= 0 \\ \mu &< 0 \end{aligned} \tag{6.4.11}$$

For any nonzero value of μ, we know that the solutions of these second-order ordinary differential equations are of the form e^{mt} and e^{nx}, respectively (see Section 1.5). The characteristic equations are

$$m^2 - \mu a^2 = 0 \tag{6.4.12}$$

$$n^2 - \mu = 0 \tag{6.4.13}$$

The roots are

$$m_1 = a\sqrt{\mu}, \quad m_2 = -a\sqrt{\mu} \tag{6.4.14}$$

$$n_1 = \sqrt{\mu}, \quad n_2 = -\sqrt{\mu} \tag{6.4.15}$$

The resulting solutions are

$$T(t) = c_1 e^{\sqrt{\mu}\, at} + c_2 e^{-\sqrt{\mu}\, at} \tag{6.4.16}$$

and

$$X(x) = c_3 e^{\sqrt{\mu}\, x} + c_4 e^{-\sqrt{\mu}\, x} \tag{6.4.17}$$

Now, consider the three cases, $\mu > 0$, $\mu = 0$, and $\mu < 0$. For $\mu > 0$, we have the result that $\sqrt{\mu}$ is a real number and the general solution is

$$u(x,t) = T(t)X(x) = (c_1 e^{\sqrt{\mu}\, at} + c_2 e^{-\sqrt{\mu}\, at})(c_3 e^{\sqrt{\mu}\, x} + c_4 e^{-\sqrt{\mu}\, x}) \tag{6.4.18}$$

which is a decaying or growing exponential. The derivative of Eq. 6.4.18 with respect to time would yield the velocity and it, too, would be growing or decaying with respect to time. This, of course, means that the kinetic energy of an element of the string would be increasing or decreasing in time, as would the total kinetic energy. However, energy remains constant; therefore, this solution violates the basic law of physical conservation of energy. The solution also does not give the desired wave motion and the boundary and initial conditions cannot be satisfied; thus, we cannot have $\mu > 0$. Similar arguments prohibit the use of $\mu = 0$. Hence, we are left with $\mu < 0$; for simplicity, let

$$\sqrt{\mu} = i\beta \tag{6.4.19}$$

That is $\mu = -\beta^2$ where β is a real number and i is $\sqrt{-1}$. For this case, Eq. 6.4.16 becomes

$$T(t) = c_1 e^{i\beta at} + c_2 e^{-i\beta at} \tag{6.4.20}$$

and Eq. 6.4.17 becomes

$$X(x) = c_3 e^{i\beta x} + c_4 e^{-i\beta x} \tag{6.4.21}$$

Using the relation

$$e^{i\theta} = \cos\theta + i\sin\theta \tag{6.4.22}$$

Eqs. 6.4.20 and 6.4.21 may be rewritten as

$$T(t) = A\sin\beta at + B\cos\beta at \tag{6.4.23}$$

and

$$X(x) = C\sin\beta x + D\cos\beta x \tag{6.4.24}$$

where A, B, C, and D are new constants. The relation of the new constants in terms of the constants c_1, c_2, c_3, and c_4 is left as an exercise.

Now that we have solutions to Eqs. 6.4.9 and 6.4.10 that are periodic in time and space, let us attempt to satisfy the boundary conditions and initial conditions given in Eqs. 6.4.2 through 6.4.5. Our solution thus far is

$$u(x,t) = (A\sin\beta at + B\cos\beta at)(C\sin\beta x + D\cos\beta x) \tag{6.4.25}$$

Up to this point, β is not yet determined.

The boundary condition $u(0, t) = 0$ states that u is zero for all t at $x = 0$; that is,

$$u(0,t) = (A\sin\beta at + B\cos\beta at)D = 0 \tag{6.4.26}$$

The only way this is possible is to have $D = 0$. Hence, we are left with

$$u(x,t) = (A\sin\beta at + B\cos\beta at)C\sin\beta x \tag{6.4.27}$$

The boundary condition $u(L, t) = 0$ states that u is zero for all t at $x = L$; this is expressed as

$$u(L,t) = (A\sin\beta at + B\cos\beta at)C\sin\beta L \tag{6.4.28}$$

This is possible if

$$\sin\beta L = 0 \tag{6.4.29}$$

For this to be true, we must have

$$\beta L = n\pi, \quad n = 1, 2, 3, \cdots \tag{6.4.30}$$

or $\beta = n\pi/L$; the quantity β is called an *eigenvalue*. When β is substituted back into $\sin\beta x$, the function $\sin n\pi x/L$ is called an *eigenfunction*. Each eigenvalue corresponding to a particular value of n produces a unique eigenfunction. Note that the $n = 0$ eigenvalue ($\mu = 0$) has already been eliminated as a possible solution, so it is not included here. The solution given in Eq. 6.4.27 may now be written as

Indeed, Eq. 6.4.10 with boundary conditions is called a differential eigenvalue problem. In this case its boundary conditions are $X(0) = X(L) = 0$.

$$u(x,t) = \left(A\sin\frac{n\pi at}{L} + B\cos\frac{n\pi at}{L}\right)C\sin\frac{n\pi x}{L} \tag{6.4.31}$$

For simplicity, let us make the substitutions

$$AC = a_n \quad \text{and} \quad BC = b_n \tag{6.4.32}$$

since each value of n may require different constants. Equation 6.4.31 is then

$u_n(x, t) = T_n(t)\, X_n(x, t)$ represents synchronous motion. That is the shape $X_n(x)$ is fixed and is modulated by a function of time, $T_n(t)$. Each particle on the string moves in or out of phase with other particles.

$$u_n(x,t) = \left(a_n \sin\frac{n\pi at}{L} + b_n \cos\frac{n\pi at}{L}\right)\sin\frac{n\pi x}{L} \tag{6.4.33}$$

where the subscript n has been added to $u(x, t)$ to allow for a different function for each value of n.

For our vibrating string, each value of n results in harmonic motion of the string with frequency $na/2L$ cycles per second (hertz). For $n = 1$ the *fundamental mode* results, and for $n > 1$ *overtones* result; see Fig. 6.9. *Nodes* are those points of the string which do not move. The velocity $\partial u_n/\partial t$ is then

$$\frac{\partial u_n}{\partial t} = \frac{n\pi a}{L}\left(a_n \cos\frac{n\pi at}{L} - b_n \sin\frac{n\pi at}{L}\right)\sin\frac{n\pi x}{L} \tag{6.4.34}$$

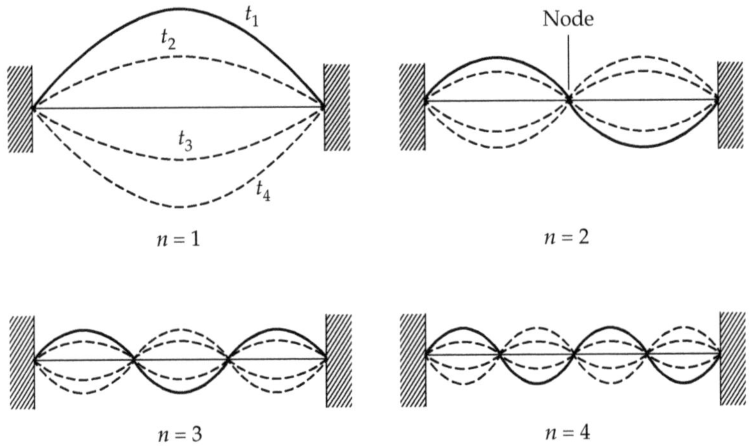

FIGURE 6.9 Harmonic motion snapshots. The solution at various values of time t is as shown.

Thus, to satisfy initial condition (6.4.4),

$$\frac{\partial u_n}{\partial t}(x,0) = \frac{n\pi a}{L} a_n \sin\frac{n\pi x}{L} = 0 \tag{6.4.35}$$

for all x, we must have $a_n = 0$. We are now left with

$$u_n(x,t) = b_n \cos\frac{n\pi at}{L}\sin\frac{n\pi x}{L} \tag{6.4.36}$$

If we are to solve our problem, we must satisfy initial condition (6.4.5),

$$u_n(x, 0) = f(x) \tag{6.4.37}$$

But, unless $f(x)$ is a multiple of sin $n\pi x/L$, no one value of n will satisfy Eq. 6.4.37. How do we then satisfy the boundary condition $u(x, 0) = f(x)$ if $f(x)$ is not a sine function?

Equation 6.4.36 is a solution of Eq. 6.4.1 and satisfies Eqs. 6.4.2 through 6.4.4 for all n ($n = 1, 2, 3, \cdots$). Hence, any linear combination of any of the solutions

$$u_n(x,t) = b_n \cos\frac{n\pi at}{L}\sin\frac{n\pi x}{L}, \qquad n = 1, 2, 3, \cdots \tag{6.4.38}$$

is also a solution, since the describing equation is linear and superposition is allowed. If we assume that for the most general function $f(x)$ we need to consider all values of n, then we should try

$$u(x,t) = \sum_{n=1}^{\infty} u_n(x,t) = \sum_{n=1}^{\infty} b_n \cos\frac{n\pi at}{L}\sin\frac{n\pi x}{L} \tag{6.4.39}$$

For the initial condition (6.4.5), we then have

$$u(x,0) = \sum_{n=1}^{\infty} b_n \sin\frac{n\pi x}{L} = f(x) \tag{6.4.40}$$

If constants b_n can be determined to satisfy Eq. 6.4.40, then we have a solution anywhere that the sum in Eq. 6.4.39 converges. The series in Eq. 6.4.40 is a *Fourier sine series*. It was presented in Section 1.10, but the essential features will be repeated here.

To find the b_n's, multiply the right side of Eq. 6.4.40 by $\sin m\pi x/L$ to give

$$\sin\frac{m\pi x}{L}\sum_{n=1}^{\infty} b_n \sin\frac{n\pi x}{L} = f(x)\sin\frac{m\pi x}{L} \tag{6.4.41}$$

Now integrate both sides of Eq. 6.4.41 from $x = 0$ to $x = L$. We may take $\sin m\pi x/L$ inside the sum, since it is a constant as far as the summation is concerned. The integral and the summation may be switched if the series converges properly. This may be done for most functions of interest in physical applications. Thus, we have

$$\sum_{n=1}^{\infty} b_n \int_0^L \sin\frac{n\pi x}{L}\sin\frac{m\pi x}{L}dx = \int_0^L f(x)\sin\frac{m\pi x}{L}dx \tag{6.4.42}$$

With the use of trigonometric identities we can verify* that

$$\int_0^L \sin\frac{n\pi x}{L}\sin\frac{m\pi x}{L}dx = \begin{cases} 0 & \text{if } m \neq n \\ \dfrac{L}{2} & \text{if } m = n \end{cases} \tag{6.4.43}$$

Hence, Eq. 6.4.42 gives us

$$b_n = \frac{2}{L}\int_0^L f(x)\sin\frac{n\pi x}{L}dx \tag{6.4.44}$$

if $f(x)$ may be expressed by

$$f(x) = \sum_{n=1}^{\infty} b_n \sin\frac{n\pi x}{L} \tag{6.4.45}$$

Equation 6.4.45 gives the Fourier sine series representation of $f(x)$ with the coefficients given by Eq. 6.4.44. Examples will illustrate the use of the above equations for particular functions $f(x)$.

---

*Use the trigonometric identities $\sin\alpha\sin\beta = 1/2[\cos(\alpha-\beta) - \cos(\alpha+\beta)]$ and $\sin^2\alpha = 1/2 - 1/2(\cos 2\alpha)$.

Example 6.3

A tight string 2 m long with $a = 30$ m/s is initially at rest but is given an initial velocity of $300 \sin 4\pi x$ from its equilibrium position. Determine the maximum displacement at the $x = \frac{1}{8}$ m location of the string.

Solution

We assume that the solution to the describing differential equation

$$\frac{\partial^2 u}{\partial t^2} = 900 \frac{\partial^2 u}{\partial x^2}$$

can be separated as

$$u(x, t) = T(t)X(x)$$

Following the procedure of the previous section, we substitute into the describing equation to obtain

$$\frac{1}{900} \frac{T''}{T} = \frac{X''}{X} = -\beta^2$$

where we have chosen the separation constant to be $-\beta^2$ so that an oscillatory motion will result. The two ordinary differential equations that result are

$$T'' + 900\beta^2 T = 0$$

$$X'' + \beta^2 X = 0$$

The general solutions to the equations above are

$$T(t) = A \sin 30\beta t + B \cos 30\beta t$$

$$X(x) = C \sin \beta x + D \cos \beta x$$

The solution for $u(x, t)$ is then

$$u(x, t) = (A \sin 30\beta t + B \cos 30\beta t)(C \sin \beta x + D \cos \beta x)$$

The end at $x = 0$ remains motionless; that is, $u(0, t) = 0$. Hence,

$$u(0, t) = (A \sin 30\beta t + B \cos 30\beta t)(0 + D) = 0$$

Thus, $D = 0$. The initial displacement $u(x, 0) = 0$. Hence,

$$u(x, 0) = (0 + B)C \sin \beta x = 0$$

Thus, $B = 0$. The solution reduces to

$$u(x, t) = AC \sin 30\beta t \sin \beta x$$

The initial velocity $\partial u/\partial t$ is given as 300 sin $4\pi x$. We then have, at $t = 0$,

$$\frac{\partial u}{\partial t} = 30\beta AC \sin \beta x = 300 \sin 4\pi x$$

This gives

$$\beta = 4\pi, \quad AC = \frac{300}{30(4\pi)} = \frac{2.5}{\pi}$$

The solution for the displacement is finally

$$u(x, t) = \frac{2.5}{\pi} \sin 120\pi t \sin 4\pi x$$

We have not imposed the condition that the end at $x = 2$ m is motionless. Insert $x = 2$ in the expression above and it is obvious that this boundary condition is satisfied; thus we have found an acceptable solution.

The maximum displacement at $x = 1/8$ m occurs when sin $120\pi t = 1$. Thus, the maximum displacement is

$$u_{\max} = \frac{2.5}{\pi} \text{ m}$$

Note that we did not find it necessary to use the general expression given by Eq. 6.4.39. We could have, but it would have required more work to obtain a solution. This happened because the initial condition was given as a sine function. Any other function would require the more general form given by Eq. 6.4.39.

Example 6.4

Determine several coefficients in the series solution for $u(x, t)$ for the vibrating string if

$$f(x) = \begin{cases} 0.1x & 0 \le x \le 1 \\ 0.2 - 0.1x & 1 < x \le 2 \end{cases}$$

The string is 2 m long. Use the boundary and initial conditions of Section 6.4.

Solution

The solution for the displacement of the string is given by Eq. 6.4.39. It is

$$u(x, t) = \sum_{n=1}^{\infty} b_n \cos \frac{n\pi a t}{2} \sin \frac{n\pi x}{2}$$

where we have used $L = 2$ m. The coefficients b_n are related to the initial displacement $f(x)$ by Eq. 6.4.44,

$$b_n = \frac{2}{2}\int_0^2 f(x) \sin \frac{n\pi x}{2} dx$$

Substituting for $f(x)$ results in

$$b_n = 0.1\int_0^1 x\sin\frac{n\pi x}{2}dx + 0.1\int_1^2 (2-x)\sin\frac{n\pi x}{2}dx$$

Performing the integrations (integration by parts* is required) gives

$$b_n = 0.1\left[-\frac{2x}{n\pi}\cos\frac{n\pi x}{2} + \frac{4}{n^2\pi^2}\sin\frac{n\pi x}{2}\right]_0^1$$
$$+0.1\left[-\frac{4}{n\pi}\cos\frac{n\pi x}{2} + \frac{2x}{n\pi}\cos\frac{n\pi x}{2} - \frac{4}{n^2\pi^2}\sin\frac{n\pi x}{2}\right]_1^2$$

By being careful in reducing this result, we have

$$b_n = \frac{0.8}{\pi^2 n^2}\sin\frac{n\pi}{2}$$

This gives several b_n's as

$$b_1 = \frac{0.8}{\pi^2},\quad b_2 = 0,\quad b_3 = -\frac{0.8}{9\pi^2},\quad b_4 = 0,\quad b_5 = \frac{0.8}{25\pi^2}$$

The solution is, finally,

$$u(x,t) = \frac{0.8}{\pi^2}\left[\cos\frac{\pi at}{2}\sin\frac{\pi x}{2} - \frac{1}{9}\cos\frac{3\pi at}{2}\sin\frac{3\pi x}{2} + \frac{1}{25}\cos\frac{5\pi at}{2}\sin\frac{5\pi x}{2} + \cdots\right]$$

We see that the amplitude of each term is getting smaller and smaller. A good approximation results if we keep several terms (say five) and simply ignore the rest. This, in fact, was done before the advent of the computer. With the computer many more terms can be retained, with more accurate numbers resulting from the calculations. A computer plot of the solution above is shown in Fig. 6.10 for $a = 100$ m/s. One hundred terms were retained.

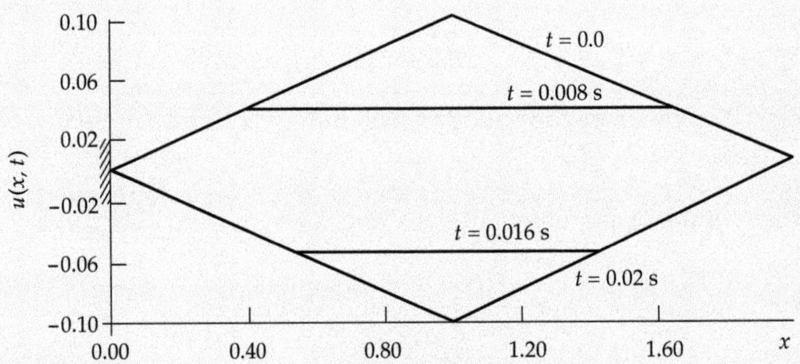

FIGURE 6.10

*We shall integrate $\int_0^\pi x\sin x\,dx$ by parts. Let $u = x$ and $dv = \sin x\,dx$. Then $du = dx$ and $v = -\cos x$. The integral is then $\int_0^\pi x\sin x\,dx = -x\cos x\Big|_0^\pi + \int_0^\pi \cos x\,dx = \pi$.

Example 6.5

A tight string, π m long and fixed at both ends, is given an initial displacement $f(x)$ and an initial velocity $g(x)$. Find an expression for $u(x, t)$.

Solution

We follow the steps of Section 6.4 and find the general solution to be

$$u(x,t) = (A\sin\beta at + B\cos\beta at)(C\sin\beta x + D\cos\beta x)$$

Using the b.c. that the left end is fixed, that is, $u(0, t) = 0$, we have $D = 0$. We also have the b.c. that $u(\pi, t) = 0$, giving

$$0 = (A \sin \beta at + B \cos \beta at)C \sin \beta\pi.$$

If we let $C = 0$, a trivial solution results, $u(x, t) = 0$. Thus, we must let

$$\beta\pi = n\pi$$

or $\beta = n$, an integer. The general solution is then

$$u_n(x,t) = (a_n \sin nat + b_n \cos nat)\sin nx$$

where the subscript n on $u_n(x, t)$ allows for a different $u(x, t)$ for each value of n. The most general $u(x, t)$ is then found by superposing all of the $u_n(x, t)$; that is,

$$u(x,t) = \sum_{n=1}^{\infty} u_n(x,t) = \sum_{n=1}^{\infty}(a_n \sin nat + b_n \cos nat)\sin nx \tag{1}$$

Now, to satisfy the initial displacement, we require that

$$u(x,0) = \sum_{n=1}^{\infty} b_n \sin nx = f(x)$$

Again, b_n are the coefficients of a Fourier sine series.

Multiply by sin mx and integrate from 0 to π. Using the results indicated in Eq. 6.4.43, we have

$$b_n = \frac{2}{\pi}\int_0^{\pi} f(x)\sin nx\, dx \tag{2}$$

Next, to satisfy the initial velocity, we must have

$$\frac{\partial u}{\partial t}(x,0) = \sum_{n=1}^{\infty} a_n an \sin nx = g(x)$$

Again, multiply by sin mx and integrate from 0 to π. Then

$$a_n = \frac{2}{an\pi}\int_0^{\pi} g(x)\sin nx\, dx \tag{3}$$

Our solution is now complete. It is given by Eq. 1 with the b_n provided by Eq. 2 and the a_n by Eq. 3. If $f(x)$ and $g(x)$ were specified, numerical values for each b_n and a_n could be obtained.

Example 6.6

A tight string, π m long, is fixed at the left end but the right end moves, with displacement 0.2 sin $15t$. Find $u(x, t)$ if the wave speed is 30 m/s and state the initial conditions if a solution using separation of variables is to be possible.

Solution

Separation of variables leads to the general solution as

$$u(x,t) = (A\sin 30\beta t + B\cos 30\beta t)(C\sin \beta x + D\cos \beta x)$$

The left end is fixed, requiring that $u(0, t) = 0$. Hence, $D = 0$. The right end moves with the displacement 0.2 sin $15t$; that is,

$$u(\pi, t) = 0.2\sin 15t = (A\sin 30\beta t + B\cos 30\beta t)C\sin \beta\pi$$

This can be satisfied if we let

$$B = 0, \qquad \beta = \frac{1}{2}, \qquad AC = 0.2$$

The resulting solution for $u(x, t)$ is

$$u(x,t) = 0.2\sin 15t \sin\frac{x}{2}$$

The initial displacement $u(x, 0)$ must be zero and the initial velocity must be

$$\frac{\partial u}{\partial t}(x,0) = 3\sin\frac{x}{2}$$

Any other set of initial conditions would not allow a solution using separation of variables.

Example 6.7

A tight string is fixed at both ends. A forcing function (this could be due to wind blowing over a wire), applied normal to the string, is given by $\mathcal{F}(t) = Km \sin \omega t$ kilograms per meter of length. Show that resonance occurs whenever $\omega = an\pi/L$.

Solution

The forcing function $\mathcal{F}(t)$ multiplied by the distance Δx can be added to the right-hand side of Eq. 6.2.8. Dividing by $m\Delta x$ results in

$$a^2\frac{\partial^2 u}{\partial x^2} = \frac{\partial^2 u}{\partial t^2} + K\sin\omega t$$

where $a^2 = P/m$. This is a nonhomogeneous partial differential equation, since the last term does not contain the dependent variable $u(x, t)$. As with ordinary differential equations that are linear, we can find a particular solution and add it to the solution of the associated homogeneous equation to form the general solution.

We assume that the effect of the forcing function will be to produce a displacement having the same frequency as the forcing function, as was the case with lumped systems. This suggests that the particular solution has the form

$$u_p(x,t) = X(x)\sin\omega t$$

Substituting this into the partial differential equation gives

$$a^2X''\sin\omega t = -X\omega^2\sin\omega t + K\sin\omega t$$

The sin ωt divides out and we are left with the ordinary differential equation

$$X'' + \frac{\omega^2}{a^2}X = K$$

The general solution to this nonhomogeneous differential equation is (see Section 1.8)

$$X(x) = c_1\sin\frac{\omega}{a}x + c_2\cos\frac{\omega}{a}x + \frac{Ka^2}{\omega^2}$$

We will force this solution to satisfy the end conditions that apply to the string. Hence,

$$X(0) = 0 = c_2 + \frac{Ka^2}{\omega^2}$$

$$X(L) = 0 = c_1\sin\frac{\omega L}{a} + c_2\cos\frac{\omega L}{a} + \frac{Ka^2}{\omega^2}$$

The equations above give

$$c_2 = -\frac{Ka^2}{\omega^2}, \qquad c_1 = \frac{\frac{Ka^2}{\omega^2}\left(\cos\frac{\omega L}{a} - 1\right)}{\sin(\omega L/a)}$$

The particular solution is then

$$u_p(x,t) = \frac{Ka^2}{\omega^2}\left[\frac{\cos\frac{\omega L}{a} - 1}{\sin(\omega L/a)}\sin\frac{\omega x}{a} - \cos\frac{\omega x}{a} + 1\right]\sin\omega t$$

The amplitude of the above becomes infinite whenever sin $\omega L/a = 0$ and cos $\omega L/a = -1$. This occurs whenever

$$\frac{\omega L}{a} = (2n-1)\pi$$

Hence, if the input frequency is such that

$$\omega = \frac{(2n-1)\pi a}{L}, \qquad n = 1, 2, 3, \cdots$$

the amplitude of the resulting motion becomes infinitely large. This equals the natural frequency corresponding to the fundamental mode or one of the significant overtones of the string, depending on the value of n. Thus, we see that a number of input frequencies can lead to resonance in the string. This is true of all phenomena modeled by the wave equation. Recall that we have neglected any type of damping.

6.5 DIFFUSION

Another class of physical problems exists that is characterized by diffusion equations. Diffusion may be likened to a spreading, smearing, or mixing. A physical system that has a high concentration of variable ϕ in volume A and a low concentration of ϕ in volume B may tend to *diffuse* so that the concentrations in A and B approach equality. This phenomenon is exhibited by the tendency of a body toward a uniform temperature. One of the most common diffusion processes that is encountered is the transfer of energy in the form of heat.

From thermodynamics we learn that heat is thermal energy in transit. It may be transmitted by conduction (when two bodies are in contact), by convection (when a body is in contact with a liquid or a gas), and by radiation (when energy is transmitted by energy waves). We shall consider the first of these mechanisms in some detail. Experimental observations have been organized to permit us to make the following two statements:

1. Heat flows in the direction of decreasing temperature.
2. The rate at which energy in the form of heat is transferred through an area is proportional to the area and to the temperature gradient normal to the area.

These statements must be expressed analytically. The heat flux through an area A oriented normal to the x axis is

$$Q = -KA\frac{\partial T}{\partial x} \tag{6.5.1}$$

where Q (watts per second, W/s) is the heat flux, $\partial T/\partial x$ is the temperature gradient normal to A, and K (W/m·s·K) is a constant of proportionality called the *thermal conductivity*. The minus sign is present since heat is transferred in the direction opposite the temperature gradient.

The energy (usually called internal energy) gained or lost by a body of mass m that undergoes a uniform temperature change ΔT is expressed as

$$\Delta E = Cm\,\Delta T \tag{6.5.2}$$

where ΔE (W) the energy change of the body and C (W/kg·K) is a constant of proportionality called the *specific heat*.

Conservation of energy is a fundamental law of nature. We shall use this law to make an energy balance on the element in Fig. 6.11. The density ρ of the element will be used to determine its mass, namely,

$$m = \rho \Delta x \Delta y \Delta z \tag{6.5.3}$$

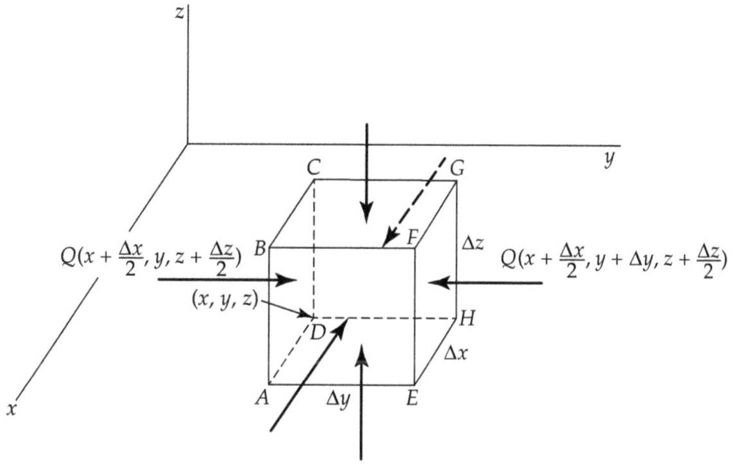

FIGURE 6.11 Element of mass.

By energy balance we mean that the net energy flowing into the element in time Δt must equal the increase in energy in the element in Δt. For simplicity, we shall assume that there are no sources inside the element. Equation 6.5.2 gives the change in energy in the element as

$$\Delta E = Cm\Delta T = C\rho \Delta x \Delta y \Delta z \Delta T \tag{6.5.4}$$

The energy that flows into the element through face $ABCD$ in Δt is, by Eq. 6.5.1,

$$\Delta E_{ABCD} = Q_{ABCD}\Delta t = -K\Delta x \Delta z \left.\frac{\partial T}{\partial y}\right|_{\substack{x+\Delta x/2 \\ y \\ z+\Delta z/2}} \Delta t \tag{6.5.5}$$

where we have approximated the temperature derivative by the value at the center of the face. The flow into the element through face $EFGH$ is

$$\Delta E_{EFGH} = K\Delta x \Delta z \left.\frac{\partial T}{\partial y}\right|_{\substack{x+\Delta x/2 \\ y+\Delta y \\ z+\Delta z/2}} \Delta t \tag{6.5.6}$$

Similar expressions are found for the other four faces. The energy balance then provides us with

$$\Delta E = \Delta E_{ABCD} + \Delta E_{EFGH} + \Delta E_{ADHE} + \Delta E_{BCGF} + \Delta E_{DHGC} + \Delta E_{BFEA} \tag{6.5.7}$$

or, using Eqs. 3.5.5, 3.5.6, and their counterparts for the x and z directions,

$$C\rho\,\Delta x\,\Delta y\,\Delta z\,\Delta T = K\,\Delta x\,\Delta z\left(\left.\frac{\partial T}{\partial y}\right|_{\substack{x+\Delta x/2\\ y+\Delta y\\ z+\Delta z/2}} - \left.\frac{\partial T}{\partial y}\right|_{\substack{x+\Delta x/2\\ y\\ z+\Delta z/2}}\right)\Delta t$$

$$+K\,\Delta y\,\Delta z\left(\left.\frac{\partial T}{\partial x}\right|_{\substack{x+\Delta x\\ y+\Delta y/2\\ z+\Delta z/2}} - \left.\frac{\partial T}{\partial x}\right|_{\substack{x\\ y+\Delta y/2\\ z+\Delta z/2}}\right)\Delta t$$

$$+K\,\Delta x\,\Delta y\left(\left.\frac{\partial T}{\partial z}\right|_{\substack{x+\Delta x/2\\ y+\Delta y/2\\ z+\Delta z}} - \left.\frac{\partial T}{\partial z}\right|_{\substack{x+\Delta x/2\\ y+\Delta y/2\\ z}}\right)\Delta t \tag{6.5.8}$$

The arrow here means "approach."

Divide both sides of the equation by $C\rho\Delta x\Delta y\Delta z\Delta t$, then let $\Delta x \to 0$, $\Delta y \to 0$, $\Delta z \to 0$, $\Delta t \to 0$; there results

$$\frac{\partial T}{\partial t} = k\left[\frac{\partial^2 T}{\partial x^2} + \frac{\partial^2 T}{\partial y^2} + \frac{\partial^2 T}{\partial z^2}\right] \tag{6.5.9}$$

where $k = K/C\rho$ is called the *thermal diffusivity* and is assumed constant. It has dimensions of square meters per second (m^2/s). Equation 6.5.9 is a *diffusion equation.*

Two special cases of the diffusion equation are of particular interest. A number of situations involve time and only one coordinate, say x, as in a long, slender rod with insulated sides. The *one-dimensional heat equation* then results. It is given by

$$\frac{\partial T}{\partial t} = k\frac{\partial^2 T}{\partial x^2} \tag{6.5.10}$$

which is a parabolic equation.

In some situations $\partial T/\partial t$ is zero and we have a steady-state condition; then we no longer have a diffusion equation, but the equation

$$\frac{\partial^2 T}{\partial x^2} + \frac{\partial^2 T}{\partial y^2} + \frac{\partial^2 T}{\partial z^2} = 0 \tag{6.5.11}$$

This equation is known as *Laplace's equation.* It is sometimes written in the shorthand form

$$\nabla^2 T = 0 \tag{6.5.12}$$

If the temperature depends only on two coordinates x and y, as in a thin rectangular plate, an elliptic equation is encountered,

$$\frac{\partial^2 T}{\partial x^2} + \frac{\partial^2 T}{\partial y^2} = 0 \tag{6.5.13}$$

Cylindrical or spherical coordinates (see Fig. 6.12) should be used in certain geometries. It is then convienient to express $\nabla^2 T$ in cylindrical coordinates as

$$\nabla^2 T = \frac{1}{r}\frac{\partial}{\partial r}\left(r\frac{\partial T}{\partial r}\right) + \frac{1}{r^2}\frac{\partial^2 T}{\partial \theta^2} + \frac{\partial^2 T}{\partial z^2} = 0 \tag{6.5.14}$$

or in spherical coordinates as

$$\nabla^2 T = \frac{1}{r^2}\frac{\partial}{\partial r}\left(r^2\frac{\partial T}{\partial r}\right) + \frac{1}{r^2\sin^2\phi}\frac{\partial^2 T}{\partial \theta^2} + \frac{1}{r^2\sin\phi}\frac{\partial}{\partial \phi}\left(\sin\phi\frac{\partial T}{\partial \phi}\right) \quad (6.5.15)$$

See Section 5.5 on cylindrical and spherical coordinates.

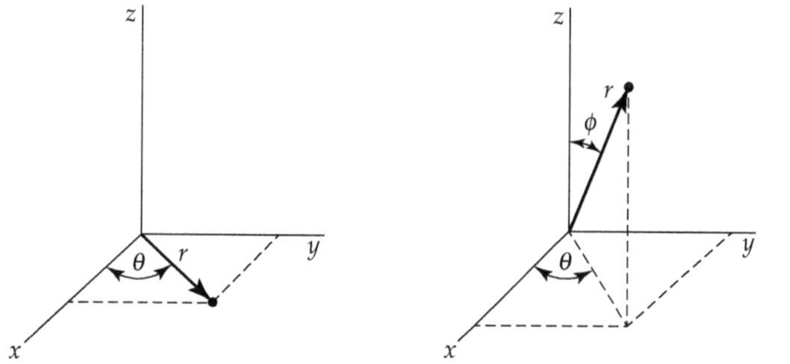

(a) Cylindrical coordinates (b) Spherical coordinates

FIGURE 6.12 Cylindrical and spherical coordinates.

Our discussion of heat transfer has included heat conduction only. Radiative and convective forms of heat transfer would necessarily lead to other partial differential equations. We have also assumed no heat sources in the volume of interest, and have assumed the conductivity K to be constant. Finally, the specification of boundary and initial conditions would make our problem statement complete. These will be reserved for the following section in which a solution to the diffusion equation is presented.

6.6 SOLUTION OF THE DIFFUSION EQUATION

This section will be devoted to a solution of the diffusion equation developed in Section 6.5. Recall that the diffusion equation is

$$\frac{\partial T}{\partial t} = k\left(\frac{\partial^2 T}{\partial x^2} + \frac{\partial^2 T}{\partial y^2} + \frac{\partial^2 T}{\partial z^2}\right) \quad (6.6.1)$$

Heat transfer will again be used to illustrate this very important phenomenon. The procedure developed for the wave equation will be used, but the solution will be quite different, owing to the presence of the first derivative with respect to time rather than the second derivative. This requires only one initial condition instead of the two required by the wave equation. We shall illustrate the solution technique with three specific situations.

6.6.1 A Long, Insulated Rod with Ends at Fixed Temperatures

A long rod, shown in Fig. 6.13, is subjected to an initial temperature distribution along its axis; the rod is insulated on the lateral surface, and the ends of the rod are kept at the same constant temperature.* The insulation prevents heat flux in the radial direction;

*We shall choose the temperature of the ends in the illustration to be 0°C. Note, however, that both ends could be held at any temperature T_0. Since it is necessary to have the ends maintained at zero, we would simply define a new variable $\theta = T - T_0$, so that $\theta = 0$ at both ends. We would then find a solution for $\theta(x, t)$ with the desired temperature given by $T(x, t) = \theta(x, t) + T_0$.

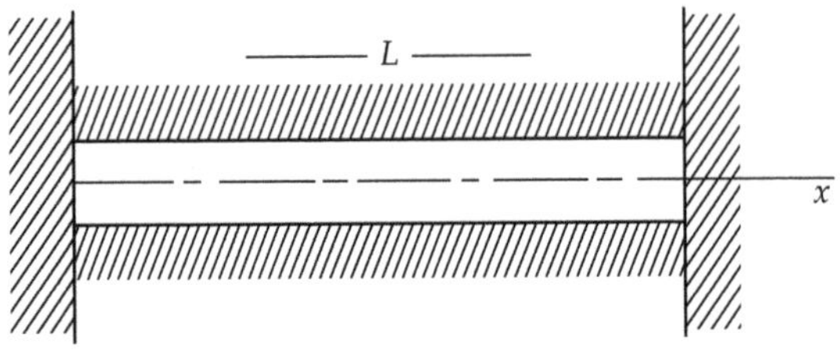

FIGURE 6.13 Heated rod.

hence, the temperature will depend on the x coordinate only. The describing equation is then the one-dimensional heat equation, given by Eq. 6.5.10, as

$$\frac{\partial T}{\partial t} = k\frac{\partial^2 T}{\partial x^2} \tag{6.6.2}$$

We shall choose to hold the ends at $T = 0°$. These boundary conditions are expressed as

$$T(0, t) = 0, \qquad T(L, t) = 0 \tag{6.6.3}$$

Let the initial temperature distribution be represented by

$$T(x, 0) = f(x) \tag{6.6.4}$$

We assume that the variables separate; that is,

$$T(x, t) = \theta(t)X(x) \tag{6.6.5}$$

Substitution of Eq. 6.6.5 into 6.6.2 yields

$$\theta' X = k\theta X'' \tag{6.6.6}$$

where $\theta' = d\theta/dt$ and $X'' = d^2X/dx^2$. This is rearranged as

$$\frac{\theta'}{k\theta} = \frac{X''}{X} \tag{6.6.7}$$

Since the left side is a function of t only and the right side is a function of x only, we set Eq. 6.6.7 equal to a constant λ (lambda). This gives

$$\theta' - \lambda k\theta = 0 \tag{6.6.8}$$

and

$$X'' - \lambda X = 0 \tag{6.6.9}$$

Separating variables in Eq. 6.6.3, we have $T(0, t) = \theta(t)X(0) = 0$ and $T(L, t) = \theta(t)X(L) = 0$, which imply that $X(0) = X(L) = 0$. These are the conditions that are needed to solve Eq. 6.6.9.

The solution of Eq. 6.6.8 is of the form

$$\theta(t) = c_1 e^{\lambda k t} \tag{6.6.10}$$

Equation 6.6.9 yields the solution

$$X(x) = c_2 e^{\sqrt{\lambda}x} + c_3 e^{-\sqrt{\lambda}x} \tag{6.6.11}$$

Again, we must decide whether

$$\lambda > 0, \qquad \lambda = 0, \qquad \lambda < 0 \tag{6.6.12}$$

For $\lambda > 0$, Eq. 6.6.10 shows that the solution has a nearly infinite temperature at large t due to exponential growth; of course, this is not physically possible. For $\lambda = 0$, the solution would be independent of time. Again our physical intuition tells us this is not expected. Therefore, we are left with $\lambda < 0$. If we can satisfy the boundary conditions, then we have found a solution. Let

$$\beta^2 = -\lambda \tag{6.6.13}$$

so that

$$\beta^2 > 0 \tag{6.6.14}$$

The solutions, Eqs. 6.6.10 and 6.6.11, may then be written as

$$\theta(t) = Ae^{-\beta^2 kt} \tag{6.6.15}$$

and

$$X(x) = B\sin\beta x + C\cos\beta x \tag{6.6.16}$$

where A, B, and C are arbitrary constants to be determined. Therefore, our solution is

$$T(x,t) = Ae^{-\beta^2 kt}\left[B\sin\beta x + C\cos\beta x\right] \tag{6.6.17}$$

The first condition of Eq. 6.6.3 implies that

$$C = 0 \tag{6.6.18}$$

Therefore, our solution reduces to

$$T(x,t) = De^{-\beta^2 kt}\sin\beta x \tag{6.6.19}$$

where $D = A \cdot B$. The second boundary condition of Eq. 6.6.3 requires that

$$\sin\beta L = 0 \tag{6.6.20}$$

This is satisfied if

$$\beta L = n\pi \quad \text{or} \quad \beta = n\pi/L, \quad n = 1, 2, 3, \ldots \tag{6.6.21}$$

The constant β is the eigenvalue, and the function $\sin n\pi x/L$ is the eigenfunction. The solution is now

Eq. 6.6.9 with its boundary conditions ($X(0) = X(L) = 0$) is a differential eigenvalue problem.

$$T(x,t) = \sum_{n=1}^{\infty} T_n(x,t) = \sum_{n=1}^{\infty} D_n e^{-kn^2\pi^2 t/L^2}\sin\frac{n\pi x}{L} \tag{6.6.22}$$

The initial condition, (6.6.4), may be satisfied at $t = 0$ if

$$T(x,0) = f(x) = \sum_{n=1}^{\infty} D_n \sin\frac{n\pi x}{L} \tag{6.6.23}$$

that is, if $f(x)$ can be expanded in a Fourier sine series. If such is the case, the coefficients will be given by (refer to Eqs. 6.4.42–6.4.44)

$$D_n = \frac{2}{L}\int_0^L f(x)\sin\frac{n\pi x}{L}\,dx \tag{6.6.24}$$

and the separation-of-variables technique is successful.

It should be noted again that not all solutions of partial differential equations can be found by separation of variables; in fact, it is only a very special set of boundary conditions that allows us to separate the variables. For example, Eq. 6.6.20 would obviously not be useful in satisfying the boundary condition $T(L, t) = 20t$. Separation of variables would then be futile. Numerical methods could be used to find a solution, or other analytical techniques not covered in this book would be necessary.

Example 6.8

A long copper rod with insulated lateral surfaces has its left end maintained at a temperature of 0°C and its right end, at $x = 2$ m, maintained at 100°C. Determine the temperature as a function of x and t if the initial condition is given by

$$T(x,0) = f(x) = \begin{cases} 100x & 0 < x < 1 \\ 100 & 1 < x < 2 \end{cases}$$

The thermal diffusivity for copper is $k = 1.14 \times 10^{-4}$ m²/s.

Solution

We again assume the variables separate as

$$T(x, t) = \theta(t)X(x)$$

with the resulting equation,

$$\frac{1}{k}\frac{\theta'}{\theta} = \frac{X''}{X} = \lambda$$

In this problem the eigenvalue $\lambda = 0$ will play an important role. The solution for $\lambda = 0$ is

$$\theta(t) = C_1, \qquad X(x) = A_1 x + B_1$$

resulting in

$$T(x, t) = C_1(A_1 x + B_1)$$

To satisfy the two end conditions $T(0, t) = 0$ and $T(2, t) = 100$, it is necessary to require $B_1 = 0$ and $A_1 C_1 = 50$. Then

$$T(x, t) = 50x \tag{1}$$

This solution is, of course, independent of time, but we will find it quite useful.

Now, we return to the case that allows for exponential decay of temperature, namely $\lambda = -\beta^2$. For this eigenvalue see Eq. 6.6.17 the solution is

$$T(x, t) = Ae^{-\beta^2 kt}\left[B \sin \beta x + C \cos \beta x\right] \tag{2}$$

We can superimpose the above two solutions, since Eq. 6.6.2 is linear, and obtain the more general solution

$$T(x, t) = 50x + Ae^{-\beta^2 kt}\left[B \sin \beta x + C \cos \beta x\right]$$

Now let us satisfy the boundary conditions. The left-end condition $T(0, t) = 0$ demands that $C = 0$. The right-end condition demands that

$$100 = 100 + A \cdot Be^{-\beta^2 kt} \sin \beta L$$

This requires that $\sin \beta L = 0$, which occurs whenever

$$\beta L = n\pi \quad \text{or} \quad \beta = n\pi/L, \qquad n = 1, 2, 3, \cdots$$

The general solution is then

$$T(x, t) = 50x + \sum_{n=1}^{\infty} D_n e^{-n^2\pi^2 kt/4} \sin \frac{n\pi x}{2}$$

using $L = 2$. Note that this satisfies the describing equation (6.6.2) and the two boundary conditions. Finally, it must satisfy the initial condition

$$f(x) = 50x + \sum_{n=1}^{\infty} D_n \sin \frac{n\pi x}{2}$$

We see that if the function $[f(x) - 50x]$ can be expanded in a Fourier sine series, then the solution will be complete. The Fourier coefficients are

$$\begin{aligned} D_n &= \frac{2}{L}\int_0^L [f(x) - 50x] \sin \frac{n\pi x}{L} dx \\ &= \frac{2}{2}\int_0^1 (100x - 50x) \sin \frac{n\pi x}{2} dx + \frac{2}{2}\int_0^2 (100 - 50x) \sin \frac{n\pi x}{2} dx \\ &= 50\left[-\frac{2x}{n\pi}\cos\frac{n\pi x}{2} + \frac{4}{n^2\pi^2}\sin\frac{n\pi x}{2}\right]_0^1 - \frac{200}{n\pi}\cos\frac{n\pi x}{2}\bigg|_1^2 \\ &\quad - 50\left[-\frac{2x}{n\pi}\cos\frac{n\pi x}{2} + \frac{4}{n^2\pi^2}\sin\frac{n\pi x}{2}\right]_1^2 \\ &= \frac{400}{n^2\pi^2}\sin\frac{n\pi}{2} \end{aligned}$$

The solution is, using $k = 1.14 \times 10^{-4}$ m²/s for copper,

$$T(x, t) = 50x + \sum_{n=1}^{\infty} \frac{40.5}{n^2} \sin \frac{n\pi}{2} e^{-2.81 \times 10^{-4} n^2 t} \sin \frac{n\pi x}{2}$$

Note that the time t is measured in seconds.

As $t \to \infty$, $T(x, t) \to 50x$, which is the steady-state temperature distribution.

6.6.2 A Long, Totally Insulated Rod

The lateral sides of the long rod of length L (see Fig. 6.13) are again insulated so that heat transfer occurs only in the x direction along the rod. The temperature in the rod is described by the one-dimensional heat equation

$$\frac{\partial T}{\partial t} = k\frac{\partial^2 T}{\partial x^2} \tag{6.6.25}$$

For this problem, we have an initial temperature distribution given by

$$T(x,0) = f(x) \tag{6.6.26}$$

Since the rod is totally insulated, the heat flux across the end faces is zero. This condition gives, with the use of Eq. 6.5.1,

$$\frac{\partial T}{\partial x}(0,t) = 0, \qquad \frac{\partial T}{\partial x}(L,t) = 0 \tag{6.6.27}$$

We assume that the variables separate,

$$T(x,t) = \theta(t)X(x) \tag{6.6.28}$$

Substitute into Eq. 6.6.25, to obtain

$$\frac{\theta'}{k\theta} = \frac{X''}{X} = -\beta^2 \tag{6.6.29}$$

where $-\beta^2$ is a negative real number. Equation 6.6.29 gives

$$\theta' = -\beta^2 k\theta \tag{6.6.30}$$

and

$$X'' + \beta^2 X = 0 \tag{6.6.31}$$

The equations have solutions in the form

$$\theta(t) = Ae^{-\beta^2 kt} \tag{6.6.32}$$

and

$$X(x) = B\sin\beta x + C\cos\beta x \tag{6.6.33}$$

The first boundary condition of (6.6.27) implies that $B = 0$, and the second requires that

$$\frac{\partial X}{\partial x}(L) = -C\beta\sin\beta L = 0 \tag{6.6.34}$$

This can be satisfied if we set

$$\sin\beta L = 0 \tag{6.6.35}$$

Hence, the eigenvalues are

$$\beta = \frac{n\pi}{L}, \quad n = 0, 1, 2, \cdots \tag{6.6.36}$$

Thus, the independent solutions are of the form

$$T_n(x,t) = a_n e^{-n^2\pi^2 kt/L^2} \cos\frac{n\pi x}{L} \tag{6.6.37}$$

where the constant a_n replaces AC. The general solution, which hopefully will satisfy the remaining initial condition, is then

$$T(x,t)=\sum_{n=0}^{\infty}a_n e^{-(n^2\pi^2k/L^2)t}\cos\frac{n\pi x}{L} \tag{6.6.38}$$

Note that we retain the $\beta = 0$ eigenvalue in the series.

The initial condition is given by Eq. 6.6.26. It demands that

$$f(x)=\sum_{n=0}^{\infty}a_n\cos\frac{n\pi x}{L} \tag{6.6.39}$$

Using trigonometric identities we can show that

$$\int_0^T\cos\frac{n\pi x}{L}\cos\frac{m\pi x}{L}dx=\begin{cases}0 & m\neq n\\ L/2 & m=n\neq 0\\ L & m=n=0\end{cases} \tag{6.6.40}$$

Multiply both sides of Eq. 6.6.39 by cos $m\pi x/L$ and integrate from 0 to L. We then have*

$$a_0=\frac{1}{L}\int_0^L f(x)dx,\quad a_n=\frac{2}{L}\int_0^L f(x)\cos\frac{n\pi x}{L}dx \tag{6.6.41}$$

The solution is finally

$$T(x,t)=\sum_{n=0}^{\infty}a_n e^{-(n^2\pi^2k/L^2)t}\cos\frac{n\pi x}{L} \tag{6.6.42}$$

Thus, the temperature distribution can be determined provided that $f(x)$ can be expanded in a Fourier cosine series.

Example 6.9

A long, laterally insulated stainless steel rod has heat generation occurring uniformly within the rod at the constant rate of 4140 $W/m^3\cdot s$. The right end is insulated and the left end is maintained at 0°C. Find an expression for $T(x, t)$ if the initial temperature distribution is

$$T(x,0)=f(x)=\begin{cases}100x & 0<x<1\\ 200-100x & 1<x<2\end{cases}$$

for the 2-m-long, 0.1-m-diameter rod. Use the specific heat C = 460 J/kg·°C, $\rho = 7820$ kg/m^3, and $k = 3.86\times10^{-6}$ m^2/s.

*Note that it is often the practice to define a_0 as $a_0=(2/L)\int_0^L f(x)dx$ and then to write the solution as $T(x,t)=a_0/2+\sum_{n=1}^{\infty}a_n e^{-n^2\pi^2kt/L^2}\cos(n\pi x/L)$. This was done in Section 1.10. Both methods are, of course, equivalent.

Solution

To find the appropriate describing equation, we must account for the heat generated in the infinitesimal element of Fig. 6.11. To Eq. 6.5.7 we would add a heat-generation term,

$$\phi(x, y, z, t)\,\Delta x\,\Delta y\,\Delta z\,\Delta t$$

where $\phi(x, y, z, t)$ is the amount of heat generated per volume per unit time. The one-dimensional heat equation would then take the form

$$\frac{\partial T}{\partial t} = k\frac{\partial^2 T}{\partial x^2} + \frac{\phi}{\rho C}$$

For the present example the describing equation is

$$\frac{\partial T}{\partial t} = k\frac{\partial^2 T}{\partial x^2} + \frac{4140}{7890 \cdot 460}$$

This nonhomogeneous, partial differential equation is solved by finding a particular solution and adding it to the solution of the homogeneous equation

$$\frac{\partial T}{\partial t} = k\frac{\partial^2 T}{\partial x^2}$$

The solution of the homogeneous equation is (see Eqs. 6.6.32 and 6.6.33)

$$T(x, t) = Ae^{-\beta^2 kt}\left[B\sin\beta x + C\cos\beta x\right]$$

The left-end boundary condition is $T(0, t) = 0$, resulting in $C = 0$. The insulated right end requires that $\partial T/\partial x\ (L, t) = 0$. This results in

$$\cos\beta L = 0$$

Thus, the quantity βL must equal $\pi/2, 3\pi/2, 5\pi/2, \cdots$. This is accomplished by using

$$\beta = \frac{(2n-1)\pi}{2L}, \quad n = 1, 2, 3, \cdots$$

The homogeneous solution is, then, using $k = 3.86 \times 10^{-6}$ and $L = 2$,

$$T(x, t) = \sum_{n=1}^{\infty} D_n e^{-2.38\times10^{-6}(2n-1)^2 t} \sin\left(\frac{2n-1}{4}\pi x\right)$$

To find the particular solution, we note that the generation of heat is independent of time. Since the homogeneous solution decays to zero with time, we anticipate that the heat-generation term will lead to a steady-state temperature distribution. Thus, we assume the particular solution to be independent of time, that is,

$$T_p(x, t) = g(x)$$

Substitute this into the describing equation, to obtain

$$0 = 3.86 \times 10^{-6} g'' + 1.15 \times 10^{-3}$$

The solution to this ordinary differential equation is

$$g(x) = -149x^2 + c_1 x + c_2$$

This solution must also satisfy the boundary condition at the left end, yielding $c_2 = 0$ and the boundary condition at the right end ($g' = 0$), giving $c_1 = 596$. The complete solution, which must now satisfy the initial condition, is

$$T(x,t) = -149x^2 + 596x + \sum_{n=1}^{\infty} D_n e^{-2.38\times10^{-6}(2n-1)^2 t} \sin\left(\frac{2n-1}{4}\pi x\right)$$

To find the unknown coefficients D_n we use the initial condition, which states that

$$f(x) = -149x^2 + 596x + \sum_{n=1}^{\infty} D_n \sin\left(\frac{2n-1}{4}\pi x\right)$$

The coefficients are then

$$\begin{aligned} D_n &= \frac{2}{2}\int_0^2 [f(x) + 149x^2 - 596x] \sin\left(\frac{2n-1}{4}\pi x\right) dx \\ &= \int_0^1 (149x^2 - 496x) \sin\left(\frac{2n-1}{4}\pi x\right) dx \\ &\quad + \int_1^2 (149x^2 - 696x + 200) \sin\left(\frac{2n-1}{4}\pi x\right) dx \end{aligned}$$

The integrals can be integrated by parts providing a complete solution.

6.6.3 Two-Dimensional Heat Conduction in a Long, Rectangular Bar

A long, rectangular bar is bounded by the planes $x = 0$, $x = a$, $y = 0$, and, $y = b$. These faces are kept at $T = 0$°C, as shown by the cross section in Fig. 6.14. The bar is heated so that

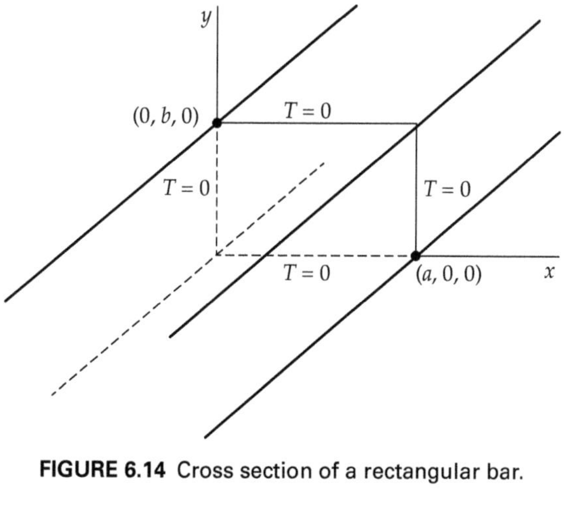

FIGURE 6.14 Cross section of a rectangular bar.

the variation in the z direction may be neglected. Thus, the variation of temperature in the bar is described by

$$\frac{\partial T}{\partial t} = k\left(\frac{\partial^2 T}{\partial x^2} + \frac{\partial^2 T}{\partial y^2}\right) \tag{6.6.43}$$

The initial temperature distribution in the bar is given by

$$T(x,y,0) = f(x,y) \tag{6.6.44}$$

We want to find an expression for $T(x, y, t)$. Hence, we assume that

$$T(x,y,t) = X(x)Y(y)\theta(t) \tag{6.6.45}$$

After Eq. 6.6.45 is substituted into Eq. 6.6.43, we find that

$$XY\theta' = k(X''Y\theta + XY''\theta) \tag{6.6.46}$$

Equation 6.6.46 may be rewritten as

$$\frac{X''}{X} = \frac{\theta'}{k\theta} - \frac{Y''}{Y} \tag{6.6.47}$$

Since the left-hand side of Eq. 6.6.47 is a function of x only and the right side is a function of t and y, we may assume that both sides equal the constant value $-\lambda$. (With experience we now anticipate the minus sign.) Therefore, we have

$$X'' + \lambda X = 0 \tag{6.6.48}$$

and

$$\frac{Y''}{Y} = \frac{\theta'}{k\theta} + \lambda \tag{6.6.49}$$

We use the same argument on Eq. 6.6.49 and set it equal to a constant $-\mu$. That is,

$$\frac{Y''}{Y} = \frac{\theta'}{k\theta} + \lambda = -\mu \tag{6.6.50}$$

This yields the two differential equations

$$Y'' + \mu Y = 0 \tag{6.6.51}$$

and

$$\theta' + (\lambda + \mu)k\theta = 0 \tag{6.6.52}$$

The boundary conditions on $X(x)$ are

$$X(0) = 0, \qquad X(a) = 0 \tag{6.6.53}$$

since the temperature is zero at $x = 0$ and $x = a$. Consequently, the solution of Eq. 6.6.48,

$$X(x) = A\sin\sqrt{\lambda}x + B\cos\sqrt{\lambda}x \tag{6.6.54}$$

reduces to

$$X(x) = A \sin \frac{n\pi x}{a} \tag{6.6.55}$$

where we have used

$$\lambda = \frac{n^2\pi^2}{a^2}, \qquad n = 1, 2, 3, \cdots \tag{6.6.56}$$

Similarly, the solution to Eq. 6.6.51 reduces to

$$Y(y) = C \sin \frac{m\pi y}{b} \tag{6.6.57}$$

where we have employed

$$\mu = \frac{m^2\pi^2}{b^2}, \qquad m = 1, 2, 3 \cdots \tag{6.6.58}$$

With the use of Eqs. 6.6.56 and 6.6.58 we find the solution of Eq. 6.6.52 to be

$$\theta(t) = De^{-\pi^2 k(n^2/a^2 + m^2/b^2)t} \tag{6.6.59}$$

Equations 6.6.55, 6.6.57 and 6.6.59 may be combined to give

$$T_{mn}(x,y,t) = A_{mn} e^{-\pi^2 k(n^2/a^2 + m^2/b^2)t} \sin \frac{n\pi x}{a} \sin \frac{m\pi y}{b} \tag{6.6.60}$$

where the constant a_{mn} replaces ACD. The most general solution is then obtained by superposition, namely,

$$T(x,y,t) = \sum_{m=1}^{\infty} \sum_{n=1}^{\infty} T_{mn}(x,y,t) \tag{6.6.61}$$

and we have

$$T(x,y,t) = \sum_{m=1}^{\infty} \sum_{n=1}^{\infty} a_{mn} e^{-\pi^2 k(n^2/a^2 + m^2/b^2)t} \sin \frac{n\pi x}{a} \sin \frac{m\pi y}{b} \tag{6.6.62}$$

This is a solution if coefficients a_{mn} can be determined so that

$$T(x,y,0) = f(x,y) = \sum_{m=1}^{\infty} \left[\sum_{n=1}^{\infty} a_{mn} \sin \frac{n\pi x}{a} \right] \sin \frac{m\pi y}{b} \tag{6.6.63}$$

We make the grouping indicated by the brackets in Eq. 6.6.63. Thus, for a given x in the range $(0, a)$, we have a Fourier series in y. [For a given x, $f(x, y)$ is a function of y only.] Therefore, the term in the brackets is the constant b_n in the Fourier sine series. Hence,

$$\begin{aligned} \sum_{n=1}^{\infty} a_{mn} \sin \frac{n\pi x}{a} &= \frac{2}{b} \int_0^b f(x,y) \sin \frac{m\pi y}{b} dy \\ &= F_m(x) \end{aligned} \tag{6.6.64}$$

The right-hand side of Eq. 6.6.64 is a series of functions of x, one for each $m = 1, 2, 3, \cdots$. Thus, Eq. 6.6.64 is a Fourier sine series for $F_m(x)$. Therefore, we have

$$a_{mn} = \frac{2}{a}\int_0^a F_m(x)\sin\frac{n\pi x}{a}dx \tag{6.6.65}$$

Substitution of Eq. 6.6.64 into Eq. 6.6.65 yields

$$a_{mn} = \frac{4}{ab}\int_0^a\int_0^b f(x,y)\sin\frac{m\pi y}{b}\sin\frac{n\pi x}{a}dydx \tag{6.6.66}$$

The solution of our problem is Eq. 6.6.62 with a_{mn} given by Eq. 6.6.66.

This problem is an example of an extension of the ideas that we have developed, to include three independent variables; the two-dimensional Fourier series representation was also utilized.

We have studied the major ideas used in the application of separation of variables to problems in rectangular coordinates; to find the solution it was, in general, necessary to expand the initial condition in a Fourier series. For other problems that are more conveniently formulated in cylindrical coordinates, we would find Bessel functions taking the place of Fourier series, and using spherical coordinates, Legendre polynomials would appear. Sections 6.7 and 6.8 will present the solutions to Laplace's equation in spherical coordinates and cylindrical coordinates, respectively.

Example 6.10

The edges of a thin plate are held at the temperatures shown in the sketch of Fig. 6.15. Determine the steady-state temperature distribution in the plate. Assume the large plate surfaces to be insulated.

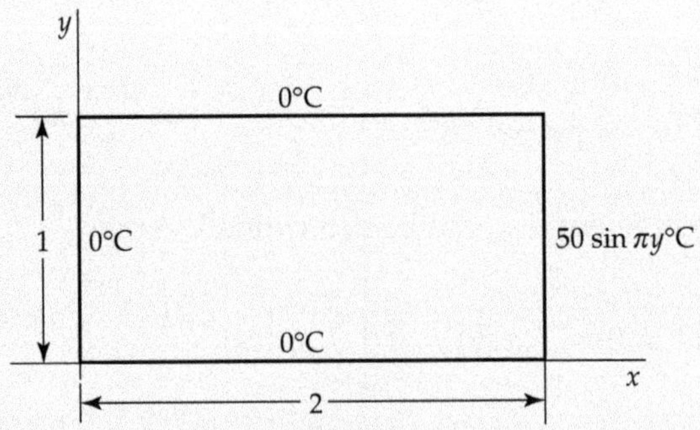

FIGURE 6.15

Solution

The describing equation is the heat equation

$$\frac{\partial T}{\partial t} = k\left(\frac{\partial^2 T}{\partial x^2} + \frac{\partial^2 T}{\partial y^2} + \frac{\partial^2 T}{\partial z^2}\right)$$

For the steady-state situation there is no variation of temperature with time; that is, $\partial T/\partial t = 0$. For a thin plate with insulated surfaces we have $\partial^2 T/\partial z^2 = 0$. Thus,

$$\frac{\partial^2 T}{\partial x^2} + \frac{\partial^2 T}{\partial y^2} = 0$$

This is Laplace's equation. Let us assume that the variables separate; that is,

$$T(x,y) = X(x)Y(y)$$

Then substitute into the describing equation to obtain

$$\frac{X''}{X} = -\frac{Y''}{Y} = \beta^2$$

where we have chosen the separation constant to be positive to allow for a sinusoidal variation* with y. The ordinary differential equations that result are

$$X'' - \beta^2 X = 0$$
$$Y'' + \beta^2 Y = 0$$

The solutions are

$$X(x) = Ae^{\beta x} + Be^{-\beta x}$$
$$Y(y) = C\sin\beta y + D\cos\,\beta y$$

The solution for $T(x, y)$ is then

$$T(x,y) = (Ae^{\beta x} + Be^{-\beta x})(C\sin\beta y + D\cos\beta y)$$

Using $T(0, y) = 0$, $T(x, 0) = 0$, and $T(x, 1) = 0$ gives

$$0 = A + B$$
$$0 = D$$
$$0 = \sin\beta$$

The final boundary condition is

$$T(2,y) = 50\sin\pi y = (Ae^{2\beta} + Be^{-2\beta})C\sin\beta y$$

From this condition we have

$$\beta = \pi$$
$$50 = C(Ae^{2\beta} + Be^{-2\beta})$$

From the equations above we can solve for the constants. We have

$$B = -A, \qquad AC = \frac{50}{e^{2\pi} - e^{-2\pi}} = 0.0934$$

Finally, the expression for $T(x, y)$ is

$$T(x,y) = 0.0934(e^{\pi x} - e^{-\pi x})\sin\pi y$$

Note that the expression above for the temperature is independent of the material properties; it is a steady-state solution.

---

*If the right-hand edge were held at a constant temperature we would also choose the separation constant so that $\cos\beta y$ and $\sin\beta y$ appear. This would allow a Fourier series to satisfy the edge condition.

6.7 ELECTRIC POTENTIAL ABOUT A SPHERICAL SURFACE

Consider that a spherical surface is maintained at an electrical potential V. The potential depends only on ϕ and is given by the function $f(\phi)$. The equation that describes the potential in the region on either side of the spherical surface is Laplace's equation (6.5.15), written in spherical coordinates (shown in Fig. 6.12) as

See Section 5.5.2 on spherical coordinates.

$$\frac{\partial}{\partial r}\left(r^2 \frac{\partial V}{\partial r}\right) + \frac{1}{\sin\phi}\frac{\partial}{\partial \phi}\left(\sin\phi \frac{\partial V}{\partial \phi}\right) = 0 \tag{6.7.1}$$

Obviously, one boundary condition requires that

$$V(r_0, \phi) = f(\phi) \tag{6.7.2}$$

The fact that a potential exists on the spherical surface of finite radius should not lead to a potential at infinite distances from the sphere; hence, we set

$$V(\infty, \phi) = 0 \tag{6.7.3}$$

We follow the usual procedure of separating variables; that is, assume that

$$V(r, \phi) = R(r)\Phi(\phi) \tag{6.7.4}$$

This leads to the equations

$$\frac{1}{R}\frac{d}{dr}\left(r^2 \frac{dR}{dr}\right) = -\frac{1}{\Phi \sin\phi}\frac{d}{d\phi}\left(\frac{d\Phi}{d\phi}\sin\phi\right) = \mu \tag{6.7.5}$$

which can be written as, letting $\cos\phi = x$, so that $\Phi = \Phi(x)$,

We used $\frac{d}{d\phi} = \frac{d}{dx}\frac{dx}{d\phi} = -\sin\phi\frac{d}{dx}$.

$$\begin{aligned} r^2R'' + 2rR' - \mu R &= 0 \\ (1-x^2)\Phi'' - 2x\Phi + \mu\Phi &= 0 \end{aligned} \tag{6.7.6}$$

The first of these is recognized as Cauchy's equation (see Section 1.11) and has the solution

$$R(r) = c_1 r^{-1/2+\sqrt{\mu+1/4}} + c_2 r^{-1/2-\sqrt{\mu+1/4}} \tag{6.7.7}$$

This is put in better form by letting $-\frac{1}{2} + \sqrt{\mu + \frac{1}{4}} = n$. Then

$$R(r) = c_1 r^n + \frac{c_2}{r^{n+1}} \tag{6.7.8}$$

The equation for Φ becomes Legendre's equation (see Section 2.3),

$$(1-x^2)\Phi'' - 2x\Phi' + n(n+1)\Phi = 0 \tag{6.7.9}$$

where n must be a positive integer for a proper solution to exist. The general solution to this equation is

$$\Phi(x) = c_3 P_n(x) + c_4 Q_n(x) \tag{6.7.10}$$

where $P_n(x)$ are Legendre polynomials and $Q_n(x)$ are Legendre functions of the second kind, defined in Eqs. 2.3.16 and 2.3.17, respectively.

Since $Q_n(x) \to \infty$ as $x \to 1$ (see Eq. 2.3.19), we set $c_4 = 0$. This results in the following solution for $V(r, x)$:

$$V(r,x) = \sum_{n=0}^{\infty} V_n(r,x) = \sum_{n=0}^{\infty} [A_n r^n P_n(x) + B_n r^{-(n+1)} P_n(x)] \tag{6.7.11}$$

Let us first consider points inside the spherical surface. The constants $B_n = 0$ if a finite potential is to exist at $r = 0$. We are left with

The solution inside the sphere and the solution outside the sphere are different.

$$V(r,x) = \sum_{n=0}^{\infty} A_n r^n P_n(x) \tag{6.7.12}$$

This equation must satisfy the boundary condition

$$V(r_0,x) = f(x) = \sum_{n=0}^{\infty} A_n r_0^n P_n(x) \tag{6.7.13}$$

The unknown coefficients A_n are found by using the property

$$\int_{-1}^{1} P_m(x)P_n(x)dx = \begin{cases} 0 & m \neq n \\ \dfrac{2}{2n+1} & m = n \end{cases} \tag{6.7.14}$$

Multiply both sides of Eq. 6.7.12 by $P_m(x)dx$ and integrate from –1 to 1. This gives

$$A_n = \frac{2n+1}{2r_0^n} \int_{-1}^{1} f(x)P_n(x)dx \tag{6.7.15}$$

For a prescribed $f(\phi)$, using $\cos\phi = x$, Eq. 6.7.12 provides us with the solution for interior points with the constants A_n given by Eq. 6.7.15.

For exterior points we require that $A_n = 0$ in Eq. 6.7.11, so the solution is bounded as $x \to \infty$. This leaves the solution

$$V(r,x) = \sum_{n=0}^{\infty} B_n r^{-(n+1)} P_n(x) \tag{6.7.16}$$

This equation must also satisfy the boundary condition

$$f(x) = \sum_{n=0}^{\infty} B_n r_0^{-(n+1)} P_n(x) \tag{6.7.17}$$

Using the property (6.7.14), the B_n's are given by

$$B_n = \frac{2n+1}{2} r_0^{n+1} \int_{-1}^{1} f(x)P_n(x)dx \tag{6.7.18}$$

If $f(x)$ is a constant we must evaluate $\int_{-1}^{1} P_n(x)dx$. Using Eq. 2.3.15 we can show that

$$\int_{-1}^{1} P_0(x)dx = 2, \quad \int_{-1}^{1} P_n(x)dx = 0, \qquad n = 1, 2, 3, \cdots \tag{6.7.19}$$

An example will illustrate the application of this presentation for a specific $f(x)$.

Example 6.11

Find the electric potential inside a spherical surface of radius r_0 if the hemispherical surface when $\pi > \phi > \pi/2$ is maintained at a constant potential V_0 and the hemispherical surface when $\pi/2 > \phi > 0$ is maintained at zero potential.

Solution

Inside the sphere of radius r_0, the solution is

$$V(r,x) = \sum_{n=0}^{\infty} A_n r^n P_n(x)$$

where $x = \cos\phi$. The coefficients A_n are given by Eq. 6.7.15,

$$\begin{aligned} A_n &= \frac{2n+1}{2r_0^n}\int_{-1}^{1} f(x)P_n(x)\,dx \\ &= \frac{2n+1}{2r_0^n}\left[\int_{-1}^{0} V_0 P_n(x)\,dx + \int_{0}^{1} 0\cdot P_n(x)\,dx\right] \\ &= \frac{2n+1}{2r_0^n} V_0 \int_{-1}^{0} P_n(x)\,dx \end{aligned}$$

where we have used $V = V_0$ for $\pi > \phi > \pi/2$ and $V = 0$ for $\pi/2 > \phi > 0$. Several A_n's can be evaluated, to give (see Eq. 2.3.15)

$$A_0 = \frac{V_0}{2},\quad A_1 = -\frac{3V_0}{4r_0},\quad A_2 = 0,\quad A_3 = \frac{7V_0}{16r_0^3},\quad A_4 = 0,\quad A_5 = -\frac{11V_0}{32r_0^5}$$

This provides us with the solution, letting $\cos\phi = x$,

$$\begin{aligned} V(r,\phi) &= A_0P_0 + A_1 rP_1 + A_2 r^2 P_2 + \cdots \\ &= V_0\left[\frac{1}{2} - \frac{3}{4}\frac{r}{r_0}\cos\phi + \frac{7}{16}\left(\frac{r}{r_0}\right)^3 P_3(\cos\phi) - \frac{11}{32}\left(\frac{r}{r_0}\right)^5 P_5(\cos\phi) + \cdots\right] \end{aligned}$$

where the Legendre polynomials are given by Eqs. 2.3.15. Note that the expression above could be used to give a reasonable approximation to the temperature in a solid sphere if the hemispheres are maintained at T_0 and zero degrees, respectively, since Laplace's equation also describes the temperature distribution in a solid body.

6.8 HEAT TRANSFER IN A CYLINDRICAL BODY

Boundary-value problems involving a boundary condition applied to a circular cylindrical surface are encountered quite often in physical situations. The solutions of such problems invariably involve Bessel functions, which were introduced in

Section 2.5. We shall use the problem of finding the steady-state temperature distribution in the cylinder shown in Fig. 6.16 as an example. Other exercises are included in the Problems.

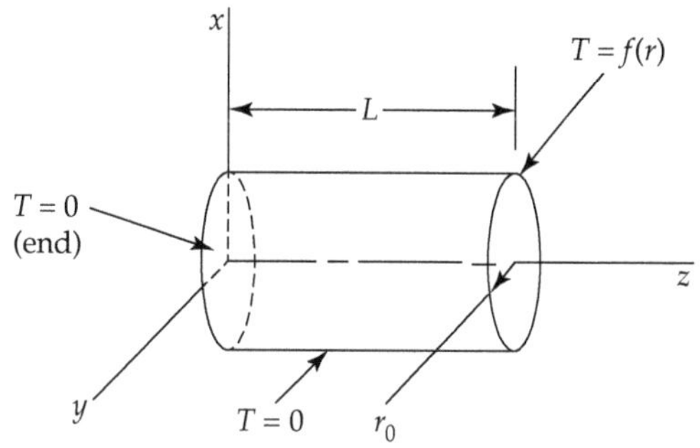

FIGURE 6.16 Circular cylinder with boundary conditions.

The partial differential equation describing the phenomenon illustrated in Fig. 6.16 is

$$\frac{\partial T}{\partial t} = k\nabla^2 T \tag{6.8.1}$$

where we have assumed constant material properties. Recall at steady state, $\frac{\partial T}{\partial t} = 0$. For a steady-state situation using cylindrical coordinates (see Eq. 6.5.14), this becomes

$$\frac{\partial^2 T}{\partial r^2} + \frac{1}{r}\frac{\partial T}{\partial r} + \frac{\partial^2 T}{\partial z^2} = 0 \tag{6.8.2}$$

where, considering the boundary conditions shown in the figure, we have assumed the temperature to be independent of θ. We assume a separated solution of the form

$$T(r,z) = R(r)Z(z) \tag{6.8.3}$$

which leads to the equations

$$\frac{1}{R}\left(R'' + \frac{1}{r}R'\right) = -\frac{Z''}{Z} = -\mu^2 \tag{6.8.4}$$

where a negative sign is chosen on the separation constant since we anticipate an exponential variation with z. We are thus confronted with solving the two ordinary differential equations

$$R'' + \frac{1}{r}R' + \mu^2 R = 0 \tag{6.8.5}$$

$$Z'' - \mu^2 Z = 0 \tag{6.8.6}$$

The solution to Eq. 6.8.6 is simply

$$Z(z) = c_1 e^{\mu z} + c_2 e^{-\mu z} \tag{6.8.7}$$

for $\mu > 0$; for $\mu = 0$, it is

$$Z(z) = c_5 z + c_6 \tag{6.8.8}$$

This solution may or may not be of use. We note that Eq. 6.8.5 is close to being Bessel's equation (2.5.1) with $\lambda = 0$. By substituting $x = \mu r$, Eq. 6.8.5 becomes

$$x^2R'' + xR' + x^2R = 0 \tag{6.8.9}$$

which is Bessel's equation with $\lambda = 0$. It possesses the general solution

$$R(x) = c_3J_0(x) + c_4Y_0(x) \tag{6.8.10}$$

where $J_0(x)$ and $Y_0(x)$ are Bessel functions of the first and second kind, respectively. We know (see Fig. 2.5) that $Y_0(x)$ is singular at $x = 0$. (This corresponds to $r = 0$.) Hence, we require that $c_4 = 0$, and the solution to our problem is

$$T(r,z) = J_0(\mu r)[Ae^{\mu z} + Be^{-\mu z}] \tag{6.8.11}$$

The temperature on the surface at $z = 0$ is maintained at zero degrees. This gives $B = -A$ from the equation above. The temperature at $r = r_0$ is also maintained at zero degrees; that is,

$$T(r_0, z) = 0 = AJ_0(\mu r_0)[e^{\mu z} - e^{-\mu z}] \tag{6.8.12}$$

The Bessel function $J_0(\mu r_0)$ has infinitely many roots that allow the equation above to be satisfied; none of these roots equal zero; thus the $\mu = 0$ eigenvalue is not of use. Let the nth root be designated μ_n. Four such roots are shown in Fig. 2.4 and are given numerically in the Appendix.

Returning to Eq. 6.8.11, our solution is now

$$T(r,z) = \sum_{n=1}^{\infty} T_n(r,z) = \sum_{n=1}^{\infty} J_0(\mu_n r)A_n[e^{\mu_n z} - e^{-\mu_n z}] \tag{6.8.13}$$

This solution should allow the final end condition to be satisfied. It is

$$T(r,L) = f(r) = \sum_{n=1}^{\infty} A_n J_0(\mu_n r)[e^{\mu_n L} - e^{-\mu_n L}] \tag{6.8.14}$$

We must now use the property that

$$\int_0^b xJ_j(\mu_n x)J_j(\mu_m x)\,dx = \begin{cases} 0 & n \neq m \\ \dfrac{b^2}{2}J_{j+1}^2(\mu_n b) & n = m \end{cases} \tag{6.8.15}$$

where the μ_n are the roots of the equation $J_j(\mu r_0) = 0$. This permits the coefficients A_n to be determined from, using $J = 0$,

$$A_n = \frac{2(e^{\mu_n L} - e^{-\mu_n L})^{-1}}{r_0^2 J_1^2(\mu_n r_0)} \int_0^{r_0} rf(r)J_0(\mu_n r)\,dr \tag{6.8.16}$$

This completes the solution. For a specified $f(r)$ for the temperature on the right end, Eq. 6.8.13 gives the temperature at any interior point if the coefficients are evaluated using Eq. 6.8.16. This process will be illustrated with an example.

Example 6.12

Determine the steady-state temperature distribution in a 2-unit-long, 4-unit-diameter circular cylinder with one end maintained at 0°C, the other end at 100r °C, and the lateral surface insulated.

Solution

Following the solution procedure outlined in the previous section, the solution is

$$T(r, z) = J_0(\mu r)[Ae^{\mu z} + Be^{-\mu z}]$$

The temperature at the base where $z = 0$ is zero. Thus, $B = -A$ and

$$T(r, z) = AJ_0(\mu r)[e^{\mu z} + e^{-\mu z}]$$

On the lateral surface where $r = 2$, the heat transfer is zero, requiring that

$$\frac{\partial T}{\partial r}(2, z) = 0 = AJ'_0(2\mu)[e^{\mu z} - e^{-\mu z}]$$

or

$$J'_0(2\mu) = 0$$

There are infinitely many values of μ that provide this condition, the first of which is $\mu = 0$. Let the nth one be μ_n, the eigenvalue. The solution corresponding to this eigenvalue is

$$T_n(r,z) = A_n J_0(\mu_n r)[e^{\mu_n z} - e^{-\mu_n z}]$$

for $\mu_n > 0$; for $\mu_1 = 0$, the solution is, using Eq. 6.8.8,

$$T_1(r, z) = A_1 z$$

The general solution is then found by superimposing all the individual solutions, resulting in

$$T(r,z) = \sum_{n=1}^{\infty} T_n(r,z) = A_1 z + \sum_{n=2}^{\infty} A_n J_0(\mu_n r)[e^{\mu_n z} - e^{-\mu_n z}]$$

The remaining boundary condition is that the end at $z = 2$ is maintained at 100r °C, that is,

$$T(r,2) = 100r = 2A_1 + \sum_{n=2}^{\infty} A_n J_0(\mu_n r)[e^{2\mu_n} - e^{-2\mu_n}]$$

We must be careful, however, and not assume that the A_n in this series are given by Eq. 6.8.16; they are not, since the roots μ_n are not to the equation $J_0(\mu r_0) = 0$, but to $J'_0(\mu r_0) = 0$. The property analogous to Eq. 6.8.15 takes the form

$$\int_0^{r_0} x J_j(\mu_n x) J_j(\mu_m x) dx = \begin{cases} 0 & n \neq m \\ \dfrac{\mu_n^2 r_0^2 - j^2}{2\mu_n^2} J_j^2(\mu_n r_0) & n = m \end{cases}$$

The key distinction with Eq. 6.8.15 is that the previous equation uses roots of $J'(\mu r_0) = 0$, not $J(\mu r_0) = 0$.

whenever μ_n are the roots of $J'_j(\mu r_0) = 0$. The coefficients A_n are then given by, using $j = 0$,

$$A_n = \frac{2(e^{2\mu_n} - e^{-2\mu_n})^{-1}}{r_0^2 J_0^2(\mu_n r_0)} \int_0^{r_0} r f(r) J_0(\mu_n r) dr$$

where $f(r) = 100r$. For the first root, $\mu_1 = 0$, the coefficient is

$$A_1 = \frac{2}{r_0^2} \int_0^{r_0} r f(r) dr$$

Some of the coefficients are, using $\mu_1 = 0$, $\mu_2 = 1.916$, $\mu_3 = 3.508$

$$A_1 = \frac{2}{2^2} \int_0^2 r(100r) dr = \frac{400}{3}$$

$$A_2 = \frac{2(e^{3.832} - e^{-3.832})^{-1}}{2^2 \times 0.403^2} \int_0^2 r(100r) J_0(1.916r) dr$$

$$= 6.68 \int_0^2 r^2 J_0(1.916r) dr = 0.951 \int_0^{3.832} x^2 J_0(x) dx$$

$$A_3 = \frac{2(e^{7.016} - e^{-7.016})^{-1}}{2^2 \times 0.300^2} \int_0^2 r(100r) J_0(3.508r) dr$$

$$= 0.501 \int_0^2 r^2 J_0(3.508r) dr = 0.0117 \int_0^{7.016} x^2 J_0(x) dx$$

The integrals above could be easily evaluated by use of a computer integration scheme. Such a scheme will be presented in Chapter 8. The solution is then

$$T(r, z) = \frac{400}{3} z + A_2 J_0(1.916r)[e^{1.916z} - e^{-1.916z}] + A_3 J_0(3.508r)[e^{3.508z} - e^{-3.508z}] + \cdots$$

6.9 GRAVITATIONAL POTENTIAL

There are a number of physical situations that are modeled by Laplace's equation. We shall choose the force of attraction of particles to demonstrate its derivation. The law of gravitation states that a lumped mass m located at the point (X, Y, Z) attracts a unit mass located at the point (x, y, z) (see Fig. 6.17), with a force directed along the line connecting the two points with magnitude given by

$$F = -\frac{km}{r^2} \tag{6.9.1}$$

where K is a positive constant and the negative sign indicates that the force acts toward the mass m. The distance between the two points is provided by the expression

$$r = \sqrt{(x - X)^2 + (y - Y)^2 + (z - Z)^2} \tag{6.9.2}$$

positive being from Q to P.

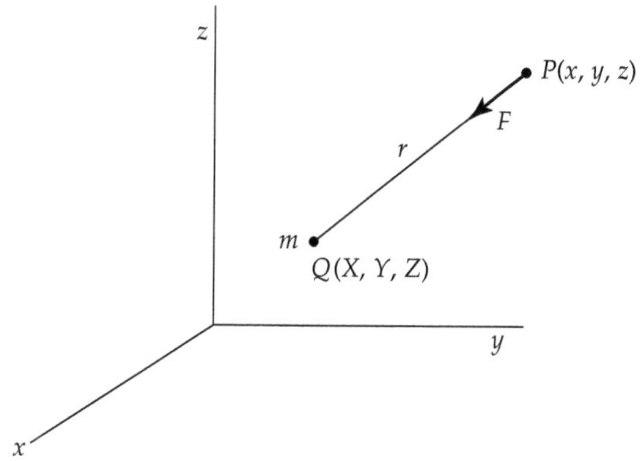

FIGURE 6.17 Gravitational attraction.

A gravitational potential ϕ can be defined as

By popular convention, ϕ is used for potential, and is not to be confused with notation for spherical coordinates of the previous section.

$$\phi = \frac{km}{r} \tag{6.9.3}$$

This allows the force F acting on a unit mass at P due to a mass at Q to be related to ϕ by the equation

$$\begin{aligned} F &= \frac{\partial \phi}{\partial r} \\ &= -\frac{km}{r^2} \end{aligned} \tag{6.9.4}$$

Now, let the mass m be fixed in space and let the unit mass move to various locations $P(x, y, z)$. The potential function ϕ is then a function of x, y, and z. If we let P move along a direction parallel to the x axis, then

$$\begin{aligned} \frac{\partial \phi}{\partial x} &= \frac{\partial \phi}{\partial r}\frac{\partial r}{\partial x} \\ &= -\frac{km}{r^2}\frac{1}{2}(2)(x - X)[(x - X)^2 + (y - Y)^2 + (z - Z)^2]^{-1/2} \\ &= -\frac{km}{r^2}\frac{x - X}{r} \\ &= F\cos\alpha = F_x \end{aligned} \tag{6.9.5}$$

where α is the angle between r and the x axis, and F_x is the projection of F in the x direction. Similarly, for the other two directions,

$$F_y = \frac{\partial \phi}{\partial y}, \qquad F_z = \frac{\partial \phi}{\partial z} \tag{6.9.6}$$

The discussion is now extended to include a distributed mass throughout a volume V. The potential $d\phi$ due to an incremental mass dm is written, following Eq. 6.9.3, as

$$d\phi = \frac{k\rho\,dV}{r} \tag{6.9.7}$$

where ρ is the density, i.e., mass per unit volume. Letting $dV = dx\,dy\,dz$, we have

$$\phi = k\iiint_V \frac{\rho\,dx\,dy\,dz}{[(x-X)^2 + (y-Y)^2 + (z-Z)^2]^{1/2}} \tag{6.9.8}$$

This is differentiated to give the force components. For example, F_x is given by

$$F_x = \frac{\partial\phi}{\partial x} = -k\iiint_V \frac{x-X}{r}\frac{\rho}{r^2}\,dx\,dy\,dz \tag{6.9.9}$$

This represents the x component of the total force exerted on a unit mass located outside the volume V at $P(x, y, z)$ due to the distributed mass in the volume V.

If we now differentiate Eq. 6.9.9 again with respect to x, we find that

$$\frac{\partial^2\phi}{\partial x^2} = -k\iiint_V \left[\frac{1}{r^3} - \frac{3(x-X)^2}{r^5}\right]\rho\,dx\,dy\,dz \tag{6.9.10}$$

We can also show that

$$\begin{aligned}\frac{\partial^2\phi}{\partial y^2} &= -k\iiint_V \left[\frac{1}{r^3} - \frac{3(y-Y)^2}{r^5}\right]\rho\,dx\,dy\,dz\\ \frac{\partial^2\phi}{\partial z^2} &= -k\iiint_V \left[\frac{1}{r^3} - \frac{3(z-Z)^2}{r^5}\right]\rho\,dx\,dy\,dz\end{aligned} \tag{6.9.11}$$

The sum of the bracketed terms inside the three integrals above is observed to be identically zero, using Eq. 6.9.2. Hence, Laplace's equation results,

$$\frac{\partial^2\phi}{\partial x^2} + \frac{\partial^2\phi}{\partial y^2} + \frac{\partial^2\phi}{\partial z^2} = 0 \tag{6.9.12}$$

or, in our shorthand notation,

$$\nabla^2\phi = 0 \tag{6.9.13}$$

Laplace's equation is also satisfied by a magnetic potential function and an electric potential function at points not occupied by magnetic poles or electric charges. We have already observed in Section 6.5 that the steady-state heat-conduction problem leads to Laplace's equation. Finally, the flow of an incompressible fluid with negligible viscous effects also leads to Laplace's equation.

We have now derived several partial differential equations that describe a variety of physical phenomena. This modeling process is quite difficult to perform on a situation that is new and different. Hopefully, the confidence gained in deriving the equations of this chapter and in finding solutions will allow the reader to derive and solve other partial differential equations arising in the multitude of application areas.

PROBLEMS

6.1 Classify each of the following equations.

(a) $$\frac{\partial^2 u}{\partial x^2} + \frac{\partial^2 u}{\partial x\,\partial y} + \frac{\partial^2 u}{\partial y^2} = 0$$

(b) $$(1 - x)\frac{\partial^2 u}{\partial x^2} + 2y\frac{\partial^2 u}{\partial x\,\partial y} + (1 + x)\frac{\partial^2 u}{\partial y^2} = 0$$

(c) $$\frac{\partial^2 u}{\partial x^2} + \sqrt{1 + \left(\frac{\partial u}{\partial x}\right)^2}\,\frac{\partial^2 u}{\partial y^2} + k\frac{\partial u}{\partial y} = G(x, y)$$

(d) $$\left(\frac{\partial u}{\partial x}\right)^2 = u(x, y)$$

(e) $$\frac{du}{dx} = u(x)$$

6.2 Verify each of the following statements.

(a) $u(x, y) = e^x \sin y$ is a solution of Laplace's equation, $\nabla^2 u = 0$.
(b) $T(x, t) = e^{-kt} \sin x$ is a solution of the parabolic heat equation, $\partial T/\partial t = k\partial^2 T/\partial x^2$.
(c) $u(x, t) = \sin \omega x \sin \omega a t$ is a solution of the wave equation, $\partial^2 u/\partial t^2 = a^2 \partial^2 u/\partial x^2$.

6.3 In arriving at the equation describing the motion of a vibrating string, the weight was assumed to be negligible. Include the weight of the string in the derivation and determine the describing equation. Classify the equation.

6.4 Derive the describing equation for a stretched string subject to gravity loading and viscous drag. Viscous drag per unit length of string may be expressed by $c(\partial u/\partial t)$; the drag force is proportional to the velocity. Classify the resulting equation.

6.5 Derive the torsional vibration equation for a circular shaft by applying the basic law which states that $I\alpha = \sum T$, where α is the angular acceleration, T is the torque ($T = GJ\theta/L$, where θ is the angle of twist of the shaft of length L and J and G are constants), and I is the mass moment of inertia ($I = k^2 m$, where the radius of gyration $k = \sqrt{J/A}$ and m is the mass of the shaft). Choose an infinitesimal element of the shaft of length Δx, sum the torques acting on it, and, using ρ as the mass density, show that the wave equation results,

$$\frac{\partial^2 \theta}{\partial t^2} = \frac{G}{\rho}\frac{\partial^2 \theta}{\partial x^2}$$

6.6 An unloaded beam will undergo vibrations when subjected to an initial disturbance. Derive the appropriate partial differential equation which describes the motion using Newton's second law applied to an infinitesimal section of the beam. Assume the inertial force to be a distributed load acting on the beam. A uniformly distributed load w is related to the vertical deflection $y(x, t)$ of the beam by $w = -EI\,\partial^4 y/\partial x^4$, where E and I are constants.

6.7 For the special situation in which $LG = RC$, show that the transmission-line equation 6.2.36 reduces to the wave equation

$$\frac{\partial^2 u}{\partial t^2} = a^2\frac{\partial^2 u}{\partial x^2}$$

if we let

$$i(x, t) = e^{-abt}u(x, t)$$

where $a^2 = 1/LC$ and $b^2 = RG$.

6.8 A tightly stretched string, with its ends fixed at the points (0, 0) and (2L, 0), hangs at rest under its own weight. The y axis points vertically upward. Find the describing equation for the position $u(x)$ of the string. Is the following expression a solution?

$$u(x) = \frac{g}{2a^2}(x - L)^2 - \frac{gL^2}{2a^2}$$

where $a^2 = P/m$. If so, show that the depth of the vertex of the parabola (i.e., the lowest point) varies directly with m (mass per unit length) and L^2, and inversely with P, the tension.

6.9 A very long string is given an initial displacement $\phi(x)$ and an initial velocity $\theta(x)$. Determine the general form of the solution for $u(x, t)$. Compare with the solution (6.3.18) and that of Example 6.1.

6.10 An infinite string with a mass of 0.03 kg/m is stretched with a force of 300 N. It is subjected to an initial displacement of cos x for $-\pi/2 < x < \pi/2$ and zero for all other x and released from rest. Determine the subsequent displacement of the string and sketch the solution for $t = 0.1$ s and 0.01 s.

6.11 Express the solution (6.4.36) in terms of the solution (6.3.10). What are f and g?

6.12 Determine the general solution for the wave equation using separation of variables assuming that the separation constant is zero. Show that this solution cannot satisfy the boundary and/or initial conditions.

6.13 Verify that

$$u(x, t) = b_n \cos\frac{\pi at}{L} \sin\frac{n\pi x}{L}$$

is a solution to Eq. 6.4.1, and the conditions 6.4.2 through 6.4.4.

6.14 Find the constants A, B, C, and D in Eqs. 6.4.23 and 6.4.24 in terms of the constants c_1, c_2, c_3, and c_4 in Eqs. 6.4.20 and 6.4.21.

6.15 Determine the relationship of the fundamental frequency of a vibrating string to the mass per unit length, the length of the string, and the tension in the string.

6.16 If, for a vibrating wire, the original displacement of the 2-m-long stationary wire is given by a) 0.1 sin $x\pi/2$, b) 0.1 sin $3\pi/2$, and c) 0.1(sin $\pi x/2$ − sin $3\pi x/2$), find the displacement function $u(x, t)$. Both ends are fixed, $P = 50$ N, and the mass per unit length is 0.01 kg/m. With what frequency does the wire oscillate? Write the eigenvalue and eigenfunction for part (a).

6.17 The initial displacement in a 2-m-long string is given by 0.2 sin πx and released from rest. Calculate the maximum velocity in the string and state its location.

6.18 A string π m long is stretched until the wave speed is 40 m/s. It is given an initial velocity of 4 sin x from its equilibrium position. Determine the maximum displacement and state its location and when it occurs.

6.19 A string 4 m long is stretched, resulting in a wave speed of 60 m/s. It is given an initial displacement of 0.2 sin $\pi x/4$ and an initial velocity of 20 sin $\pi x/4$. Find the solution representing the displacement of the string.

6.20 A 4-m-long stretched string with $a = 20$ m/s is fixed at each end.

(a) The string is started off by an initial displacement $u(x, 0) = 0.2 \sin \pi x/4$. The initial velocity is zero. Determine the solution for $u(x, t)$.

(b) Suppose that we wish to generate the same string vibration as in part (a) (a standing half-sine wave with the same amplitude), but we want to start with a zero-displacement, non-zero-velocity condition. That is, $u(x, 0) = 0$, $\partial u/\partial t(x, 0) = g(x)$. What should $g(x)$ be?

(c) For $u(x, 0) = 0.1 \sin \pi x/4$ and $\partial u/\partial t(x, 0) = 10 \sin \pi x/4$, what are the arbitrary constants? What is the maximum displacement value $u_{max}(x, t)$, and where does it occur?

6.21 Suppose that a tight string is subjected to the following conditions: $u(0, t) = 0$, $u(L, t) = 0$, $\partial u/\partial t(x, 0) = 0$, $u(x, 0) = k$. Calculate the first three nonzero terms of the solution $u(x, t)$.

6.22 A string π m long is started into motion by giving the middle one-half an initial velocity of 20 m/s. The string is stretched until the wave speed is 60 m/s. Determine the resulting displacement of the string as a function of x and t.

6.23 The right end of a 6-m-long wire, which is stretched until the wave speed is 60 m/s, is continually moved with the displacement 0.5 cos $4\pi t$. What is the maximum amplitude of the resulting displacement?

6.24 The wind is blowing over some suspension cables on a bridge, causing a force that is approximated by the function 0.02 sin $21\pi t$. Is resonance possible if the force in the cable is 40,000 N, the cable has a mass of 10 kg/m, and it is 15 m long?

6.25 A circular shaft π m long is fixed at both ends. The middle of the shaft is twisted through an angle α, the remainder of the shaft through an angle proportional to the distance from the nearest end, and then the shaft is released from rest. Determine the subsequent motion expressed as $\theta(x, t)$. Problem 6.5 gives the appropriate wave equation.

6.26 Modify Eq. 6.5.9 to account for internal heat generation within the rod. The rate of heat generation is denoted ϕ (W/m$^3 \cdot$ s).

6.27 Allow the sides of a long, slender circular rod to transfer heat by convection. The convective rate of heat loss is given by $Q = hA(T - T_f)$, where h (W/m$^2 \cdot$ s $\cdot$ °C) is the convection coefficient, A is the surface area, and T_f is the temperature of the surrounding fluid. Derive the describing partial differential equation. (*Hint:* Apply an energy balance to an elemental slice of the rod.)

6.28 The tip of a 2-m-long slender rod with lateral surface insulated is dipped into a hot liquid at 200°C. What differential equation would describe the temperature? After a long time, what would be the temperature distribution in the rod if the other end is held at 0°C? The lateral surfaces of the rod are insulated.

6.29 The conductivity K in the derivation of Eq. 6.3.10 was assumed constant. Let K be a function of x and let C and ρ be constants. Write the appropriate describing equation.

6.30 Write the one-dimensional heat equation that would be used to determine the temperature in a) a flat circular disk with the flat surfaces insulated, and b) in a sphere with initial temperature a function of r only.

6.31 Determine the steady-state temperature distribution in a) a flat circular disc with sides held at 100°C with the flat surfaces insulated, and b) a sphere with the outer surface held at 100°C.

6.32 The initial temperature in a 10-m-long iron rod is 300 sin $\pi x/10$, with both ends being held at zero temperature. Determine the times necessary for the midpoint of the rod to reach 200, 100, and 50, respectively. The material constant $k = 1.7 \times 10^{-5}$ m^2/s. The lateral surfaces are insulated.

6.33 A 1-m-long, 50-mm-diameter aluminum rod, with lateral surfaces insulated, is initially at a temperature of 200(1 + sin πx). Calculate the rate at which the rod is transferring heat at the left end initially and after 600 s if both ends are maintained at 200°C. For aluminum, $K = 200$ W/m $\cdot$ °C and $k = 8.6 \times 10^{-5}$ m^2/s. (*Hint:* Let $\theta(x, t) = T(x, t) - 200$.)

6.34 The initial temperature distribution in a 2-m-long brass bar is given by

$$f(x) = \begin{cases} 50x & 0 < x < 1 \\ 100 - 50x & 1 < x < 2 \end{cases}$$

Both ends are maintained at zero temperature. Determine the solution for $T(x, t)$. How long would you predict it would take for the center of the rod to reach a temperature of 10°C? The material constant $k = 2.9 \times 10^{-5}$ m^2/s. The lateral surfaces are insulated.

6.35 The initial temperature distribution in a 2-m-long steel rod is given by

$$f(x) = \begin{cases} 50x & 0 < x < 1 \\ 100 - 50x & 1 < x < 2 \end{cases}$$

The rod is completely insulated. Determine the temperature distribution in the rod and predict the temperature that the rod will eventually attain. $k = 3.9 \times 10^{-6}$ m^2/s.

6.36 A 2-m-long aluminum bar, with lateral surfaces insulated, is given the initial temperature distribution $f(x) = 50\,x^2$. The left end of the bar is maintained at 0°C and the right end at 200°C. Determine the subsequent temperature distribution in the bar. $k = 8.6 \times 10^{-5}$ m^2/s.

6.37 The infinite slab of Fig. 6.17 is initially at temperature $f(x)$. The face at $x = 0$ is held at $T = 0$°C. Determine the temperature $T(x, t)$ of the slab for $t > 0$.

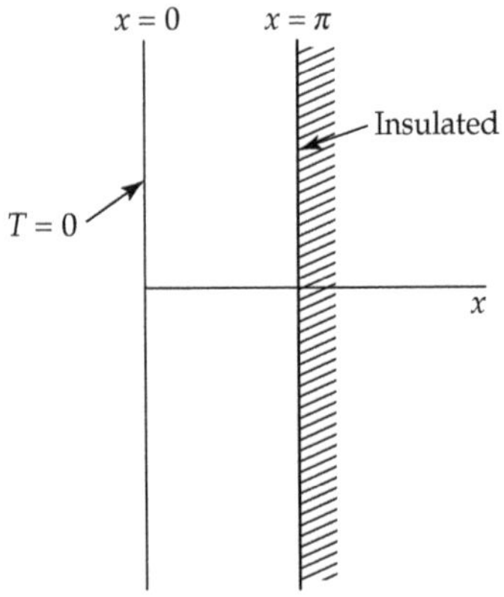

FIGURE 6.17

6.38 The aluminum slab in Problem 6.37 is given the initial temperature distribution

$$f(x) = \begin{cases} 100 & 0 < x < \pi/2 \\ 0 & \pi/2 < x < \pi \end{cases}$$

Estimate the rate of heat transfer per square meter from the left face at $t = 10^4$ s if $k = 8.6 \times 10^{-5}$ m²/s and $K = 200$ W/m · °C.

6.39 Heat generation occurs within a 4-m-long copper rod at the variable rate of $2000(4x - x^2)$ W/m³ · s. Both ends are maintained at 0°C. $C = 380$ J/kg · °C, $\rho = 8940$ kg/m³, and $k = 1.14 \times 10^{-4}$ m²/s.

(a) Find the steady-state solution for the temperature distribution in the rod.

(b) Find the transient temperature distribution in the rod if the initial temperature was constant at 100°C. Just set up the integral for the Fourier coefficients; do not integrate.

6.40 Find the steady-state temperature distribution in a 1-m² slab if three sides are maintained at 0°C and the remaining side (at $y = 1$ m) is held at $100 \sin \pi x$ °C. All other surfaces are insulated.

6.41 Three edges of a thin 1-m by 2-m plate are held at 0°C, while the fourth edge, at $y = 1$ m, is held at 100°C. All other surfaces are insulated. Determine an expression for the temperature distribution in the plate.

6.42 Find the steady-state temperature distribution in a 2 m-square slab if three sides are maintained at 100°C and the remaining side (at $x = 2$ m) is held at 200°C. The two flat surfaces are insulated.

6.43 The temperature of a spherical surface 0.2 m in diameter is maintained at a temperature of 250°C. This surface is interior to a very large mass. Find an expression for the temperature distribution inside and outside the surface.

6.44 The temperature on the surface of a 1-m-diameter sphere is $100 \cos \phi$ °C. What is the temperature distribution inside the sphere?

6.45 Find the potential field between two concentric spheres if the potential of the outer sphere is maintained at $V = 100$ and the potential of the inner sphere is maintained at zero. The radii are 2 m and 1 m, respectively.

6.46 A right circular cylinder is 1 m long and 2 m in diameter. Its left end and lateral surface are maintained at a temperature of 0°C and its right end at 100°C. Find an expression for its temperature at any interior point. Calculate the first three coefficients in the series expansion.

6.47 Determine the solution for the temperature as a function of r and t in a circular cylinder of radius r_0 with insulated (or infinitely long) ends if the initial temperature distribution is a function $f(r)$ of r only and the lateral surface is maintained at 0°C. See Eq. 6.5.14.

6.48 An aluminum circular cylinder 50 mm in diameter with ends insulated is initially at 100°C. Approximate the temperature at the center of the cylinder after 2 s if the lateral surface is kept at 0°C. For aluminum, $k = 8.6 \times 10^{-5}$ m²/s.

6.49 A circular cylinder 1 m in radius is completely insulated and has an initial temperature distribution $100r$ °C. Find an expression for the temperature as a function of r and t. Write integral expressions for at least three coefficients in the series expansion.

6.50 Differentiate Eq. 6.9.8 and show that Eq. 6.9.9 results. Also verify Eq. 6.9.10.

7 Complex Variables

Outline

7.1 Introduction
7.2 Complex Numbers
7.3 Elementary Functions
7.4 Analytic Functions
7.5 Complex Integration
7.5.1 Green's Theorem
7.5.2 Cauchy's Integral Theorem
7.5.3 Cauchy's Integral Formula
7.6 Series
7.6.1 Taylor Series
7.6.2 Laurent Series
7.7 Residues

Problems

7.1 INTRODUCTION

The use of complex numbers has been illustrated when the solution of a particular equation involves the quantity $\sqrt{-1}$. The solution of the differential equation describing the motion of a spring–mass system is an example. Various relationships have been used involving complex numbers; in this chapter these relationships will be discussed in detail. Additional information will be presented that is often useful in the solution of the equations encountered in the modeling of physical phenomena.

In addition to the discussion of complex numbers, complex variables will also be presented. Complex variables differ from complex numbers in that the real and imaginary parts of a quantity are functions of the independent variables. Various physical quantities can be expressed in terms of complex variables. These are encountered in such areas as fluids mechanics, solid mechanics, electrical fields, and heat transfer. The fact that a variable can be complex introduces many additional mathematical techniques. In this chapter we shall introduce some of the basic ideas upon which a theory of complex variables has been built. We will in no way exhaust this topic; one should consult a textbook on the subject for a more complete treatment.

7.2 COMPLEX NUMBERS

The solution of algebraic equations such as

$$z^2 - 12z + 52 = 0 \tag{7.2.1}$$

M. C. Potter and B. F. Feeny, *Mathematical Methods for Engineering and Science*,
https://doi.org/10.1007/978-3-031-26151-0_7

has led to the introduction of complex numbers.* The solution to the equation above is

$$z = 6 \pm 4\sqrt{-1} \tag{7.2.2}$$

which is not a real number. If we were solving for a distance z, the solution above would not make sense. Yet, in some situations this solution is acceptable. The solution (7.2.2) is written as the *complex number*

$$z = 6 \pm 4i \tag{7.2.3}$$

where $i = \sqrt{-1}$. The quantity 6 is the *real part* and the quantity ±4 is the *imaginary part* of the complex number. In general, the complex number z will be written as

$$z = x + iy \tag{7.2.4}$$

where x and y are the rectangular coordinates shown in Fig. 7.1. The x axis is the *real axis* and the y axis is the *imaginary axis*.

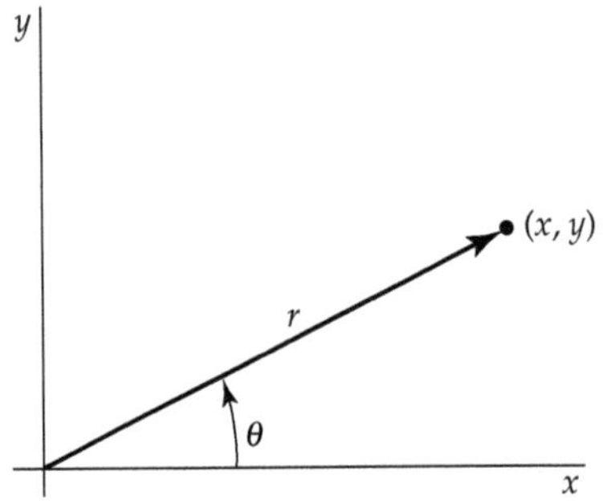

FIGURE 7.1 The complex plane.

In terms of polar coordinates (r, θ) the variables x and y are

$$x = r\cos\theta, \quad y = r\sin\theta \tag{7.2.5}$$

The complex variable z is then written as

$$z = r(\cos\theta + i\sin\theta) \tag{7.2.6}$$

The quantity r is the *absolute value* of z and is

$$r = |z| = \sqrt{x^2 + y^2} \tag{7.2.7}$$

The angle θ, measured in radians and positive in the counterclockwise sense, is the *argument* of z given by

$$\theta = \tan^{-1}\frac{y}{x} \tag{7.2.8}$$

The signs of x and y determine the quadrant and hence define the principal value of θ.

Obviously, there are an infinite number of θ's satisfying (7.2.8) at intervals of 2π radians. We shall make the usual choice of limiting θ to the interval $0 \le \theta < 2\pi$ for its *principal value*.†

---

*The term "complex number" was introduced by Carl F. Gauss in the early nineteenth century.
†Another interval for the principal value which is used in some texts is $-\pi < \theta < \pi$.

A complex number is *pure imaginary* if the real part is zero. It is *real* if the imaginary part is zero. The *conjugate* of the complex number z is denoted by $\bar{z}$; it is found by changing the sign on the imaginary part of z, that is,

$$\bar{z} = x - iy \tag{7.2.9}$$

The conjugate is useful in manipulations involving complex numbers. An interesting observation and often useful result is that the product of a complex number and its conjugate is real. This follows from

$$\begin{aligned} z\bar{z} &= (x+iy)(x-iy) \\ &= x^2 - i^2y^2 \\ &= x^2 + y^2 \end{aligned} \tag{7.2.10}$$

Note that $z\bar{z} = |z|^2$.

where we have used $i^2 = -1$. The imaginary part of $z\bar{z}$ is obviously zero.

The addition, subtraction, multiplication, or division of two complex numbers $z_1 = x_1 + iy_1$ and $z_2 = x_2 + iy_2$ is accomplished as follows:

$$\begin{aligned} z_1 + z_2 &= (x_1 + iy_1) + (x_2 + iy_2) \\ &= (x_1 + x_2) + i(y_1 + y_2) \end{aligned} \tag{7.2.11}$$

$$\begin{aligned} z_1 - z_2 &= (x_1 + iy_1) - (x_2 + iy_2) \\ &= (x_1 - x_2) + i(y_1 - y_2) \end{aligned} \tag{7.2.12}$$

$$\begin{aligned} z_1 z_2 &= (x_1 + iy_1)(x_2 + iy_2) \\ &= (x_1x_2 - y_1y_2) + i(x_1y_2 + x_2y_1) \end{aligned} \tag{7.2.13}$$

$$\begin{aligned} \frac{z_1}{z_2} &= \frac{x_1 + iy_1}{x_2 + iy_2} = \frac{x_1 + iy_1}{x_2 + iy_2}\frac{x_2 - iy_2}{x_2 - iy_2} \\ &= \frac{x_1x_2 + y_1y_2}{x_2^2 + y_2^2} + i\frac{x_2y_1 - x_1y_2}{x_2^2 + y_2^2} \end{aligned} \tag{7.2.14}$$

Note that the conjugate of z_2 was used to form a real number in the denominator of z_1/z_2.

Addition and subtraction is illustrated graphically in Fig. 7.2. From the parallelogram formed by the addition of the two complex numbers, we observe that

$$|z_1 + z_2| \le |z_1| + |z_2| \tag{7.2.15}$$

This inequality will be quite useful in later considerations.

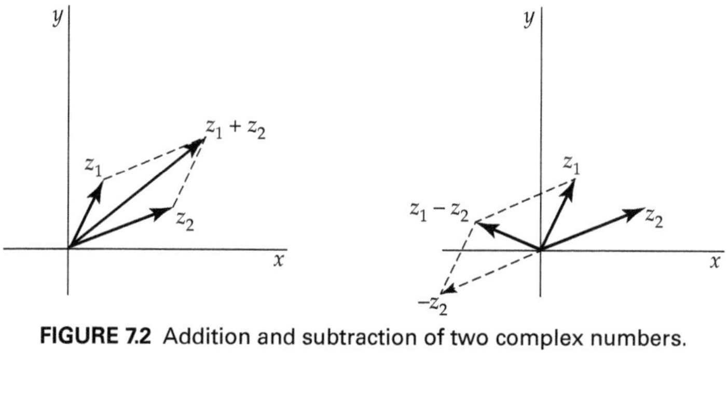

FIGURE 7.2 Addition and subtraction of two complex numbers.

The multiplication of two complex numbers is particularly useful when expressed in polar form. The complex numbers z_1 and z_2 are

$$z_1 = r_1(\cos\theta_1 + i\sin\theta_1), \quad z_2 = r_2(\cos\theta_2 + i\sin\theta_2) \tag{7.2.16}$$

so that

$$\begin{aligned} z_1 z_2 &= r_1 r_2[(\cos\theta_1\cos\theta_2 - \sin\theta_1\sin\theta_2) + i(\sin\theta_1\cos\theta_2 + \sin\theta_2\cos\theta_1)] \\ &= r_1 r_2[\cos(\theta_1 + \theta_2) + i\sin(\theta_1 + \theta_2)] \end{aligned} \tag{7.2.17}$$

where we have used appropriate trigonometric identities. From the above, the following generalized rule is stated:

$$z_1 z_2 \cdots z_n = r_1 r_2 \cdots r_n[\cos(\theta_1 + \theta_2 + \cdots + \theta_n) + i\sin(\theta_1 + \theta_2 + \cdots + \theta_n)] \tag{7.2.18}$$

Letting $z_1 = z_2 = z_3 = \cdots = z_n$, we have the important relationship

$$z^n = r^n(\cos n\theta + i\sin n\theta) \tag{7.2.19}$$

From Eq. 7.2.17 we see that (refer to Eq. 7.2.7)

$$|z_1 z_2| = |z_1||z_2| \tag{7.2.20}$$

since

$$|\cos(\theta_1 + \theta_2) + i\sin(\theta_1 + \theta_2)| = 1 \tag{7.2.21}$$

Also, note that the argument of $z_1 z_2$ is $\theta_1 + \theta_2$. Similarly, division yields

$$\frac{z_1}{z_2} = \frac{r_1}{r_2}[\cos(\theta_1 - \theta_2) + i\sin(\theta_1 - \theta_2)] \tag{7.2.22}$$

showing that

$$\left|\frac{z_1}{z_2}\right| = \frac{|z_1|}{|z_2|} \tag{7.2.23}$$

The nth *root* of a complex number is denoted by $z^{1/n}$. In polar form we can write

$$z = r[\cos(\theta + 2k\pi) + i\sin(\theta + 2k\pi)], \quad k = 0, 1, 2, \cdots \tag{7.2.24}$$

where the additional multiples of 2π radians are necessary when finding the nth root. Referring to Eq. 7.2.19, we have

$$z^{1/n} = r^{1/n}\left[\cos\left(\frac{\theta + 2k\pi}{n}\right) + i\sin\left(\frac{\theta + 2k\pi}{n}\right)\right] \tag{7.2.25}$$

The nth root of unity would be

$$1^{1/n} = \cos\frac{2\pi k}{n} + i\sin\frac{2\pi k}{n} \tag{7.2.26}$$

where $r = 1$ and $\theta = 0°$. For $n = 4$, k would take on the values of 0, 1, 2, and 3 and four distinct values would result. For $k = 4$ a result identical to $k = 0$ would result and no different root is obtained; hence, additional k values would not be considered. Note that, in general, there will be n distinct values for the nth root.

Obviously, we can combine the above to find $z^{m/n}$, where m and n are integers; that is, m/n is a real rational number. Again, there will be n distinct values for $z^{m/n}$. A complex number raised to a power that is not a real rational number will be considered in the next section.

Finally, we note the rather obvious fact that if two complex numbers are equal, the real parts and imaginary parts are equal, respectively. Hence, an equation written in terms of complex variables includes two real equations, one found by equating the real parts from each side of the equation and the other found by equating the imaginary parts.

Example 7.1

Express the complex number $3 + 6i$ in polar form, and also divide it by $2 - 3i$.

Solution

To express $3 + 6i$ in polar form we must determine r and θ. We have

$$r = \sqrt{x^2 + y^2} = \sqrt{3^2 + 6^2} = 6.708$$

The angle θ is found, in degrees, to be

$$\theta = \tan^{-1}\frac{6}{3} = 63.43°$$

In polar form we have

$$3 + 6i = \underline{6.708(\cos 63.43° + i \sin 63.43°)}$$

The desired division is

$$\frac{3+6i}{2-3i} = \frac{3+6i}{2-3i}\frac{2+3i}{2+3i}$$

$$= \frac{6-18+i(12+9)}{4+9} = \frac{1}{13}(-12+21i) = \underline{-0.9231+1.615i}$$

Example 7.2

Determine (a) $(3 + 4i)^2$, and (b) $(3 + 4i)^{1/3}$.

Solution

The number is expressed in polar form, using $r = \sqrt{3^2 + 4^2} = 5$ and $\theta = \tan^{-1}\frac{4}{3} = 53.13°$ as

$$3 + 4i = 5(\cos 53.13° + i \sin 53.13°)$$

a) To determine $(3+4i)^2$ we use Eq. 7.2.19 and find

$$\begin{aligned}(3+4i)^2 &= 5^2(\cos 2\times 53.13° + i\sin 2\times 53.13°)\\ &= 25(-0.280+0.960i)\\ &= -7+24i\end{aligned}$$

We could also simply form the product

$$\begin{aligned}(3+4i)^2 &= (3+4i)(3+4i)\\ &= 9-16+12i+12i = \underline{-7+24i}\end{aligned}$$

b) There are three distinct roots that must be determined when evaluating $(3+4i)^{1/3}$. They are found by expressing $(3+4i)^{1/3}$ as

$$(3+4i)^{1/3} = 5^{1/3}\left(\cos\frac{53.13+360k}{3} + i\sin\frac{53.13+360k}{3}\right)$$

The first root is then, using $k=0$,

$$\begin{aligned}(3+4i)^{1/3} &= 5^{1/3}(\cos 17.71° + i\sin 17.71°)\\ &= 1.710(0.9526+0.3042i)\\ &= \underline{1.629+0.5202i}\end{aligned}$$

The second root is, using $k=1$,

$$\begin{aligned}(3+4i)^{1/3} &= 5^{1/3}(\cos 137.7° + i\sin 137.7°)\\ &= 1.710(-0.7397+0.6729i)\\ &= \underline{-1.265+1.151i}\end{aligned}$$

The third and final root is, using $k=2$,

Our angles are expressed in degrees, not radians, in this example.

$$\begin{aligned}(3+4i)^{1/3} &= 5^{1/3}(\cos 257.7° + i\sin 257.7°)\\ &= 1.710(-0.2129-0.9771i)\\ &= \underline{-0.3641-1.671i}\end{aligned}$$

If we chose $k=3$, we would simply repeat the first root. The various roots are positioned in the diagram in Fig. 7.3.

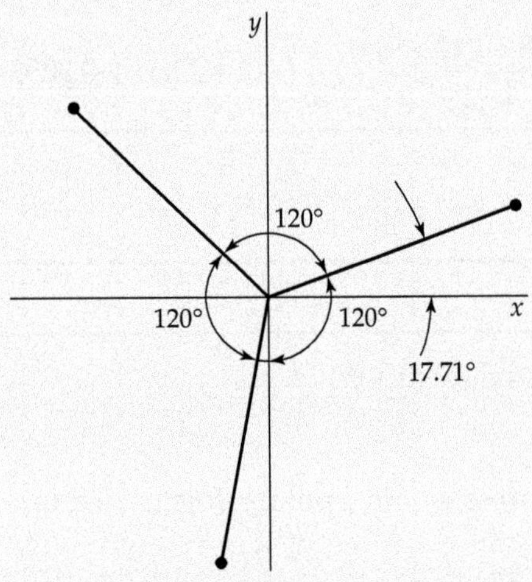

FIGURE 7.3

Example 7.3

What is the curve represented by $|z/(z-1)| = 2$?

Solution
First, using Eq. 7.2.23, we can write

$$\left|\frac{z}{z-1}\right| = \frac{|z|}{|z-1|}$$

Then, recognizing that the magnitude squared of a complex number is the real part squared plus the imaginary part squared, we have

$$\frac{|z|^2}{|z-1|^2} = \frac{x^2+y^2}{(x-1)^2+y^2}$$

where we have used $z - 1 = x - 1 + iy$. The desired equation is then

$$\frac{x^2+y^2}{(x-1)^2+y^2} = 4$$

or

$$x^2 + y^2 = 4(x-1)^2 + 4y^2$$

This can be written as

$$\left(x - \frac{4}{3}\right)^2 + y^2 = \frac{4}{9}$$

which is the equation of a circle of radius $\frac{2}{3}$ with center at $\left(\frac{4}{3}, 0\right)$.

7.3 ELEMENTARY FUNCTIONS

Some of the most common elementary functions will now be considered. Some interesting properties are observed when the functions are written in terms of the complex variable z. We shall require that the functions reduce to the forms encountered in our studies of real variables if the imaginary part is set equal to zero. A *polynomial function* is

$$f(z) = a_n z^n + a_{n-1} z^{n-1} + \cdots + a_1 z + a_0 \tag{7.3.1}$$

where the coefficients $a_n, \ldots, a_1, a_0$ may be complex and n is a positive integer. The equation above can be extended to the infinite series, a *power series*,

$$f(z) = \sum_{n=0}^{\infty} a_n z^n \tag{7.3.2}$$

The ratio test is used, as in real variables, to determine if the series converges. The radius of convergence is found from

$$\frac{1}{R} = \lim_{x\to\infty}\left|\frac{a_{n+1}}{a_n}\right| \tag{7.3.3}$$

Then, for the series to converge, we require that

$$|z| < R \tag{7.3.4}$$

We will define e^z, $\sin z$, and $\cos z$ in terms of convergent power series to be

$$e^z = 1 + z + \frac{z^2}{2!} + \cdots \tag{7.3.5}$$

$$\sin z = z - \frac{z^3}{3!} + \frac{z^5}{5!} - \cdots \tag{7.3.6}$$

$$\cos z = 1 - \frac{z^2}{2!} + \frac{z^4}{4!} - \cdots \tag{7.3.7}$$

so that the series expressions reduce to those of real variables when the imaginary part of z vanishes. The expressions above can be manipulated to give

$$\sin z = \frac{e^{iz} - e^{-iz}}{2i}, \qquad \cos z = \frac{e^{iz} + e^{-iz}}{2} \tag{7.3.8}$$

Also, we note that, letting $z = i\theta$, from Eq. 7.3.5 we have

$$\begin{aligned} e^{i\theta} &= 1 + i\theta - \frac{\theta^2}{2!} - i\frac{\theta^3}{3!} + \frac{\theta^4}{4!} + \frac{i\theta^5}{5!} + \cdots \\ &= \left(1 - \frac{\theta^2}{2!} + \frac{\theta^4}{4!} - \cdots\right) + i\left(\theta - \frac{\theta^3}{3!} + \frac{\theta^5}{5!} - \cdots\right) \\ &= \cos\theta + i\sin\theta \end{aligned} \tag{7.3.9}$$

This leads to a very useful expression for the complex variable z. In polar form, $z = r(\cos\theta + i\sin\theta)$, so Eq. 7.3.9 allows us to write

$$z = re^{i\theta} \tag{7.3.10}$$

This form is quite useful in obtaining powers and roots of z and in various other operations involving complex numbers.

The *hyperbolic sine* and *cosine* are defined as

$$\sinh z = \frac{e^z - e^{-z}}{2}, \qquad \cosh z = \frac{e^z + e^{-z}}{2} \tag{7.3.11}$$

With the use of Eqs. 7.3.8 we see that

$$\begin{aligned} \sinh iz &= i\sin z, & \sin iz &= i\sinh z \\ \cosh iz &= \cos z, & \cos iz &= \cosh z \end{aligned} \tag{7.3.12}$$

We can then separate the real and imaginary parts from sin z and cos z, with the use of trigonometric identities, as follows:

$$\begin{aligned}\sin z &= \sin(x + iy)\\ &= \sin x \cos iy + \sin iy \cos x\\ &= \sin x \cosh y + i \sinh y \cos x\end{aligned} \tag{7.3.13}$$

$$\begin{aligned}\cos z &= \cos(x + iy)\\ &= \cos x \cos iy - \sin x \sin iy\\ &= \cos x \cosh y - i \sin x \sinh y\end{aligned} \tag{7.3.14}$$

The *natural logarithm* of a complex number, denoted by ln z, is also of interest. It is expressed as

$$w = \ln z \tag{7.3.15}$$

and must satisfy the equation

$$z = e^w \tag{7.3.16}$$

Express z in polar form and write Eq. 7.3.15 as

$$\begin{aligned}w = \ln z &= \ln(re^{i\theta})\\ &= \ln r + \ln e^{i\theta}\\ &= \ln r + i\theta\end{aligned} \tag{7.3.17}$$

This result is then substituted into Eq. 7.3.16 to obtain the relationship $z = re^{i\theta}$, which shows that Eq. 7.3.17 is an acceptable expression for ln z. From Eq. 7.3.17 we see that ln z is multivalued, since θ can take on an infinite number of values using multiples of 2π. If the imaginary part of z is zero so that $z = x$, we desire that ln z reduce to that of real variables. Then $\theta = 0$ only, and θ must be restricted from taking on values of multiples of 2π. Hence, we restrict θ to be in the interval $0 \le \theta < 2\pi$ and use only the principal value of ln z.

We are now in a position to consider the general power of a complex number, that is, $w = z^a$. Writing $z = e^{\ln z}$ (see Eq. 7.3.16), we express w as

$$w = e^{a \ln z} \tag{7.3.18}$$

The quantity ln z can then be expressed as a complex number using Eq. 7.3.17 and the result follows. An example will illustrate this procedure.

Finally, in our discussion of elementary functions, we shall include the inverse trigonometric functions and inverse hyperbolic functions. Let

$$w = \sin^{-1} z \tag{7.3.19}$$

Then, using Eq. 7.3.8,

$$z = \sin w = \frac{e^{iw} - e^{-iw}}{2i} \tag{7.3.20}$$

Rearranging and multiplying by e^{iw} gives

$$e^{2iw} - 2ize^{iw} - 1 = 0 \tag{7.3.21}$$

This quadratic equation (let $e^{iw} = \phi$, so that $\phi^2 - 2iz\,\phi - 1 = 0$) is solved to yield

$$e^{iw} = iz + (1 - z^2)^{1/2} \tag{7.3.22}$$

Note that the quantity $(1 - z^2)^{1/2}$ contains the usual two roots $\pm\sqrt{1 - x^2}$ when $z = x$. By taking the logarithm of both sides of Eq. 7.3.22 we find that

$$\begin{aligned} w &= \sin^{-1} z \\ &= -i \ln\left[iz + (1 - z^2)^{1/2}\right] \end{aligned} \tag{7.3.23}$$

This expression is double-valued because of the square root and is multivalued because of the logarithm. Two principal values result for each complex number z except for $z = 1$, in which case the square-root quantity is zero. In a similar manner we can find expressions for the other inverse functions. They are listed in the following:

$$\begin{aligned} \sin^{-1} z &= -i \ln\left[iz + (1 - z^2)^{1/2}\right] \\ \cos^{-1} z &= -i \ln\left[z + (z^2 - 1)^{1/2}\right] \\ \tan^{-1} z &= \frac{i}{2} \ln \frac{1 - iz}{1 + iz} \end{aligned} \tag{7.3.24}$$

$$\begin{aligned} \sinh^{-1} z &= \ln\left[z + (1 + z^2)^{1/2}\right] \\ \cosh^{-1} z &= \ln\left[z + (z^2 - 1)^{1/2}\right] \\ \tanh^{-1} z &= \frac{1}{2} \ln \frac{1 + z}{1 - z} \end{aligned} \tag{7.3.25}$$

An important observation from the expressions of this section is that the definition of the derivative

$$\frac{df}{dz} = \lim_{\Delta z \to 0} \frac{f(z + \Delta z) - f(z)}{\Delta z} \tag{7.3.26}$$

leads to the same formulas as those derived for real variables, e.g., $d/dz(e^z) = e^z$.

Example 7.4

Find the three roots of $(5 + 12i)^{1/3}$ using the exponential form of the complex number.

Solution

The complex number is written as

$$5 + 12i = 13e^{1.176i}$$

where the angle θ in radians is $\tan^{-1} \dfrac{12}{5} = 1.176$. The first root is

$$\begin{aligned} (5 + 12i)^{1/3} &= 13^{1/3} e^{(1.176/3)i} = 13^{1/3}\left(\cos\frac{1.176}{3} + i \sin\frac{1.176}{3}\right) \\ &= 2.351(0.9241 + 0.3820i) \\ &= \underline{2.173 + 0.8981i} \end{aligned}$$

The third distinct root is found by adding 4π to 1.177 and then evaluating

$$13^{1/3}\left(\cos\frac{1.176+4\pi}{3}+i\sin\frac{1.176+4\pi}{3}\right)$$

Example 7.5

Using $z = 3 - 4i$, find the value or principal value of (a) e^{iz}, (b) e^{-iz}, (c) $\sin z$, and (d) $\ln z$.

Solution

a) $$\begin{aligned} e^{i(3-4i)} &= e^{4+3i} = e^4 e^{3i} \\ &= 54.60(\cos 3 + i\sin 3) \\ &= 54.60(-0.990 + 0.1411i) \\ &= -54.05 + 7.704\,i \end{aligned}$$

b) $$\begin{aligned} e^{-i(3-4i)} &= e^{-4-3i} = e^{-4}e^{-3i} \\ &= 0.01832[\cos(-3) + i\sin(-3)] \\ &= 0.01832[-0.990 - 0.1411i] \\ &= -0.01814 - 0.002585i \end{aligned}$$

c) $$\begin{aligned} \sin(3-4i) &= \frac{e^{i(3-4i)} - e^{-i(3-4i)}}{2i} \\ &= \frac{-54.05 + 7.704i - (-0.01814 - 0.002585i)}{2i} \\ &= 3.853 + 27.01i \end{aligned}$$

d) $$\begin{aligned} \ln(3-4i) &= \ln r + i\theta \\ &= \ln 5 + i\tan^{-1}\frac{-4}{3} \\ &= 1.609 + 5.356i \end{aligned}$$

where the angle θ is expressed in radians.

Example 7.6

Determine the principal value of the complex quantity $(2+i)^{1-i}$.

Solution

We write the complex quantity in the form

$$(2+i)^{1-i} = e^{(1-i)\ln(2+i)}$$

Carrying out the indicated operations, we have

$$\begin{aligned}(2+i)^{1-i} &= e^{(1-i)\ln(2+i)}\\ &= e^{(1-i)(\ln\sqrt{5}+0.4636i)} = e^{1.2683-0.3411i}\\ &= e^{1.2683}(\cos 0.3411 - i\sin 0.3411)\\ &= 3.555(0.9424 - 0.3345i)\\ &= 3.350 - 1.189i\end{aligned}$$

Example 7.7

What is the value of z so that $\sin z = 10$?

Solution

We write $\sin z$ in exponential form as

$$\sin z = \frac{e^{iz} - e^{-iz}}{2i} = 10$$

Writing $z = x + iy$ so that

$$e^{iz} = e^{i(x+iy)} = e^{-y+ix} = e^{-y}(\cos x + i\sin x)$$

we have

$$e^{-y}(\cos x + i\sin x) - e^{y}(\cos x - i\sin x) = 20i$$

Equating real and imaginary parts results in

$$\begin{aligned}\cos x(e^{-y} - e^{y}) &= 0\\ \sin x(e^{-y} - e^{y}) &= 20\end{aligned}$$

If we set the quantity in parentheses in the first equation equal to zero, the second equation cannot be satisfied. Hence, we set $\cos x = 0$ so that $x = \pi/2$. The second equation then gives

$$e^{-y} + e^{y} = 20$$

or

$$y = 2.993 \quad \text{or} \quad -2.993$$

The two principal values are then

$$z = \frac{\pi}{2} + 2.993i \quad \text{and} \quad z = \frac{\pi}{2} - 2.993i$$

7.4 ANALYTIC FUNCTIONS

The definition of the derivative of a function $f(z)$ is

$$f'(z) = \lim_{\Delta z \to 0} \frac{f(z+\Delta z) - f(z)}{\Delta z} \tag{7.4.1}$$

It is important to note that in the limiting process as $\Delta z \to 0$ there are an infinite number of paths that Δz can take. Some of these are sketched in Fig.7.4. For a derivative to exist we demand that $f'(z)$ be unique as $\Delta z \to 0$, *regardless of the path chosen.* In real variables this restriction on the derivative was not necessary since only one path was used, along the x axis only. Let us illustrate the importance of this demand with the function

$$f(z) = \bar{z} = x - iy \tag{7.4.2}$$

The quotient in the definition of the derivative using $\Delta z = \Delta x + i\,\Delta y$ is

$$\frac{f(z+\Delta z) - f(z)}{\Delta z} = \frac{[(x+\Delta x) - i(y+\Delta y)] - (x - iy)}{\Delta x + i\Delta y} \tag{7.4.3}$$

First, let $\Delta y \to 0$ and then let $\Delta x \to 0$. Then, the quotient is +1. Next, let $\Delta x \to 0$ and then let $\Delta y \to 0$. Now the quotient is –1. Obviously, we obtain two different values for each of the paths. Actually, there would be a different value for the quotient for each value of the slope of the line along which Δz approaches zero. The derivative is not unique; hence, we say that the derivative does not exist. We shall now derive the necessary conditions which allow us to identify whether or not a function possesses a derivative.

Let us assume now that the derivative $f'(z)$ does exist; it is unique. The real and imaginary parts of $f(z)$ are denoted by $u(x, y)$ and $v(x, y)$, respectively; that is,

$$f(z) = u(x,y) + iv(x,y) \tag{7.4.4}$$

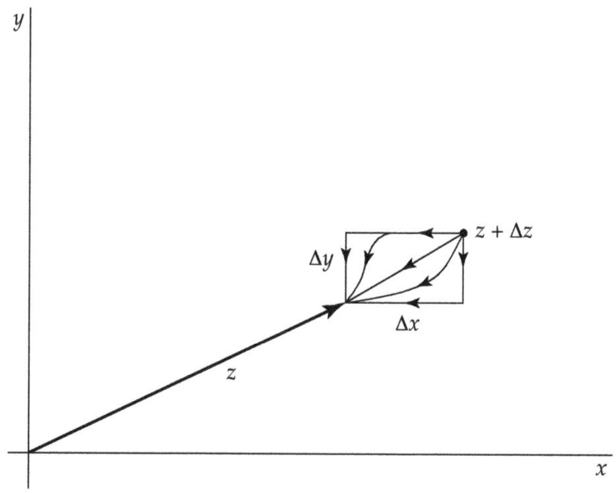

FIGURE 7.4 Various paths for Δz to approach zero.

First, let $\Delta y = 0$ so that $\Delta z \to 0$ parallel to the x axis. From Eq. 7.4.1 with $\Delta z = \Delta x$,

$$
\begin{aligned}
f'(z) &= \lim_{\Delta x \to 0} \frac{u(x+\Delta x, y) + iv(x+\Delta x, y) - u(x,y) - iv(x,y)}{\Delta x} \\
&= \lim_{\Delta x \to 0} \left[\frac{u(x+\Delta x, y) - u(x,y)}{\Delta x} + i \frac{v(x+\Delta x, y) - v(x,y)}{\Delta x} \right] \\
&= \frac{\partial u}{\partial x} + i \frac{\partial v}{\partial x}
\end{aligned} \tag{7.4.5}
$$

Next, let $\Delta x = 0$ so that $\Delta z \to 0$ parallel to the y axis. Then, using $\Delta z = i\,\Delta y$,

$$
\begin{aligned}
f'(z) &= \lim_{\Delta y \to 0} \frac{u(x, y+\Delta y) + iv(x, y+\Delta y) - u(x,y) - iv(x,y)}{i\,\Delta y} \\
&= \lim_{\Delta y \to 0} \left[\frac{u(x, y+\Delta y) - u(x,y)}{i\,\Delta y} + \frac{v(x, y+\Delta y) - v(x,y)}{\Delta y} \right] \\
&= -i \frac{\partial u}{\partial y} + \frac{\partial v}{\partial y}
\end{aligned} \tag{7.4.6}
$$

For the derivative to exist, it is necessary that these two expressions for $f'(z)$ be equal. Hence,

$$
\frac{\partial u}{\partial x} + i \frac{\partial v}{\partial x} = -i \frac{\partial u}{\partial y} + \frac{\partial v}{\partial y} \tag{7.4.7}
$$

Setting the real parts and the imaginary parts equal to each other, respectively, we find that

$$
\frac{\partial u}{\partial x} = \frac{\partial v}{\partial y}, \qquad \frac{\partial u}{\partial y} = -\frac{\partial v}{\partial x} \tag{7.4.8}
$$

the famous *Cauchy-Riemann equations.* We have derived these equations by considering only two possible paths along which $\Delta z \to 0$. It can be shown (we shall not do so in this text) that no additional relationships are necessary to ensure the existence of the derivative. If the Cauchy-Riemann equations are satisfied at a point $z = z_0$, then the derivative $f'(z_0)$ exists. If $f'(z)$ exists at $z = z_0$ and at every point in a neighborhood of z_0, then the function $f(z)$ is said to be *analytic* at z_0. If $f(z)$ is analytic at all points in a neighborhood of z_0 but is not analytic at z_0, then z_0 is called a *singular point.* An *analytic function* is a function that is analytic at all points in the region of interest; the region must not contain any singular points. Functions of this sort are encountered quite often when modeling physical phenomena.

From Eqs. 7.4.8 we can write*

$$
\frac{\partial^2 u}{\partial x^2} = \frac{\partial^2 v}{\partial x\, \partial y}, \qquad \frac{\partial^2 u}{\partial y^2} = -\frac{\partial^2 v}{\partial x\, \partial y} \tag{7.4.9}
$$

*We have interchanged the order of differentiation since the second partial derivatives are assumed to be continuous.

and

$$\frac{\partial^2 v}{\partial y^2} = \frac{\partial^2 u}{\partial x\,\partial y}, \quad \frac{\partial^2 v}{\partial x^2} = -\frac{\partial^2 u}{\partial x\,\partial y} \tag{7.4.10}$$

From Eqs. 7.4.9 we see that

$$\frac{\partial^2 u}{\partial x^2} + \frac{\partial^2 u}{\partial y^2} = 0 \tag{7.4.11}$$

and from Eqs. 7.4.10,

$$\frac{\partial^2 v}{\partial x^2} + \frac{\partial^2 v}{\partial y^2} = 0 \tag{7.4.12}$$

The real and imaginary parts of an analytic function satisfy Laplace's equation. Functions that satisfy Laplace's equation are called *harmonic functions.* Hence, $u(x, y)$ and $v(x, y)$ are harmonic functions. Two functions that satisfy Laplace's equation and the Cauchy–Riemann equations are known as *conjugate harmonic functions.* If one of the conjugate harmonic functions is known, the other can be found by using the Cauchy–Riemann equations. This will be illustrated by an example.

If $u(x, y)$ is harmonic, the conjugate harmonic function $v(x, y)$ is that which makes $f(z) = u(x, y) + iv(x, y)$ analytic.

Finally, let us show that constant u lines are normal to constant v lines if $u + iv$ is an analytic function. From the chain rule of calculus

$$du = \frac{\partial u}{\partial x}dx + \frac{\partial u}{\partial y}dy \tag{7.4.13}$$

Along a constant u line, $du = 0$. Hence,

$$\frac{dy}{dx} = -\frac{\partial u/\partial x}{\partial u/\partial y} \tag{7.4.14}$$

Along a constant v line, $dv = 0$, giving

$$\frac{dy}{dx} = -\frac{\partial v/\partial x}{\partial v/\partial y} \tag{7.4.15}$$

But, using the Cauchy–Riemann equations,

$$-\frac{\partial u/\partial x}{\partial u/\partial y} = \frac{\partial v/\partial y}{\partial v/\partial x} \tag{7.4.16}$$

The slope of the constant u line is the negative reciprocal of the slope of the constant v line. Hence, the lines are orthogonal. This property is useful in sketching constant u and v lines, as in fluid fields or electrical fields.

Example 7.8

Determine if and where the functions $z\bar{z}$ and z^2 are analytic ($z = x + iy$).

Solution

The function $z\bar{z}$ is written as

$$\begin{aligned} f(z) = z\bar{z} &= (x + iy)(x - iy) \\ &= x^2 + y^2 \end{aligned}$$

and is a real function only; its imaginary part is zero. That is, using $f = u + iv$,

$$u = x^2 + y^2, \qquad v = 0$$

The Cauchy-Riemann equations give

$$\frac{\partial u}{\partial x} = \frac{\partial v}{\partial y} \quad \text{or} \quad 2x = 0$$

$$\frac{\partial u}{\partial y} = -\frac{\partial v}{\partial x} \quad \text{or} \quad 2y = 0$$

Hence, we see that x and y must each be zero for the Cauchy–Riemann equations to be satisfied. This is true at the origin but not in the neighborhood (however small) of the origin. Thus, the derivative does not exist and the function $z\bar{z}$ is not analytic anywhere.

Now consider the function z^2. It is

$$\begin{aligned} f(z) = z^2 &= (x + iy)(x + iy) \\ &= x^2 - y^2 + i2xy \end{aligned}$$

The real and imaginary parts are

$$u = x^2 - y^2, \quad v = 2xy$$

The Cauchy–Reimann equations give

$$\frac{\partial u}{\partial x} = \frac{\partial v}{\partial y} \quad \text{or} \quad 2x = 2x$$

$$\frac{\partial u}{\partial y} = \frac{\partial v}{\partial x} \quad \text{or} \quad -2y = -2y$$

We see that these equations are satisfied at all points in the xy plane. Hence, the function z^2 is analytic everywhere.

Example 7.9

The real function $u(x, y) = Ax + By$ obviously statisfies Laplace's equation. Find its conjugate harmonic function, and write the function $f(z)$.

Solution

The conjugate harmonic function, denoted $v(x, y)$, is related to $u(x, y)$ by the Cauchy–Riemann equations. Thus,

$$\frac{\partial u}{\partial x} = \frac{\partial v}{\partial y} \quad \text{or} \quad \frac{\partial v}{\partial y} = A$$

The solution for v is

$$v = Ay + g(x)$$

where $g(x)$ is an unknown function to be determined. The relationship above must satisfy the other Cauchy-Riemann equation,

$$\frac{\partial v}{\partial x} = -\frac{\partial u}{\partial y} \quad \text{or} \quad \frac{dg}{dx} = -B$$

Hence,

The conjugate harmonic function has an arbitrary constant (but is otherwise unique).

$$g(x) = -Bx + C$$

where C is a constant of integration, to be determined by an imposed condition. Finally, the conjugate harmonic function $v(x, y)$ is

$$v(x, y) = Ay - Bx + C$$

The function $f(z)$ is then

$$\begin{aligned} f(z) &= Ax + By + i(Ay - Bx + C) \\ &= A(x + iy) - Bi(x + iy) + iC \\ &= (A - iB)z + iC \\ &= K_1 z + K_2 \end{aligned}$$

where K_1 and K_2 are complex constants.

7.5 COMPLEX INTEGRATION

There are a number of applications in which line integrals are of interest. The real integral between two limits a and b can be thought of as a line integral. Occasionally, there exists a singular point between the limits of integration of a real integral; the line integral in the complex plane is especially useful is such situations. There are also some proofs of basic properties of complex functions that can be accomplished with complex line integrals.

7.5.1 Green's Theorem

There is an important relationship that allows us to transform a line integral into an area integral for lines and areas in the xy plane. It is often referred to as *Green's theorem,* which states that

$$\oint_C u\,dx - v\,dy = -\iint_R \left(\frac{\partial v}{\partial x} + \frac{\partial u}{\partial y} \right) dx\,dy \tag{7.5.1}$$

where C is a closed curve surrounding the region R. The curve C is traversed with the region R always to the left.

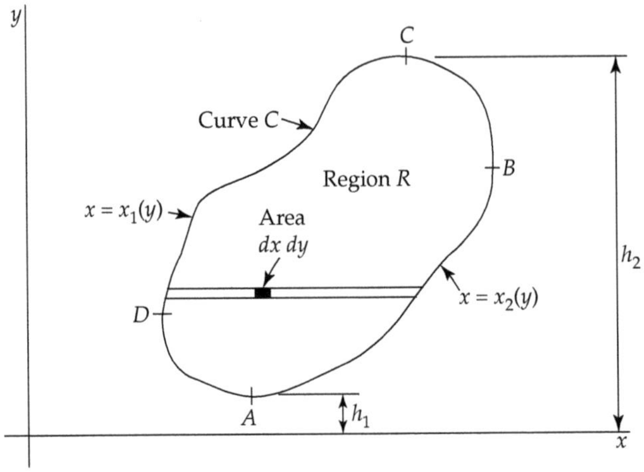

FIGURE 7.5 Curve C surrounding region R in Green's theorem.

To prove Green's theorem, consider the curve C surrounding the region R in Fig. 7.5. Let us investigate the first part of the area integral in Green's theorem. It can be written as

$$\iint_R \frac{\partial v}{\partial x} dx\, dy = \int_{h_1}^{h_2} \int_{x_1(y)}^{x_2(y)} \frac{\partial v}{\partial x} dx\, dy$$

$$= \int_{h_1}^{h_2} [v(x_2, y) - v(x_1, y)] dy$$

$$= \int_{h_1}^{h_2} v(x_2, y) dy + \int_{h_2}^{h_1} v(x_1, y) dy \qquad (7.5.2)$$

The first integral on the right-hand side is the line integral of $v(x, y)$ taken along the path ABC from A to C and the second integral is the line integral of $v(x, y)$ taken along the path ADC from C to A. Note that the region R is on the left. Hence, we can write

$$\iint_R \frac{\partial v}{\partial x} dx\, dy = \oint_C v(x, y) dy \qquad (7.5.3)$$

Similarly, we can show that

$$\iint_R \frac{\partial u}{\partial y} dx\, dy = -\oint_C u(x, y) dx \qquad (7.5.4)$$

and Green's theorem is proved. It should be noted that Green's theorem may be applied to a multiply connected region by appropriately cutting the region, as shown in Fig. 7.6. This makes a simply connected region* from the original multiply connected region. The contribution from the cuts is zero, since each cut is traversed twice, in opposite directions, thereby making no net contribution to the line integral.

*A simply connected region is one in which any closed curve contained in the region can be shrunk to zero without passing through points not in the region. A circular ring (like a washer) is not simply connected.

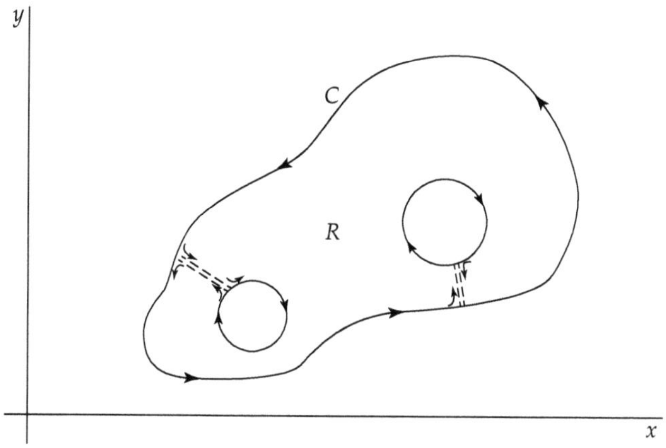

FIGURE 7.6 Multiply connected region.

Example 7.10

Verify Green's theorem by integrating the quantities $u = x + y$ and $v = 2y$ around the unit square shown in Fig. 7.7.

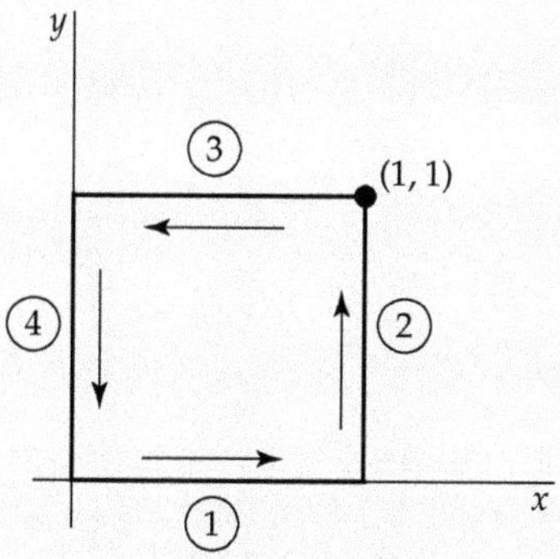

FIGURE 7.7

Solution

Let us integrate around the closed curve C formed by the four sides of the square. We have

$$\begin{aligned}\oint u\,dx - v\,dy &= \int_{①} u\,dx - v\,dy + \int_{②} u\,dx - v\,dy + \int_{③} u\,dx - v\,dy \\ &\quad + \int_{④} u\,dx - v\,dy \\ &= \int_0^1 x\,dx + \int_0^1 -2y\,dy + \int_1^0 (x+1)\,dx + \int_0^1 -2y\,dy\end{aligned}$$

where along 1, $dy = 0$ and $y = 0$; along 2, $dx = 0$; along 3, $dy = 0$ and $y = 1$; and along 4, $dx = 0$. The equation above is integrated to give

$$\oint u\,dx - v\,dy = \frac{1}{2} - 1 - \left(\frac{1}{2} + 1\right) + 1 = -1$$

Now, using Green's theorem, let us evaluate the double integral

$$-\iint\left(\frac{\partial v}{\partial x}+\frac{\partial u}{\partial y}\right)dx\,dy$$

Using $\partial v/\partial x = 0$ and $\partial u/\partial y = 1$, there results

$$-\iint\left(\frac{\partial v}{\partial x}+\frac{\partial u}{\partial y}\right)dx\,dy = -\iint(1)\,dx\,dy = -\text{area} = -1$$

For the functions $u(x, y)$ and $v(x, y)$ of this example we have verified Green's theorem.

7.5.2 Cauchy's Integral Theorem

Let us now investigate the line integral, $\oint_C f(z)dz$ where $f(z)$ is an analytic function within the simply connected region R enclosed by the curve C. We can write

$$\begin{aligned}\oint_C f(z)dz &= \oint_C (u+iv)(dx+i\,dy)\\ &= \oint_C [(u\,dx - v\,dy) + i(v\,dx + u\,dy)]\\ &= \oint_C [(u\,dx - v\,dy) + i\oint_C (v\,dx + u\,dy)]\end{aligned} \qquad (7.5.5)$$

Green's theorem allows us to transform the above to

$$\oint_C f(z)\,dz = -\iint_R\left(\frac{\partial v}{\partial x}+\frac{\partial u}{\partial y}\right)dx\,dy - i\iint_R\left(-\frac{\partial u}{\partial x}+\frac{\partial v}{\partial y}\right)dx\,dy \qquad (7.5.6)$$

Using the Cauchy–Riemann equations (7.4.8) we arrive at *Cauchy's integral theorem,*

$$\oint_C f(z)dz = 0 \qquad (7.5.7)$$

in which $f(z)$ must be an analytic function, and C must be a closed curve surrounding a simply connected region.

If we divide the closed curve C into two parts, as shown in Fig. 7.8, Cauchy's integral theorem can be written as

$$\begin{aligned}\oint_C f(z)dz &= \underset{\text{along } C_1}{\int_a^b} f(z)dz + \underset{\text{along } C_2}{\int_b^a} f(z)dz\\ &= \underset{\text{along } C_1}{\int_a^b} f(z)dz - \underset{\text{along } C_2}{\int_a^b} f(z)dz = 0\end{aligned} \qquad (7.5.8)$$

where the negative sign appears since we have exchanged the limits on the last integral. Thus, we have

$$\underset{\text{along } C_1}{\int_a^b} f(z)dz = \underset{\text{along } C_2}{\int_a^b} f(z)dz \qquad (7.5.9)$$

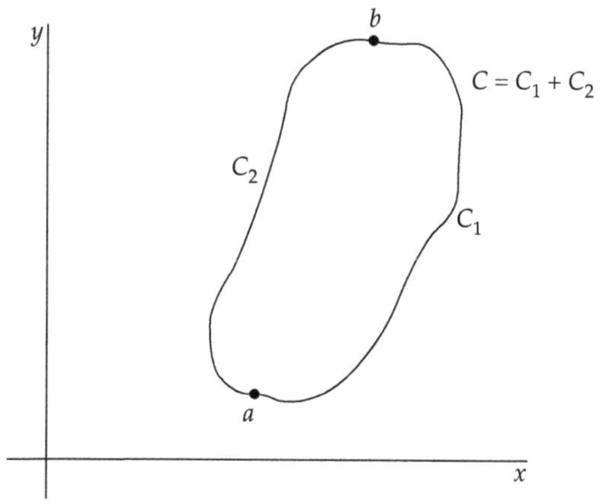

FIGURE 7.8 Two paths from *a* to *b* enclosing a simply connected region.

showing that the value of a line integral between two points is *independent of the path* provided that the integrand is an analytic function. It follows that $f(z)\,dz$ is an exact differential of a function $F(z)$; that is,

$$dF(z) = f(z)dz \tag{7.5.10}$$

or

$$F(z) = \int f(z)dz \tag{7.5.11}$$

Thus, if $f(z)$ is analytic, we can write

$$\int_a^b f(z)dz = F(b) - F(a) \tag{7.5.12}$$

We observe that contour C_1 and contour C_2 in Fig. 7.9 can be considered equivalent if

$$\oint_{C_1} f(z)dz = \oint_{C_2} f(z)dz \tag{7.5.13}$$

This is indeed the case provided that $f(z)$ is analytic on C_1 and C_2 and in the region between C_1 and C_2. This follows by cutting the region as shown and then applying Cauchy's integral theorem to the resulting single curve C. It is often useful to do this, since the integration process is usually simplified by using a circular path C_2.

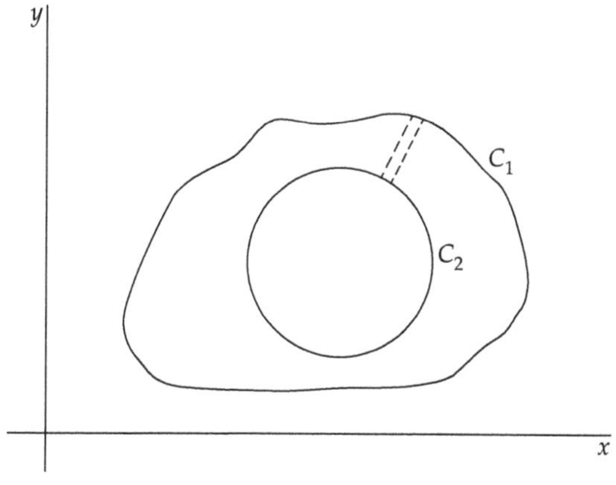

FIGURE 7.9 Equivalent contours C_1 and C_2.

Example 7.11

Using the analytic function $f(z) = z^2$, show that the integration $\int f(z)dz$ is independent of the path by integrating from the origin to the point (1, 1) along a) the straight line connecting the two points, and b) the x axis to the point (1, 0) and vertically to the point (1, 1).

Solution

The function $f(z) = z^2$ is written as

$$\begin{aligned} f(z) &= z^2 = (x + iy)^2 \\ &= x^2 - y^2 + i2xy \end{aligned}$$

Multiply this by $dz = dx + i\,dy$, to obtain

$$\begin{aligned} f(z)dz &= (x^2 - y^2 + i2xy)(dx + i\,dy) \\ &= (x^2 - y^2)dx - 2xy\,dy + i[2xy\,dx + (x^2 - y^2)dy] \end{aligned}$$

a) The integral along the line $x = y$ from origin to (1, 1) is then

$$\int_{0,0}^{1,1} \{\cancelto{0}{(x^2 - y^2)}dx - 2y^2dy + i[2x^2dx + \cancelto{0}{(x^2 - y^2)}dy]\} = -\frac{2}{3} + \frac{2}{3}i$$

b) Now, integrating along the x axis, we have, with $y = 0$ and $dy = 0$,

$$\int_{0,0}^{1,0} \{(x^2 - \cancelto{0}{y^2})dx - \cancelto{0}{2xy}\,dy + i[\cancelto{0}{2xy}\,dx + (x^2 - y^2)\cancelto{0}{dy}]\} = \frac{1}{3}$$

Along the vertical line $x = 1$ and $dx = 0$. We have

$$\begin{aligned} &\int_{0,0}^{1,1} \{(x^2 - y^2)\cancelto{0}{dx} - 2(1)y\,dy + i[2(1)y\,\cancelto{0}{dx} + (1^2 - y^2)dy]\} \\ &= -1 + i\left(1 - \frac{1}{3}\right) \\ &= -1 + \frac{2}{3}i \end{aligned}$$

Adding the above two numbers together gives the result $-\frac{2}{3} + \frac{2}{3}i$, the same result we obtained using the line $x = y$, thereby verifying Cauchy's integral theorem.

Example 7.12

Evaluate the integral $\oint dz/z$ around the unit circle with center at the origin.

Solution

The integral $\oint dz/z$ can be written as $\oint f(z)dz$, where $f(z) = 1/z$. This is not analytic at the origin, where $z = 0$; thus, Cauchy's integral theorem is not applicable. Let us use polar coordinates and write, with $r = 1$,

$$z = e^{i\theta} \quad \text{and} \quad dz = ie^{i\theta}\, d\theta$$

where we have noted that $r = 1$ for a unit circle with center at the origin. We then have

$$\begin{aligned}\oint \frac{dz}{z} &= \int_0^{2\pi} \frac{ie^{i\theta}\, d\theta}{e^{i\theta}} \\ &= \int_0^{2\pi} i\, d\theta \\ &= \underline{2\pi i}\end{aligned}$$

This is an important integration technique and will be used quite often in the remainder of this chapter.

Example 7.13

Evaluate the integral $\oint dz/z^2$ around the unit circle with center at the origin.

Solution

Since the function $f(z) = 1/z^2$ is not analytic at the origin, we cannot use Cauchy's integral theorem. Hence, we write, letting $r = 1$,

$$z = e^{i\theta} \quad \text{and} \quad dz = ie^{i\theta}\, d\theta$$

We then have

$$\begin{aligned}\oint \frac{dz}{z^2} &= \int_0^{2\pi} \frac{ie^{i\theta} d\theta}{e^{2i\theta}} \\ &= \int_0^{2\pi} ie^{-i\theta}\, d\theta = -e^{i\theta}\Big|_0^{2\pi} \\ &= -(1-1) \\ &= 0\end{aligned}$$

Note that even though the integrand is not analytic, the integral is still zero. If the integrand is analytic, the integral $\oint f(z)dz$ must be zero; if the integrand is not analytic, as in this example, then we may perform the integration to evaluate the integral. It may or may not be zero.

Example 7.14

Evaluate the integral $\oint f(z)dz$ around the circle of radius 2 with center at the origin, as sketched in Fig. 7.10, if $f(z) = 1/(z-1)$.

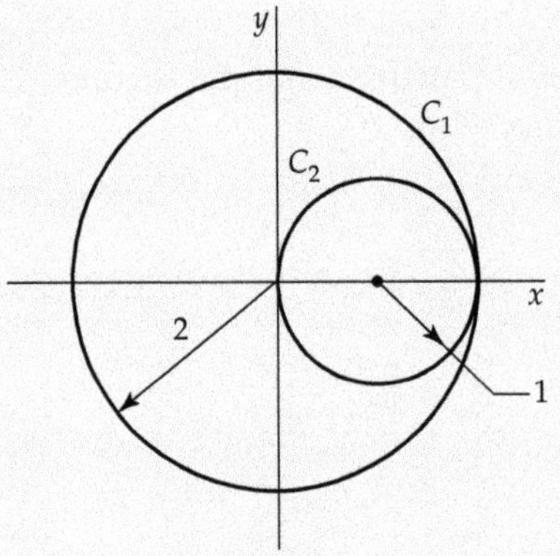

FIGURE 7.10

Solution

The given function $f(z)$ is not analytic at $z = 1$, a point inside the integration curve. Thus, Cauchy's integral theorem cannot be used. It would be relatively difficult to proceed directly with the integration. Instead, we use the suggestion associated with Fig. 7.9 and place a unit circle around the singular point, as shown. Since the function $f(z)$ is analytic in the region between these two circles, we employ Eq. 7.5.13 and evaluate the integral around the unit circle with center at $z = 1$. For this circle we let

$$z - 1 = e^{i\theta} \quad \text{and} \quad dz = ie^{i\theta}\, d\theta$$

where θ is now measured with respect to a radius emanating from $z = 1$. The integral becomes

$$\begin{aligned}\oint_{C_1} \frac{dz}{z-1} &= \oint_{C_2} \frac{dz}{z-1} \\ &= \int_0^{2\pi} \frac{ie^{i\theta}\, d\theta}{e^{i\theta}} \\ &= 2\pi i\end{aligned}$$

Observe that this integration is independent of the radius of the circle with center at $z = 1$; a circle of any radius would serve our purpose. Often we choose a circle of radius ϵ, a very small radius. Also, note that the integration around any curve enclosing the point $z = 1$, whether it is a circle or not, would give the value $2\pi i$.

7.5.3 Cauchy's Integral Formula

An important application of Section 7.5.2 is illustrated as we attempt to evaluate the integral $\oint_C f(z)/(z - z_0)\, dz$ where the point z_0 is included in the region R surrounded by the curve C (Fig. 7.11). The function $f(z)$ is analytic in R and the integrand $f(z)/(z - z_0)$

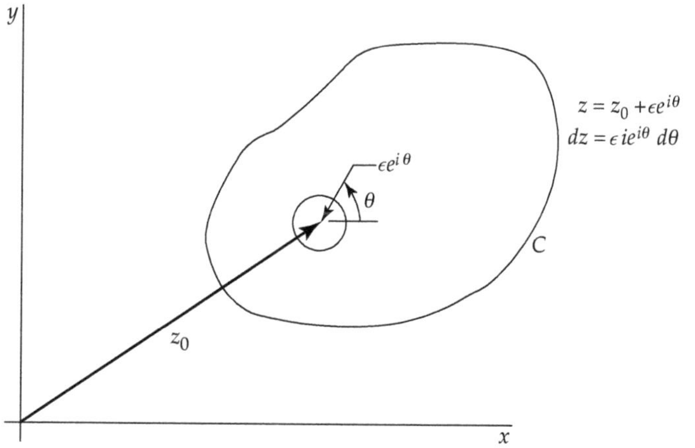

FIGURE 7.11 Small-circle equivalent to the curve *C*.

is analytic everywhere in R except at the point z_0, a singular point. We shall choose an equivalent curve around which to integrate. It is a small circle of radius ϵ. Hence,

$$\begin{aligned}\oint_C \frac{f(z)}{z - z_0} dz &= \oint_{\text{circle}} \frac{f(z)}{z - z_0} dz \\ &= \int_0^{2\pi} \frac{f(z_0 + \epsilon e^{i\theta})}{\epsilon e^{i\theta}} \epsilon e^{i\theta} i\, d\theta \\ &= \int_0^{2\pi} f(z_0 + \epsilon e^{i\theta}) i\, d\theta \end{aligned} \tag{7.5.14}$$

Now, we let the radius of the circle become extremely small ($\epsilon \to 0$) so that $f(z_0 + \epsilon e^{i\theta}) \to f(z_0)$, a constant. Finally,

$$\begin{aligned}\oint_C \frac{f(z)}{z - z_0} dz &= f(z_0) \int_0^{2\pi} i\, d\theta \\ &= 2\pi i f(z_0)\end{aligned} \tag{7.5.15}$$

This is *Cauchy's integral formula*, usually written as

$$f(z_0) = \frac{1}{2\pi i} \oint_C \frac{f(z)}{z - z_0} dz \tag{7.5.16}$$

We can obtain an expression for the derivative of $f(z)$ at z_0 by using Cauchy's integral formula in the definition of a derivative as follows:

$$\begin{aligned}f'(z_0) &= \lim_{\Delta z_0 \to 0} \frac{f(z_0 + \Delta z_0) - f(z_0)}{\Delta z_0} \\ &= \lim_{\Delta z_0 \to 0} \frac{1}{\Delta z_0} \left[\frac{1}{2\pi i} \oint_C \frac{f(z) dz}{z - z_0 - \Delta z_0} - \frac{1}{2\pi i} \oint_C \frac{f(z)}{z - z_0} dz \right]\end{aligned}$$

$$= \lim_{\Delta z_0 \to 0} \frac{1}{\Delta z_0} \left[\frac{1}{2\pi i} \oint_C f(z) \left(\frac{1}{z - z_0 - \Delta z_0} - \frac{1}{z - z_0} \right) dz \right]$$

$$= \lim_{\Delta z_0 \to 0} \frac{1}{\Delta z_0} \left[\frac{\Delta z_0}{2\pi i} \oint_C \frac{f(z)dz}{(z - z_0 - \Delta z_0)(z - z_0)} \right]$$

$$= \frac{1}{2\pi i} \oint_C \frac{f(z)}{(z - z_0)^2} dz \tag{7.5.17}$$

In a like manner we can show that

$$f''(z_0) = \frac{2!}{2\pi i} \oint_C \frac{f(z)}{(z - z_0)^3} dz \tag{7.5.18}$$

or, in general,

$$f^{(n)}(z_0) = \frac{n!}{2\pi i} \oint_C \frac{f(z)}{(z - z_0)^{n+1}} dz \tag{7.5.19}$$

Thus, we have established the fact that analytic functions possess derivatives of all orders; also, all derivatives of analytic functions are analytic. This is quite different from our experience with real variables, where we have encountered functions that possess, perhaps, first and second derivatives at a particular point, but yet the third derivative is not defined.

Cauchy's integral formula (7.5.16) allows us to determine the value of an analytic function at any point z_0 interior to a simply connected region by integrating around a curve C surrounding the region. Only values of the function on the boundary are used. Thus, we note that if an analytic function is prescribed on the entire boundary of a simply connected region, the function and all its derivatives can be determined at all interior points. We can write Eq. 7.5.16 in the alternative form

$$f(z) = \frac{1}{2\pi i} \oint_C \frac{f(w)}{w - z} dw \tag{7.5.20}$$

where z is any interior point such as that shown in Fig. 7.12. The complex variable w is simply a dummy variable of integration that disappears in the integration process. Cauchy's integral formula is often used in this form.

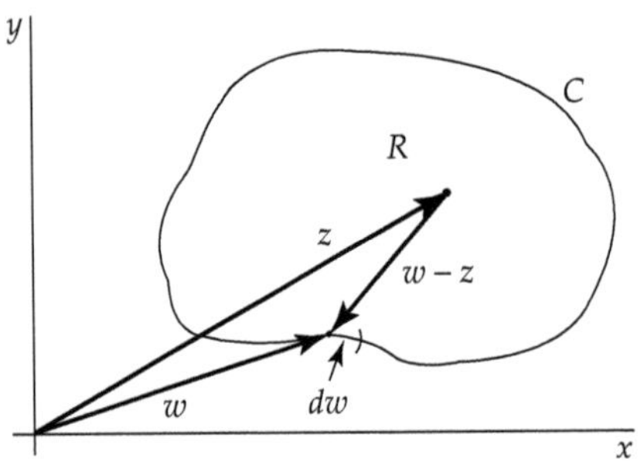

FIGURE 7.12 Integration variables for Cauchy's integral theorem.

Example 7.15

Find the value of the integral $\oint z^2/(z^2-1)dz$ around the unit circle with center at (a) $z=1$, (b) $z=-1$, and (c) $z=\frac{1}{2}$, drawn in Fig. 7.13 with the solid curve.

Note that $z^2/(z^2-1)$ has singular points at $z=1$ and $z=-1$.

Solution

Using Cauchy's integral formula, Eq. 7.5.16, we must make sure that $f(z)$ is analytic in the unit circle, and that z_0 lies within the circle.

a) With the center of the unit circle at $z=1$, we write

$$\oint \frac{z^2}{z^2-1}dz = \oint \frac{z^2/(z+1)}{z-1}dz$$

where we recognize that

$$f(z) = \frac{z^2}{z+1}$$

This function is analytic at $z=1$ and in the unit circle. Hence, at that point

$$f(1) = \frac{1}{2}$$

and we have

$$\oint \frac{z^2}{z^2-1}dz = 2\pi i\left(\frac{1}{2}\right) = \underline{\pi i}$$

b) With the center of the unit circle at $z=-1$, we write

$$\oint \frac{z^2}{z^2-1}dz = \oint \frac{z^2/(z-1)}{z+1}dz$$

where

$$f(z) = \frac{z^2}{z-1} \quad \text{and} \quad f(-1) = -\frac{1}{2}$$

There results

$$\oint \frac{z^2}{z^2-1}dz = 2\pi i\left(-\frac{1}{2}\right) = \underline{-\pi i}$$

c) Rather than integrating around the unit circle with center at $z=\frac{1}{2}$, we can integrate around any curve enclosing the point $z=1$, for example, that displayed in Fig. 7.13

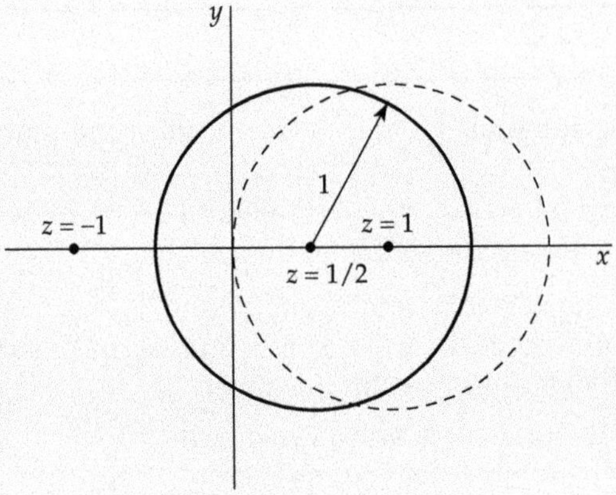

FIGURE 7.13

with the dashed line, just so the curve does not enclose the other singular point at $z = -1$. This is stated by Eq. 7.5.13. Obviously, the unit circle of part (a) is an acceptable alternative curve. Hence,

$$\oint \frac{z^2}{z^2 - 1} dz = \underline{\pi i}$$

Example 7.16

Evaluate the integrals

$$\text{a)} \oint \frac{z^2 + 1}{(z-1)^2} dz \quad \text{and} \quad \text{b)} \oint \frac{\cos z}{z^3} dz$$

around the circle $|z| = 2$.

Solution

a) Using Eq. 7.5.17, we can write the integral as

$$\oint \frac{z^2 + 1}{(z-1)^2} dz = 2\pi i f'(1)$$

where

$$f(z) = z^2 + 1 \quad \text{and} \quad f'(z) = 2z$$

Then

$$f'(1) = 2$$

The value of the integral is then determined to be

$$\oint \frac{z^2 + 1}{(z-1)^2} dz = 2\pi i(2)$$
$$= \underline{4\pi i}$$

b) For this integral, using Eq. 7.5.18 we have

$$\oint \frac{\cos z}{z^3} dz = \frac{2\pi i}{2!} f''(0)$$

where

$$f(z) = \cos z \quad \text{and} \quad f''(z) = -\cos z$$

At the origin

$$f''(0) = -1$$

The integral is then

$$\oint \frac{\cos z}{z^3} dz = \frac{2\pi i}{2!}(-1)$$
$$= -\pi i$$

7.6 SERIES

7.6.1 Taylor Series

The representation of analytic functions by an infinite series is basic in the application of complex variables. Let us show that if $f(z)$ is analytic at the point $z = a$, then $f(z)$ can be represented by a series of powers of $z - a$.

Expand the quantity $(w - z)^{-1}$ in the series

$$\frac{1}{w-z} = \frac{1}{(w-a)-(z-a)} = \frac{1}{w-a}\left[\frac{1}{1-\dfrac{z-a}{w-a}}\right]$$
$$= \frac{1}{w-a}\left[1 + \frac{z-a}{w-a} + \left(\frac{z-a}{w-a}\right)^2 + \cdots\right] \tag{7.6.1}$$

where we have used

$$\frac{1}{1-x} = 1 + x + x^2 + \cdots \tag{7.6.2}$$

Recall that Eq. 7.6.2 is called the geometric series.

We can now substitute the expansion of $1/(w - z)$ in Cauchy's integral formula (7.5.20) to obtain

$$f(z) = \frac{1}{2\pi i}\oint_C \frac{f(w)}{w-a}\left[1 + \frac{z-a}{w-a} + \frac{(z-a)^2}{(w-a)^2} + \cdots\right] dw$$
$$= \frac{1}{2\pi i}\oint_C \frac{f(w)}{w-a} dw + \frac{z-a}{2\pi i}\oint_C \frac{f(w)}{(w-a)^2} dw + \frac{(z-a)^2}{2\pi i}\oint_C \frac{f(w)}{(w-a)^3} dw + \cdots \tag{7.6.3}$$

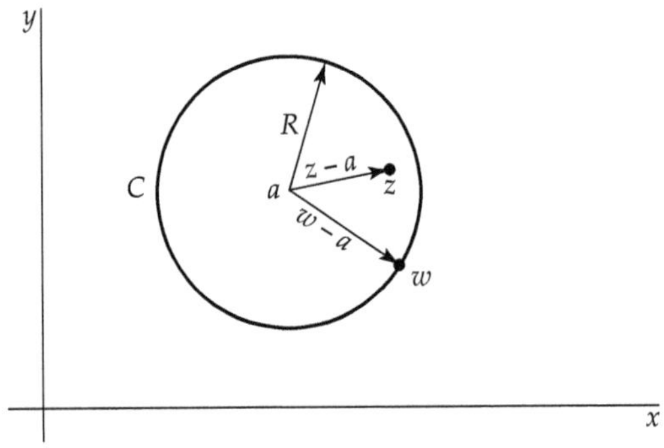

FIGURE 7.14 Circular region of convergence for the Taylor series.

Now, Cauchy's integral formula (7.5.20) and the expressions for the derivatives (7.5.17) through (7.5.19) can be substituted into the expression above to give us a series expansion for the analytic function $f(z)$,

$$f(z) = f(a) + f'(a)(z-a) + f''(a)\frac{(z-a)^2}{2!} + f'''(a)\frac{(z-a)^3}{3!} + \cdots \tag{7.6.4}$$

This is the famous *Taylor series*, which is used so often in applications. Note that the coefficients of powers of $(z - a)$ involve the integrals shown in Eq. 7.6.3. The curve C surrounds a simply connected region in which there are no singularities. The curve C is chosen to be a circle with radius $R = |w - a|$, (see Fig. 7.14). The radius R must not be larger than the distance from $z = a$ to the nearest singular point. If it were, the circle would include a point at which $f(z)$ would not be analytic; this is not allowed in Cauchy's integral formula. The largest radius R for which $f(z)$ is analytic throughout its interior is called the *radius of convergence.*

The discussion above applies equally well to a circular region about the origin, $a = 0$. The resultant series expression

$$f(z) = f(0) + zf'(0) + \frac{z^2}{2!}f''(0) + \cdots \tag{7.6.5}$$

is sometimes called a *Maclaurin series.*

7.6.2 Laurent Series

There are many applications in which we wish to expand a function $f(z)$ in a series about a point $z = a$, which is a singular point. Consider the annulus shown in Fig. 7.15a. The function $f(z)$ which will be expanded in a series is analytic in the annular region; however, there may be singular points inside the smaller circle or outside the larger circle. The possibility of a singular point inside the smaller circle bars us from expanding in a Taylor series, since the function $f(z)$ must be analytic at all interior points. We can apply Cauchy's integral formula to the multiply connected region by cutting the region as shown in Fig. 7.15b, thereby forming a simply connected region bounded by the curve C'. Cauchy's integral formula is then

$$\begin{aligned} f(z) &= \frac{1}{2\pi i}\oint_{C'}\frac{f(w)}{w-z}dw \\ &= \frac{1}{2\pi i}\oint_{C_2}\frac{f(w)}{w-z}dw - \frac{1}{2\pi i}\oint_{C_1}\frac{f(w)}{w-z}dw \end{aligned} \tag{7.6.6}$$

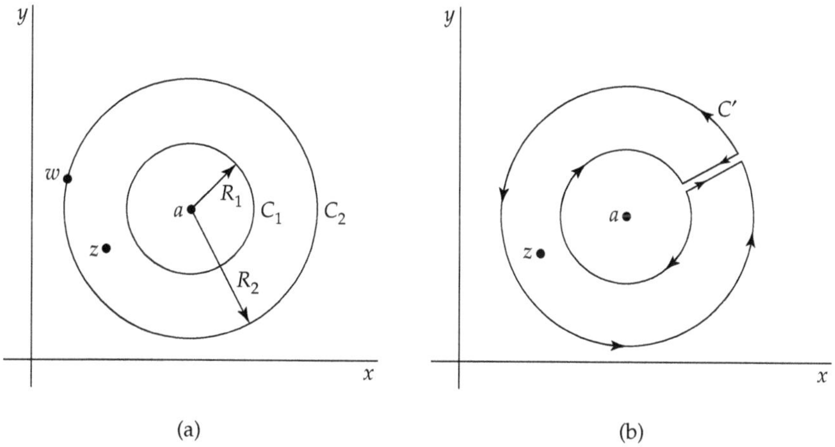

FIGURE 7.15 Annular region inside of which a singular point exists.

where C_1 and C_2 are both traversed in the counterclockwise direction. The negative sign results because the direction of integration was reversed on C_1. Now, let us express the quantity $(w - z)^{-1}$ in the integrand of Eq. 7.6.6 in a form that results in positive powers of $(z - a)$ in the C_2 integration, and that results in negative powers in the C_1 integration. If no singular points exist inside C_1, then the coefficients of the negative powers will all disappear and a Taylor series will reappear. Doing this, we have

$$f(z) = \frac{1}{2\pi i}\oint_{C_2} \frac{f(w)}{w-a}\left[\frac{1}{1-\dfrac{z-a}{w-a}}\right]dw + \frac{1}{2\pi i}\oint_{C_1} \frac{f(w)}{z-a}\left[\frac{1}{1-\dfrac{w-a}{z-a}}\right]dw \tag{7.6.7}$$

By using the expression (7.6.2) the equation above becomes

$$\begin{aligned} f(z) &= \frac{1}{2\pi i}\oint_{C_2} \frac{f(w)}{w-a}\left[1+\frac{z-a}{w-a}+\left(\frac{z-a}{w-a}\right)^2+\cdots\right]dw \\ &\quad + \frac{1}{2\pi i}\oint_{C_1} \frac{f(w)}{z-a}\left[1+\frac{w-a}{z-a}+\left(\frac{w-a}{z-a}\right)^2+\cdots\right]dw \end{aligned} \tag{7.6.8}$$

This is a series expression, often written as

$$f(z) = a_0 + a_1(z-a) + a_2(z-a)^2 + \cdots + b_1(z-a)^{-1} + b_2(z-a)^{-2} + \cdots \tag{7.6.9}$$

where

$$a_n = \frac{1}{2\pi i}\oint_{C_2} \frac{f(w)}{(w-a)^{n+1}}dw, \quad b_n = \frac{1}{2\pi i}\oint_{C_1} f(w)(w-a)^{n-1}\,dw \tag{7.6.10}$$

The series expression (7.6.9) is a *Laurent series*. The integral expression for a_n must not be related to the derivatives as in Eqs. 7.5.17 through 7.5.19, since $f(z)$ is not analytic throughout the entire interior of the circle C_2. Note, however, that if $f(z)$ is analytic in the circle C_1, the integrand in the integral for b_n is everywhere analytic, requiring the b_n's all be zero, a direct application of Cauchy's integral theorem, and a Taylor series would result.

The integral expressions (7.6.10) for the coefficients in the Laurent series are not normally used to find the coefficients. It is known that the series expansion is unique; hence, elementary techniques are usually used to find the Laurent series. This will be illustrated with an example. The region of convergence may be found, in most cases, by putting the desired $f(z)$ in the form $1/(1-z^*)$ so that $|z^*| < 1$ for convergence, a result of Eq. 7.3.4.

Example 7.17

Use the Taylor series representation of $f(z)$ and find a series expansion about the origin for (a) $f(z) = \sin z$, (b) $f(z) = e^z$, and (c) $f(z) = 1/(1-z)^m$.

Solution

a) To use Eq. 7.6.4, we must evaluate the derivatives at $z = 0$. They are

$$f'(0) = \cos 0 = 1, \quad f''(0) = -\sin 0 = 0,$$
$$f'''(0) = -\cos 0 = -1, \quad \text{etc.}$$

The Taylor series is then, with $f(z) = \sin z$,

$$\begin{aligned}\sin z &= \sin 0 + 1\cdot(z-0) + 0\cdot\frac{(z-0)^2}{2!} - 1\cdot\frac{(z-0)^3}{3!} + \cdots \\ &= z - \frac{z^3}{3!} + \frac{z^5}{5!} - \cdots\end{aligned}$$

This series is valid for all z since no singular point exists in the xy plane.

b) For the second function the derivatives are

$$f'(0) = e^0 = 1, \quad f''(0) = e^0 = 1, \quad f'''(0) = e^0 = 1, \cdots$$

The Taylor series for $f(z) = e^z$ is then

$$\begin{aligned}e^z &= e^0 + 1\cdot z + 1\cdot\frac{z^2}{2!} + 1\cdot\frac{z^3}{3!} + \cdots \\ &= 1 + z + \frac{z^2}{2!} + \frac{z^3}{3!} + \cdots\end{aligned}$$

This series is valid for all z.

c) We determine the derivatives to be

$$f'(z) = m(1-z)^{-m-1}, \quad f''(z) = m(m+1)(1-z)^{-m-2},$$
$$f'''(z) = m(m+1)(m+2)(1-z)^{-m-3}, \cdots$$

Substitute into a Taylor series to obtain

$$\begin{aligned}\frac{1}{(1-z)^m} &= \frac{1}{(1-0)^m} + m(1-0)^{-m-1}z + m(m+1)(1-0)^{-m-2}\frac{z^2}{2!} + \cdots \\ &= 1 + mz + m(m+1)\frac{z^2}{2!} + m(m+1)(m+2)\frac{z^3}{3!} + \cdots\end{aligned}$$

This series converges for $|z| < 1$ and does not converge for $|z| \geq 1$ since a singular point exists at $z = 1$. Using $m = 1$, an often used expression for $1/(1-z)$ results:

$$\frac{1}{1-z} = 1 + z + z^2 + z^3 + \cdots$$

Example 7.18

Find the Taylor series representation of $\ln(1+z)$ by noting that

$$\frac{d}{dz}\ln(1+z) = \frac{1}{1+z}$$

Solution

First, let us write the Taylor series expansion of $1/(1+z)$. It is, using the results of Example 7.17(c),

$$\frac{1}{1+z} = \frac{1}{1-(-z)} = 1 - z + z^2 - z^3 + \cdots$$

Now, we can perform the integration

$$\int d[\ln(1+z)] = \int \frac{1}{1+z}\, dz$$

using the series expansion of $1/(1+z)$ to obtain

$$\ln(1+z) = \int \frac{1}{1+z} dz = z - \frac{z^2}{2} + \frac{z^3}{3} - \frac{z^4}{4} + \cdots + C$$

The constant of integration $C = 0$, since when $z = 0$, $\ln(1) = 0$. The power-series expansion is finally

$$\ln(1+z) = z - \frac{z^2}{2} + \frac{z^3}{3} - \cdots$$

This series is valid for $|z| < 1$ since a singularity exists at $z = -1$.

Example 7.19

Determine the Taylor series expansion of

$$f(z) = \frac{1}{z^2 - 3z + 2}$$

about the origin.

Solution

First, represent the function $f(z)$ as partial fractions; that is,

$$\frac{1}{z^2-3z+2}=\frac{1}{(z-2)(z-1)}$$
$$=\frac{1}{z-2}-\frac{1}{z-1}$$

The series representations are then, using the results of Example 7.17(c),

$$\frac{1}{z-1}=-\frac{1}{1-z}=-(1+z+z^2+\cdots)$$
$$\frac{1}{z-2}=-\frac{1}{2}\left(\frac{1}{1-z/2}\right)=-\frac{1}{2}\left[1+\frac{z}{2}+\left(\frac{z}{2}\right)^2+\left(\frac{z}{2}\right)^3+\cdots\right]$$
$$=-\frac{1}{2}\left[1+\frac{z}{2}+\frac{z^2}{4}+\frac{z^3}{8}+\cdots\right]$$

Finally, the difference of the two series provides

$$\frac{1}{z^2-3z-2}=\frac{1}{2}+\frac{3}{4}z+\frac{7}{8}z^2+\frac{15}{16}z^3+\cdots$$

We could also have multiplied the two series together to obtain the same result.

Example 7.20

Find the Taylor series expansion of

$$f(z)=\frac{1}{z^2-9}$$

by expanding about the point $z = 1$.

Solution

We write the function $f(z)$ in partial fractions as

$$\frac{1}{z^2-9}=\frac{1}{(z-3)(z+3)}=\frac{1}{6}\left(\frac{1}{z-3}\right)-\frac{1}{6}\left(\frac{1}{z+3}\right)$$
$$=-\frac{1}{6}\left[\frac{1}{2-(z-1)}\right]-\frac{1}{6}\left[\frac{1}{4+(z-1)}\right]$$
$$=-\frac{1}{12}\left[\frac{1}{1-\frac{z-1}{2}}\right]-\frac{1}{24}\left[\frac{1}{1-\left(-\frac{z-1}{4}\right)}\right]$$

Now, we can expand in a Taylor series as

$$\frac{1}{z^2-9} = -\frac{1}{12}\left[1+\frac{z-1}{2}+\left(\frac{z-1}{2}\right)^2+\left(\frac{z-1}{2}\right)^3+\cdots\right]$$

$$-\frac{1}{24}\left[1-\frac{z-1}{4}+\left(\frac{z-1}{4}\right)^2-\left(\frac{z-1}{4}\right)^3+\cdots\right]$$

$$= -\frac{1}{8}-\frac{1}{32}(z-1)-\frac{3}{128}(z-1)^2-\frac{5}{512}(z-1)^3+\cdots$$

The nearest singularity is at the point $z = 3$; hence, the radius of convergence is 2; that is $|z - 1| < 2$. This is also obtained from the first ratio since $|(z - 1)/2| < 1$ or $|z - 1| < 2$. The second ratio is convergent if $|-(z - 1)/4| < 1$ or $|z - 1| < 4$; thus, it is the first ratio that limits the radius of convergence.

Example 7.21

What is the Laurent series expansion of

$$f(z) = \frac{1}{z^2-3z+2}$$

valid in each of the shaded regions shown in Fig. 7.16?

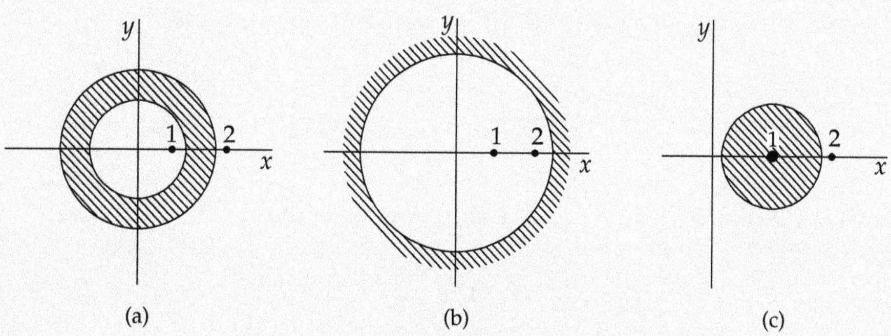

FIGURE 7.16

Solution

a) To obtain a Laurent series expansion in the shaded region of Fig. 7.16a, we expand about the origin. We express the ratio in partial fractions as

$$\frac{1}{z^2-3z+2} = \frac{1}{(z-2)(z-1)} = \frac{1}{z-2}-\frac{1}{z-1}$$

$$= -\frac{1}{2}\left(\frac{1}{1-z/2}\right)-\frac{1}{z}\left(\frac{1}{1-1/z}\right)$$

The first fraction has a singularity at $z/2 = 1$ and can be expanded in a Taylor series that converges if $(z/2) < 1$ or $|z| < 2$. The second fraction has a singularity at $1/z = 1$ and can be expanded in a Laurent series that converges if $|1/z| < 1$ or $|z| > 1$. The two fractions are expressed in the appropriate series as

$$-\frac{1}{2}\left(\frac{1}{1-z/2}\right) = -\frac{1}{2}\left[1+\frac{z}{2}+\left(\frac{z}{2}\right)^2+\left(\frac{z}{2}\right)^3+\cdots\right]$$

$$= -\frac{1}{2}-\frac{z}{4}-\frac{z^2}{8}-\frac{z^3}{16}-\cdots$$

$$-\frac{1}{z}\left(\frac{1}{1-1/z}\right) = -\frac{1}{z}\left[1+\frac{1}{z}+\left(\frac{1}{z}\right)^2+\left(\frac{1}{z}\right)^3+\cdots\right]$$

$$= -\frac{1}{z}-\frac{1}{z^2}-\frac{1}{z^3}-\frac{1}{z^4}-\cdots$$

where the first series is valid for $|z| < 2$ and the second series for $|z| > 1$. Adding the two expressions above yields the Laurent series

$$\frac{1}{z^2-3z+2} = \cdots-\frac{1}{z^3}-\frac{1}{z^2}-\frac{1}{z}-\frac{1}{2}-\frac{z}{4}-\frac{z^2}{8}-\frac{z^3}{16}-\cdots$$

which is valid in the region $1 < |z| < 2$.

b) In the region exterior to the circle $|z| = 2$ of Fig. 7.16b, we expand $1/(z-1)$ as

$$\frac{1}{z-1} = \frac{1}{z}\left(\frac{1}{1-1/z}\right) = \frac{1}{z}+\frac{1}{z^2}+\frac{1}{z^3}+\cdots$$

which is valid if $|1/z| < 1$ or $|z| > 1$. But, now we write

$$\frac{1}{z-2} = \frac{1}{z}\left(\frac{1}{1-2/z}\right) = \frac{1}{z}\left[1+\frac{2}{z}+\left(\frac{2}{z}\right)^2+\left(\frac{2}{z}\right)^3+\cdots\right]$$

$$= \frac{1}{z}+\frac{2}{z^2}+\frac{4}{z^3}+\frac{8}{z^4}+\cdots$$

which is valid if $|2/z| < 1$ or $|z| > 2$. The two series expansions above are thus valid for $|z| > 2$, and we have the Laurent series

$$\frac{1}{z^2-3z+2} = \frac{1}{z^2}+\frac{3}{z^3}+\frac{7}{z^4}+\frac{15}{z^5}+\cdots$$

valid in the region $|z| > 2$.

c) To obtain a series expansion in the region $0 < |z-1| < 1$ of Fig. 7.16c, we expand about the point $z = 1$ and obtain

$$\frac{1}{z^2-3z+2} = \frac{1}{z-1}\left(-\frac{1}{2-z}\right) = \frac{1}{z-1}\left[\frac{-1}{1-(z-1)}\right]$$

$$= \frac{-1}{z-1}[1+(z-1)+(z-1)^2+(z-1)^3+\cdots]$$

$$= -\frac{1}{z-1} - 1 + (z-1) + (z-1)^2 + \cdots$$

This Laurent series is valid if $0 < |z-1| < 1$.

7.7 RESIDUES

In this section we shall present a technique that is especially useful when evaluating certain types of real integrals [i.e., of the form $\int f(x)\,dx$] that are encountered when solving problems that arise in various physical situations. Suppose that a function $f(z)$ is singular at the point $z = a$ and is analytic at all other points within some circle with center at $z = a$. Then $f(z)$ can be expanded in the Laurent series (see Eq. 7.6.9)

$$f(z) = \frac{b_m}{(z-a)^m} + \frac{b_{m-1}}{(z-a)^{m-1}} + \cdots + \frac{b_2}{(z-a)^2} + \frac{b_1}{z-a} + a_0 + a_1(z-a) + \cdots \qquad (7.7.1)$$

The function $f(z)$ is said to have a *pole of order m* at $z = a$. If the Laurent series for $f(z)$ contains an infinite number of negative powers of $(z-a)$, then $z = a$ is an *essential singularity* of $f(z)$. From the expressions (7.6.10) for the coefficients we see that

$$b_1 = \frac{1}{2\pi i}\oint_{C_1} f(w)\,dw \qquad (7.7.2)$$

Hence, the integral of a function $f(z)$ about some connected curve surrounding a singular point is given by

$$\oint_{C_1} f(z)\,dz = 2\pi i b_1 \qquad (7.7.3)$$

where b_1 is the coefficient of the $(z-a)^{-1}$ term in the Laurent series expansion at the point $z = a$. The quantity b_1 is called the *residue* of $f(z)$ at $z = a$. Thus, to find the integral of a function about a singular point, we simply find the Laurent series expansion and use the relationship (7.7.3). An actual integration is not necessary. If more than one singularity exists within the closed curve C, we make it simply connected by cutting it as shown in Fig. 7.17. Then an application of Cauchy's integral theorem gives

$$\oint_C f(z)\,dz + \oint_{C_1} f(z)\,dz + \oint_{C_2} f(z)\,dz + \oint_{C_3} f(z)\,dz = 0 \qquad (7.7.4)$$

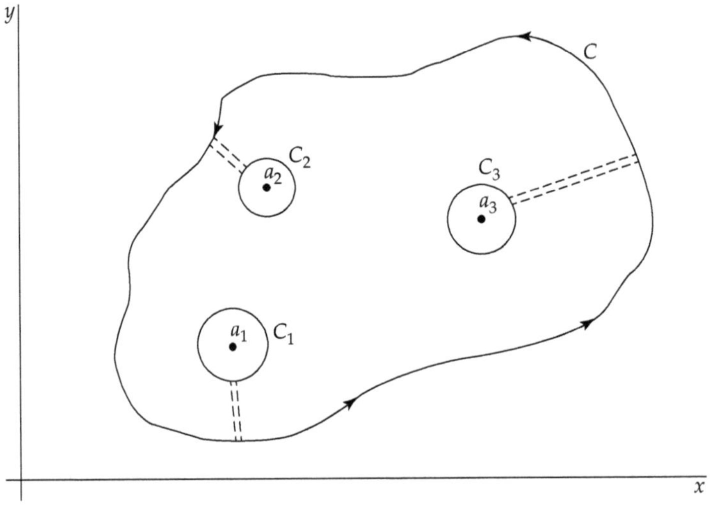

FIGURE 7.17 Integration about a curve that surrounds singular points.

since $f(z)$ is analytic at all points in the region outside the small circles and inside C. If we reverse the direction of integration on the integrals around the circles, there results

$$\oint_C f(z)dz = \oint_{C_1} f(z)dz + \oint_{C_2} f(z)dz + \oint_{C_3} f(z)dz \tag{7.7.5}$$

In terms of the residues at the points, we have Cauchy's *residue theorem*,

$$\oint_C f(z)dz = 2\pi i[(b_1)_{a_1} + (b_1)_{a_2} + (b_1)_{a_3}] \tag{7.7.6}$$

where the b_1's are coefficients of the $(z-a)^{-1}$ terms of the Laurent series expansions at each of the points.

Another technique, which may be less tedious than finding the Laurent series expansion, often used to find the residue at a particular singular point, is to multiply the Laurent series (7.7.1) by $(z-a)^m$, to obtain

$$\begin{aligned}(z-a)^m f(z) &= b_m + b_{m-1}(z-a) + \cdots \\ &+ b_1(z-a)^{m-1} + a_0(z-a)^m + a_1(z-a)^{m+1} + \cdots\end{aligned} \tag{7.7.7}$$

Now, if the series above is differentiated $(m-1)$ times and we let $z = a$, the residue results; that is,

$$b_1 = \frac{1}{(m-1)!}\left[\frac{d^{m-1}}{dz^{m-1}} g(z)\right]_{z=a} = \frac{1}{(m-1)!}\left\{\frac{d^{m-1}}{dz^{m-1}}[(z-a)^m f(z)]\right\}_{z=a} \tag{7.7.8}$$

Obviously, the order of the pole (or singularity) must be known before this method is useful. If $m = 1$, no differentiation is required and the residue results from $[(z-a)f(z)]_{z=a}$.

The residue theorem can be used to evaluate certain real integrals. Several examples will be presented here. Consider the real integral

$$I = \int_0^{2\pi} g(\cos\theta, \sin\theta)d\theta \tag{7.7.9}$$

where $g(\cos\theta, \sin\theta)$ is a rational* function of $\cos\theta$ and $\sin\theta$ with no singularities in the interval $2\pi > \theta \geq 0$. Let us make the substitution

$$e^{i\theta} = z \tag{7.7.10}$$

resulting in

$$\begin{aligned} \cos\theta &= \frac{1}{2}(e^{i\theta} + e^{-i\theta}) = \frac{1}{2}\left(z + \frac{1}{z}\right) \\ \sin\theta &= \frac{1}{2i}(e^{i\theta} - e^{-i\theta}) = \frac{1}{2i}\left(z - \frac{1}{z}\right) \\ d\theta &= \frac{dz}{ie^{i\theta}} = \frac{dz}{iz} \end{aligned} \tag{7.7.11}$$

As θ ranges from 0 to 2π, the complex variable z moves around the unit circle, as shown in Fig. 7.18, in the counterclockwise sense. The real integral now takes the form

$$I = \oint_C \frac{f(z)}{iz} dz \tag{7.7.12}$$

The residue theorem can be applied to the integral above once $f(z)$ is known. All residues inside the unit circle must be accounted for.

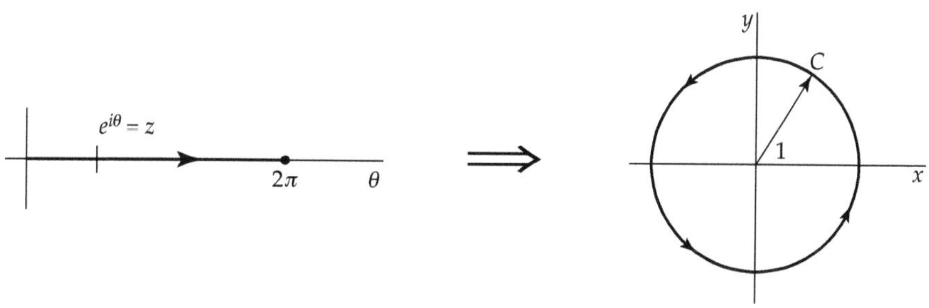

FIGURE 7.18 Paths of integration.

A second real integral that can be evaluated using the residue theorem is the integral

$$I = \int_{-\infty}^{\infty} f(x)dx \tag{7.7.13}$$

where $f(x)$ is the rational function

$$f(x) = \frac{p(x)}{q(x)} \tag{7.7.14}$$

and $q(x)$ has no zeros and is of degree at least 2 greater than $p(x)$. Consider the corresponding integral

$$I_1 = \oint_C f(z)dz \tag{7.7.15}$$

*Recall that a rational function can be expressed as the ratio of two polynomials.

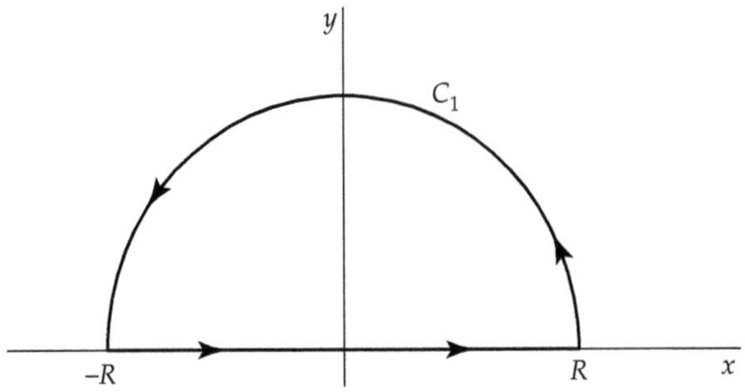

FIGURE 7.19 Path of integration.

where C is the closed path shown in Fig. 7.19. If C_1 is the semicircular part of curve C, Eq. 7.7.15 can be written as

$$I_1 = \int_{C_1} f(z)dz + \int_{-R}^{R} f(x)dx = 2\pi i \sum_{n=1}^{N} (b_1)_n \tag{7.7.16}$$

where Cauchy's residue theorem has been used. N in this equation represents the number of singularities in the upper half-plane contained within the semicircle. Let us now show that

$$\int_{C_1} f(z)dz = 0 \tag{7.7.17}$$

as $R \to \infty$. Using Eq. 7.7.14 and the restriction that $q(z)$ is of degree at least 2 greater than $p(z)$, we have

$$|f(z)| = \left|\frac{p(z)}{q(z)}\right| = \frac{|p(z)|}{|q(z)|} \sim \frac{1}{R^2} \tag{7.7.18}$$

Then there results*

$$\left|\int_{C_1} f(z)dz\right| \le \left|f(z)\right|_{\max} \pi R \sim \frac{1}{R} \tag{7.7.19}$$

As the radius R of the semicircle approaches ∞, we see that

$$\int_{C_1} f(z)dz \to 0 \tag{7.7.20}$$

---

*The integral is approximated by $\sum_{n=1}^{N} f(z_n)\Delta z_n$, where $\sum_{n=1}^{N} \Delta z_n$ represents the length of the line segments, which together approximate the length of the curve C_1; the number of line segments N should be large for a good approximation. Generalizing on Eq. 7.2.15, we observe that

$$\left|\sum_{n=1}^{N} f(z_n)\Delta z_n\right| \le |f(z_1)||\Delta z_1| + |f(z_2)|\,|\Delta z_2| + \cdots + |f(z_N)||\Delta z_N| \le M \sum_{n=1}^{N} \Delta z_n = ML$$

where M is the maximum value of $f(z)$ on the curve and L is the length of the curve. Thus, $\left|\int_{C_1} f(z)dz\right| \le \underset{\max}{f(z)|}\, L$.

Finally,

$$\int_{-\infty}^{\infty} f(x)dx = 2\pi i \sum_{n=1}^{N} (b_1)_n \tag{7.7.21}$$

where the b_1's include the residues of $f(z)$ at all singularities in the upper half-plane.

A third real integral that may be evaluated using the residue theorem is

$$I = \int_{-\infty}^{\infty} f(x)\sin mx\,dx \quad \text{or} \quad I = \int_{-\infty}^{\infty} f(x)\cos mx\,dx \tag{7.7.22}$$

Consider the complex integral

$$I_1 = \oint_C f(z)e^{imz}\,dz \tag{7.7.23}$$

where m is assumed to be a positive quantity and C is the curve of Fig. 7.19. Using the relationships

$$|e^{imz}| = |e^{imx}||e^{-my}| = e^{-my} \le 1 \tag{7.7.24}$$

if we limit ourselves to the upper half-plane so that $y \ge 0$; m is considered to be positive. We then have

$$|f(z)e^{imz}| = |f(z)||e^{imz}| \le |f(z)| \tag{7.7.25}$$

The remaining steps follow as in the previous example for $\int_{-\infty}^{\infty} f(z)dz$ using Fig. 7.19, resulting in

$$\int_{-\infty}^{\infty} f(x)e^{imx}dx = 2\pi i \sum_{n=1}^{N} (b_1)_n \tag{7.7.26}$$

where the b_1's include the residues of $[f(z)e^{imz}]$ at all singularities in the upper half-plane. The integral of interest in Eq. 7.7.22 would follow from either the real or imaginary part of Eq. 7.7.26.

Example 7.22

Find the value of the following integrals, where C is the circle $|z| = 2$.

a) $\oint_C \frac{\cos z}{z^3} dz$ b) $\oint_C \frac{dz}{z^2+1}$

c) $\oint_C \frac{z^2-2}{z(z-1)(z+4)} dz$ d) $\oint_C \frac{z}{(z-1)^3(z+3)} dz$

Solution

a) We expand the function $\cos z$ as

$$\cos z = 1 - \frac{z^2}{2!} + \frac{z^4}{4!} - \cdots$$

The integrand is then

$$\frac{\cos z}{z^3} = \frac{1}{z^3} - \frac{1}{2z} + \frac{z}{4!} + \cdots$$

The residue, the coefficient of the 1/z term, is

$$b_1 = -\frac{1}{2}$$

Thus, the value of the integral is

$$\oint_C \frac{\cos z}{z} dz = 2\pi i\left(-\frac{1}{2}\right) = \underline{-\pi i}$$

b) The integrand is factored as

$$\frac{1}{z^2+1} = \frac{1}{(z+i)(z-i)}$$

Two singularities exist inside the circle of interest. The residue at each singularity is found to be

$$(b_1)_{z=i} = (z-i)\frac{1}{(z+i)(z-i)}\bigg|_{z=i} = \frac{1}{2i}$$

$$(b_1)_{z=-i} = (z+i)\frac{1}{(z+i)(z-i)}\bigg|_{z=-i} = -\frac{1}{2i}$$

The value of the integral is

$$\oint_C \frac{dz}{z^2+1} = 2\pi i\left(\frac{1}{2i} - \frac{1}{2i}\right) = \underline{0}$$

Actually, this is the value of the integral around every curve that encloses the two poles.

c) There are two poles of order 1 in the region of interest, one at $z = 0$ and the other at $z = 1$. The residue at each of these poles is

$$(b_1)_{z=0} = z\frac{z^2-2}{z(z-1)(z+4)}\bigg|_{z=0} = \frac{1}{2}$$

$$(b_1)_{z=1} = (z-1)\frac{z^2-2}{z(z-1)(z+4)}\bigg|_{z=1} = -\frac{1}{5}$$

The integral is evaluated to be

$$\oint_C \frac{z^2-2}{z(z-1)(z+4)} dz = 2\pi i\left(\frac{1}{2} - \frac{1}{5}\right) = \underline{3\pi i/5}$$

d) There is one pole in the circle $|z| = 2$, a pole of order 3. The residue at that pole is (see Eq. 7.7. 8)

$$b_1 = \frac{1}{2!}\frac{d^2}{dz^2}\left[(z-1)^3 \frac{z}{(z-1)^3(z+3)}\right]_{z=1}$$

$$= \frac{1}{2}\frac{-6}{(z+3)^3}\bigg|_{z=1} = -\frac{3}{64}$$

The value of the integral is then

$$\oint_C \frac{z}{(z-1)^3(z+3)}dz = 2\pi i\left(-\frac{3}{64}\right) = \underline{-0.2945i}$$

Example 7.23

Evaluate the real integral $\int_0^{2\pi} d\theta/(2+\cos\theta)$.

Solution

Using Eqs. 7.7.11, the integral becomes

$$\int_0^{2\pi} \frac{d\theta}{2+\cos\theta} = \oint_C \frac{dz/iz}{2+\frac{1}{2}\left(z+\frac{1}{z}\right)} = -2i\oint \frac{dz}{z^2+4z+1}$$

where C is the unit circle. The roots of the denominator are found to be

$$z = -2 \pm \sqrt{3}$$

Hence, there is a zero at $z = -0.2679$ and at $z = -3.732$. The first of these zeros is located in the unit circle, so we must determine the residue at that zero; the second is outside the unit circle, so we forget it. To find the residue, write the integrand as partial fractions

$$\frac{1}{z^2+4z+1} = \frac{1}{(z+0.2679)(z+3.732)} = \frac{0.2887}{z+0.2679} + \frac{-0.2887}{z+3.732}$$

The residue at the singularity in the unit circle is then the coefficient of the $(z+0.2679)^{-1}$ term. It is 0.2887. Thus, the value of the integral is, using the residue theorem,

$$\int_0^{2\pi} \frac{d\theta}{2+\cos\theta} = -2i(2\pi i \times 0.2887) = \underline{3.628}$$

Example 7.24

Evaluate the real integral $\int_0^\infty dx/(1+x^2)$. Note that the lower limit is zero.

Solution
We consider the complex function $f(z) = 1/(1+z^2)$. Two poles exist at the points where

$$1 + z^2 = 0$$

They are

$$z_1 = i \quad \text{and} \quad z_2 = -i$$

The first of these roots lies in the upper half-plane. The residue there is

$$(b_1)_{z=i} = (z-i)\frac{1}{(z-i)(z+i)}\bigg|_{z=i} = \frac{1}{2i}$$

The value of the integral is then (refer to Eq. 7.7.21)

$$\int_{-\infty}^{\infty} \frac{dx}{1+x^2} = 2\pi i\left(\frac{1}{2i}\right) = \pi$$

Since the integrand is an even function,

$$\int_0^\infty \frac{dx}{1+x^2} = \frac{1}{2}\int_{-\infty}^{\infty} \frac{dx}{1+x^2}$$

Hence,

$$\int_0^\infty \frac{dx}{1+x^2} = \underline{\pi/2}$$

Example 7.25

Determine the value of the real integrals

$$\int_{-\infty}^{\infty} \frac{\cos x}{1+x^2}dx \quad \text{and} \quad \int_{-\infty}^{\infty} \frac{\sin x}{1+x^2}dx$$

Solution
To evaluate the given integrals we consider the integral

$$I_1 = \oint_C \frac{e^{iz}}{1+z^2}dz$$

Considering the curve C to surround the upper half-plane, we must locate all the singularities inside C. The quantity $(1 + z^2)$ has zeros as $z = \pm i$. One of these points is in the upper half-plane. The residue at $z = i$ is

$$(b_1)_{z=i} = (z-i)\frac{e^{iz}}{1+z^2}\bigg|_{z=i} = \frac{e^{-1}}{2i} = -0.1839i$$

The value of the integral is then

$$\int_{-\infty}^{\infty} \frac{e^{ix}}{1+x^2}dx = 2\pi i(-0.1839i) = 1.188$$

The integral can be rewritten as

$$\int_{-\infty}^{\infty} \frac{e^{ix}}{1+x^2}dx = \int_{-\infty}^{\infty} \frac{\cos x}{1+x^2}dx + i\int_{-\infty}^{\infty} \frac{\sin x}{1+x^2}dx$$

Equating real and imaginary parts, we have

$$\int_{-\infty}^{\infty} \frac{\cos x}{1+x^2}dx = \underline{1.118} \quad \text{and} \quad \int_{-\infty}^{\infty} \frac{\sin x}{1+x^2}dx = \underline{0}$$

The result with $\sin x$ should not be surprising, since $(1 + x^2)$ is symmetric and $\sin x$ is skew-symmetric. This would produce a result for negative x that would be negative to the result for positive x. The two numbers would cancel, yielding zero, as above.

PROBLEMS

7.1 Determine the angle θ, in degrees and radians, which is necessary to write each of the following complex numbers in polar form.

a) $4+3i$ b) $-4+3i$
c) $4-3i$ d) $-4-3i$

7.2 For the complex number $z = 3-4i$, find each of the following terms.

a) z^2 b) $z\bar{z}$
c) $z\sqrt{z}$ d) $\left|\dfrac{z+1}{z-1}\right|$
e) $(z+1)(z-i)$ f) $|z^2|$
g) $(z-i)^2/(z-1)^2$ h) z^4
i) $z^{1/2}$ j) $z^{1/3}$
k) $z^{2/5}$ l) $\left|\dfrac{z^{1/2}}{z^{1/3}}\right|$

7.3 Determine the roots of each of the following (express in the form $a+ib$).

a) $1^{1/5}$ b) $-16^{1/4}$
c) $i^{1/3}$ d) $9^{1/2}$

7.4 Show that each of the equations represents a circle.

a) $|z|=4$ b) $|z-2|=2$
c) $|(z-1)/(z+1)|=3$

7.5 Find the equation of each of the curves represented by the following.

a) $|(z-1)/(z+1)|=3$
b) $|(z+i)/(z-i)|=2$

7.6 Identify the region represented by $|z-2| \le x$.

7.7 Show that $\sin z = (e^{iz}-e^{-iz})/2i$ and $\cos z = (e^{iz}-e^{-iz})/2$ using Eqs. 7.3.5 through 7.3.7.

7.8 Express each of the following complex numbers in exponential form. (See Eq. 7.3.10.)

a) -2 b) $2i$
c) $-2i$ d) $3+4i$
e) $e-12i$ f) $-3-4i$
g) $-5+12i$ h) $0.213-2.15i$

7.9 Using $z = (\pi/2)-i$, show that Eq. 7.3.8 yields the same result as Eq. 7.3.13 for sin z.

7.10 Find the value of e^z for each of the following values of z. (Express in the form $a+bi$.)

a) $\dfrac{\pi}{2}i$ b) $2i$
c) $-\dfrac{\pi}{4}i$ d) $4\pi i$
e) $2+\pi i$ f) $-1-\dfrac{\pi}{4}i$

7.11 Find each of the following quantities using Eq. 7.3.10.

a) $1^{1/5}$ b) $(1-i)^{1/4}$
c) $(-1)^{1/3}$ d) $(2+i)^3$
e) $(3+4i)^4$ f) $\sqrt{2-i}$

7.12 For the value $z = \pi/2-(\pi/4)i$, find each of the following terms.

a) e^{iz} b) $\sin z$
c) $\cos z$ d) $\sinh z$
e) $\cosh z$ f) $|\sin z|$
g) $|\tan z|$

7.13 Find the principal value of the ln z for each of the following values for z.

a) i b) $3+4i$
c) $4-3i$ d) $-5+12i$
e) ei f) -4
g) e^t

7.14 Using the relationship that $z^a = e^{\ln z^a} = e^{a \ln x}$, find the principal value of each of the following values of a.

a) i^t b) $(3+4i)^{(1-i)}$
c) $(4-3i)^{(2+i)}$ d) $(1+i)^{(1+i)}$
e) $(-1-i)^{-1/2}$

7.15 Find the values or principal values of z for each of the following.

a) $\sin z = 2$ b) $\cos z = 4$
c) $e^z = -3$ d) $\sin z = -2i$
e) $\cos z = -2$

7.16 Show that each of the following is true.

a) $\cos^{-1} z = -i\ln[z+(z^2-1)^{1/2}]$

b) $\sinh^{-1} z = \ln[z+(1+z^2)^{1/2}]$

7.17 For $z = 2 - i$, evaluate each of the following.

a) $\sin^{-1} z$ **b)** $\tan^{-1} z$

c) $\cosh^{-1} z$

7.18 Compare the derivatives of each of the following functions $f(z)$ using Eq. 7.4.5 with those obtained using Eq. 7.4.6.

a) z^2 **b)** $\bar{z}$

c) $\dfrac{1}{z+2}$ **d)** $(z-1)^{1/2}$

e) $\ln(z-1)$ **f)** e^z

g) $\bar{z}z$

7.19 Express a complex function in polar coordinates as $f(z) = u(r, \theta) + iv(r, \theta)$ and:

a) Show that the Cauchy-Riemann equations can be expressed as

$$\frac{\partial u}{\partial r} = \frac{1}{r}\frac{\partial v}{\partial \theta}, \quad \frac{\partial v}{\partial r} = -\frac{1}{r}\frac{\partial u}{\partial \theta}$$

Hint: Sketch Δz using polar coordinates; then, note that for $\Delta\theta = 0$, $\Delta z = \Delta r(\cos\theta + i\sin\theta)$, and for $\Delta r = 0$, $\Delta z = r\,\Delta\theta(-\sin\theta + i\cos\theta)$.

b) Derive Laplace's equation for polar coordinates.

c) Find the conjugate harmonic-function associated with $u(r, \theta) = \ln r$. Sketch some constant u and v lines.

7.20 Show that each of the following functions is harmonic and find the conjugate harmonic function. Also, write the analytic function $f(z)$.

a) xy **b)** $x^2 - y^2$

c) $e^y \sin x$ **d)** $\ln(x^2 + y^2)$

7.21 Sketch the constant u lines and constant v lines for parts (a) and (b) in Problem 7.20. Show that the u lines and v lines intersect at right angles.

7.22 Integrate each of the following functions around the closed curve indicated and compare with the double integral of Eq. 7.5.1.

a) $u = y$, $v = x$ around the unit square as in Example 7.10.

b) $u = y$, $v = -x$ around the unit circle with center at the origin.

c) $u = x^2 - y^2$, $v = -2xy$ around the triangle with vertices at (0, 0) (2, 0), (2, 2).

d) $u = x + 2y$, $v = x^2$ around the triangle of part (c).

e) $u = y^2$, $v = -x^2$ around the unit circle with center at the origin.

7.23 To show that line integrals are, in general, dependent on the limits of integration, evaluate each of the following line integrals.

a) $\int_{0,0}^{2,2} (x - iy)\,dz$ along a straight line connecting the two points.

$\int_{0,0}^{2,2} (x - iy)dz$ along the x axis to the point (2, 0) and then vertical.

b) $\int_{0,0}^{0,2} (x^2 + y^2)\,dz$ along the y axis.

$\int_{0,0}^{0,2} (x^2 + y^2)\,dz$ along the x axis to the point (2, 0), then along a circular arc.

7.24 To verify that the line integral of an analytic function is independent of the path, evaluate each of the following.

a) $\int_{0,0}^{2,2} z\,dz$ along a straight line connecting the two points.

$\int_{0,0}^{2,2} z\,dz$ along the z axis to the point (2, 0) and then vertical.

b) $\int_{0,0}^{0,2} z^2\,dz$ along the x axis to the point (2, 0) and then along a circular arc.

$\int_{0,0}^{0,2} z^2\,dz$ along the y axis.

7.25 Evaluate $\oint f(z)\,dz$ for each of the following functions, where the path of integration is the unit circle with center at the origin.

a) e^z **b)** $\sin z$

c) $1/z^3$ **d)** $\dfrac{1}{z-2}$

e) $1/\bar{z}$ **f)** $\dfrac{1}{z^2 - 5z + 6}$

7.26 Evaluate $\oint f(z)\,dz$ by direct integration using each of the following functions, when the path of integration is the circle with radius 4, center at the origin.

a) $1/z$ b) $\dfrac{1}{z^2-5z+6}$

c) $\dfrac{1}{z-1}$ d) $\dfrac{1}{z^2-4}$

e) z^2+1/z^2 f) $\dfrac{z}{z-1}$

7.27 Find the value of each of the following integrals around the circle $|z|=2$ using Cauchy's integral formula.

a) $\oint \dfrac{\sin z}{z}dz$

b) $\oint \dfrac{e^z}{z-1}dz$

c) $\oint \dfrac{z}{z^2+4z+3}dz$

d) $\oint \dfrac{z^2-1}{z^3-z^2+9z-9}dz$

e) $\oint \dfrac{\cos z}{z-1}dz$

f) $\oint \dfrac{z^2}{z+i}dz$

7.28 Evaluate the integral $\oint (z-1)/(z^2+1)dz$ around each of the following curves.

a) $|z-i|=1$
b) $|z+i|=1$
c) $|z|=1/2$
d) $|z-1|=1$
e) $|z|=2$
f) the ellipse $2x^2+(y+1)^2=1$

7.29 If the curve C is the circle $|z|=2$, determine each of the following.

a) $\oint_C \dfrac{\sin z}{z^2}dz$ b) $\oint_C \dfrac{z-1}{(z+1)^2}dz$

c) $\oint_C \dfrac{z^2}{(z-1)^3}dz$ d) $\oint_C \dfrac{\cos z}{(z-1)^2}dz$

e) $\oint_C \dfrac{e^z}{(z-i)^2}dz$ f) $\oint_C \dfrac{\sinh z}{z^4}dz$

7.30 Using the Taylor series Eq. 7.6.4, find the series expansion about the origin for each of the following functions. State the radius of convergence.

a) $\cos z$ b) $\dfrac{1}{1+z}$

c) $\ln(1+z)$ d) $\dfrac{z-1}{z+1}$

e) $\cosh z$ f) $\sinh z$

7.31 For the function $1/(z-2)$, determine the Taylor series expansion about each of the following points. Use the known series expansion for $1/(1-z)$. State the radius of covergence for each.

a) $z=0$ b) $z=1$
c) $z=i$ d) $z=-1$
e) $z=3$ f) $z=-2i$

7.32 Using known series expansions, find the Taylor series expansion about the origin of each of the following.

a) $\dfrac{1}{1-z^2}$ b) $\dfrac{z-1}{1+z^3}$

c) $\dfrac{z^2+3}{2-z}$ d) $\dfrac{1}{z^2-3z-4}$

e) e^{-z^2} f) e^{2-z}
g) $\sin \pi z$ h) $\sin z^2$

i) $\dfrac{\sin z}{1-z}$ j) $e^z \cos z$

k) $\tan z$ l) $\dfrac{\sin z}{e^{-z}}$

7.33. What is the Taylor series expansion about the origin for each of the following?

a) $\int_0^z e^{-w^2}dw$ b) $\int_0^z \sin w^2\, dw$

c) $\int_0^z \dfrac{\sin w}{w}dw$ d) $\int_0^z \cos w^2\, dw$

7.34 Find the Taylor series expansion about the origin of $f(z)=\tan^{-1}z$ by recognizing that $f'(z)=1/(1+z^2)$.

7.35 Determine the Tailor series expansion of each of the following functions about the point $z=a$.

a) $e^z, a=1$ b) $\dfrac{1}{1-z}, a=2$

c) $\sin z, a=\dfrac{\pi}{2}$ d) $\ln z, a=1$

e) $\dfrac{1}{z^2-z-2}, a=0$ f) $\dfrac{1}{z^2}, a=1$

7.36 Expand each of the following functions in a Laurent series about the origin, convergent in the region $0<|z|<R$. State the radius of convergence R.

a) $\dfrac{1}{z^2}\sin z$ b) $\dfrac{1}{z^2 - 2z}$

c) $\dfrac{1}{z(z^2 + 3z + 2)}$ d) $\dfrac{e^{z-1}}{z}$

7.37 For each of the following functions, find all Taylor series and Laurent series expansions about the point $z = a$ and state the region of convergence for each.

a) $\dfrac{1}{z}$, $a = 1$ b) $e^{1/z}$, $a = 0$

c) $\dfrac{1}{1-z}$, $a = 0$ d) $\dfrac{1}{1-z}$, $a = 1$

e) $\dfrac{1}{1-z}$, $a = 2$ f) $\dfrac{1}{z(z-1)}$, $a = 0$

g) $\dfrac{z}{1-z^2}$, $a = 1$ h) $\dfrac{1}{z^2+1}$, $a = i$

i) $\dfrac{1}{(z+1)(z-2)}$, $a = 0$

j) $\dfrac{z}{(z+1)(z-2)}$, $a = -1$

k) $\dfrac{1}{(z+1)(z-2)}$, $a = 2$

7.38 Find the residue of each of the following functions at each pole.

a) $\dfrac{1}{z^2+4}$ b) $\dfrac{z}{z^2+4}$

c) $\dfrac{1}{z^2}\sin 2z$ d) $\dfrac{e^z}{(z-1)^2}$

e) $\dfrac{\cos z}{z^2 + 2z + 1}$ f) $\dfrac{z^2+1}{z^2 + 3z + 2}$

7.39 Evaluate each of the following integrals around the circle $|z| = 2$.

a) $\oint \dfrac{e^z}{z^4} dz$ b) $\oint \dfrac{\sin z}{z^3} dz$

c) $\oint \dfrac{z^2}{1-z} dz$ d) $\oint \dfrac{z+1}{z+i} dz$

e) $\oint \dfrac{z\,dz}{z^2 + 4z + 3}$ f) $\oint \dfrac{dz}{4z^2 + 9}$

g) $\oint \dfrac{e^{1/z}}{z} dz$ h) $\oint \dfrac{\sin z}{z^3 - z^2} dz$

i) $\oint e^z \tan z\,dz$ j) $\oint \dfrac{z^2+1}{z(z+1)^3} dz$

k) $\oint \dfrac{\sinh z}{z^2+1} dz$ l) $\oint \dfrac{\cosh z}{z^2+z} dz$

7.40 Determine the value of each of the following real integrals.

a) $\displaystyle\int_0^{2\pi} \frac{\sin\theta}{1+\cos\theta} d\theta$ b) $\displaystyle\int_0^{2\pi} \frac{d\theta}{(2+\cos\theta)^2}$

c) $\displaystyle\int_0^{2\pi} \frac{d\theta}{5 - 4\cos\theta}$ d) $\displaystyle\int_0^{2\pi} \frac{d\theta}{\cos\theta + 2\sin\theta}$

e) $\displaystyle\int_0^{2\pi} \frac{\sin 2\theta\, d\theta}{5 + 4\cos\theta}$ f) $\displaystyle\int_0^{2\pi} \frac{\sin 2\theta\, d\theta}{5 - 4\cos\theta}$

7.41 Evaluate each of the following integrals.

a) $\displaystyle\int_{-\infty}^{\infty} \frac{dx}{1+x^4}$ b) $\displaystyle\int_0^{\infty} \frac{x^2 dx}{(1+x^2)^2}$

c) $\displaystyle\int_{-\infty}^{\infty} \frac{1+x}{1+x^3} dx$ d) $\displaystyle\int_{-\infty}^{\infty} \frac{x^2 dx}{x^4 + x^2 + 1}$

e) $\displaystyle\int_0^{\infty} \frac{x^2 dx}{1+x^6}$ f) $\displaystyle\int_{-\infty}^{\infty} \frac{dx}{x^4 + 3x + 2}$

7.42 Find the value of each of the following integrals.

a) $\displaystyle\int_{-\infty}^{\infty} \frac{\cos 2x}{1+x^2} dx$ b) $\displaystyle\int_{-\infty}^{\infty} \frac{\cos x}{(1+x^2)^2} dx$

c) $\displaystyle\int_{-\infty}^{\infty} \frac{x\sin x}{1+x^2} dx$ d) $\displaystyle\int_0^{\infty} \frac{\cos x}{1+x^4} dx$

e) $\displaystyle\int_{-\infty}^{\infty} \frac{x\sin x}{x^2 + 3x + 2} dx$ f) $\displaystyle\int_0^{\infty} \frac{\cos 4x}{(1+x^2)^2} dx$

7.43 Find the value of $\int_{-\infty}^{\infty} dx/(x^4 - 1)$ following the technique using the path of integration of Fig. 7.19, but integrate around the two poles on the x axis by considering the path of integration shown in Fig. 7.20.

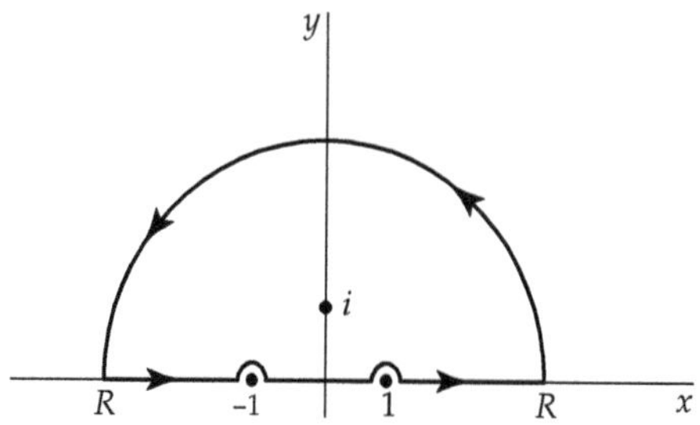

FIGURE 7.20

8 Numerical Methods

Outline

8.1 Introduction
8.2 Finite-Difference Operators
8.3 The Differential Operator Related to the Difference Operators
8.4 Truncation Error
8.5 Numerical Integration
8.6 Numerical Interpolation
8.7 Roots of Equations
8.8 Initial-Value Problems—Ordinary Differential Equations
8.8.1 Taylor's Method
8.8.2 Euler's Method
8.8.3 Adams' Method
8.8.4 Runge–Kutta Methods
8.8.5 Direct Method
8.9 Higher-Order Equations
8.10 Boundary-Value Problems—Ordinary Differential Equations
8.10.1 Iterative Method
8.10.2 Superposition
8.10.3 Simultaneous Equations
8.11 Numerical Stability
8.12 Numerical Solution of Partial Differential Equations
8.12.1 The Diffusion Equation
8.12.2 The Wave Equation
8.12.3 Laplace's Equation

Problems

8.1 INTRODUCTION

In previous chapters analytical solution techniques to both ordinary and partial differential equations were presented. More often than not, problems are encountered for which the describing differential equations are extremely difficult, if not impossible, to solve analytically. Fortunately, since the latter part of the 1950s, the digital computer has become an increasingly useful tool for solving differential equations, whether they be ordinary or partial, linear or nonlinear, homogeneous or nonhomogeneous, or first order or tenth order. It is, of course, not always simple to solve a differential equation, or a set of differential equations, using numerical methods. A numerical technique can be very intricate and difficult to understand, requiring substantial computer capability. Some techniques exist only in the literature or in advanced texts on the subject. We will, however, present several of the simplest methods for solving both ordinary and partial differential equations.

M. C. Potter and B. F. Feeny, *Mathematical Methods for Engineering and Science*,
https://doi.org/10.1007/978-3-031-26151-0_8

This chapter is intended to present some fundamental ideas in numerical methods and is not meant to be exhaustive. Textbooks on the subject should be consulted for a more complete treatment. Some sample computer programs will be presented; however, it is assumed that the reader is capable of using the computer, so the numerical methods outlined can be applied to the solution of real problems.

The numerical solution to a problem is quite different from the analytical solution. The analytical solution provides the value of the dependent variable for any value of the independent variable; that is, for the simple spring–mass system the analytical solution is

$$y(t) = A \sin \omega t + B \cos \omega t \tag{8.1.1}$$

We can choose any value of t and determine the displacement of the mass. Eq. 8.1.1 is a solution of*

$$\ddot{y} + \omega^2 y = 0 \tag{8.1.2}$$

If Eq. 8.1.2 were solved numerically, the time interval of interest would be divided into a predetermined number of increments, not necessarily of equal length. Initial conditions would be necessary to "start" the solution at $t = 0$; then the solution would be "generated" by solving numerically for the dependent variable y at each incremental step. This is done by using one of a host of numerical methods, any of which allows one to predict the value of the dependent variable at the $(i + 1)$ increment knowing its value at the ith increment [and possibly the $(i - 1)$ and $(i - 2)$ increments, depending on the method chosen]. The derivatives of the dependent variable may also be required in this process. After the solution is completed, the results are presented either in graphical or tabular form.

For a sufficiently small step size the solution to Eq. 8.1.2 closely approximates the solution given by Eq. 8.1.1. However, problems are encountered which are fairly common in numerical work. After one "debugs" a computer program, which may turn one's hair gray prematurely, a numerical solution may become "unstable"; that is, as the solution is progressing from one step to the next, the numerical results may begin to oscillate in an uncontrolled manner. This is referred to as a *numerical instability.* If the step size is changed, the stability characteristic changes. The objective is to choose an appropriate step size such that the solution is reasonably accurate and such that no instability results.

The problem of *truncation error,* which will be discussed in a later section, exists when a series of computational steps is truncated prematurely with the hope that the truncated terms are negligible. Truncation error depends on each method used and is minimized by retaining additional terms in the series of computational steps, or by choosing a different numerical technique with less truncation error.

Another problem that always exists in numerical work is that of *round-off error.* Numerical computations are rounded off† to a particular number of digits at each step in a numerical process, whether it be a *fixed-point system,* in which numbers are expressed with a fixed number of decimal places (e.g., 0.1734, 69.3712), or a *floating-point system,*

---

*In this chapter we shall often use the notation $\dot{y} = dy/dt$, $\dot{y} = dy/dx$, or $y' = dy/dx$.

†The general rule for rounding off is best reviewed by giving examples. If we round off to three digits, $62.55 \rightarrow 62.6$, $62.45 \rightarrow 62.4$, $0.01724 \rightarrow 0.0172$, $0.017251 \rightarrow 0.0173$, and $99.97 \rightarrow 100$.

in which numbers are expressed with a fixed number of significant digits (e.g., $3.22 \times 10^4, 5.00 \times 10^{-10}$). Round-off error accumulates in computations and thus increases with an increasing number of steps. Consequently, we are limited in the number of steps in solving a particular problem if we wish to minimize round-off error.

Usually, various choices in step size are used and the numerical results compared. The best solution is then chosen. Of course, the larger the step size, the shorter the computer time required, which leads to savings in computer costs. Thus, one must choose a small-enough step size to guarantee accurate results: not so small as to give excessive round-off error, and not so small as to incur high computer costs.

Another restraint in the numerical solution of problems is the size of the computer. A computer has only a particular number of "cells" in which information can be stored; for example, to store the number 2, or the number 2.00, one cell is used. In any numerical solution the total number of cells necessary to solve the problem must not exceed the "memory" of the computer in which information is stored. In the past, this was a definite limitation of computers; now, only in the solution of very complex problems does one experience a lack of memory space on the large computers.

8.2 FINITE-DIFFERENCE OPERATORS

A knowledge of the finite-difference operators is helpful to an understanding of numerical methods. The difference operator is analogous to the differential df of differential equations. The most often used difference operator is the *forward difference operator* Δ, which when operating on a function $f(x)$ yields

$$\Delta f_i = f_{i+1} - f_i \tag{8.2.1}$$

The quantity f_{i+1} is the value of $f(x)$ at x_{i+1}, and f_i is the value of $f(x)$ at x_i (see Fig. 8.1). In this chapter all the increments will be of equal size, namely Δx. In addition to the forward difference operator, two other difference operators are used. The *backward difference operator* ∇ operates on $f(x)$ to give

$$\nabla f_i = f_i - f_{i-1} \tag{8.2.2}$$

The *central difference operator* δ operating on $f(x)$ gives

$$\delta f_i = f_{i+1/2} - f_{i-1/2} \tag{8.2.3}$$

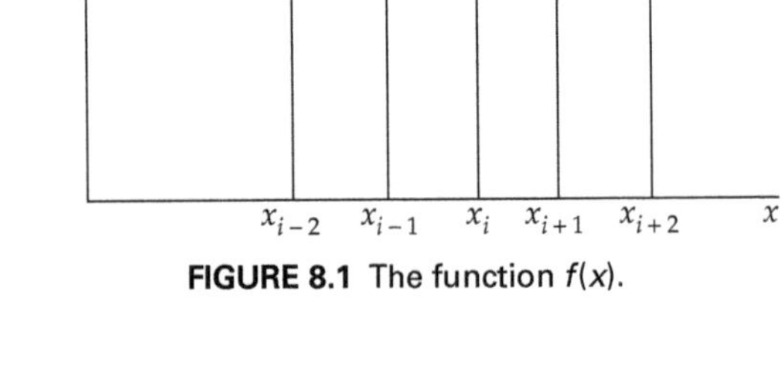

FIGURE 8.1 The function $f(x)$.

The differences described on the previous page are referred to as the *first differences*. The second forward difference is

$$
\begin{aligned}
\Delta^2 f_i &= \Delta(\Delta f_i) \\
&= \Delta(f_{i+1} - f_i) \\
&= \Delta f_{i+1} - \Delta f_i \\
&= f_{i+2} - f_{i+1} - f_{i+1} + f_i \\
&= f_{i+2} - 2f_{i+1} + f_i
\end{aligned} \tag{8.2.4}
$$

The second backward difference is

$$\nabla^2 f_i = f_i - 2f_{i-1} + f_{i-2} \tag{8.2.5}$$

Similarly, the second central difference is

$$
\begin{aligned}
\delta^2 f_i &= \delta(f_{i+1/2} - f_{i-1/2}) \\
&= f_{i+1} - f_i - f_i + f_{i-1} \\
&= f_{i+1} - 2f_i + f_{i-1}
\end{aligned} \tag{8.2.6}
$$

Continuing to the third differences, we have

$$\Delta^3 f_i = f_{i+3} - 3f_{i+2} + 3f_{i+1} - f_i \tag{8.2.7}$$

$$\nabla^3 f_i = f_i - 3f_{i-1} + 3f_{i-2} - f_{i-3} \tag{8.2.8}$$

$$\delta^3 f_i = f_{i+3/2} - 3f_{i-1/2} + 3f_{i-1/2} - f_{i-3/2} \tag{8.2.9}$$

Another useful operator is the operator E, defined by

$$Ef_i = f_{i+1} \tag{8.2.10}$$

It then follows that

$$
\begin{aligned}
E^{-1} f_i &= f_{i-1} \\
E^{1/2} f_i &= f_{i+1/2} \\
E^{-1/2} f_i &= f_{i-1/2} \\
E^2 f_i &= f_{i+2}
\end{aligned} \tag{8.2.11}
$$

The E operator can be related to the difference operators by observing that

$$
\begin{aligned}
\Delta f_i &= f_{i+1} - f_i \\
&= Ef_i - f_i \\
&= (E - 1) f_i
\end{aligned} \tag{8.2.12}
$$

We see then that the operator Δ operating on f_i is equal to $(E - 1)$ operating on f_i. We then conclude that

$$\Delta = E - 1 \tag{8.2.13}$$

Similarly, we can show that

$$\begin{aligned} \nabla &= 1 - E^{-1} \\ \delta &= E^{1/2} - E^{-1/2} \end{aligned} \tag{8.2.14}$$

Rewritten, we have

$$\begin{aligned} E &= \Delta + 1 \\ E^{-1} &= 1 - \nabla \end{aligned} \tag{8.2.15}$$

We can easily verify, by using the definitions, that

$$\nabla E = E\nabla = \Delta = \delta E^{1/2} \tag{8.2.16}$$

Another operator, the *averaging operator* μ, is defined by

$$\mu = \frac{1}{2}(E^{1/2} + E^{-1/2}) \tag{8.2.17}$$

Results in tabular form are presented in Table 8.1. Note that after the operators have been separated from the function they operate on, we can treat them as algebraic quantities. We can manipulate them into various expressions to give the desired form of the answer. This will be illustrated with Examples and Problems. A word of caution is also in order, namely that the operators operate on a function, such as ∇f_i. This order must be retained since $\nabla f_i \neq f_i \nabla$.

TABLE 8.1 The Difference Operators

$\Delta f_i = f_{i+1} - f_i$	$\Delta^2 f_i = f_{i+2} - 2f_{i+1} + f_i$	$\Delta^3 f_i = f_{i+3} - 3f_{i+2} + 3f_{i+1} - f_i$
$\nabla f_i = f_i - f_{i-1}$	$\nabla^2 f_i = f_i - 2f_{i-1} + f_{i-2}$	$\nabla^3 f_i = f_i - 3f_{i-1} + 3f_{i-2} - f_{i-3}$
$\delta f_i = f_{i+1/2} - f_{i-1/2}$	$\delta^2 f_i = f_{i+1} - 2f_i + f_{i-1}$	$\delta^3 f_i = f_{i+3/2} - 3f_{i+1/2} + 3f_{i-1/2} - f_{i-3/2}$
$Ef_i = f_{i+1}$	$E^2 f_i = f_{i+2}$	$E^3 f_i = f_{i+3}$
$\mu f_i = \frac{1}{2}(f_{i+1/2} + f_{i-1/2})$		

Example 8.1

Derive the relationships: a) $\Delta = \delta^2/2 + \delta\sqrt{1+\delta^2/4}$, b) $\nabla = -\delta^2/2 + \delta\sqrt{1+\delta^2/4}$, and c) $\mu = \sqrt{1+\delta^2/4}$.

Solution

a) The definition of the central difference operator is

$$\begin{aligned} \delta f_i &= f_{i+1/2} - f_{i-1/2} \\ &= (E^{1/2} - E^{-1/2}) f_i \end{aligned}$$

Hence,

$$\delta = E^{1/2} - E^{-1/2}$$

Using $E = 1 + \Delta$, we have

$$\delta = \sqrt{1+\Delta} - \frac{1}{\sqrt{1+\Delta}}$$

Squaring both sides gives

$$\delta^2 = 1 + \Delta + \frac{1}{1+\Delta} - 2$$

or

$$\delta^2 + 2 = \frac{(1+\Delta)^2 + 1}{1+\Delta}$$

Put this in the standard quadratic form,

$$\Delta^2 - \delta^2\Delta - \delta^2 = 0$$

The quadratic formula gives, using the positive root,

$$\begin{aligned}\Delta &= \frac{\delta^2}{2} + \frac{1}{2}\sqrt{\delta^4 + 4\delta^2} \\ &= \frac{\delta^2}{2} + \delta\sqrt{1 + \frac{\delta^2}{4}}\end{aligned}$$

b) Similarly, using $E^{-1} = 1 - \nabla$, we find that

$$\delta = \frac{1}{\sqrt{1-\nabla}} - \sqrt{1-\nabla}$$

After writing this in the standard quadratic form, the positive root is

$$\nabla = -\frac{\delta^2}{2} + \delta\sqrt{1 + \frac{\delta^2}{4}}$$

c) Now, μ can be written as

$$\mu = \frac{1}{2}(E^{1/2} + E^{-1/2})$$

or, squaring both sides,

$$4\mu^2 = E + 2 + E^{-1}$$

Also, if we square the initial expression in this Example for δ, we have

$$\begin{aligned}\delta^2 &= E - 2 + E^{-1} \\ &= E + 2 + E^{-1} - 4 \\ &= 4\mu^2 - 4\end{aligned}$$

Thus,

$$\mu^2 = 1 + \frac{\delta^2}{4}$$

or

$$\mu = \sqrt{1 + \frac{\delta^2}{4}}$$

We could also write

$$\Delta = \frac{\delta^2}{2} + \delta\mu, \quad \nabla = -\frac{\delta^2}{2} + \delta\mu$$

8.3 THE DIFFERENTIAL OPERATOR RELATED TO THE DIFFERENCE OPERATORS

We shall now relate the various operators to the *differential operator* $D = d/dx$. In this process the Taylor series is used; it is

$$f(x+h) = f(x) + h\frac{df}{dx} + \frac{h^2}{2!}\frac{d^2 f}{dx^2} + \frac{h^3}{3!}\frac{d^3 f}{dx^3} + \cdots \tag{8.3.1}$$

where the derivatives are evaluated at x and the step size $\Delta x = h$. This is written, using the difference notation, as

$$f_{i+1} = f_i + hf_i' + \frac{h^2}{2!}hf_i'' + \frac{h^3}{3!}hf_i''' + \cdots \tag{8.3.2}$$

where the primes denote differentiation with respect to the independent variable. The higher-order derivatives are written as

$$D^2 = \frac{d^2}{dx^2}, \qquad D^3 = \frac{d^3}{dx^3}, \cdots \tag{8.3.3}$$

Then Eq. 8.3.2 can be written as

$$Ef_i = \left[1 + hD + \frac{h^2D^2}{2!} + \frac{h^3D^3}{3!} + \cdots\right] f_i \tag{8.3.4}$$

We recognize that the quantity in brackets is (see Eq. 2.1.6)

$$e^{hD} = 1 + hD + \frac{h^2D^2}{2!} + \frac{h^3D^3}{3!} + \cdots \tag{8.3.5}$$

which leads to

$$Ef_i = e^{hD} f_i \tag{8.3.6}$$

This relates the operator E to the operator D as

$$E = e^{hD} \tag{8.3.7}$$

Making the substitution

$$E = \Delta + 1 \tag{8.3.8}$$

we have

$$\Delta = e^{hD} - 1$$
$$= hD + \frac{h^2D^2}{2!} + \frac{h^3D^3}{3!} + \cdots \tag{8.3.9}$$

You can also obtain this result by using $\Delta f_i = f_{i+1} - f_i$ and Eq. 8.3.4.

The second forward difference is found by squaring the equation above, to obtain

$$\Delta^2 = \left(hD + \frac{h^2D^2}{2} + \frac{h^3D^3}{6} + \cdots\right)^2$$
$$= h^2D^2 + h^3D^3 + \frac{7}{12}h^4D^4 + \cdots \tag{8.3.10}$$

To find D in terms of Δ we take the natural logarithm of both sides of Eq. 8.3.7 to obtain

$$D = \frac{1}{h}\ln E = \frac{1}{h}\ln(1+\Delta) \tag{8.3.11}$$

In series form, we have

$$\ln(1+\Delta) = \Delta - \frac{\Delta^2}{2} + \frac{\Delta^3}{3} - \cdots \tag{8.3.12}$$

We have finally related the differential operator to the forward difference operator; there results

$$D = \frac{1}{h}\left(\Delta - \frac{\Delta^2}{2} + \frac{\Delta^3}{3} - \cdots\right) \tag{8.3.13}$$

Squaring both sides yields

$$D^2 = \frac{1}{h^2}\left(\Delta^2 - \Delta^3 + \frac{11}{12}\Delta^4 - \frac{5}{6}\Delta^5 - \cdots\right) \tag{8.3.14}$$

The central and backward difference operators can be related to the differential operator by using Eq. 8.3.7 to write

$$E^{1/2} = e^{hD/2}, \qquad E^{-1/2} = e^{-hD/2}, \qquad E^{-1} = e^{-hD} \tag{8.3.15}$$

The resulting expressions will be included in the Examples and Problems. They are summarized in Table 8.2.

The results above are used to express the first derivative of the function $f(x)$ at x_i as

$$Df_i = \frac{df_i}{dx} = \frac{1}{h}\left(\Delta f_i - \frac{\Delta^2}{2}f_i + \frac{\Delta^3}{3}f_i - \cdots\right) \tag{8.3.16}$$

The second derivative is

$$D^2 f_i = \frac{d^2 f_i}{dx^2} = \frac{1}{h^2}\left(\Delta^2 f_i - \Delta^3 f_i + \frac{11}{12}\Delta^4 f_i - \cdots\right) \tag{8.3.17}$$

Higher-order derivatives can be generated.

TABLE 8.2 Relationships Between the Operators

$$\Delta = E - 1 \qquad \delta = E^{1/2} - E^{-1/2} \qquad \nabla = -\frac{\delta^2}{2} + \delta\sqrt{1 + \frac{\delta^2}{4}}$$

$$\nabla = 1 - E^{-1} \qquad 2\mu\delta = E - E^{-1}$$

$$2\mu = E^{1/2} + E^{-1/2} \qquad \Delta = \frac{\delta^2}{2} + \delta\sqrt{1 + \frac{\delta^2}{4}} \qquad \mu = \sqrt{1 + \frac{\delta^2}{4}}$$

$$D = \frac{1}{h}\left[\Delta - \frac{\Delta^2}{2} + \frac{\Delta^3}{3} - \cdots\right] = \frac{1}{h}\left[\nabla + \frac{\nabla^2}{2} + \frac{\nabla^3}{3} + \cdots\right] = \frac{\mu}{h}\left[\delta - \frac{\delta^3}{6} + \frac{\delta^5}{30} - \cdots\right]$$

$$D^2 = \frac{1}{h^2}\left[\Delta^2 - \Delta^3 + \frac{11}{12}\Delta^4 - \cdots\right] = \frac{1}{h^2}\left[\nabla^2 + \nabla^3 + \frac{11}{12}\nabla^4 + \cdots\right]$$

$$= \frac{1}{h^2}\left[\delta^2 - \frac{\delta^4}{12} + \frac{\delta^6}{90} - \cdots\right]$$

$$D^3 = \frac{1}{h^3}\left[\Delta^3 - \frac{3}{2}\Delta^4 + \frac{7}{4}\Delta^5 - \cdots\right] = \frac{1}{h^3}\left[\nabla^3 + \frac{3}{2}\nabla^4 + \frac{7}{4}\nabla^5 + \cdots\right]$$

$$= \frac{\mu}{h^3}\left[\delta^3 - \frac{\delta^5}{4} + \frac{7}{120}\delta^7 - \cdots\right]$$

$$D^4 = \frac{1}{h^4}\left[\Delta^4 - 2\Delta^5 + \frac{17}{6}\Delta^6 - \cdots\right] = \frac{1}{h^4}\left[\nabla^4 + 2\nabla^5 + \frac{17}{6}\nabla^6 + \cdots\right]$$

$$= \frac{1}{h^4}\left[\delta^4 - \frac{\delta^6}{6} + \frac{7}{240}\delta^8 - \cdots\right]$$

$$\Delta = hD + \frac{h^2}{2}D^2 + \frac{h^3}{6}D^3 + \cdots \qquad \Delta^2 = h^2D^2 + h^3D^3 + \frac{7}{12}h^4D^4 + \cdots$$

$$\Delta^3 = h^3D^3 + \frac{3}{2}h^4D^4 + \frac{5}{4}h^5D^5 + \cdots$$

$$\nabla = hD - \frac{h^2}{2}D^2 + \frac{h^3}{6}D^3 + \cdots \qquad \nabla^2 = h^2D^2 - h^3D^3 + \frac{7}{12}h^4D^4 + \cdots$$

$$\nabla^3 = h^3D^3 - \frac{3}{2}h^4D^4 + \frac{5}{4}h^5D^5 + \cdots$$

$$\mu\delta = hD + \frac{h^3D^3}{6} + \frac{h^5D^5}{120} + \cdots \qquad \delta^2 = h^2D^2 + \frac{h^4D^4}{12} + \frac{h^6D^6}{360} + \cdots$$

$$\mu\delta^3 = h^3D^3 + \frac{h^5D^5}{4} + \frac{h^7D^7}{40} + \cdots$$

Example 8.2

Relate the differential operator D to the central difference operator δ by using Eq. 8.3.13 and the result of Example 8.1.

Solution

We can use the relationship $\Delta = \delta^2/2 + \delta\sqrt{1+\delta^2/4}$ (see Example 8.1). Expand $(1+\delta^2/4)^{1/2}$ in a series using the binomial theorem* to give

$$\Delta = \frac{\delta^2}{2} + \delta\left(1 + \frac{\delta^2}{8} - \frac{\delta^4}{128} + \cdots\right)$$
$$= \delta + \frac{\delta^2}{2} + \frac{\delta^3}{8} - \frac{\delta^5}{128} + \cdots$$

Substitute this into Eq. 8.3.13 to find

$$D = \frac{1}{h}\left[\delta + \frac{\delta^2}{2} + \frac{\delta^3}{8} - \frac{\delta^5}{128} + \cdots - \frac{1}{2}\left(\delta + \frac{\delta^2}{2} + \frac{\delta^3}{8} - \frac{\delta^5}{128} + \cdots\right)^2\right.$$
$$\left. + \frac{1}{3}\left(\delta + \frac{\delta^2}{2} + \frac{\delta^3}{8} - \frac{\delta^5}{128} + \cdots\right)^3 + \cdots\right]$$
$$= \frac{1}{h}\left(\delta - \frac{\delta^3}{24} + \frac{3\delta^5}{640} - \cdots\right)$$

This expression would allow us to relate Df_i to quantities such as $f_{i+1/2}$, $f_{i-1/2}$, $f_{i+3/2}$, $f_{i-5/2}$, and so on. It is more useful to introduce the averaging operator μ so that quantities with integer subscripts result such as f_{i+1}, f_{i-1}, f_{i+2}, etc. From Example 8.1 we have $\mu = \sqrt{1+\delta^2/4}$. Expressing this as a series, we can write

$$\mu = 1 + \frac{\delta^2}{8} - \frac{\delta^4}{128} + \cdots$$

Now, we can write D as

$$D = \frac{\mu}{h}\left(\delta - \frac{\delta^3}{24} + \frac{3\delta^5}{640} - \cdots\right)\frac{1}{\mu}$$
$$= \frac{\mu}{h}\,\frac{\delta - \dfrac{\delta^3}{24} + \dfrac{3\delta^5}{640} - \cdots}{1 + \dfrac{\delta^2}{8} - \dfrac{\delta^4}{128} + \cdots}$$

*The binomial theorem is written

$$(a+b)^n = a^n + na^{n-1}b + n(n-1)a^{n-2}b^2/2! + \cdots$$

Dividing one series by the other, we finally have

$$D = \frac{\mu}{h}\left(\delta - \frac{\delta^3}{6} + \frac{\delta^5}{30} - \cdots\right)$$

Using this expression Df_i would contain only integer subscripts; this would allow use on a computer.

Example 8.3

Relate the central difference operator to the differential operator by starting with a Taylor series.

Solution

A Taylor series can be written as

$$f\left(x + \frac{h}{2}\right) = f(x) + \frac{h}{2}f'(x) + \left(\frac{h}{2}\right)^2 \frac{1}{2!}f''(x) + \left(\frac{h}{2}\right)^3 \frac{1}{3!}f'''(x) + \cdots$$

where the primes denote differentiation with respect to x. In difference notation, we have

$$\begin{aligned} f_{i+1/2} &= f_i + \frac{h}{2}f_i' + \frac{h^2}{8}f_i'' + \frac{h^3}{48}f_i''' + \cdots \\ &= \left(1 + \frac{hD}{2} + \frac{h^2D^2}{8} + \frac{h^3D^3}{48} + \cdots\right)f_i \end{aligned}$$

Similarly,

$$\begin{aligned} f_{i-1/2} &= f_i - \frac{h}{2}f_i' + \frac{h^2}{8}f_i'' - \frac{h^3}{48}f_i''' + \cdots \\ &= \left(1 - \frac{hD}{2} + \frac{h^2D^2}{8} - \frac{h^3D^3}{48} + \cdots\right)f_i \end{aligned}$$

Subtracting gives

$$\begin{aligned} \delta f_i &= f_{i+1/2} - f_{i-1/2} \\ &= \left(hD + \frac{h^3D^3}{24} + \frac{h^5D^5}{1920} + \cdots\right)f_i \end{aligned}$$

Finally,

$$\delta = hD + \frac{h^3D^3}{24} + \frac{h^5D^5}{1920} + \cdots$$

This could also have been obtained by using Eqs. 8.2.14 and 8.3.7. We would have

$$\delta = E^{1/2} - E^{-1/2} = e^{hD/2} - e^{-hD/2}$$

Expanding the exponentials as in Eq. 8.3.5, there results

$$\delta = \left(1 + \frac{hD}{2} + \frac{h^2D^2}{8} + \frac{h^3D^3}{48} + \cdots\right) - \left(1 - \frac{hD}{2} + \frac{h^2D^2}{8} - \frac{h^3D^3}{48} + \cdots\right)$$
$$= hD + \frac{h^3D^3}{24} + \frac{h^5D^5}{1920} + \cdots$$

8.4 TRUNCATION ERROR

We obviously cannot use all the terms in the infinite series when representing a derivative in finite-difference form, as in Eqs. 8.3.16 and 8.3.17. The series is truncated and the sum of the omitted terms is the *truncation error.* It is quite difficult to determine the sum of the omitted terms; instead, we estimate the magnitude of the first term truncated in the series. Since each term is smaller than the preceding term, we call the magnitude of the first truncated term the *order of magnitude* of the error. Its primary function is to allow a comparison of formulas. If the magnitude of the first truncated term of one formula is smaller than that of another formula, we assume the first formula to be more accurate.

If the first term truncated is $\Delta^2 f_i/2$ in Eq. 8.3.16, the order of magnitude of the error would be of order h, written symbolically as $e = o(h)$, since from Eq. 8.3.10 Δ^2 is of order h^2 and in Eq. 8.3.16 we divide $\Delta^2 f_i/2$ by h. Hence, we can express the first derivative of a function, with $e = o(h)$, as

$$Df_i = \frac{1}{h}\Delta f_i$$
$$= \frac{1}{h}(f_{i+1} - f_i), \qquad e = o(h) \tag{8.4.1}$$

If a smaller error were desired, then an additional term would be maintained and

$$Df_i = \frac{1}{h}(\Delta f_i - \Delta^2 f_i/2)$$
$$= \frac{1}{h}[f_{i+1} - f_i - \frac{1}{2}(f_{i+2} - 2f_{i+1} + f_i)]$$
$$= \frac{1}{2h}(-f_{i+2} + 4f_{i+1} - 3f_i), \qquad e = o(h^2) \tag{8.4.2}$$

This, of course, requires additional information, the value of $f(x)$ at x_{i+2}.

The second derivative would be approximated, with $e = o(h)$, by

$$D^2 f_i = \frac{1}{h^2}\Delta^2 f_i$$
$$= \frac{1}{h^2}(f_{i+2} - 2f_{i+1} + f_i), \qquad e = o(h) \tag{8.4.3}$$

Note that the Δ^3-term was omitted. It is of order h^3; but it is divided by h^2, hence $e = o(h)$. Maintaining an additional term in Eq. 8.3.17, the second derivative, with $e = o(h^2)$, would be

$$
\begin{aligned}
D^2 f_i &= \frac{1}{h^2}\left(\Delta^2 f_i - \Delta^3 f_i\right) \\
&= \frac{1}{h^2}(f_{i+2} - 2f_{i+1} + f_i - f_{i+3} + 3f_{i+2} - 3f_{i+1} - f_i) \\
&= \frac{1}{h^2}(-f_{i+3} + 4f_{i+2} - 5f_{i+1} + 2f_i), \qquad e = o(h^2)
\end{aligned} \tag{8.4.4}
$$

Results in tabular form are presented in Table 8.3.

The error analysis above is meaningful only when the phenomenon of interest occurs over a time duration of order unity or over a length of order unity. If we are studying a phenomenon that occurs over a long time T, then the time increment Δt could be quite

TABLE 8.3 The Derivatives in Finite-Difference Form

Forward	Backward	Central
$e = o(h)$		
$Df_i = \frac{1}{h}(f_{i+1} - f_i)$	$\frac{1}{h}(f_i - f_{i-1})$	
$D^2 f_i = \frac{1}{h^2}(f_{i+2} - 2f_{i+1} + f_i)$	$\frac{1}{h^2}(f_i - 2f_{i-1} + f_{i-2})$	
$D^3 f_i = \frac{1}{h^3}(f_{i+3} - 3f_{i+2} + 3f_{i+1} - f_i)$	$\frac{1}{h^3}(f_i - 3f_{i-1} + 3f_{i-2} - f_{i-3})$	
$D^4 f_i = \frac{1}{h^4}(f_{i+4} - 4f_{i+3} + 6f_{i+2} - 4f_{i+1} + f_i)$	$\frac{1}{h^4}(f_i - 4f_{i-1} + 6f_{i-2} - 4f_{i-3} + f_{i-4})$	
$e = o(h^2)$		
$Df_i = \frac{1}{2h}(-f_{i+2} + 4f_{i+1} - 3f_i)$	$\frac{1}{2h}(3f_i - 4f_{i-1} + f_{i-2})$	$\frac{1}{2h}(f_{i+1} - f_{i-1})$
$D^2 f_i = \frac{1}{2h^2}(-f_{i+3} + 4f_{i+2} - 5f_{i+1} + 2f_i)$	$\frac{1}{h^2}(2f_i - 5f_{i-1} + 4f_{i-2} - f_{i-3})$	$\frac{1}{h^2}(f_{i+1} - 2f_i + f_{i-1})$
$D^3 f_i = \frac{1}{2h^3}(-3f_{i+4} + 14f_{i+3} - 24f_{i+2} + 18f_{i+1} - 5f_i)$	$\frac{1}{2h^3}(5f_i - 18f_{i-1} + 24f_{i-2} - 14f_{i-3} + 3y_{i-4})$	$\frac{1}{2h^3}(f_{i+2} - 2f_{i+1} + 2f_{i-1} - f_{i-2})$
$D^4 f_i = \frac{1}{h^4}(-2f_{i+5} + 11f_{i+4} - 24f_{i+3} + 26f_{i+2} - 14f_{i+1} + 3f_i)$	$\frac{1}{h^4}(3f_i - 14f_{i-1} + 26f_{i-2} - 24f_{i-3} + 11f_{i-4} - 2f_{i-5})$	$\frac{1}{h^4}(f_{i+2} - 4f_{i+1} + 6f_i - 4f_{i-1} + f_{i-2})$

large if a reasonable number of steps are considered; or if the phenomenon occurs over a large length L, the length increment Δx could be quite large. For example, the deflection of a 300-m-high smokestack on a power plant could be reasonably calculated with increments of 3 m. We would not then say that the error gets larger with each term truncated, i.e., $o(h^3) > o(h^2)$. The same reasoning is applied to phenomena of very short duration or lengths. Then h is extremely small and the truncation error would appear to be much smaller than it actually is. The quantity that determines the error is actually the step size involved when the time duration or the length scale is of order unity; hence, to determine the error when large or small scales are encountered, we first "normalize" on the independent variable so that the phenomenon occurs over a duration or length of order unity. That is, we consider the quantity h/T or h/L to determine the order of the error. The expressions for error in Eqs. 8.4.1, 8.4.2, and 8.4.3 are based on the assumption that the time duration or length scale is of order 1.

Example 8.4

Find an expression for the second derivative using central differences with $e = o(h^4)$.

Solution

D^2 in terms of central differences is found by squaring the expression given in Example 8.2 for D in terms of δ. It is

$$
\begin{aligned}
D^2 &= \left[\frac{1}{h}\left(\delta - \frac{\delta^3}{24} + \frac{3\delta^5}{640} - \cdots\right)\right]^2 \\
&= \frac{1}{h^2}\left(\delta^2 - \frac{\delta^4}{12}\right), \qquad e = o(h^4)
\end{aligned}
$$

This expression can also be found in Table 8.2. Now, we have

$$
\begin{aligned}
D^2 f_i &= \frac{1}{h^2}\left(\delta^2 f_i - \frac{1}{12}\delta^4 f_i\right) \\
&= \frac{1}{h^2}\left[f_{i+1} - 2f_i + f_{i-1} - \frac{1}{12}(f_{i+2} - 4f_{i+1} + 6f_i - 4f_{i-1} + f_{i-2})\right] \\
&= \frac{1}{12h^2}(-f_{i+2} + 16f_{i+1} - 30f_i + 16f_{i-1} - f_{i-2}), \quad e = o(h^4)
\end{aligned}
$$

The relationship for $\delta^4 f_i$ is part of Problem 8.1.

Example 8.5

Write the differential equation

$$
\ddot{y} - \frac{C}{M}\dot{y} + \frac{K}{M}y = A\sin\omega t
$$

in difference notation using forward differences with $e = o(h^2)$.

Solution

The first derivative is given by Eq. 8.4.2 as

$$\dot{y}_i = \frac{1}{2h}(-y_{i+2} + 4y_{i+1} - 3y_i)$$

The second derivative, found by maintaining the first two terms in Eq. 8.3.17, with $e = o(h^2)$, is given by Eq. 8.4.4. The equation is then written in difference form as

$$\frac{1}{h^2}(-y_{i+3} + 4y_{i+2} - 5y_{i+1} + 2y_i) - \frac{C}{2Mh}(-y_{i+2} + 4y_{i+1} - 3y_i) + \frac{K}{M}y_i$$
$$= A\sin\omega t_i$$

This is the difference equation of the given differential equation. By letting $i = 0$, y_3 can be related to y_2, y_1, y_0, and t_0. This would be the first value of the dependent variable $y(t)$ that could be found by the difference equation. But we do not know the values of y_2 and y_1. (The value of y_0 would be known from an initial condition.) Thus, a "starting technique" is necessary to find the values y_1 and y_2. This will be presented in Section 8.9.

8.5 NUMERICAL INTEGRATION

The symbol used for differentiating a function is $D = d/dx$; we will now show that the process of integrating is the inverse of the process of differentiating. This is written as

$$D^{-1}f(x) = \int f(x)dx \tag{8.5.1}$$

or, between the limits of x_i and x_{i+1}, this is

$$\int_{x_i}^{x_{i+1}} f(x)dx = D^{-1}f(x)\Big|_{x_i}^{x_{i+1}}$$
$$= D^{-1}(f_{i+1} - f_i) = \frac{E-1}{D}f_i \tag{8.5.2}$$

Relating this to the forward difference operator, we use Eqs. 8.2.13 and 8.3.13 to obtain

$$\int_{x_i}^{x_{i+1}} f(x)dx = \frac{\Delta}{\frac{1}{h}\left(\Delta - \frac{\Delta^2}{2} + \frac{\Delta^3}{3} - \cdots\right)}f_i$$
$$= h\left(1 + \frac{\Delta}{2} - \frac{\Delta^2}{12} + \frac{\Delta^3}{24} - \cdots\right)f_i \tag{8.5.3}$$

If we neglect the Δ^2-term and the higher-order terms in the parentheses above, there results

$$\int_{x_i}^{x_{i+1}} f(x)dx = h\left(1 + \frac{\Delta}{2}\right)f_i$$
$$= \frac{h}{2}(f_{i+1} + f_i), \qquad e = o(h^3) \tag{8.5.4}$$

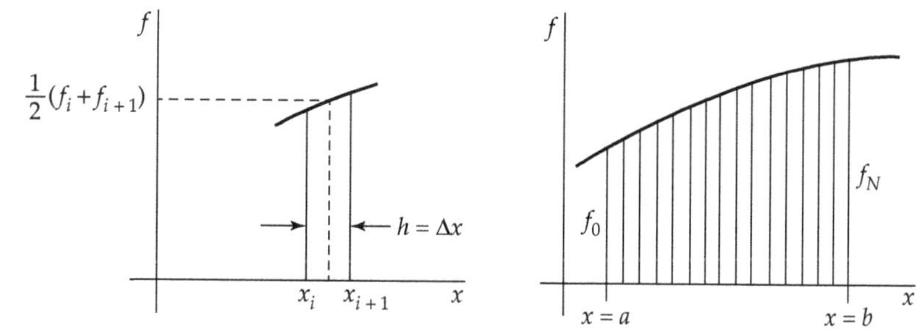

FIGURE 8.2 The integral of $f(x)$.

It is simply the average of $f(x)$ between x_{i+1} and x_i multiplied by the step size Δx (see Fig. 8.2). It definitely is an approximation to the integral of $f(x)$ between x_i and x_{i+1}. The smaller the step size h, the closer the approximation to the integral. The error results from the neglected $o(h^2)$ term multiplied by h; it is $e = o(h^3)$.

To obtain the integral from $x = a$ to $x = b$, we simply add up all the areas to arrive at

$$\begin{aligned}\int_a^b f(x)dx &\cong \frac{h}{2}[(f_0 + f_1) + (f_1 + f_2) + (f_2 + f_3) + \cdots \\ &\quad + (f_{N-2} + f_{N-1}) + (f_{N-1} + f_N)] \\ &= \frac{h}{2}(f_0 + 2f_1 + 2f_2 + \cdots + 2f_{N-1} + f_N) \qquad (8.5.5)\end{aligned}$$

where $N = (b - a)/h$. This is the *trapezoidal rule* of integration. Each element in the interval from a to b contains an error $e = o(h^3)$. Hence, assuming the interval to be of order unity, that is, $b - a = o(1)$, it follows that $N = o(1/h)$. The order of magnitude of the total error in the integration formula 8.5.5 is then $N \times o(h^3)$, or $o(h^2)$.

We can also determine the numerical approximation to the integral between x_i and x_{i+2} as

$$\begin{aligned}\int_{x_i}^{x_{i+1}} f(x)\,dx &= D^{-1}f(x)\Big|_{x_i}^{x_{i+2}} = D^{-1}(f_{i+2} - f_i) \\ &= \frac{E^2 - 1}{D} f_i \\ &= \frac{2\Delta + \Delta^2}{(1/h)(\Delta - \Delta^2/2 + \Delta^3/3 - \cdots)} f_i \\ &= h\left(2 + 2\Delta + \frac{\Delta^2}{3} - \frac{\Delta^4}{90} + \cdots\right) f_i \qquad (8.5.6)\end{aligned}$$

There results, keeping terms up through Δ^2, so that we do not go outside the limits of integration,

$$\begin{aligned}\int_{x_i}^{x_{i+2}} f(x)dx &= h(2 + 2\Delta + \Delta^2/3)f_i \\ &= \frac{h}{3}(f_{i+2} + 4f_{i+1} + f_i), \qquad e = o(h^5) \qquad (8.5.7)\end{aligned}$$

The error in this formula is, surprisingly, of order $o(h^5)$ because of the absence of the Δ^3-term. This small error makes this a popular integration formula. The integral from $x = a$ to $x = b$ is then

$$\int_a^b f(x)dx \cong \frac{h}{3}[(f_0 + 4f_1 + f_2) + (f_2 + 4f_3 + f_4) + \cdots + (f_{N-4} + 4f_{N-3} + f_{N-2}) + (f_{N-2} + 4f_{N-1} + f_N)]$$

$$= \frac{h}{3}(f_0 + 4f_1 + 2f_2 + 4f_3 + 2f_4 + \cdots + 2f_{N-2} + 4f_{N-1} + f_N) \qquad (8.5.8)$$

where $N = (b - a)/h$ and N is an even integer. This is *Simpson's one-third rule* of integration. The integral has been approximated by $N/2$ pairs of elements, each pair having $e = o(h^5)$. Since $N = o(1/h)$ it follows that the order of the error for the formula (8.5.8) is $e = (N/2) \times o(h^5) = o(h^4)$. Note that the factor 2 does not change the order of the error.

Similarly, we have

$$\int_{x_i}^{x_{i+1}} f(x)dx = \frac{3h}{8}(f_{i+3} + 3f_{i+2} + 3f_{i+1} + f_i), \quad e = o(h^5) \qquad (8.5.9)$$

The integration formula is then

$$\int_a^b f(x)dx = \frac{3h}{8}[(f_0 + 3f_1 + 3f_2 + f_3) + (f_3 + 3f_4 + 3f_5 + f_6) + \cdots + (f_{N-3} + 3f_{N-2} + 3f_{N-1} + f_N)]$$

$$= \frac{3h}{8}(f_0 + 3f_1 + 3f_2 + 2f_3 + 3f_4 + 3f_5 + 2f_6 + \cdots + 2f_{N-3} + 3f_{N-2} + 3f_{N-1} + f_N) \qquad (8.5.10)$$

where $N = (b - a)/h$ and N is divisible by 3. This is *Simpson's three-eights rule* of integration. The error is found to be of order $o(h^4)$, essentially the same as Simpson's one-third rule.

If we desired the integral in backward difference form, for example, $\int_{x_{i-2}}^{x_i} f(x)dx$, we would have chosen to express E and D in terms of backward differences; if $\int_{x_{i-2}}^{x_{i+2}} f(x)dx$ were desired, central differences would be chosen. Examples of these will be included in the Problems and the Examples.

Numerical integration is important; in addition to providing us with the values of integrals, it is often used in the numerical solution of differential equations.

It is possible to establish error bounds on the numerical integration process, a more exact error analysis than the order of magnitude. Let us first consider the trapezoidal rule of integration. The error e involved is (see Eq. 8.5.5)

$$e = \frac{h}{2}(f_0 + 2f_1 + 2f_2 + \cdots + f_N) - \int_a^b f(x)dx \qquad (8.5.11)$$

We will find the error for only the first interval, letting the step size h be a variable, as shown in Fig. 8.3. Using the relationship above, the error in this single strip is

We can also obtain this from Eq. 8.5.4 with $x_i = a$ and $x_{i+1} = t$.

$$e(t) = \frac{t - a}{2}[f(a) + f(t)] - \int_a^t f(x)dx \qquad (8.5.12)$$

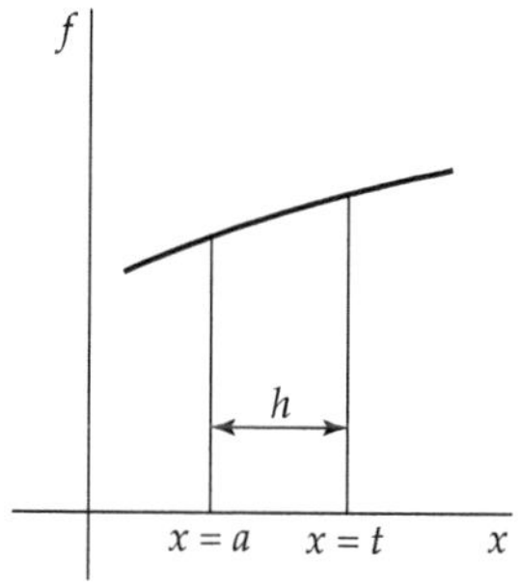

FIGURE 8.3 Variables with element used in the error analysis.

Differentiate this equation with respect to t to obtain

$$e'(t) = \frac{1}{2}[f(a) + f(t)] + \frac{t-a}{2} f'(t) - f(t) \tag{8.5.13}$$

where we have used Liebnitz' rule of integration from calculus to obtain

$$\frac{d}{dt}\int_a^t f(x)dx = f(t) \tag{8.5.14}$$

Again, differentiate to find

$$e''(t) = \frac{t-a}{2} f''(t) \tag{8.5.15}$$

Thus, the maximum value of e'' is obtained if we replace $f''(t)$ with its maximum value in the interval, and the minimum value results when $f''(t)$ is replaced with its minimum value. This is expressed by

$$\frac{t-a}{2} f''_{\text{min}} \le e''(t) \le \frac{t-a}{2} f''_{\text{max}} \tag{8.5.16}$$

Now, let us integrate to find the bounds on the error $e(t)$. Integrating once gives

$$\frac{(t-a)^2}{4} f''_{\text{min}} \le e'(t) \le \frac{(t-a)^2}{4} f''_{\text{max}} \tag{8.5.17}$$

A second integration results in

$$\frac{(t-a)^3}{12} f''_{\text{min}} \le e(t) \le \frac{(t-a)^3}{12} f''_{\text{max}} \tag{8.5.18}$$

In terms of the step size, the error for this first step is

$$\frac{h^3}{12} f''_{\text{min}} \le e \le \frac{h^3}{12} f''_{\text{max}} \tag{8.5.19}$$

But there are N steps in the interval of integration from $x = a$ to $x = b$. Assuming that each step has the same bounds on its error, the total accumulated error is N times that of a single step,

$$\frac{h^3}{12} N f''_{\text{min}} \le e \le \frac{h^3}{12} N f''_{\text{max}} \tag{8.5.20}$$

where f''_{min} and f''_{max} are the smallest and largest second derivatives, respectively, in the interval of integration.

A similar analysis, using Simpson's one-third rule, leads to an error bounded by

$$\frac{h^5}{180}Nf_{\min}^{(\mathrm{iv})} \le e \le \frac{h^5}{180}Nf_{\max}^{(\mathrm{iv})} \tag{8.5.21}$$

There are some integration approaches that can be quickly and easily performed in MATLAB. Let us first re-examine Eq. 8.5.3. If we neglect terms of order Δ and higher in Eq. 8.5.3, we have

$$\int_{x_i}^{x_{i+1}} f(x)dx \cong hf_i \tag{8.5.22}$$

which is a very coarse numerical integration method classified as a rectangular rule, in which the trapezoidal strips in Fig. 8.2 are replaced with rectangular strips. An integral over many steps, with constant step size h, would lead to

$$\int_{x_i}^{x_{i+1}} f(x)dx \cong h\sum_{i=1}^{n} f_i \tag{8.5.23}$$

This is a sum of rectangles of width h and height f_i, such that if a rectangle is located between x_i and x_{i+1}, then the height of this rectangle is defined from its left edge, that is, f_i. Integration by Eq. 8.5.23 is perhaps too simple, but it lends itself to very quick first-approximation MATLAB calculations.

Suppose the x coordinate is divided into small equal intervals of size h from a to b, and we aim to approximate the area of $f(x) = \tanh(5x)$. If `a`, `b`, and `h` have already been assigned values, then we get a quick estimate from these lines of code:

```
>> x = a:h:(b-h); % an array of values not including the endpoint
>> f = tanh(5*x);
>> area = sum(f)*h;
```

For example, if `h = 0.01`, `a = 0`, and `b = 0.5`, there are 50 thin rectangles, and the approximated integral is `area = 0.3577`, while the true value is 0.3627.

If the x coordinate is not divided into equal intervals from a to b, we can make a minor enhancement to the sum. First we need the interval sizes (rectangle widths), and then we construct the sum so that each rectangle has the appropriate rectangle width. If an array `x` is already defined between `a`, `b`, then we get a quick estimate from these lines:

```
>> N = length(x)
>> h = x(2:N) - x(1:N-1); % an array of length N-1 of interval sizes
>> f = tanh(5*x(1:N-1));
>> area = f*h';             % An inner product between row vector f and
                            % column vector h' (transpose of h).
                            % This inner product performs the sum.
```

A more accurate option, if the function is known, is the built-in MATLAB function `integral(f,a,b)`. In this case, the mathematical function to be integrated is first established and identified with a "function handle". The established function is passed into the arguments of the built-in MATLAB function, with the specified endpoints.

The function `integral` uses an adaptive internal code to meet default tolerances. The code below provides an example, for the case in which `x` is an undefined function or variable.

```
>> clear x                   % Make sure x is undefined.
>> f = @(x)tanh(5*x);        % The @ is a "function handle"
>> area = integral(f,a,b); % so the function can be passed.
```

MATLAB has a built-in function that can do an integration resulting in a sampled function, not just a final area. The MATLAB function `cumtrapz` provides a calculation of a cumulative integral by using the trapezoidal rule (Eq. 8.5.4). If `f` is a sampled mathematical function (or sampled data), `cumtrapz` produces a sampled function of the numerical integral of `f`, over a specified sampled interval from a to b. The numerical integral is performed in the context of the area of the curve between a and b. That means, if the result of `cumtrapz` is plotted versus `x`, it starts with a value of `0` at `x(1)`, i.e., at $x = a$. A sample set of lines is below.

```
>> x = a:h:(b-h);  % An array of values not including the endpoint.
>> f = tanh(5*x); % An array of function evaluations.
>> fintegral = cumtrapz(x,f);
```

The resulting variable `fintegral` has the same dimensions as arrays `f` and `x`.

Example 8.6

Find an approximate value for $\int_0^2 x^{1/3}dx$ using the trapezoidal rule of integration with eight increments.

Solution

The formula for the trapezoidal rule of integration is given by Eq. 8.5.5. It is, using $h = \frac{2}{8} = \frac{1}{4}$,

$$\begin{aligned}\int_0^2 x^{1/3}dx &\cong \frac{1}{8}(f_0 + 2f_1 + 2f_2 + \cdots + 2f_7 + 2f_8)\\ &= \frac{1}{8}\left[0 + 2\left(\frac{1}{4}\right)^{1/3} + 2\left(\frac{2}{4}\right)^{1/3} + \cdots + 2\left(\frac{7}{4}\right)^{1/3} + 2^{1/3}\right]\\ &= \frac{1}{8}[2(.63 + .794 + .909 + 1.0 + 1.077 + 1.145 + 1.205) + 1.26]\\ &= \underline{1.85}\end{aligned}$$

This compares with the exact value of

$$\int_0^2 x^{1/3}dx = \frac{3}{4}x^{4/3}\bigg|_0^2 = \frac{3}{4}(2)^{4/3} = 1.89$$

Example 8.7

Derive the integration formula using central differences with the largest error.

Solution

The integral of interest would be $\int_{x_{i-1}}^{x_{i+1}} f(x)dx$. In difference notation it would be expressed as

$$\int_{x_{i-1}}^{x_{i+1}} f(x)dx = D^{-1}(f_{i+1} - f_{i-1}) = \frac{(E - E^{-1})}{D} f_i$$
$$= \frac{(E^{1/2} + E^{-1/2})(E^{1/2} - E^{-1/2})}{D} f_i$$
$$= \frac{\delta 2\mu}{D} f_i$$

using the results of Example 8.1. With the appropriate expression from Table 8.2, we have

$$\int_{x_{i-1}}^{x_{i+1}} f(x)dx = \frac{2\mu\delta}{(\mu/h)(\delta - \delta^3/6 + \delta^5/30 - \cdots)} f_i$$

Dividing, we get, neglecting terms of $o(h^4)$ in the series expansion,

$$\int_{x_{i-1}}^{x_{i+1}} f(x)dx = 2h\left(1 + \frac{\delta^2}{6}\right) f_i$$
$$= \frac{h}{3}(f_{i+1} + 4f_i + f_{i-1}), \qquad e = o(h^5)$$

Note that it would not be correct to retain the δ^4-term in the above since it would result in f_{i+2} and f_{i-2}, quantities outside the limits of integration. The integration formula is then

$$\int_a^b f(x)dx = \frac{h}{3}[(f_0 + 4f_1 + f_2) + (f_2 + 4f_3 + f_4) + \cdots$$
$$+ (f_{N-4} + 4f_{N-3} + f_{N-2}) + (f_{N-2} + 4f_{N-1} + f_N)]$$
$$= \frac{h}{3}(f_0 + 4f_1 + 2f_2 + 4f_3 + 2f_4 + \cdots + 2f_{N-2} + 4f_{N-1} + f_N)$$

This is identical to Simpson's one-third rule.

8.6 NUMERICAL INTERPOLATION

We often desire information at points other than a multiple of Δx, or at points other than at the entries in a table of numbers. The value desired is f_{i+n}, where n is not an integer but some fractional number such as $\frac{2}{3}$ (see Fig. 8.4). This is simply written as

$$E^n f_i = f_{i+n} \tag{8.6.1}$$

In terms of the forward difference operator Δ, we have

$$(1+\Delta)^n f_i = f_{i+n} \tag{8.6.2}$$

or, by using the binomial theorem, we have

$$(1+\Delta)^n = 1 + n\Delta + \frac{n(n-1)}{2}\Delta^2 + \frac{n(n-1)(n-2)}{6}\Delta^3 + \cdots \tag{8.6.3}$$

There results

$$f_{i+n} = \left[1 + n\Delta + \frac{n(n-1)}{2}\Delta^2 + \frac{n(n-1)(n-2)}{6}\Delta^3 + \cdots\right] f_i \tag{8.6.4}$$

Neglecting terms of order higher than Δ^3, this becomes

$$f_{i+n} = \left[f_i + n(f_{i+1} - f_i) + \frac{n(n-1)}{2}(f_{i+2} - 2f_{i+1} + f_i) + \frac{n(n-1)(n-2)}{6}(f_{i+3} - 3f_{i+2} + 3f_{i+1} - f_i)\right] \tag{8.6.5}$$

If we desired f_{i-n}, where n is a fraction, we would use backward differences to obtain

$$f_{i-n} = \left[f_i - n(f_i - f_{i-1}) + \frac{n(n-1)}{2}(f_i - 2f_{i-1} + f_{i-2}) - \frac{n(n-1)(n-2)}{6}(f_i - 3f_{i-1} + 3f_{i-2} - f_{i-3})\right] \tag{8.6.6}$$

This formula would be used to interpolate for a value near the end of a set of numbers.

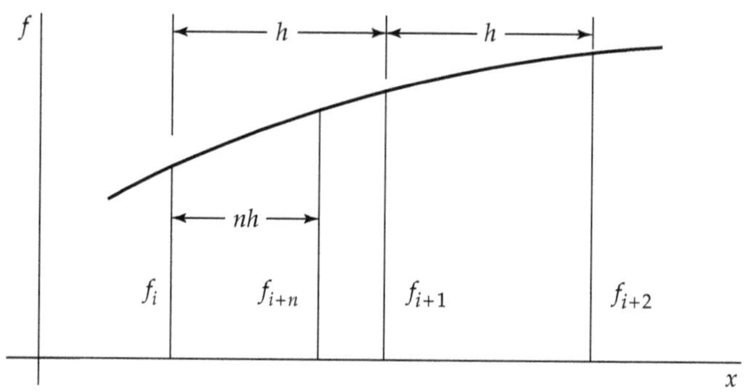

FIGURE 8.4 Numerical interpolation.

Example 8.8

Find the value for the Bessel function $J_0(x)$ at $x = 2.06$ using numerical interpolation with (a) $e = o(h^2)$ and (b) $e = o(h^3)$. Use forward differences and four significant numbers.

Solution

a) Using Eq. 8.6.4 with $e = o(h^2)$, we have

$$\begin{aligned} f_{i+n} &= (1+n\Delta)f_i \\ &= f_i + n(f_{i+1} - f_i) \end{aligned}$$

The table for Bessel functions in the Appendix is given with $h = 0.1$. For our problem,

$$n = \frac{0.06}{0.1} = 0.6$$

The interpolated value is then, using the ith term corresponding to $x = 2.0$,

$$\begin{aligned} J_0(2.06) = f_{i+0.6} &= 0.2239 + 0.6(0.1666 - 0.2239) \\ &= \underline{0.2583} \end{aligned}$$

This is a *linear interpolation*, the method used most often when interpolating for values in a table.

b) Now, let us determine a more accurate value for $J_0(2.06)$. Equation 8.6.4 with $e = o(h^3)$ is

$$\begin{aligned} f_{i+n} &= \left[1 + n\Delta + \frac{1}{2}(n)(n-1)\Delta^2\right] f_i \\ &= f_i + n(f_{i+1} - f_i) + \frac{n(n-1)}{2}(f_{i+2} - 2f_{i+1} + f_i) \end{aligned}$$

Again, using $n = 0.6$, we have

$$\begin{aligned} J_0(0.06) = f_{i+0.6} &= 0.2239 + 0.6(0.1666 - 0.2239) \\ &\quad + \frac{0.6(0.6-1)}{2}(0.1104 - 2\times 0.1666 + 0.2239) \\ &= \underline{0.2582} \end{aligned}$$

Note that the linear interpolation was not valid to four significant numbers; the next-order interpolation scheme was necessary to obtain the fourth significant number.

8.7 ROOTS OF EQUATIONS

It is often necessary to find roots of equations, that is, the values of x for which $f(x) = 0$. This was encountered when solving the characteristic equation of ordinary differential equations with constant coefficients. It may also be necessary to find roots of equations when using numerical methods in solving differential equations. We will study one technique that is commonly used in locating roots; it is *Newton's method,* sometimes called the *Newton–Raphson method.* We make a guess at the root, say $x = x_0$. Using this value of x_0 we calculate $f(x_0)$ and $f'(x_0)$ from the given equation

$$f(x) = 0 \tag{8.7.1}$$

Then, a Taylor series with $e = o(h^2)$ is used to predict an improved value for the root. Using Eq. 8.3.1, this is written as

$$f(x_0 + h) = f(x_0) + hf'(x_0) \tag{8.7.2}$$

We presume that $f(x_0)$ will not be zero, since we only guessed at the root. What we desire from Eq. 8.7.2 is that $f(x_0 + h) = 0$; then $x_1 = x_0 + h$ will be our next guess for the root. Setting $f(x_0 + h) = 0$ and solving for h, we have

$$h = -\frac{f(x_0)}{f'(x_0)} \tag{8.7.3}$$

The next guess is then

$$x_1 = x_0 - \frac{f(x_0)}{f'(x_0)} \tag{8.7.4}$$

Adding another iteration gives a third guess as

$$x_2 = x_1 - \frac{f(x_1)}{f'(x_1)} \tag{8.7.5}$$

or, in general,

$$x_{n+1} = x_n - \frac{f(x_n)}{f'(x_n)} \tag{8.7.6}$$

This process can be visualized by considering the function $f(x)$ displayed in Fig. 8.5. Let us search for the root x_r shown. Assume that the first guess x_0 is too small, so that $f(x_0)$ is negative as shown and $f'(x_0)$ is positive. The first derivative $f'(x_0)$ is equal to $\tan \alpha$. Then, from Eq. 8.7.3,

$$\tan \alpha = \frac{-f(x_0)}{h} \tag{8.7.7}$$

where h is the horizontal leg on the triangle shown. The next guess can then be observed to be

$$x_1 = x_0 + h \tag{8.7.8}$$

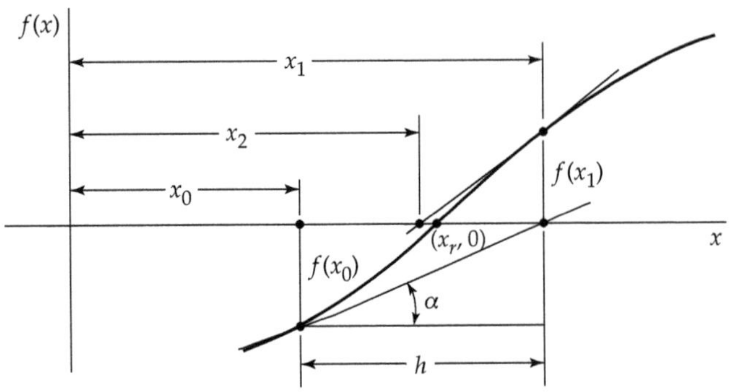

FIGURE 8.5 Newton's method.

Repeating the steps above gives x_2 as shown. A third iteration can be added to the figure with x_3 being very close to x_r. It is obvious that this iteration process converges to the root x_r. However, there are certain functions $f(x)$ for which the initial guess must be very close to a root for convergence to result. An example of this kind of function is shown in Fig. 8.6a. An initial guess outside the small increment Δx will lead to one of the other two roots shown and not to x_r. The root x_r would be quite difficult to find using Newton's method.

Another type of function for which Newton's method may give trouble is shown in Fig. 8.6b. By making the guess x_0, following Newton's method, the first iteration would yield x_1. The next iteration could yield a value x_2 close to the initial guess x_0. The process would just repeat itself indefinitely. To avoid an infinite loop of this nature, we should set a maximum number of iterations for our calculations.

One last word of caution is in order. Note from Eq. 8.7.3 that if we ever guess a point on the curve where $f'(x_0) = 0$, or approximately zero, then h is undefined or extremely large and the process will not work. A new guess may be attempted, or we may use Taylor series with $e = o(h^3)$, neglecting the first derivative term; in that case,

$$f(x_0 + h) = f(x_0) + \frac{h^2}{2} f''(x_0) \tag{8.7.9}$$

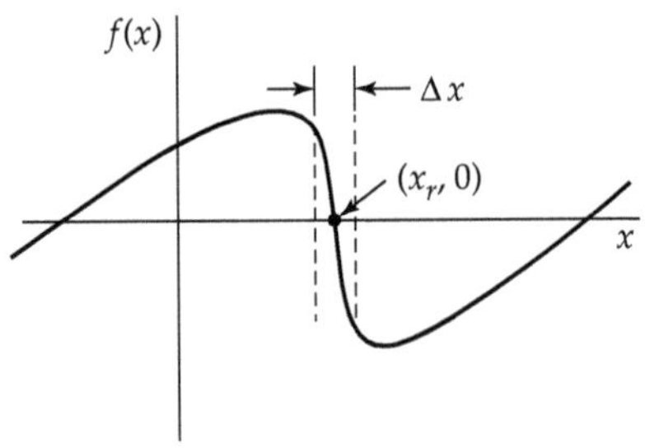

(a)

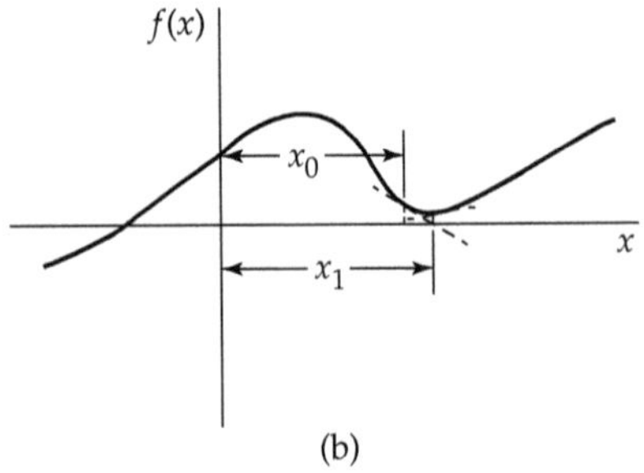

(b)

FIGURE 8.6 Examples for which Newton's method gives trouble.

Setting $f(x_0 + h) = 0$, we have

$$h^2 = -\frac{2f(x_0)}{f''(x_0)} \tag{8.7.10}$$

This step would then be substituted into the iteration process in place of the step in which $f'(x_0) \cong 0$.

Example 8.9

Find at least one root of the equation $x^5 - 10x + 100 = 0$. Carry out four iterations from an initial guess.

Solution

The function $f(x)$ and its first derivative are

$$f(x) = x^5 - 10x + 100$$
$$f'(x) = 5x^4 - 10$$

Note that the first derivative is zero at $x^4 = 2$. This gives a value $x = \pm\sqrt[4]{2}$. So, let's keep away from these points of zero slope. A positive value of $x > 1$ is no use since $f(x)$ will always be positive, so let's try $x_0 = -2$. At $x_0 = -2$, we have

$$f(x_0) = 88$$
$$f'(x_0) = 70$$

For the first iteration, Eq. 8.7.4 gives

$$x_1 = -2.0 - \frac{88}{70} = -3.26$$

Using this value, we have

$$x_2 = -3.26 - \frac{-235}{555} = -2.84$$

The third iteration gives

$$x_3 = -2.84 - \frac{-56.1}{315} = -2.66$$

Finally, the fourth iteration results in

$$x_4 = -2.66 - \frac{-6}{240} = \underline{-2.64}$$

If a more accurate value of the root is desired, a fifth iteration would be necessary. Obviously, a computer would zero in on this root with extreme accuracy in just a few iterations.

Please note that the first derivative was required when applying Newton's method. There are situations in which the first derivative is very difficult, if not impossible, to write explicitly. For those situations we would form the first derivative using a numerical expression such as that given by Eq. 8.4.2.

8.8 INITIAL-VALUE PROBLEMS—ORDINARY DIFFERENTIAL EQUATIONS

One of the most important and useful applications of numerical analysis is in the solution of differential equations, both ordinary and partial differential equations. There are two common problems encountered when finding the numerical solution to a differential equation. The first is: When one finds a numerical solution, is the solution acceptable; that is, is it sufficiently close to the exact solution? If one has an analytical solution, this can easily be checked; but for a problem for which an analytical solution is not known, one must be careful in concluding that a particular numerical solution is acceptable. When extending a solution from x_i to x_{i+1}, a truncation error is incurred, as discussed in Section 8.4, and as the solution is extended across the interval of interest, this error accumulates to give an accumulated truncation error. After, say, 100 steps, this error must be sufficiently small so as to give acceptable results. Obviously, all the various methods give different accumulated error. Usually, a method is chosen that requires a minimum number of steps, requiring the shortest possible computer time, yet one that does not give excessive error.

The second problem often encountered in numerical solutions to differential equations is the instability of numerical solutions. The actual solution to the problem of interest is stable (well behaved), but the errors incurred in the numerical solution are magnified in such a way that the numerical solution is obviously incompatable with the actual solution. This often results in a wildly oscillating solution in which extremely large variations occur in the dependent variable from one step to the next. When this happens, the numerical solution is unstable. By changing the step size or by changing the numerical method, a stable numerical solution can usually be found.

A numerical method that gives accurate results and is stable with the least amount of computer time often requires that it be "started" with a somewhat less accurate method and then continued with a more accurate technique. There are, of course, a host of starting techniques and methods that are used to continue a solution. We shall consider only a few methods; the first will not require starting techniques and will be the most inaccurate. However, the various methods do include the basic ideas of numerical solution of differential equations, and hence are quite important.

We shall initially focus our attention on solving first-order equations, since, generally speaking, it is possible to solve a system of first-order equations instead of a higher-order equation. Hence, the methods outlined in this section can be used to solve higher-order equations.

Many of the examples in which ordinary differential equations describe the phenomenon of interest involve time as the independent variable; thus, we shall use time t in place of the independent variable x of the preceding sections. Naturally, the difference operators are used as defined, with t substituted for x.

Any first-order equation can be put in the form

$$\dot{y} = f(y, t) \tag{8.8.1}$$

where $\dot{y} = dy/dt$. If a solution (y_t and $\dot{y}_t$) at t_i is known, then Eq. 8.8.1 can be used to give the solution (y_{i+1} and $\dot{y}_{i+1}$) at t_{i+1}. We shall assume that the necessary condition is given at a particular time t_0.

8.8.1 Taylor's Method

A simple technique for solving a first-order differential equation is to use a Taylor series, which if written in difference notation is

$$y_{i+1} = y_i + h\dot{y}_i + \frac{h^2}{2}\ddot{y}_i + \frac{h^3}{6}\dddot{y}_i + \cdots \tag{8.8.2}$$

where h is the step size $(t_{i+1} - t_i)$. This obviously requires several derivatives, depending on the order of the terms truncated. These derivatives are found by differentiating the equation

$$\dot{y} = f(y, t) \tag{8.8.3}$$

Since we consider the function f to depend on the two variables y and t, where $y = y(t)$, we must be careful when differentiating with respect to t. For example, consider $\dot{y} = 2y^2t$. Then to find $\ddot{y}$ we must differentiate a product and use the chain rule to give

$$\ddot{y} = 4y\dot{y}t + 2y^2 \tag{8.8.4}$$

and, differentiating again,

$$\dddot{y} = 4yt + 4\dot{y}\dot{y}^2t + 8y\dot{y} \tag{8.8.5}$$

Higher-order derivatives follow.

By knowing an initial condition, y_0 at $t = t_0$, the first derivative $\dot{y}_0$ is calculated from the given differential equation and $\ddot{y}_0$ and $\dddot{y}_0$ from equations similar to Eqs. 8.8.4 and 8.8.5. The value y_1 at $t = t_1$ then follows by putting $i = 0$ in Eq. 8.8.2 which is truncated appropriately. The derivatives, at $t = t_1$, are then calculated from Eqs. 8.8.3, 8.8.4, and 8.8.5. This procedure is continued to the maximum t that is of interest.

This method can also be used to solve higher-order equations simply by expressing the higher-order equation as a set of first-order equations and proceeding with a simultaneous solution of the set of equations. We address this in detail in Section 8.9.

8.8.2 Euler's Method

Euler's method results from the definition of the derivative as

$$\frac{dy}{dt} = \lim_{\Delta t \to 0} \frac{\Delta y}{\Delta t} \tag{8.8.6}$$

We then have the approximation

$$\Delta y \cong \dot{y}\,\Delta t \tag{8.8.7}$$

or, in difference notation,

$$y_{i+1} = y_i + h\dot{y}_i, \qquad e = o(h^2) \tag{8.8.8}$$

This is immediately recognized as the first-order approximation of Taylor's method; thus, we would expect for the same step size that more accurate results would occur if Taylor's method is used, retaining higher-order terms. Euler's method is, of course, simpler to use, since we do not have to compute the higher derivatives at each point. It could also be used to solve higher-order equations, as will be illustrated later.

8.8.3 Adams' Method

Adams' method is one of the multitude of the more accurate methods that exists. It illustrates another technique of solving first-order differential equations.

The Taylor series allows us to write

$$\begin{aligned} y_{i+1} &= y_i + \left(hD + \frac{h^2D^2}{2} + \frac{h^3D^3}{6} + \cdots \right) y_i \\ &= y_i + \left(1 + \frac{hD}{2} + \frac{h^2D^2}{6} + \cdots \right) hDy_i \end{aligned} \tag{8.8.9}$$

Let us neglect terms of order h^5 and greater so that $e = o(h^5)$. Then, writing D in terms of ∇ (see Table 8.2), we have

$$\begin{aligned} y_{i+1} &= y_i + h\left[1 + \frac{1}{2}\left(\nabla + \frac{\nabla^2}{2} + \frac{\nabla^3}{3} + \cdots \right) + \frac{1}{6}(\nabla^2 + \nabla^3 + \cdots) + \frac{1}{24}(\nabla^3 + \cdots) \right] Dy_i \\ &= y_i + h\left(1 + \frac{\nabla}{2} + \frac{5\nabla^2}{12} + \frac{3\nabla^3}{8} \right) Dy_i, \qquad e = o(h^5) \end{aligned} \tag{8.8.10}$$

Using the notation, $Dy_i = \dot{y}_i$, the equation above can be put in the form (using Table 8.1)

$$y_{i+1} = y_i + \frac{h}{24}(55\dot{y}_i - 59\dot{y}_{i-1} + 37\dot{y}_{i-2} - 9\dot{y}_{i-3}), \qquad e = o(h^5) \tag{8.8.11}$$

Adams' method uses the above expression to predict y_{i+1} in terms of previous information. This method requires several starting values, which could be obtained by Taylor's or Euler's methods, usually using smaller step sizes to maintain accuracy. Note that the first value obtained by Eq. 8.8.11 would be y_4. Thus, we must use a different technique to find y_1, y_2, and y_3. If we were to use Adams' method with $h = 0.1$ we could choose Taylor's method with $e = o(h^3)$ and use $h = 0.02$ to find the starting values so that the same accuracy as that of Adams' method results. We would then apply Taylor's method for 15 steps and use every fifth value for y_1, y_2, and y_3 to be used in Adams' method. Equation 8.8.11 is then used to continue the solution. The method is quite accurate, however, since $e = o(h^5)$.

Adams' method can be used to solve a higher-order equation by writing the higher-order equation as a set of first-order equations, or by differentiating Eq. 8.8.11 to give the higher-order derivatives. One such derivative would be

$$\dot{y}_{i+1} = \dot{y}_i + \frac{h}{24}(55\ddot{y}_i - 59\ddot{y}_{i-1} + 37\ddot{y}_{i-2} - 9\ddot{y}_{i-3}) \tag{8.8.12}$$

Others follow naturally.

8.8.4 Runge–Kutta Methods

In order to produce accurate results using Taylor's method, derivatives of higher order must be evaluated. This may be difficult, or the higher-order derivatives may be inaccurate. Adams' method requires several starting values, which may be obtained by less accurate methods, resulting in larger truncation error than desirable. Methods that require only the first-order derivative and give results with the same order of truncation error as Taylor's method maintaining the higher-order derivatives, are called the

Runge–Kutta methods. Estimates of the first derivative must be made at points within each interval $t_i \le t \le t_{i+1}$. The prescribed first-order equation is used to provide the derivative at the interior points. The Runge–Kutta method with $e = o(h^3)$ will be developed and methods with $e = o(h^4)$ and $e = o(h^5)$ will simply be presented with no development.

Let us again consider the first-order equation $\dot{y} = f(y, t)$. All Runge–Kutta methods utilize the approximation

$$y_{i+1} = y_i + h\phi_i \tag{8.8.13}$$

where ϕ_i is an approximation to the slope in the interval $t_i \le t \le t_{i+1}$. Certainly, if we used $\phi_i = f_i$, the approximation for y_{i+1} would be too large for the curve in Fig. 8.7; and, if we used $\phi_i = f_{i+1}$, the approximation would be too small. Hence, the correct ϕ_i needed to give the exact y_{i+1} lies in the interval $f_i \le \phi_i \le f_{i+1}$. The trick is to find a technique that will give a good approximation to the correct slope ϕ_i. Let us assume that

$$\phi_i = a\xi_i + b\eta_i \tag{8.8.14}$$

where

$$\xi_i = f(y_i, t_i) \tag{8.8.15}$$

$$\eta_i = f(y_i + qh\xi_i, t_i + ph) \tag{8.8.16}$$

The quantities a, b, p, and q are constants to be established later.

A good approximation for η_i is found by expanding in a Taylor series, neglecting higher-order terms, as

$$\begin{aligned}\eta_i &= f(y_i,t_i) + \frac{\partial f}{\partial y}(y_i,t_i)\Delta y + \frac{\partial f}{\partial t}(y_i,t_i)\Delta t + o(h^2)\\ &= f_i + qhf_i\frac{\partial f}{\partial y}(y_i,t_i) + ph\frac{\partial f}{\partial t}(y_i,t_i) + o(h^2)\end{aligned} \tag{8.8.17}$$

where we have used $\Delta y = qhf_i$ and $\Delta t = ph$, as required by Eq. 8.8.16. Equation 8.8.13 then becomes, using $\xi_i = f_i$,

$$\begin{aligned}y_{i+1} &= y_i + h\phi_i\\ &= y_i + h(a\xi_i + b\eta_i)\\ &= y_i + h(af_i + bf_i) + h^2\left[bqf_i\frac{\partial f}{\partial y}(y_i,t_i) + bp\frac{\partial f}{\partial t}(y_i,t_i)\right] + o(h^3)\end{aligned} \tag{8.8.18}$$

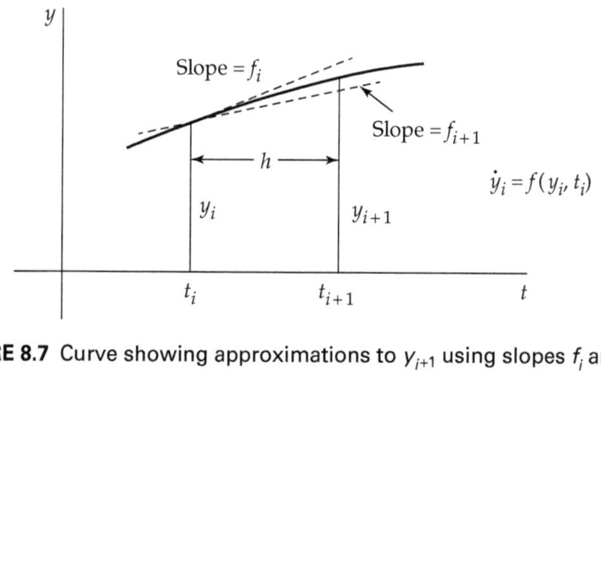

FIGURE 8.7 Curve showing approximations to y_{i+1} using slopes f_i and f_{i+1}.

where we have substituted for ξ_i and η_i from Eqs. 8.8.15 and 8.8.17, respectively. Expand y_i in a Taylor series, with $e = o(h^3)$, as

$$\begin{aligned} y_{i+1} &= y_i + h\dot{y}_i + \frac{h^2}{2}\ddot{y}_i \\ &= y_i + hf(y_i, t_i) + \frac{h^2}{2}\dot{f}(y_i, t_i) \end{aligned} \tag{8.8.19}$$

Now, using the chain rule,

$$\begin{aligned} \dot{f} &= \frac{\partial f}{\partial y}\frac{dy}{dt} + \frac{\partial f}{\partial t}\frac{dt}{dt} \\ &= \dot{y}\frac{\partial f}{\partial y} + \frac{\partial f}{\partial t} \\ &= f\frac{\partial f}{\partial y} + \frac{\partial f}{\partial t} \end{aligned} \tag{8.8.20}$$

Thus, we have

$$y_{i+1} = y_i + hf(y_i, t_i) + \frac{h^2}{2}\left[f_i \frac{\partial f}{\partial y}(y_i, t_i) + \frac{\partial f}{\partial t}(y_i, t_i) \right] \tag{8.8.21}$$

Comparing this with Eq. 8.8.18, we find that (equating terms in like powers of h)

$$\begin{aligned} a + b &= 1 \\ bq &= \frac{1}{2} \\ bp &= \frac{1}{2} \end{aligned} \tag{8.8.22}$$

These three equations contain four unknowns, hence one of them is arbitrary. It is customary to choose $b = \frac{1}{2}$ or $b = 1$. For $b = \frac{1}{2}$, we have $a = \frac{1}{2}$, $q = 1$, and $p = 1$. Then our approximation for y_{i+1} from Eq. 8.8.18 becomes

$$\begin{aligned} y_{i+1} &= y_i + h(a\xi_i + b\eta_i) \\ &= y_i + \frac{h}{2}[f(y_i, t_i) + f(y_i + hf_i, t_i + h)], \qquad e = o(h^3) \end{aligned} \tag{8.8.23}$$

For $b = 1$, we have $a = 0$, $q = \frac{1}{2}$, and $p = \frac{1}{2}$, there results

$$\begin{aligned} y_{i+1} &= y_i + h\eta_i \\ &= y_i + hf\left(y_i + \frac{h}{2}f_i, t_i + \frac{h}{2} \right), \qquad e = o(h^3) \end{aligned} \tag{8.8.24}$$

Knowing y_i, t_i, and $\dot{y}_i = f_i$ we can now calculate y_{i+1} with the same accuracy obtained using Taylor's method that required us to know $\ddot{y}_i$.

The Runge–Kutta method, with $e = o(h^4)$, could be developed in a similar manner. First, the function ϕ_i would be assumed to have the form

$$\phi_i = a\xi_i + b\eta_i + c\zeta_i \tag{8.8.25}$$

where

$$\begin{aligned} \xi_i &= f(y_i, t_i) \\ \eta_i &= f(y_i + ph\xi_i, t_i + ph) \\ \zeta_i &= f[y_i + sh\eta_i + (r-s)h\xi_i, t_i + rh] \end{aligned} \tag{8.8.26}$$

Equating coefficients of the Taylor series expansions would result in two arbitrary coefficients. The common choice results in $a = \frac{1}{6}$, $b = \frac{2}{3}$, and $c = \frac{1}{6}$. We then have

$$y_{i+1} = y_i + \frac{h}{6}(\xi_i + 4\eta_i + \zeta_i), \qquad e = o(h^4) \tag{8.8.27}$$

with

$$\begin{aligned} \xi_i &= f(y_i, t_i) \\ \eta_i &= f\left(y_i + \frac{h}{2}\xi_i, t_i + \frac{h}{2}\right) \\ \zeta_i &= f(y_i + 2h\eta_i - h\xi_i, t_i + h) \end{aligned} \tag{8.8.28}$$

The Runge–Kutta method with $e = o(h^5)$ is perhaps the most widely used method for solving ordinary differential equations. One such method results in

$$y_{i+1} = y_i + \frac{h}{6}[\xi_i + (2-\sqrt{2})\eta_i + (2+\sqrt{2})\zeta_i + \omega_i], \qquad e = o(h^5) \tag{8.8.29}$$

where

$$\begin{aligned} \xi_i &= f(y_i, t_i) \\ \eta_i &= f\left(y_i + \frac{h}{2}\xi_i, t_i + \frac{h}{2}\right) \\ \zeta_i &= f\left[y_i + \frac{h}{\sqrt{2}}(\xi_i - \eta_i) - \frac{h}{2}(\xi_i - 2\eta_i), t_i + \frac{h}{2}\right] \\ \omega_i &= f\left[y_i - \frac{h}{\sqrt{2}}(\eta_i - \zeta_i) + h\zeta_i, t_i + h\right] \end{aligned} \tag{8.8.30}$$

Another method with $e = o(h^5)$ that is widely used gives

$$y_{i+1} = y_i + \frac{h}{6}(\xi_i + 2\eta_i + 2\zeta_i + \omega_i), \qquad e = o(h^5) \tag{8.8.31}$$

where

$$\begin{aligned} \xi_i &= f(y_i, t_i) \\ \eta_i &= f\left(y_i + \frac{h}{2}\xi_i, t_i + \frac{h}{2}\right) \\ \zeta_i &= f\left(y_i + \frac{h}{2}\eta_i, t_i + \frac{h}{2}\right) \\ \omega_i &= f(y_i + h\zeta_i, t_i + h) \end{aligned} \tag{8.8.32}$$

In all the methods above no information is needed other than the initial condition. For example, y_1 is approximated by using y_0, ξ_0, η_0, and so on. The quantities are found from the given equation with no differentiation required. These reasons, combined with the accuracy of the Runge–Kutta methods, make them extremely popular.

8.8.5 Direct Method

The final method that will be discussed is seldom used because of its inaccuracy, but it is easily understood and follows directly from the expressions for the derivatives presented in Section 8.4. It also serves to illustrate the method used to solve partial differential equations. Let us again use as an example the first-order differential equation

$$\dot{y} = 2y^2 t \tag{8.8.33}$$

Then, from Table 8.3, with $e = o(h^2)$, and using forward differences, we have

$$\dot{y}_i = \frac{dy_i}{dt} = Dy_i = \frac{1}{2h}(-y_{i+2} + 4y_{i+1} - 3y_i) \tag{8.8.34}$$

Substitute this directly into Eq. 8.8.33 to obtain

$$\frac{1}{2h}(-y_{i+2} + 4y_{i+1} - 3y_i) = 2y_i^2 t_i \tag{8.8.35}$$

This is rearranged to give

$$y_{i+2} = 4y_{i+1} - (3 + 4hy_i t_i)y_i, \qquad e = o(h^2) \tag{8.8.36}$$

Using $i = 0$, we can determine y_2 if we know y_1 and y_0. This requires a starting technique to find y_1. We could use Euler's method, since that also has $e = o(h^2)$.

This method can easily be used to solve higher-order equations. We simply substitute from Table 8.3 for the higher-order derivatives and find an equation similar to Eq. 8.8.36 to advance the solution.

Let us now work some examples using the techniques of this section.

Example 8.10

Use Euler's method to solve $\dot{y} + 2yt = 4$ if $y(0) = 0.2$. Compare with Taylor's method, $e = o(h^3)$. Use $h = 0.1$. Carry the solution out for four time steps.

Solution

In Euler's method we must have the first derivative at each point; it is given by

$$\dot{y}_i = 4 - 2y_i t_i$$

The solution is approximated at each point by

$$y_{i+1} = y_i + h\dot{y}_i$$

For the first four steps there results

$$\begin{aligned}
t_0 &= 0: & y_0 &= 0.2 \\
t_1 &= 0.1: & y_1 &= y_0 + h\dot{y}_0 = 0.2 + 0.1 \times 4 = \underline{0.6} \\
t_2 &= 0.2: & y_2 &= y_1 + h\dot{y}_1 = 0.6 + 0.1 \times 3.88 = \underline{0.988} \\
t_3 &= 0.3: & y_3 &= y_2 + h\dot{y}_2 = 0.988 + 0.1 \times 3.60 = \underline{1.35} \\
t_4 &= 0.4: & y_4 &= y_3 + h\dot{y}_3 = 1.35 + 0.1 \times 3.19 = \underline{1.67}
\end{aligned}$$

Using Taylor's method with $e = o(h^3)$, we approximate y_{i+1} using

$$y_{i+1} = y_i + h\dot{y}_i + \frac{h^2}{2}\ddot{y}_i$$

Thus we see that we need $\ddot{y}$. It is found by differentiating the given equation, providing us with

$$\ddot{y}_i = -2\dot{y}_i t_i - 2y_i$$

Progressing in time as in Euler's method, there results

$$\begin{aligned}
t_0 &= 0: \quad y_0 = 0.2 \\
t_1 &= 0.1: \quad y_1 = y_0 + h\dot{y}_0 + \frac{h^2}{2}\ddot{y}_0 = 0.2 + 0.1 \times 4 + 0.005 \times (-0.4) = \underline{0.598} \\
t_2 &= 0.2: \quad y_2 = y_1 + h\dot{y}_1 + \frac{h^2}{2}\ddot{y}_1 = 0.598 + 0.1 \times 3.88 + 0.005 \times (-1.97) = \underline{0.976} \\
t_3 &= 0.3: \quad y_3 = y_2 + h\dot{y}_2 + \frac{h^2}{2}\ddot{y}_2 = \underline{1.32} \\
t_4 &= 0.4: \quad y_4 = y_3 + h\dot{y}_3 + \frac{h^2}{2}\ddot{y}_3 = \underline{1.62}
\end{aligned}$$

Example 8.11

Use a Runge–Kutta method with $e = o(h^5)$ and solve $\dot{y} + 2yt = 4$ if $y(0) = 0.2$ using $h = 0.1$. Carry out the solution for two time steps.

Solution

We will choose Eq. 8.8.31 to illustrate the Runge–Kutta method. The first derivative is used at various points interior to each interval. It is found from

$$\dot{y}_i = 4 - 2y_i t_i$$

To find y_1 we must know y_0, ξ_0, η_0, ζ_0, and ω_0. They are

$$\begin{aligned}
y_0 &= 0.2 \\
\xi_0 &= 4 - 2y_0 t_0 = 4 - 2 \times 0.2 \times 0 = 4 \\
\eta_0 &= 4 - 2\left(y_0 + \frac{h}{2}\xi_0\right)\left(t_0 + \frac{h}{2}\right) = 4 - 2\left(0.2 + \frac{0.1}{2} \times 4\right)\left(\frac{0.1}{2}\right) = 3.96 \\
\zeta_0 &= 4 - 2\left(y_0 + \frac{h}{2}\eta_0\right)\left(t_0 + \frac{h}{2}\right) = 4 - 2\left(0.2 + \frac{0.1}{2} \times 3.96\right)\left(\frac{0.1}{2}\right) = 3.96 \\
\omega_0 &= 4 - 2(y_0 + h\zeta_0)(t_0 + h) = 4 - 2(0.2 + 0.1 \times 3.96)(0.1) = 3.88
\end{aligned}$$

Thus,

$$\begin{aligned}
y_1 &= y_0 + \frac{h}{6}(\xi_0 + 2\eta_0 + 2\zeta_0 + \omega_0) \\
&= 0.2 + \frac{0.1}{6}(3.96 + 7.92 + 7.92 + 3.88) = \underline{0.595}
\end{aligned}$$

To find y_2 we calculate

$$\xi_1 = 4 - 2y_1t_1 = 4 - 2\times 0.595\times 0.1 = 3.88$$

$$\eta_1 = 4-2\left(y_1+\frac{h}{2}\xi_1\right)\left(t_1+\frac{h}{2}\right) = 4-2\left(0.595+\frac{0.1}{2}\times 3.88\right)\left(0.1+\frac{0.1}{2}\right) = 3.76$$

$$\zeta_1 = 4-2\left(y_1+\frac{h}{2}\eta_1\right)\left(t_1+\frac{h}{2}\right) = 4-2\left(0.595+\frac{0.1}{2}\times 3.76\right)\left(0.1+\frac{0.1}{2}\right) = 3.77$$

$$\omega_1 = 4-2(y_1+h\zeta_1)(t_1+h) = 4-2(0.595+0.1\times 3.76)(0.1+0.1) = 3.61$$

Finally,

$$y_2 = y_1 + \frac{h}{6}(\xi_1 + 2\eta_1 + 2\zeta_1 + \omega_1)$$

$$= 0.595 + \frac{0.1}{6}(3.88 + 7.52 + 7.54 + 3.61) = \underline{0.971}$$

Additional values follow. Note that the procedure above required no starting values and no higher-order derivatives, but still $e = o(h^5)$.

Example 8.12

Use the direct method to solve the equation $\dot{y} + 2yt = 4$ if $y(0) = 0.2$ using $h = 0.1$. Use forward differences with $e = o(h^2)$. Carry out the solution for four time steps.

Solution

Using the direct method we substitute for $\dot{y}_i = Dy_i$ from Table 8.3 with $e = o(h^2)$. We have

$$\frac{1}{2h}(-y_{i+2} + 4y_{i+1} - 3y_i) + 2y_it_i = 4$$

Rearranging, there results

$$y_{i+2} = 4y_{i+1} - (3 - 4ht_i)y_i - 8h$$

The first value that we can find with the formula above is y_2. Hence, we must find y_1 by some other technique. Use Euler's method to find y_1. It is

$$y_1 = y_0 + h\dot{y}_0 = 0.2 + 0.1\times 4 = \underline{0.6}$$

We now can use the direct method to find

$$
\begin{aligned}
y_2 &= 4y_1 - (3 - 4ht_0)y_0 - 8h \\
&= 4 \times 0.6 - (3 - 0) \times 0.2 - 8 \times 0.1 = \underline{1.0} \\
y_3 &= 4y_2 - (3 - 4ht_1)y_1 - 8h \\
&= 4 \times 1.0 - (3 - 4 \times 0.1 \times 0.1) \times 0.6 - 8 \times 0.1 = \underline{1.424} \\
y_4 &= 4y_3 - (3 - 4ht_2)y_2 - 8h \\
&= 4 \times 1.424 - (3 - 4 \times 0.1 \times 0.2) \times 1.0 - 8 \times 0.1 = \underline{1.976}
\end{aligned}
$$

These results are, of course, less accurate than those obtained using Taylor's method or the Runge-Kutta method in the preceding two examples.

8.9 HIGHER-ORDER EQUATIONS

Taylor's method can be used to solve higher-order differential equations without representing them as a set of first-order differential equations. Consider the third-order equation

$$\dddot{y} + 4t\ddot{y} + 5y = t^2 \tag{8.9.1}$$

with three required initial conditions imposed at $t = 0$, namely, y_0, $\dot{y}_0$, and $\ddot{y}_0$. Thus, at $t = 0$ all the necessary information is known and y_1 can be found from the Taylor series, with $e = o(h^3)$,

$$y_1 = y_0 + h\dot{y}_0 + \frac{h^2}{2}\ddot{y}_0 \tag{8.9.2}$$

To find y_2 the derivatives $\dot{y}_1$ and $\ddot{y}_1$ would be needed. To find them we differentiate the Taylor series to get

$$
\begin{aligned}
\dot{y}_1 &= \dot{y}_0 + h\ddot{y}_0 + \frac{h^2}{2}\dddot{y}_0 \\
\ddot{y}_1 &= \ddot{y}_0 + h\dddot{y}_0 + \frac{h^2}{2}\left(\frac{d^4y}{dt^4}\right)_0
\end{aligned}
\tag{8.9.3}
$$

The third derivative, $\dddot{y}_0$, is then found from Eq. 8.9.1 and $(d^4y/dt^4)_0$ is found by differentiating Eq. 8.9.1. We can then proceed to the next step and continue during the interval of interest.

Instead of using Taylor's method directly, we could have written Eq. 8.9.1 as the following set of first-order equations:

$$
\begin{aligned}
\dot{y}_i &= u_i \\
\dot{u}_i &= v_i \\
\dot{v}_i &= -4t_i v_i - 5y_i + t_i^2
\end{aligned}
\tag{8.9.4}
$$

The last of these equations results from substituting the first two into Eq. 8.9.1. The initial conditions specified at $t = 0$ would be y_0, $u_0 = \dot{y}_0$, and $v_0 = \ddot{y}_0$. If Euler's method were used we would have, at the first step,

$$v_1 = v_0 + \dot{v}_0 h = v_0 + (-4t_0 v_0 - 5y_0 + t_0^2)h$$
$$u_1 = u_0 + \dot{u}_0 h = u_0 + v_0 h \qquad (8.9.5)$$
$$y_1 = y_0 + \dot{y}_0 h = y_0 + u_0 h, \quad e = o(h^2)$$

With the values at $t_0 = 0$ known we could perform the calculations. This procedure is continued for all additional steps. Other methods for solving the first-order equations could also be used.

We also could have chosen the direct method by expressing Eq. 8.9.1 in finite-difference notation using the information contained in Table 8.3. For example, the forward-differencing relationships could be used to express Eq. 8.9.1 as

$$y_{i+3} - 3y_{i+2} + 3y_{i+1} - y_i + 4t_i h(y_{i+2} - 2y_{i+1} + y_i) + 5y_i h^3 = t_i^2 h^3, \qquad e = o(h) \qquad (8.9.6)$$

This is rewritten as

$$y_{i+3} = (3 - 4t_i h)y_{i+2} - (3 - 8t_i h)y_{i+1} + (1 - 4t_i h - 5h^3)y_i + t_i^2 h^3 \qquad (8.9.7)$$

For $i = 0$, this becomes, using the initial condition $y = y_0$ at $t_0 = 0$,

$$y_3 = 3y_2 - 3y_1 + 1(1 - 5h^3)y_0 \qquad (8.9.8)$$

To find y_3 we would need the starting values y_1 and y_2. They could be found by using Euler's method. Equation 8.9.7 would then be used until all values of interest are determined.

A decision that must be made when solving problems numerically is how small the step size should be. The phenomenon of interest usually has a time scale T, or a length L, associated with it. T would be the time necessary for a complete cycle of a periodic phenomenon, or the time required for a transient phenomenon to disappear; see Fig. 8.8. The length L may be the distance between telephone poles or the size of a capillary tube. What is necessary is that $h \ll T$ or $h \ll L$. If the numerical results using the various techniques or smaller step sizes differ considerably, this usually implies that h is not sufficiently small.

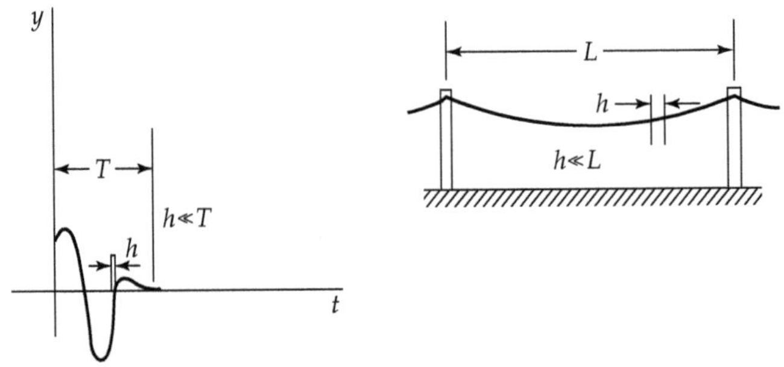

FIGURE 8.8 Examples of how the step size h should be chosen.

Example 8.13

Solve the differential equation $\ddot{y} - 2ty = 5$ with initial conditions $y(0) = 2$ and $\dot{y}(0) = 0$. Use Adams' method with $h = 0.2$ using Taylor's method with $e = o(h^3)$ to start the solution. Carry out the solution for five time steps.

Solution

Adams' method predicts the dependent variables at a forward step to be

$$y_{i+1} = y_i + \frac{h}{24}(55\dot{y}_i - 59\dot{y}_{i-1} + 37\dot{y}_{i-2} - 9\dot{y}_{i-3})$$

Differentiating this expression results in

$$\dot{y}_{i+1} = \dot{y}_i + \frac{h}{24}(55\ddot{y}_i - 59\ddot{y}_{i-1} + 37\ddot{y}_{i-2} - 9\ddot{y}_{i-3})$$

These two expressions can be used with $i \geq 3$; hence, we need a starting technique to give y_1, y_2, and y_3. We shall use Taylor's method with $e = o(h^3)$ to start the solution. Taylor's method uses

$$y_{i+1} = y_i + h\dot{y}_i + \frac{h^2}{2}\ddot{y}_i$$

$$\dot{y}_{i+1} = \dot{y}_i + h\ddot{y}_i + \frac{h^2}{2}\dddot{y}_i$$

This requires the second and third derivatives; the second derivative is provided by the given differential equation,

$$\ddot{y}_i = 5 + 2t_i y_i$$

The third derivative is found by differentiating the above and is

$$\dddot{y}_i = 2y_i + 2t_i\dot{y}_i$$

Taylor's method provides the starting values.

$$y_1 = y_0 + h\dot{y}_0 + \frac{h^2}{2}\ddot{y}_0 = \underline{2.1}$$

$t_1 = 0.2$:

$$\dot{y}_1 = \dot{y}_0 + h\ddot{y}_0 + \frac{h^2}{2}\dddot{y}_0 = 1.08$$

$$y_2 = y_1 + h\dot{y}_1 + \frac{h^2}{2}\ddot{y}_1 = \underline{2.43}$$

$t_2 = 0.4$:

$$\dot{y}_2 = \dot{y}_1 + h\ddot{y}_1 + \frac{h^2}{2}\dddot{y}_1 = 2.34$$

$$y_3 = y_2 + h\dot{y}_2 + \frac{h^2}{2}\ddot{y}_2 = \underline{3.04}$$

$t_3 = 0.6$:

$$\dot{y}_3 = \dot{y}_2 + h\ddot{y}_2 + \frac{h^2}{2}\dddot{y}_2 = 3.86.$$

Now Adams' method can be used to predict additional values:

$$y_4 = y_3 + \frac{h}{24}(55\dot{y}_3 - 59\dot{y}_2 + 37\dot{y}_1 - 9\dot{y}_0) = \underline{3.99}$$

$t_4 = 0.8$:

$$\dot{y}_4 = \dot{y}_3 + \frac{h}{24}(55\ddot{y}_3 - 59\ddot{y}_2 + 37\ddot{y}_1 - 9\ddot{y}_0) = 5.84$$

$$y_5 = y_4 + \frac{h}{24}(55\dot{y}_4 - 59\dot{y}_3 + 37\dot{y}_2 - 9\dot{y}_1) = \underline{5.41}$$

t_4 was used to evaluate $\dot{y}_4$, needed for y_5. t_5 was used to obtain $\dot{y}_5$, needed for y_6, and so on.

$t_5 = 1.0$:

$$\dot{y}_5 = \dot{y}_4 + \frac{h}{24}(55\ddot{y}_4 - 59\ddot{y}_3 + 37\ddot{y}_2 - 9\ddot{y}_1) = 8.51$$

Additional values can be found similarly.

Extension 8.13.1: Compare the values y_4 and y_5 from Adams' method above with values found by extending Taylor's method.

8.13.2: Write the differential equation in central difference form with $e = o(h^2)$ and show that

$$y_{i+1} = 5h^2 + 2(1 - t_i h^2)y_i - y_{i-1}$$

Use this expression to find y_2, y_3, and y_4. Compare with the above.

Example 8.14

Solve the differential equation $\ddot{y} - 2ty = 5$ with initial conditions $y(0) = 2$ and $\dot{y}(0) = 0$. Use the direct method with $e = o(h^2)$, using forward differences and $h = 0.2$. Start the solution with the values from Example 8.13 and carry out five time steps.

Solution

We write the differential equation in difference form using the relationships of Table 8.3. There results

$$\frac{1}{h^2}(-y_{i+3} + 4y_{i+2} - 5y_{i+1} + 2y_i) - 2t_i y_i = 5$$

This is rearranged as

$$y_{i+3} + 4y_{i+2} - 5y_{i+1} + (2 - 2t_i h^2)y_i - 5h^2$$

Letting $i = 0$, the first value that we can find from the equation above is y_3. Thus, we need to use a starting method to find y_1 and y_2. From Example 8.13 we have $y_1 = \underline{2.1}$ and $y_2 = \underline{2.43}$. Now we can use the equation of this example to find y_3. It is, letting $i = 0$,

$$y_3 = 4y_2 - 5y_1 + (2 - 2\overset{0}{\not{t_0}}h^2)y_0 - 5h^2 = \underline{3.02}$$

Two additional values are found as follows:

$$y_4 = 4y_3 - 5y_2 + (2 - 2t_1h^2)y_1 - 5h^2 = \underline{3.90}$$
$$y_5 = 4y_4 - 5y_3 + (2 - 2t_2h^2)y_2 - 5h^2 = \underline{5.08}$$

This method is, of course, less accurate than the method of the previous example. It is, however, easier to use, and if a smaller step size were chosen, more accurate numbers would result.

Example 8.15

Solve the differential equation $\ddot{y} - 2ty = 5$ using MATLAB. Set the initial conditions as $y(0) = 2$ and $\dot{y}(0) = 0$, and plot the results for y, $\dot{y}$, and $\ddot{y}$. (a) Write a code using Adams' method with 44 time steps at $h = .04$, and use Taylor's method with $e = o(h^3)$ using $h = 0.02$ to start the solution. (b) Write a code that uses the Runge-Kutta method based on the built-in MATLAB function `ode45`, using 44 time steps at $h = .04$. (c) Plot the results for both the Adams' method and the Runge-Kutta method.

Solution

a) The MATLAB script for Adam's method, started with Taylor's method, is below.

```
% Solve an ODE using Adams' method.
% The ODE is y" - 2t y = 5.
%
clear
y(1) = 2.0;
dy(1) = 0.0;
d2y(1) = 5.0;
t = 0.0;
h = 0.02;
%
% Solve d2y - 2 t y = 5 for y having the initial value of 1 and dy being
% initially equal to 0.
%
% First use Taylor's method with h = 0.02 to find the starting values.
for i = 1:6
    t = t+ 0.02;
```

```
    y(i+1) = y(i) + h*dy(i) + (h*h/2)*d2y(i);
    dy(i+1) = dy(i)+h*d2y(i) + (h*h/2)*(2*t*dy(i) + 2*y(i));
    d2y(i+1) = 5 + 2*t*y(i+1);
end
% Keep the time samples with h = 0.04, and index accordingly.
for i=1:4
    y(i) =  y(2*i-1);
    dy(i) = dy(2*i-1);
    d2y(i) = d2y(2*i-1);
end
% Now use Adam's method.
h = 0.04;
t = h*(0:44);
for i = 4:44
    y(i + 1) = y(i) + (h/24)*(55*dy(i) - 59*dy(i -1) + 37*dy(i-2) - 9*dy(i-3));
    dy(i+1)=dy(i)+(h/24)*(55*d2y(i)-59*d2y(i-1)+37*d2y(i-2)-9*d2y(i-↙
3));
    d2y(i+1) = 5 + 2*(t(i)+h)*y(i+1);
end
%
% Plot y, dy, and d2y versus time.
figure(1); plot(t,y,'b'); hold on; plot(t, dy, 'b--'); plot(t, d2y,'b-.'); hold off;
set(gca,'FontSize',14);
xlabel('t','fontsize',18); ylabel('solutions','fontsize',18);
lgd = legend('y','dy/dt', 'd^2y/dt^2','Location','northwest');
lgd.FontSize = 16;
```

b) The MATLAB script using `ode45` and plotting commands is below.

```
% Solve an ODE using ode45 (Runge-Kutta).
%The ODE is y" - 2t y = 5.
%
clear
% Set up the time samples
h = 0.04;
t = h*(0:44);
% specify the initial conditions
y0 = 2.0;
dy0 = 0.0;
x0 = [y0 dy0];

% Call the integration program, ode45
% This command finds the ODE in first-order form in a separate file called
% diffeq.m , which is in the same directory as the current Matlab path.
% Remember to set the path to the desired directory.
[t,x] = ode45(@diffeq,t,x0);
```

```
% Note that variable x contains the numerical solutions for y and dy/dt,
% sampled at times t, organized as two columns in x.
% The first column, x(:,1), contains the time samples of y.
% The second column, x(:,2), contains time samples of dy/dt.
%
% Since ode45 does not need to keep information about d^2 y/ dt^2, we get it
% from the ode:
d2y = 5 + 2*t.*x(:,1);

% Plot y, dy, and d2y versus time.
figure(2);plot(t,x(:,1),'b');hold on;plot(t,x(:,2),'b--');plot(t,d2y,'b-.');hold off;
set(gca,'FontSize',14);
xlabel('t','fontsize',18); ylabel('solutions','fontsize',18);
lgd = legend('y', 'dy/dt', 'd^2y/dt^2','Location','northwest');
lgd.FontSize =16;
```

The built-in MATLAB function `ode45` calls a user-defined function, here named `diffeq`, containing the ODE written as a set of first-order equations. The function is in a separate file. The file name is the same as the function name, i.e., `diffeq.m`.

```
function xdot = diffeq(t,x)
% function name MUST match the file name, e.g. diffeq.m
% Write the ODE as a set of first-order ODEs.
xdot = zeros(2,1);
xdot(1) = x(2):
xdot(2) = 2*t*x(1)+5;
return
```

c) The solution from Adams' method is on the left, and from Runge-Kutta on the right. The difference in this example is visually imperceptible.

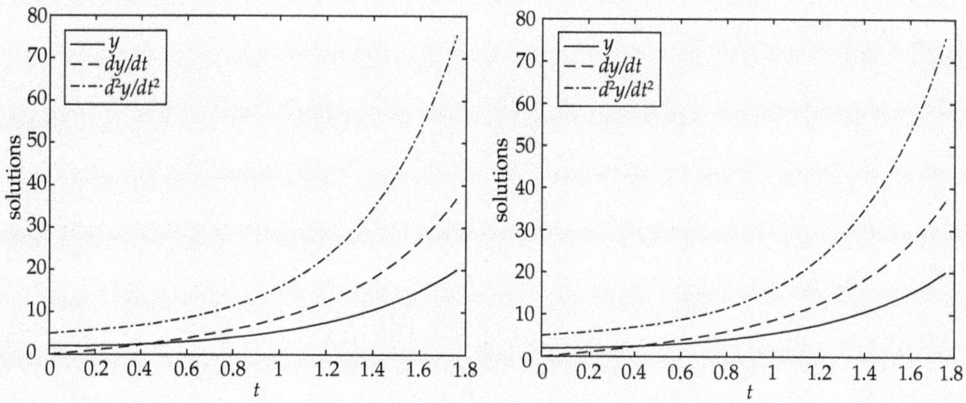

8.10 BOUNDARY-VALUE PROBLEMS—ORDINARY DIFFERENTIAL EQUATIONS

The initial-value problem, for which all the necessary conditions are given at a particular point or instant, was considered in the previous section. Now we shall consider problems for which the conditions are given at two different positions. For example, in the hanging string problem, information for the second-order describing equation is known at $x = 0$ and $x = L$. It is a boundary-value problem. Boundary-value problems are very common in physical applications; thus, several techniques to solve them will be presented.

8.10.1 Iterative Method

Suppose that we are solving the second-order differential equation

$$\ddot{y} + 3x\dot{y} + (x^2 - 1)y = \sin\frac{\pi x}{4} \tag{8.10.1}$$

This requires that two conditions be given; let them be $y = 0$ at $x = 0$, and $y = 0$ at $x = 6$. Because the conditions are given at two different values of the independent variable, it is a boundary-value problem. Now, if we knew $\dot{y}$ at $x = 0$ it would be an initial-value problem and Taylor's (or any other) method could be used. So, let's *assume* a value for $\dot{y}_0$ and proceed as though it is an initial-value problem. Then, when $x = 6$ is reached, the boundary condition there requires that $y = 0$. Of course, in general, this condition will not be satisfied and the procedure must be repeated with another guess for $\dot{y}_0$. An interpolation (or extrapolation) scheme could be employed to zero in on the correct $\dot{y}_0$. The procedure works for both linear and nonlinear equations.

An interpolation scheme that can be employed when using a computer is derived by using a Taylor series with $e = o(h^2)$. We consider the value of y at $x = 6$, let's call it y_N, to be the dependent variable and y_0 to be the independent variable. Then, using $\dot{y}_0^{(1)}$ and $\dot{y}_0^{(2)}$ to be the first two guesses leading to the values $y_N^{(1)}$ and $y_N^{(2)}$, respectively, we anticipate by linear interpolation that

$$y_N^{(3)} = y_N^{(2)} + \frac{y_N^{(2)} - y_N^{(1)}}{\dot{y}_0^{(2)} - \dot{y}_0^{(1)}}\left[\dot{y}_0^{(3)} - \dot{y}_0^{(2)}\right] \tag{8.10.2}$$

The quantity $y_N^{(3)}$ will hopefully be zero. So we set it to zero and calculate a new guess to be

$$\dot{y}_0^{(3)} = \dot{y}_0^{(2)} - \frac{\dot{y}_0^{(2)} - \dot{y}_0^{(1)}}{y_N^{(2)} - y_N^{(1)}} y_N^{(2)} \tag{8.10.3}$$

Using this value for $\dot{y}_0^{(3)}$, we calculate $y_N^{(3)}$. If it is not sufficiently close to zero, we go through another iteration and find a new value $y_N^{(4)}$. Each additional value should be nearer to zero and the iterations are stopped when y_N is sufficiently small.

8.10.2 Superposition

For a linear equation we can use the principle of superposition. Consider Eq. 8.10.1 with the same boundary conditions. Completely ignore the given boundary conditions and choose any arbitrary set of intitial conditions, for example, $y^{(1)}(0) = 1$ and $\dot{y}^{(1)}(0) = 0$. This leads to a solution $y^{(1)}(x)$. Now, change the initial conditions to, for example, $y^{(2)}(0) = 0$ and $\dot{y}^{(2)}(0) = 1$. The solution $y^{(2)}(x)$ would follow. The solutions are now superposed, made possible because of the linear equation, to give the desired solution as

$$y(t) = c_1 y^{(1)}(x) + c_2 y^{(2)}(x) + y_p(x) \tag{8.10.4}$$

The actual boundary conditions are then used to determine c_1 and c_2.

If a third-order equation were being solved, then three arbitrary, but different, sets of initial conditions would lead to three constants to be determined by the boundary conditions. The method for solving the initial-value problems could be any of those described earlier.

A word of caution is necessary. We must be careful when we choose the two sets of initial conditions. They must be chosen so that the two solutions generated are, in fact, independent. If either of the arbitrary constants c_1 and c_2 is determined to be zero, the solutions are not independent. Other sets of initial conditions may then be attempted in search of two independent solutions.

8.10.3 Simultaneous Equations

Let's write Eq. 8.10.1 in finite-difference form for each step in the interval of interest. The equations are written for each value of i and then all the equations are solved simultaneously. There are a sufficient number of equations to equal the unknowns y_i. In finite-difference form, using forward differences with $e = o(h)$, Eq. 8.10.1 is

$$\frac{1}{h^2}(y_{i+2} - 2y_{i+1} + y_i) + \frac{3x_i}{h}(y_{i+1} - y_i) + (x_i^2 - 1)y_i = \sin\frac{\pi}{4}x_i \tag{8.10.5}$$

Now write the equations for each value of i, $i = 0$ to $i = N$. Using $x_0 = 0$, $x_1 = h$, $x_2 = 2h$, and so on, and choosing $h = 1.0$ so that $N = 6$, there results

$$\begin{aligned}
y_2 - 2y_1 + y_0 + 3x_0(y_1 - y_0) + (x_0^2 - 1)y_0 &= \sin\frac{\pi}{4}x_0 \\
y_3 - 2y_2 + y_1 + 3x_1(y_2 - y_1) + (x_1^2 - 1)y_1 &= \sin\frac{\pi}{4}x_1 \\
y_4 - 2y_3 + y_2 + 3x_2(y_3 - y_2) + (x_2^2 - 1)y_2 &= \sin\frac{\pi}{4}x_2 \\
y_5 - 2y_4 + y_3 + 3x_3(y_4 - y_3) + (x_3^2 - 1)y_3 &= \sin\frac{\pi}{4}x_3 \\
y_6 - 2y_5 + y_4 + 3x_4(y_5 - y_4) + (x_4^2 - 1)y_4 &= \sin\frac{\pi}{4}x_4
\end{aligned} \tag{8.10.6}$$

Now, with $y_0 = y_6 = 0$ and $x_0 = 0$, there results

$$
\begin{aligned}
-2y_1 + y_2 \qquad\qquad &= 0 \\
-2y_1 + y_2 + y_3 \qquad &= \sin\frac{\pi}{4} \\
-2y_2 + 4y_3 + y_4 \qquad &= \sin\frac{\pi}{2} \\
7y_4 + \quad y_5 &= \sin\frac{3\pi}{4} \\
4y_4 + 10y_5 &= 0
\end{aligned}
\qquad (8.10.7)
$$

There are five equations which can be solved to give the five unknowns $y_1, y_2, y_3, y_4,$ and y_5. If the number of steps is increased to 100, there would be 99 equations to solve simultaneously. A computer would then be used to solve the algebraic equations.

Equations 8.10.7 are often written in matrix form as

$$
\begin{pmatrix} -2 & 1 & 0 & 0 & 0 \\ -2 & 1 & 1 & 0 & 0 \\ 0 & -2 & 4 & 1 & 0 \\ 0 & 0 & 0 & 7 & 1 \\ 0 & 0 & 0 & 4 & 10 \end{pmatrix}
\begin{pmatrix} y_1 \\ y_2 \\ y_3 \\ y_4 \\ y_5 \end{pmatrix}
=
\begin{pmatrix} 0 \\ \sin\pi/4 \\ \sin\pi/2 \\ \sin 3\pi/4 \\ 0 \end{pmatrix}
\qquad (8.10.8)
$$

or, using matrix notation, as

$$A_{ij} y_i = B_j \qquad (8.10.9)$$

The solution is written as

$$y_i = A_{ij}^{-1} B_j \qquad (8.10.10)$$

where A_{ij}^{-1} is the inverse of A_{ij}. It is found using techniques presented in matrix theory. Usually, matrix inversion routines are available in computing center libraries.

Example 8.16

Solve the boundary-value problem defined by the differential equation $\ddot{y} - 10y = 0$ with boundary conditions $y(0) = 0.4$ and $y(1.2) = 0$. Choose six steps to illustrate the procedure using Taylor's method with $e = o(h^3)$.

Solution

The superposition method will be used to illustrate the numerical solution of the linear equation. We can choose any arbitrary initial conditions, so choose, for the first solution $y_0^{(1)} = 1$ and $\dot{y}_0^{(1)} = 0$. The solution then proceeds as follows for the $y^{(1)}$ solution.

Using $h = 0.2$ and $\ddot{y} = 10\dot{y}$, we have

$$y_0 = 1.0 \qquad \ddot{y}_0 = 10y_0 = 10$$
$$\dot{y}_0 = 0.0 \qquad \dddot{y}_0 = 10\dot{y}_0 = 0$$
$$y_1 = y_0 + h\dot{y}_0 + \frac{h^2}{2}\ddot{y}_0 = 1.2$$
$$\dot{y}_1 = \dot{y}_0 + h\ddot{y}_0 + \frac{h^2}{2}\dddot{y}_0 = 2.0$$
$$y_2 = y_1 + h\dot{y}_1 + \frac{h^2}{2}\ddot{y}_1 = 1.84$$
$$\dot{y}_2 = \dot{y}_1 + h\ddot{y}_1 + \frac{h^2}{2}\dddot{y}_1 = 4.8$$
$$y_3 = 3.17 \qquad y_4 = 5.69 \qquad y_5 = 10.37 \qquad y_6 = 18.96$$
$$\dot{y}_3 = 9.44 \qquad \dot{y}_4 = 17.7 \qquad \dot{y}_5 = 32.6$$

To find $y^{(2)}$ we choose a different set of initial values, say $y_0^{(2)} = 0.0$ and $\dot{y}_0^{(2)} = -1$. Then proceeding as before we find $y^{(2)}$ to be given by

$$y_0 = 0.0$$
$$\dot{y}_0 = -1.0$$
$$y_1 = y_0 + h\dot{y}_0 + \frac{h^2}{2}\ddot{y}_0 = -0.2$$
$$\dot{y}_1 = \dot{y}_0 + h\ddot{y}_0 + \frac{h^2}{2}\dddot{y}_0 = -1.2$$
$$y_2 = -0.48 \qquad y_3 = -1.184 \qquad y_4 = -2.135 \qquad y_5 = -3.89 \qquad y_6 = -7.12$$
$$\dot{y}_2 = -1.84 \qquad \dot{y}_3 = -3.57 \qquad \dot{y}_4 = -6.65 \qquad \dot{y}_5 = -12.25$$

Combine the two solutions with the usual superposition technique and obtain

$$y = Ay^{(1)} + By^{(2)}$$

The actual boundary conditions require that

$$0.4 = A(1.0) + B(0.0)$$
$$0 = A(18.96) + B(-7.12)$$

Thus,

$$A = 0.4 \quad \text{and} \quad B = 1.064$$

The solution is then

$$y = 0.4y^{(1)} + 1.064y^{(2)}$$

The solution with independent solutions $y^{(1)}$ and $y^{(2)}$ are tabulated below.

x	0	0.2	0.4	0.6	0.8	1.0	1.2
$y^{(1)}$	1.0	1.2	1.84	3.17	5.69	10.37	18.96
$y^{(2)}$	0.0	−0.20	−0.48	−1.184	−2.135	−3.89	−7.12
y	0.4	0.268	0.224	0.01	0.005	0.002	0.0

Extension 8.16.1: Choose a different set of initial conditions for $y^{(2)}$ and show that the combined solution remains essentially unchanged. Choose the conditions $y_0^{(2)} = 1.0$ and $\dot{y}^{(2)} = 1.0$.

8.11 NUMERICAL STABILITY

In numerical calculations the quantity of interest, the dependent variable, being calculated is dependent on all previous calculations made of this quantity, its derivatives, and the step size. Truncation and round-off errors are contained in each calculated value. If the change in the dependent variable is small for small changes in the independent variable, then the solution is stable. There are times, however, when small step changes in the independent variable lead to large changes in the dependent variable so that a condition of instability exists. It is usually possible to detect such instability, since such results will usually violate physical reasoning. It is possible to predict numerical instabilities for linear equations. It is seldom done, though, since by changing the step size or the numerical method, instabilities can usually be avoided. With the present capacity of high-speed computers, stability problems, if ever encountered, can usually be eliminated by using smaller and smaller step sizes.

When solving partial differential equations with more than one independent variable, stability may be influenced by controlling the relationship between the step sizes chosen. For example, in solving a second-order equation with t and x the independent variables, if a numerical instability results, attempts would be made to eliminate the instability by changing the relationship between Δx and Δt. If this is not successful, a different numerical technique may eliminate the instability.

There are problems, though, for which direct attempts at a numerical solution lead to numerical instability, even though various step sizes are attempted and various methods utilized. This type of problem can often be solved by either using multiple precision (instead of the usual number of significant digits, a computer is capable of solving problems using two or even three times the number of significant digits) or by employing a specially devised technique.

8.12 NUMERICAL SOLUTION OF PARTIAL DIFFERENTIAL EQUATIONS

In Chapter 6 an analytical technique was presented for solving partial differential equations, the separation-of-variables technique. When a solution to a partial differential equation is being sought, one should always attempt to separate the variables even though the ordinary differential equations that result may not lead to an analytical solution directly. It is always advisable to numerically solve a set of ordinary differential equations instead of the original partial differential equation. There are occasions, however, when either the equation will not separate or the boundary conditions will not admit a separated solution. For example, in the heat-conduction problem the general solution, assuming the variables separate, for the long rod was

$$T(x, t) = e^{-k\beta^2 t}[A \sin \beta x + B \cos \beta x] \tag{8.12.1}$$

Attempt to satisfy the end condition that $T(0, t) = 100t$, instead of $T(0, t) = 0$ as was used in Chapter 6. This would require that

$$T(0, t) = 100t = Be^{-k\beta^2 t} \tag{8.12.2}$$

The constant B cannot be chosen to satisfy the condition; hence, the solution 8.12.1 is not acceptable. We may then turn to a numerical solution of the original partial differential equation.

8.12.1 The Diffusion Equation

We can solve the diffusion problem numerically for a variety of boundary and initial conditions. The diffusion equation for a rod of length L, assuming no heat losses from the lateral surfaces, is

$$\frac{\partial T}{\partial t} = a \frac{\partial^2 T}{\partial x^2} \tag{8.12.3}$$

Let the conditions be generalized as $T(0, t) = f(t)$, $T(L, t) = g(t)$, $T(x, 0) = F(x)$. The function $T(x, t)$ is written in difference notation as T_{ij}; this represents the temperature at $x = x_i$ and $t = t_j$. If we hold the time fixed (this is done by keeping j unchanged), the second derivative with respect to x becomes, using a central difference method with $e = o(h^2)$, referring to Table 8.3,

$$\frac{\partial^2 T}{\partial x^2}(x_i, t_i) = \frac{1}{h^2}(T_{i+1,j} - 2T_{i,j} + T_{i-1,j}) \tag{8.12.4}$$

where h is the step size Δx.

The time step Δt is chosen as k. Then, using forward differences on the time derivative, with $e = o(k)$,

$$\frac{\partial T}{\partial t}(x_i, t_j) = \frac{1}{k}(T_{i,j+1} - 2T_{i,j}) \tag{8.12.5}$$

The diffusion equation (8.12.3) is then written, in difference notation,

$$\frac{1}{k}(T_{i,j+1} - T_{i,j}) = \frac{a}{h^2}(T_{i+1,j} - 2T_{i,j} + T_{i-1,j}) \tag{8.12.6}$$

or, by rearranging,

$$T_{i,j+1} = \frac{ka}{h^2}(T_{i+1,j} - 2T_{i,j} + T_{i-1,j}) + T_{i,j} \tag{8.12.7}$$

The given boundary conditions, in difference notation, are

$$T_{0,j} = f(t_j), \qquad T_{N,j} = g(t_j), \qquad T_{i,0} = F(x_i) \tag{8.12.8}$$

where N is the total number of x steps.

A stability analysis, or some experimenting on the computer, would show that a numerical solution may be unstable unless the time step and displacement step satisfy the criterion $ak/h^2 \leq \frac{1}{2}$.

The solution proceeds by determining the temperature at $t_0 = 0$ at the various x locations: that is, $T_{0,0}, T_{1,0}, T_{2,0}, T_{3,0}, \ldots, T_{N,0}$ from the given $F(x_i)$. Equation 8.12.7 then allows us to calculate, at lime $t_1 = k$, the values $T_{0,1}, T_{1,1}, T_{2,1}, T_{3,1}, \ldots, T_{N,1}$. This process is continued to $t_2 = 2k$ to give $T_{0,2}, T_{1,2}, T_{2,2}, T_{3,2}, \ldots, T_{N,2}$ and repeated for all additional t_j's of interest. By choosing the number N of x steps sufficiently large, a satisfactory approximation to the actual temperature distribution should result.

The boundary conditions and the initial condition are given by the conditions (8.12.8). If an insulated boundary condition were imposed at $x = 0$, then $\partial T/\partial x(0, t) = 0$. Using a forward difference this would be expressed by

$$\frac{1}{h}(T_{1,j} - T_{0,j}) = 0 \tag{8.12.9}$$

or

$$T_{1,j} = T_{0,j} \tag{8.12.10}$$

For an insulated boundary at $x = L$, using a backward difference, we would have

$$T_{N,j} = T_{N-1,j} \tag{8.12.11}$$

8.12.2 The Wave Equation

The same technique can be applied to solve the wave equation as was used in the solution of the diffusion equation. We express the wave equation

$$\frac{\partial^2 u}{\partial t^2} = a^2 \frac{\partial^2 u}{\partial x^2} \tag{8.12.12}$$

using central differences for both derivatives, as

$$\frac{1}{k^2}(u_{i,j+1} - 2u_{i,j} + u_{i,j-1}) = \frac{a^2}{h^2}(u_{i+1,j} - 2u_{i,j} + u_{i-1,j}) \tag{8.12.13}$$

This is rearranged as

$$u_{i,j+1} = \frac{a^2 k^2}{h^2}(u_{i+1,j} - 2u_{i,j} + u_{i-1,j}) + 2u_{i,j} - u_{i,j-1} \tag{8.12.14}$$

The boundary and initial conditions are

$$\begin{aligned} u(0, t) &= f(t), \qquad & u(L, t) &= g(t) \\ u(x, 0) &= F(x), \qquad & \frac{\partial u}{\partial t}(x, 0) &= G(x) \end{aligned} \tag{8.12.15}$$

which, if written in difference notation, are

$$u_{0,j} = f(t_j), \qquad u_{N,j} = g(t_j)$$
$$u_{i,0} = F(x_i), \qquad u_{i,1} = u_{i,0} + kG(x_i) \tag{8.12.16}$$

where N is the total number of x steps. The values $u_{i,0}$ and $u_{i,1}$ result from the initial conditions. The remaining values $u_{i,2}$, $u_{i,3}$, $u_{i,4}$, etc., result from Eq. 8.12.14. Hence we can find the numerical solution for $u(x, t)$. Instability is usually avoided in the numerical solution of the wave equation if $ak/h \le 1$.

8.12.3 Laplace's Equation

It is again convenient if Laplace's equation

$$\frac{\partial^2 T}{\partial x^2} + \frac{\partial^2 T}{\partial y^2} = 0 \tag{8.12.17}$$

is written in difference notation using central differences. It is then

$$\frac{1}{h^2}(T_{i+1,j} - 2T_{i,j} + T_{i-1,j}) + \frac{1}{k^2}(T_{i,j+1} - 2T_{i,j} + T_{i,j-1}) = 0 \tag{8.12.18}$$

Solving for $T_{i,j}$, and letting $h = k$ (which is not necessary but convenient), we have

$$T_{i,j} = \frac{1}{4}[T_{i,j+1} + T_{i,j-1} + T_{i+1,j} + T_{i-1,j}] \tag{8.12.19}$$

Using central differences, we see that the temperature at a particular mesh point is the average of the four neighboring temperatures. For example (see Fig. 8.9),

$$T_{8,5} = \frac{1}{4}[T_{8,6} + T_{8,4} + T_{9,5} + T_{7,5}] \tag{8.12.20}$$

The Laplace equation requires that the dependent variable (or its derivative) be specified at all points surrounding a particular region of interest. A typical set of boundary conditions, for a rectangular region of interest, would be

$$T(0, y) = f(y), \qquad T(W, y) = g(y)$$
$$T(x, 0) = F(x), \qquad T(x, H) = G(x) \tag{8.12.21}$$

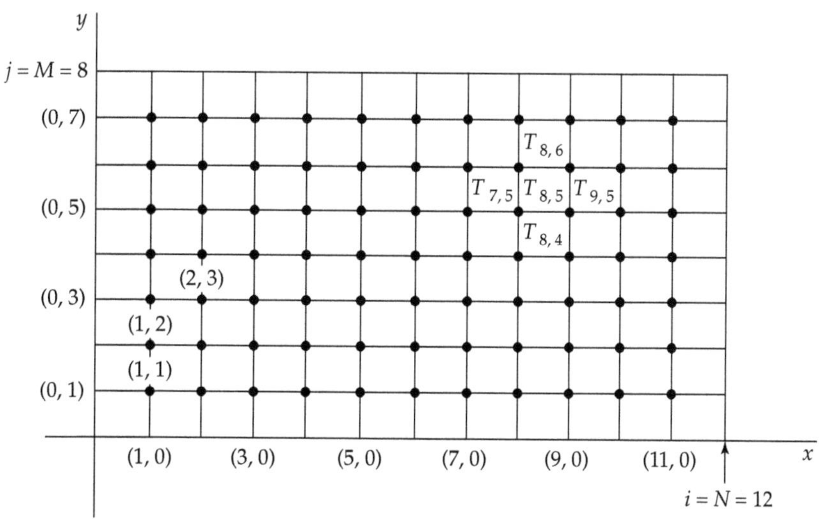

FIGURE 8.9 Typical mesh for the solution of Laplace's equation.

where W and H are the dimensions of the rectangle. In difference notation these conditions are

$$T_{0,j} = f(y_i), \quad T_{N,j} = g(y_i)$$
$$T_{i,0} = F(x_i), \quad T_{i,M} = G(x_i) \tag{8.12.22}$$

The temperature is known at all the boundary mesh points, and with a 12 by 8 mesh, shown in Fig. 8.9, Eq. 8.12.19 would give 77 algebraic equations. These equations, which include the 77 unknowns $T_{i,j}$ at each of the interior points, can then be solved simultaneously. Of course, a computer would be used to solve the set of simultaneous equations; computer programs are generally available to accomplish this.

It is also possible, and often necessary, to specify that no heat transfer occurs across a boundary, so that the temperature gradient is zero. This would, of course, change the conditions (8.12.22).

It should be pointed out that there is a simple technique, especially useful before the advent of the computer, that gives a quick approximation to the solution of Laplace's equation. It is a *relaxation* method. In this method the temperatures at every interior mesh point are guessed at. Then Eq. 8.12.19 is used in a systematic manner by starting, say, at the (1, 1) element, averaging for a new value and working across the first horizontal row, then going to the (1, 2) element and working across the second horizontal row, always using the most recently available values. This is continued until the values at every interior point of the complete mesh of elements are changed from the guessed values. A second iteration is then accomplished by recalculating every temperature again starting at the (1, 1) element. This iteration process is continued until the value at each point converges to a particular number that does not significantly change with successive iterations. This can be done by hand for a fairly large number of mesh points and hence can provide a quick approximation to the solution of Laplace's equation.

Example 8.17

A 1-m-long, laterally insulated rod, originally at 60°C, is subjected at one end to 500°C. Estimate the temperature in the rod as a function of time if the ends are held at 500°C and 60°C, respectively. The diffusivity is 2×10^{-6} m²/s. Use five displacement steps with a time step of 4 ks.

Solution

The diffusion equation describes the heat-transfer phenomenon, hence the difference equation (8.12.7) is used to estimate the temperature at successive times. At time $t = 0$ we have

$$T_{0,0} = 500, T_{1,0} = 60, T_{2,0} = 60, T_{3,0} = 60, T_{4,0} = 60, T_{5,0} = 60$$

The left end will be maintained at 500°C and the right end at 60°C. These boundary conditions are expressed in difference form as $T_{0,j} = 500$ and $T_{5,j} = 60$. Using Eq. 8.12.7 we have, at $t = 4$ ks,

$$T_{i,1} = \frac{ak}{h^2}(T_{i+1,0} - 2T_{i,0} + T_{i-1,0}) + T_{i,0}$$

Letting i assume the values 1, 2, 3, and 4 successively, we have, with $ak/h^2 = \frac{1}{5}$,

$$T_{1,1} = \frac{1}{5}(T_{2,0} - 2T_{1,0} + T_{0,0}) + T_{1,0} = 148$$

$$T_{2,1} = \frac{1}{5}(T_{3,0} - 2T_{2,0} + T_{1,0}) + T_{2,0} = 60$$

$$T_{3,1} = T_{4,1} = T_{5,1} = 60$$

At $t = 8$ ks, there results

$$T_{i,2} = \frac{1}{5}(T_{i+1,1} - 2T_{i,1} + T_{i-1,1}) + T_{i,1}$$

For the various values of i, we have

$$T_{1,2} = \frac{1}{5}(T_{2,1} - 2T_{1,1} + T_{0,1}) + T_{1,1} = 201$$

$$T_{2,2} = \frac{1}{5}(T_{3,1} - 2T_{2,1} + T_{1,1}) + T_{2,1} = 78$$

$$T_{3,2} = T_{4,2} = T_{5,2} = 60$$

At $t = 12$ ks, the temperature is

$$T_{i,3} = \frac{1}{5}(T_{i+1,2} - 2T_{i,2} + T_{i-1,2}) + T_{i,2}$$

yielding

$$T_{1,3} = 236, \quad T_{2,3} = 99, \quad T_{3,3} = 64, \quad T_{4,3} = T_{5,3} = 60$$

At $t = 16$ ks, there results

$$T_{i,4} = \frac{1}{5}(T_{i+1,3} - 2T_{i,3} + T_{i-1,3}) + T_{i,3}$$

giving

$$T_{1,4} = 261, \quad T_{2,4} = 119, \quad T_{3,4} = 70, \quad T_{4,4} = 61, \quad T_{5,4} = 60$$

At $t = 20$ ks, we find that

$$T_{1,5} = 281, \quad T_{2,5} = 138, \quad T_{3,5} = 78, \quad T_{4,5} = 63, \quad T_{5,5} = 60$$

Temperatures at future times follow. The temperatures will eventually approach a linear distribution as predicted by the steady-state solution.

Extension 8.17.1: Solve for the steady-state temperature $T_{1,\infty}$.

8.17.2: Predict the time necessary for the temperature at $x = 0.4$ to reach 145°C.

Example 8.18

A tight 6-m-long string is set in motion by releasing the string from rest, as shown in Fig. 8.10. Find an appropriate solution for the deflection using increments of 1 m and time steps of 0.01 s. The wave speed is 100 m/s.

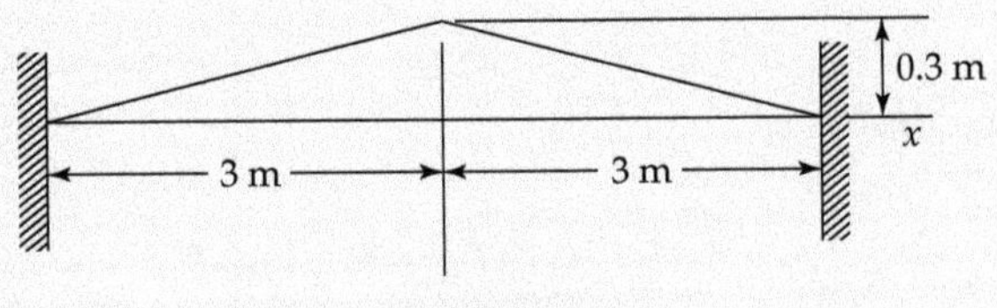

FIGURE 8.10

Solution

The initial displacement is given by

$$u(x,0) = \begin{cases} 0.1x & 0 < x < 3 \\ 0.1(6-x) & 3 < x < 6 \end{cases}$$

In difference form we have

$$u_{0,0} = 0,\ u_{1,0} = 1.0,\ u_{2,0} = 2.0,\ u_{3,0} = 3.0,\ u_{4,0} = 2.0,\ u_{5,0} = 0.1,\ u_{6,0} = 0$$

The initial velocity is zero since the string is released from rest. Using the appropriate condition listed in Eq. 8.12.16 with $G(x) = 0$, we have at $t = 0.01$,

$$u_{0,1} = 0,\ u_{1,1} = 0.1,\ u_{2,1} = 0.2,\ u_{3,1} = 0.3,\ u_{4,1} = 0.2,\ u_{5,1} = 0.1,\ u_{6,1} = 0$$

Now, we can use Eq. 8.12.14, which marches the solution forward in time and obtain, with $a^2k^2/h^2 = 1$,

$$u_{i,j+1} = u_{i+1,j} + u_{i-1,j} + u_{i,j-1}$$

This yields the following solution.

At $t = 0.02$:

$$\begin{aligned} u_{0,2} &= 0 \\ u_{1,2} &= u_{2,1} + u_{0,1} - u_{1,0} = 0.2 + 0 - 0.1 = 0.1 \\ u_{2,2} &= u_{3,1} + u_{1,1} - u_{2,0} = 0.3 + 0.1 - 0.2 = 0.2 \\ u_{3,2} &= u_{4,1} + u_{2,1} - u_{3,0} = 0.2 + 0.2 - 0.3 = 0.1 \\ u_{4,2} &= u_{5,1} + u_{3,1} - u_{4,0} = 0.1 + 0.3 - 0.2 = 0.2 \\ u_{5,2} &= u_{6,1} + u_{4,1} - u_{5,0} = 0 + 0.2 - 0.1 = 0.1 \\ u_{6,2} &= 0 \end{aligned}$$

At $t = 0.03$:

$$u_{0,3} = 0,\ u_{1,3} = 0.1,\ u_{2,3} = 0.0,\ u_{3,3} = 0.1,\ u_{4,3} = 0.0,\ u_{5,3} = 0.1,\ u_{6,3} = 0$$

At $t = 0.04$:

$$u_{0,4} = 0,\ u_{1,4} = -0.1,\ u_{2,4} = 0.0,\ u_{3,4} = -0.1,\ u_{4,4} = 0.0,\ u_{5,4} = -0.1,\ u_{6,4} = 0$$

At $t = 0.05$:

$$u_{0,5} = 0,\ u_{1,5} = 0.1,\ u_{2,5} = -0.2,\ u_{3,5} = -0.1,\ u_{4,5} = -0.2,\ u_{5,5} = -0.1,\ u_{6,5} = 0$$

At $t = 0.06$:

$$u_{0,6} = 0, \ u_{1,6} = -0.1, \ u_{2,6} = -0.2, \ u_{3,6} = -0.3, \ u_{4,6} = -0.2, \ u_{5,6} = -0.1, \ u_{6,6} = 0$$

Two observations are made from the results above. First, the solution remains symmetric, as it should. Second, the numerical results are significantly in error; it should be noted, however, that at $t = 0.06$ s we have completed one half a cycle. This is exactly as it should be, since the frequency is $a/2L$ cycles/second (see Fig. 6.9), and thus it takes $2L/a = 0.12$ s to complete one cycle. Substantially smaller length increments and time steps are necessary to obtain a solution that approximates the actual solution. An acceptable solution would result for this problem if we used 100 length increments and a time step size chosen so that $ak/h = 1$.

To see the numerical error, compare with Example 6.4, which is the same problem with different dimensions; it therefore qualitatively demonstrates the true behavior.

Example 8.19

Solve the problem presented in Example 8.18 using MATLAB. Use 90 spatial increments. Choose $k = 0.0006$ s. Evaluate the parameter $r = ak/h$ and to check whether the calculation should be stable. Plot every 10th time step and every 10th spatial location for one half cycle.

Solution

The MATLAB script and output, including the plotted solution, are below.

```
%
% Finite difference solution to the string equation for one half cycle.
%
% i = index for x, j = index for time, and u(x_i,t_j) is displacement
% Parameters are
clear
a = 100;          % wave speed, m/s
k = 0.0006;       % time step
L = 6;            % length of the string
% Compute the time for one-half cycle, as the time it takes for the
T = L/a;          % wave to travel across string one way.
% Sample the x-space and t-space (including fixed endpoints).
Nt = T/k+1;       % number of time steps. +1 adds the first time point
                  % of the next half cycle for plot visualization.
t = k*(0:Nt);     % length of t is Nt+1
Nx = 90;          % number of spatial intervals. Use an even number.
h = L/Nx;         % spatial increment
x = 0:h:L;        % length of x is Nx+1
%
% Stability assessment
r = a*k/h;
if r <= 1
   disp (['The calculation should be numerically stable, r = ', num2str(r)]);
else
```

```
    disp (['The calculation may be numerically unstable, r = ',num2str(r)]);
end
%
% Initial conditions: Set up initial displacement
for i = 1:(Nx/2)   % even Nx
    u(i,1) = 0.1*x(i);
end
for i = (Nx/2+1):(Nx+1);
    u(i,1) = 0.1*(6.0−x(i));
end
%
% Boundary conditions: u(0,t) = u(L,t) = 0.
for j = 1:Nt
    u(1,j) = 0;
    u(Nx +1,j) = 0;
end
%
% Apply difference equation
%
% At time = 0.0006 sec, u(1,2) = u(1,1), (approximately) since the initial
% velocity is zero.
for i = 1:Nx
    u(i,2)=u(i,1);
end
for j = 2:Nt
    for i = 2:Nx
        u(i,j+1) = (a*k/h)^2*(u(i+1,j) − 2*u(i,j) + u(i−1,j)) + 2*u(i,j) − u(i, j−1);
    end
end
% Notice that if r = (ak/h)^2 = 1, the above equation simplifies (two
% terms cancel), which in turn makes it easier if doing hand calculations. With
% the computer, it is not so important. If r > 1 the computation is numerically
% unstable.
%
% Plot snapshots of the displacement solution for every 10th time sample.
%
inc = 10;
figure(1); plot(x,u(:,1:inc:Nt));
set(gca:'FontSize',14);
title('Snapshots of string vibration','FontSize',18);
xlabel('x','fontsize',18); ylabel('u','fontsize',18);
axis([min(x) max(x) min(min(u)) max(max(u))]);
%
% Plot one half cycle of motion for every 10th point on half the string.
%
figure(2); plot(t,u(1:inc:Nx/2,:)');
set(gca,'FontSize',14);
title('Points on string versus time','fontSize',18);
xlabel('t','fontsize',18); ylabel('u','fontsize',18);
axis([min(t) max(t) min(min(u)) max(max(u))]);
```

The text output from this MATLAB script is

```
>> The calculation should be numerically stable, r = 0.9
```

The script produces the figures below. The plot on the left shows snapshots of one-half cycle of animated string vibration. The plot on the right shows one-half cycle of motion for selected points on the string.

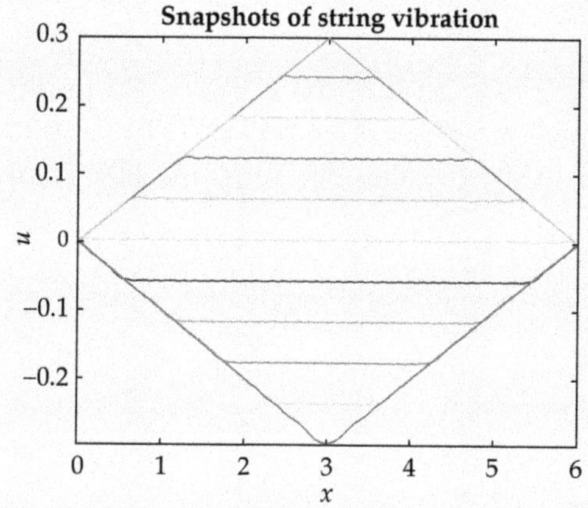

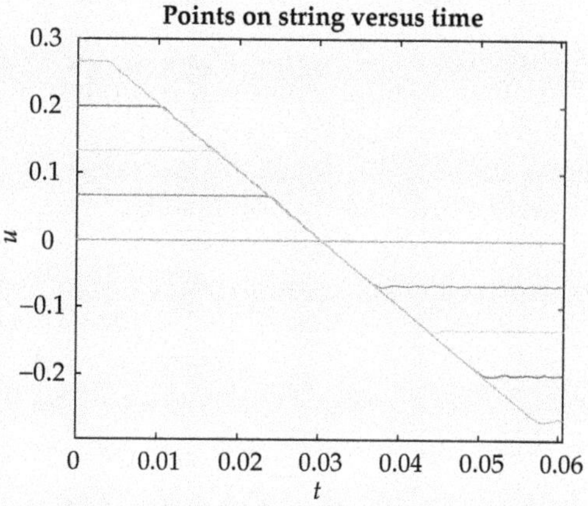

Example 8.20

A 50- by 60-mm flat plate, insulated on both flat surfaces, has its edges maintained at 0, 100, 200, and 300°C, in that order, going counterclockwise. Using the relaxation method, determine the steady-state temperature at each grid point, using a 10 × 10 mm grid.

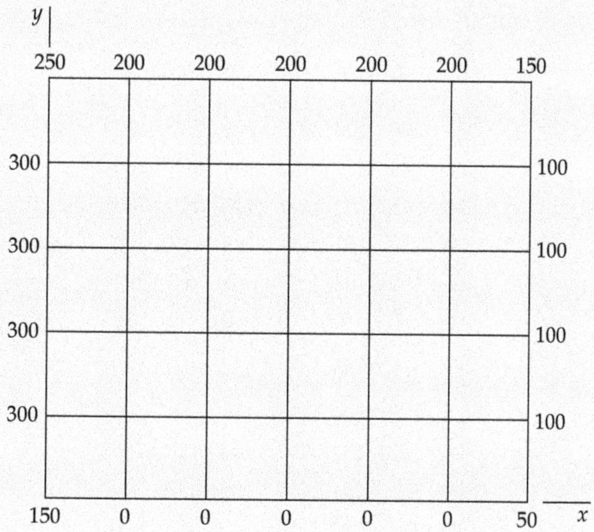

Solution

The grid is set up as shown. Note that the corner temperatures are assumed to be the average of the neighboring two temperatures. The actual solution does not involve the corner temperatures. We start assuming a temperature at each grid point; the more accurate our assumption, the fewer iterations required for convergence. Let us assume the following:

250	200	200	200	200	200	150
300	290	270	240	200	150	100
300	280	250	220	180	130	100
300	200	150	100	100	100	100
300	50	50	50	50	50	100
150	0	0	0	0	0	50

The first iteration comes by applying Eq. 8.12.19 to each of the interior grid points above. Starting at the lower left interior grid point (50°C) and continuing to the second row up (200°C), using the corrected values in the process, the following improved temperature distribution results:

250	200	200	200	200	200	150
300	259	232	207	180	152	100
300	267	230	196	161	130	100
300	217	163	135	117	110	100
300	138	84	58	52	63	100
100	0	0	0	0	0	50

We continue the iterations until there is no significant change in additional iterations. Three more iterations are listed below:

250	200	200	200	200	200	150
300	246	216	194	168	147	100
300	252	212	180	127	121	100
300	220	170	138	118	104	100
300	150	93	70	62	68	100
100	0	0	0	0	0	50

250	200	200	200	200	200	150
300	240	209	186	168	148	100
300	244	202	165	140	122	100
300	219	167	136	108	99	100
300	153	98	77	66	68	100
100	0	0	0	0	0	50

250	200	200	200	200	200	150
300	237	204	183	168	148	100
300	240	194	162	140	122	100
300	216	164	128	108	99	100
300	154	100	76	63	66	100
100	0	0	0	0	0	50

Note that in the last three iterations, the maximum change in temperature from one iteration to the next is 34, 15, and 8°C, respectively. Two more iterations should result in a steady-state temperature distribution, accurate to within about 1°C.

PROBLEMS

8.1 Derive expressions for $\Delta^4 f_i$, $\nabla^4 f_i$, and $\delta^4 f_i$.

8.2 Show that $\nabla \Delta f_i = \delta^2 f_i$.

8.3 Show that all of the difference operators commute with one another, i.e., $\Delta E = E\Delta$.

8.4 Verify that $\nabla E = \Delta = \delta E^{1/2}$.

8.5 Prove that $E^{-1/2} = \mu - \delta/2$ and that $\mu\delta f_i = \frac{1}{2}(f_{i+1} - f_{i-1})$. Also, find an expression for $\mu\delta^3 f_i$.

8.6 Use the binomial theorem $(a + x)^n = a^n + na^{n-1}x + n(n-1)a^{n-2}x^2/2! + n(n-1)(n-2)a^{n-3}x^3/3! + \cdots$ and find a series expression for Δ in terms of δ.

8.7 Show that $2\mu\delta = \nabla + \Delta$. Also express $(E^2 - E^{-2})$ in terms of δ and μ.

8.8 Verify the following expressions by squaring the appropriate series.

$$\Delta^2 = h^2D^2 + h^3D^3 + \frac{7}{12}h^4D^4 + \cdots$$

$$D^2 = \frac{1}{h^2}\left(\Delta^2 - \Delta^3 + \frac{11}{12}\Delta^4 - \frac{5}{6}\Delta^5 + \cdots\right)$$

8.9 Relate the backward difference operator ∇ to the differential operator D using h as the step size. Also find ∇^2 in terms of D, and D^2 in terms of ∇. Check Table 8.2 for the correct expressions.

8.10 Find an expression for $\mu\delta^3$ in terms of D. Use the results of Example 8.3.

8.11 Derive an expression for d^2f/dx^2 in difference notation using backward differences with $e = o(h)$.

8.12 Express df/dx in difference notation using forward differences with $e = o(h^3)$.

8.13 Derive the expression for d^3f/dx^3 in difference notation using central differences with $e = o(h^2)$.

8.14 Determine the value of $d/dx(\text{erf } x)$ at $x = 1.6$ using the tables in the Appendix. Employ central differences with a) $e = o(h^2)$, and b) $e = o(h^4)$. Check with the exact value obtained analytically. Use five significant numbers.

8.15 Find the value of $d^2/dx^2\, J_1(x)$ at $x = 2.0$ using the tables in the Appendix. Employ a) forward differences, b) backward differences, and c) central differences, all with $e = o(h^2)$.

8.16 Express the value of the integral of $f(x)$ from x_{i-2} to x_i using backward differences.

8.17 Approximate the value of the integral of $f(x) = x^2$ from $x = 0$ to $x = 6$ using six steps. Use a) the trapezoidal rule, and b) Simpson's one-third rule. Compare with the actual value found by integrating. Then, for part (a) show that the error falls within the limits established by Eq. 8.5.20.

8.18 Determine an approximate value for the integral $\int_0^9 x^2 \sin(\pi x/6)\, dx$ using a) the trapezoidal rule, and b) Simpson's three-eights rule. Use nine steps.

8.19 Determine a value for $\int_0^2 j_0(x)\, dx$ applying a) the trapezoidal rule, and b) Simpson's one-third rule, using ten steps.

8.20 Find an expression for the integral of $f(x)$ from x_{i-2} to x_{i+2} using central differences. Using this expression, determine a formula for the integral $\int_a^b f(x)\, dx$. What is the order of magnitude of the error?

8.21 We desire the value of $J_0(x)$ at $x = 7.24$ using the information in the Appendix. Approximate its value using a forward-difference interpolation formula with a) $e = o(h^2)$, b) $e = o(h^3)$, and c) $e = o(h^4)$. d) For part (b), compare the forward interpolation value with the backward interpolation value.

8.22 Find an approximate value for erf (0.523) by interpolating from the numerical values given in the Appendix using a forward-difference interpolation formula with a) $e = o(h^2)$, and b) $e = o(h^3)$.

8.23 Find approximate values for the roots of the equation $x^3 + 5x^2 - 6x - 2 = 0$. The roots are in the neighborhood of a) –6, b) –1, and c) 1. Carry out the iteration process until three significant numbers are obtained.

8.24 Determine to three significant numbers at least one of the roots of the equation $x^3 + 6x - 2 = 0$.

8.25 Find the root of the equation $\cos x = 2x$.

8.26 Locate the root of the equation $\tan x = 2 + 10x$.

8.27 Solve the differential equation $2y\dot{y} - 3t^2 = 0$ if $y = 2$ at $t = 0$. Use a) Euler's method, and b) Taylor's method with $e = o(h^3)$. Use five steps between $t = 0$ and $t = 2$. c) Compare with the exact solution.

8.28 Find a numerical solution between $t = 0$ and $t = 1$ to $\dot{y} + 4ty - 4t^2 = 1$ with $y(0) = 6$. Use Euler's method with a) five step sizes. b) Compare with the exact solution.

8.29 Determine the numerical solution between $t = 0$ and $t - 1$ to $\dot{y} + y = 0$ if at $t = 0$, $y = 0$, and $\dot{y} = 10$. Solve two first-order equations with five steps. a) Use Euler's method. b) Compare with the exact solution.

8.30 Derive an Adams' method with the order of magnitude of the error of $o(h^4)$.

8.31 Determine the maximum height a ball reaches if it is thrown upward at 40 m/s. Assume the drag force to be given by $0.0012\dot{y}^2$. The ball's mass is 0.2 kg. Does the equation $\ddot{y} + 0.006\dot{y}^2 + 9.81 = 0$ describe the motion of the ball? Solve the problem numerically using a) Euler's method, b) Taylor's method with $e = o(h^3)$, and c) Adams' method. A computer should be used and the number of steps should be varied so as to yield acceptable results. Compare the three methods using the same number of steps for each. (To estimate the time required, eliminate the $\dot{y}^2$ term and determine t_{max}. This will give an approximation to the total time required from which an appropriate h can be chosen.) Can you determine an exact solution to the given equation?

8.32 Apply the Runge–Kutta method using Eq. 8.8.24 to Problem 8.27.

8.33 Using the Runge–Kutta method with $e = o(h^4)$, find a solution for Problem 8.28. Solve for y_1 and y_2.

8.34 Use five steps and solve $\ddot{y} + 4y = 0$ from $t = 0$ to $t = 1$, if at $t = 0$, $y = 0$, and $\dot{y} = 10$. a) Use Taylor's method with $e = o(h^3)$. b) Express the differential equation in finite-difference form with $e = o(h)$ (see Table 8.3) and find the approximate values for y_i. c) Compare with the exact solution.

8.35 Solve the equation $\dddot{y} + 4t\ddot{y} + 5y = t^2$ if at $t = 0$, $y = 0$, $\dot{y} = -2$, and $\ddot{y} = 0$. Use a) Taylor's method with $e = o(h^3)$, and b) the equation expressed in finite difference form with $e = o(h^2)$. Choose various step sizes and present the solution from $t = 0$ to $t = 10$. A computer is recommended.

8.36 Assume a tight telephone wire and show that the equation describing $y(x)$ is $d^2y/dx^2 - b = 0$. Express b in terms of the tension P in the wire, the mass per unit length m of the wire, and gravity g. The boundary conditions are $y = 0$ at $x = 0$ and $x = L$. Solve the problem numerically with $b = 10^{-4}$ m^{-1} and $L = 20$ m and solve for the maximum sag. Use a) the iterative method, and b) the simultaneous equation method. Five steps are sufficient to illustrate the procedures using Euler's method.

8.37 Assume a loose hanging wire; then $d^2y/dx^2 - b[1 + (dy/dx)^2]^{1/2} = 0$ describes the resulting curve, a catenary. Can the superposition method be used? Using the iterative method, determine the maximum sag if $b = 10^{-3}$ m^{-1}. The boundary conditions are $y = 0$ at $x = 0$ and $y = 40$ m at $x = 100$ m. Using the computer, find an approximation to the minimum number of steps necessary for accuracy of three significant figures.

8.38 A bar connects two bodies of temperatures 150°C and 0°C, respectively. Heat is lost by the surface of the bar to the 30°C surrounding air and is conducted from the hot to the colder body. The describing equation is $d^2T/dx^2 - 0.01\,(T - 30) = 0$. Calculate the temperature in the 2-m-long bar. Use the superposition method with five steps. Compare with the exact solution.

8.39 A laterally insulated fin 2.5 m long connects two large bodies. One body, originally at 60°C, is suddenly heated to 600°C. The other is maintained at 60°C. Calculate the time necessary for the center of the bar to reach a temperature of 90°C. Use five x steps and time increments of 5 ks. The diffusivity is $a = 1.0 \times 10^{-5}$ m²/s. Use linear interpolation.

8.40 The initial temperature of a laterally insulated steel rod is 0°C. If the end at $x = 2$ m is insulated and the end at $x = 0$ is suddenly subjected to a temperature of 200°C, predict the temperature, using five displacement increments and four time steps of 20 ks. For steel, use $a = 4.0 \times 10^{-6}$ m²/s.

8.41 A 6-m-long wire, fixed at both ends, is given an initial displacement of $u_{1,0} = 0.1$, $u_{2,0} = 0.2$, $u_{3,0} = 0.3$, $u_{4,0} = 0.4$, and $u_{5,0} = 0.2$ at the displacement steps. The initial velocity is zero, expressed in difference form as $u_{i,1} = u_{i,0}$. Predict the displacement at three future time steps, using $k = 0.1$ s if the wave speed a is 10 m/s.

8.42 Using five increments in a 1-m-long tight string, find an approximate solution for the displacement if it is released from rest with the displacement

$$u(x,0) = \begin{cases} 0.1x & 0 < x < 0.4 \\ 0.02 & 0.4 < x < 0.6 \\ 0.1(1-x) & 0.6 < x < 1 \end{cases}$$

Assume a wave speed of 50 m/s and use a time step of 0.004 s. Find the approximate solution at three additional time steps.

8.43 A 4-m by 5-m plate is divided into 1-m squares. One long side is maintained at 100°C and the other at 200°C. The short sides are maintained at 300°C and 0°C. The flat surfaces are insulated. Predict the steady-state temperature distribution in the plate using the relaxation method.

8.44 Solve Problem 8.43 with the 100°C changed to a linear distribution varying from 0 to 300°C. The 0°C corner is adjacent to the 0°C side.

Bibliography

AYRES, F., JR., *Matrices*, Schaum Publishing Co., New York, 1962.

CHURCHILL, R. V., *Fourier Series and Boundary Value Problems*, McGraw-Hill Book Company, New York, 1941.

HILDEBRAND, F. B., *Advanced Calculus for Application*, Prentice-Hall, Inc., Englewood Cliffs, N.J., 1976.

HOVANESSIAN, S. A., and PIPES, L. A., *Digital Computer Methods in Engineering*, McGraw-Hill Book Company, New York, 1969.

ISAACSON, E., and KELLER, H. B., *Analysis of Numerical Methods*, John Wiley & Sons, Inc., New York, 1966.

KREYSZIG, E., *Advanced Engineering Mathematics*, John Wiley & Sons, Inc., New York, 1972.

RAINVILLE, E. D., *Intermediate Differential Equations*, John Wiley & Sons, Inc., New York, 1943.

RAO, N. N., *Basic Electromagnetics with Applications*, Prentice-Hall, Inc., Englewood Cliffs, N.J., 1972.

SPIEGEL, M. R., *Laplace Transforms*, Schaum Publishing Co., New York, 1965.

SPIEGEL, M. R., *Vector Analysis*, Schaum Publishing Co., New York, 1959.

WYLIE, C. R., JR., *Advanced Engineering Mathematics*, McGraw-Hill Book Company, New York, 1951.

M. C. Potter and B. F. Feeny, *Mathematical Methods for Engineering and Science*,
https://doi.org/10.1007/978-3-031-26151-0

Appendix A

TABLE A1 United States Engineering Units, SI Units, and Their Conversion Factors

Quantity	Engineering Units (U.S. System)	International System (SI)[a]	Conversion Factor
Length	inch	millimeter	1 in. = 25.4 mm
	foot	meter	1 ft = 0.3048 m
	mile	kilometer	1 mi = 1.609 km
Area	square inch	square centimeter	1 in.^{2} = 6.452 cm^2
	square foot	square meter	1 ft^2 = 0.09290 m^2
Volume	cubic inch	cubic centimeter	1 in.^{3} = 16.39 cm^3
	cubic foot	cubic meter	1 ft^3 = 0.02832 m^3
	gallon		1 gal = 0.004546 m^3
Mass	pound-mass	kilogram	1 lb_m = 0.4536 kg
	slug		1 slug = 14.61 kg
Density	pound/cubic foot	kilogram/cubic meter	1 $\text{lb}_\text{m}/\text{ft}^3$ = 16.02 kg/m^3
Force	pound-force	newton	1 lb = 4.448 N
Work or torque	foot-pound	newton-meter	1 ft-lb = 1.356 N · m
Pressure	pound/square inch	newton/square meter	1 psi = 6895 N/m^2
	pound/square foot		1 psf = 47.88 N/m^2
Temperature	degree Fahrenheit	degree Celsius	$°F = \frac{9}{5}\ °C + 32$
	degree Rankine	degree Kelvin	$°R = \frac{9}{5}\ °K$
Energy	British thermal unit	joule	1 Btu = 1055 J
	calorie		1 cal = 4.186 J
	foot-pound		1 ft-lb = 1.356 J
Power	horsepower	watt	1 hp = 745.7 W
	foot-pound/second		1 ft-lb/s = 1.356 W
Velocity	foot/second	meter/second	1 fps = 0.3048 m/s
Acceleration	foot/second squared	meter/second squared	1 ft/s^2 = 0.3048 m/s^2
Frequency	cycle/second	hertz	1 cps = 1.000 Hz

[a]The reversed initials in this abbreviation come from the French form of the name: *Système International.*

M. C. Potter and B. F. Feeny, *Mathematical Methods for Engineering and Science*,
https://doi.org/10.1007/978-3-031-26151-0

TABLE A2 Gamma Function, $\Gamma(\alpha) = \int_0^\infty e^{-t} t^{\alpha-1} dt$

α	$\Gamma(\alpha)$	α	$\Gamma(\alpha)$	α	$\Gamma(\alpha)$	α	$\Gamma(\alpha)$	α	$\Gamma(\alpha)$
1.00	1.000000	1.20	0.918169	1.40	0.887264	1.60	0.893516	1.80	0.931384
1.01	0.994326	1.21	0.915577	1.41	0.886764	1.61	0.894681	1.81	0.934076
1.02	0.988844	1.22	0.913106	1.42	0.886356	1.62	0.895924	1.82	0.936845
1.03	0.983550	1.23	0.910755	1.43	0.886036	1.63	0.897244	1.83	0.939690
1.04	0.978438	1.24	0.908521	1.44	0.885805	1.64	0.898642	1.84	0.942612
1.05	0.973504	1.25	0.906403	1.45	0.885661	1.65	0.900117	1.85	0.945611
1.06	0.968744	1.26	0.904397	1.46	0.885604	1.66	0.901668	1.86	0.948687
1.07	0.964152	1.27	0.902503	1.47	0.885633	1.67	0.903296	1.87	0.951840
1.08	0.959725	1.28	0.900719	1.48	0.885747	1.68	0.905001	1.88	0.955071
1.09	0.955459	1.29	0.899042	1.49	0.885945	1.69	0.906782	1.89	0.958380
1.10	0.951351	1.30	0.897471	1.50	0.886227	1.70	0.908639	1.90	0.961766
1.11	0.947395	1.31	0.896004	1.51	0.886592	1.71	0.910572	1.91	0.965231
1.12	0.943590	1.32	0.894640	1.52	0.887039	1.72	0.912580	1.92	0.968774
1.13	0.939931	1.33	0.893378	1.53	0.887568	1.73	0.914665	1.93	0.972397
1.14	0.936416	1.34	0.892215	1.54	0.888178	1.74	0.916826	1.94	0.976099
1.15	0.933041	1.35	0.891151	1.55	0.888869	1.75	0.919062	1.95	0.979881
1.16	0.929803	1.36	0.890184	1.56	0.889639	1.76	0.921375	1.96	0.983742
1.17	0.926700	1.37	0.889313	1.57	0.890490	1.77	0.923763	1.97	0.987685
1.18	0.923728	1.38	0.888537	1.58	0.891420	1.78	0.926227	1.98	0.991708
1.19	0.920885	1.39	0.887854	1.59	0.892428	1.79	0.928767	1.99	0.995813
1.20	0.918169	1.40	0.887264	1.60	0.893516	1.80	0.931384	2.00	1.000000

Polynomial approximation[a]:

$$\Gamma(x+1) = 1 - 0.577191652x + 0.988205891x^2 - 0.897056937x^3 + 0.918206857x^4 - 0.756704078x^5 + 0.482199394x^6 - 0.193527818x^7 + 0.035868343x^8 + \epsilon$$

$$|\epsilon| \leq 3 \times 10^{-7}$$

[a] From C. Hastings, Jr., *Approximations for Digital Computers*, Princeton University Press, Princeton, N.J., 1955.

TABLE A3 Error Function, erf $x = \frac{2}{\sqrt{\pi}}\int_0^x e^{-t^2}\,dt$

x	Erf(x)	x	Erf(x)	x	Erf(x)
0.00	0.000000	0.68	0.663782	1.36	0.945561
0.02	0.022564	0.70	0.677801	1.38	0.949016
0.04	0.045111	0.72	0.691433	1.40	0.952285
0.06	0.067621	0.74	0.704678	1.42	0.955376
0.08	0.090078	0.76	0.717537	1.44	0.958296
0.10	0.112463	0.78	0.730010	1.46	0.961053
0.12	0.134758	0.80	0.742101	1.48	0.963654
0.14	0.156947	0.82	0.753811	1.50	0.966105
0.16	0.179012	0.84	0.765143	1.52	0.968413
0.18	0.200936	0.86	0.776100	1.54	0.970586
0.20	0.222703	0.88	0.786688	1.56	0.972628
0.22	0.244296	0.90	0.796908	1.58	0.974547
0.24	0.265700	0.92	0.806768	1.60	0.976348
0.26	0.286900	0.94	0.816271	1.62	0.978038
0.28	0.307880	0.96	0.825424	1.64	0.979622
0.30	0.328627	0.98	0.834232	1.66	0.981105
0.32	0.349126	1.00	0.842701	1.68	0.982493
0.34	0.369365	1.02	0.850838	1.70	0.983791
0.36	0.389330	1.04	0.858650	1.72	0.985003
0.38	0.409010	1.06	0.866144	1.74	0.986135
0.40	0.428392	1.08	0.873326	1.76	0.987190
0.42	0.447468	1.10	0.880205	1.78	0.988174
0.44	0.466225	1.12	0.886788	1.80	0.989091
0.46	0.484655	1.14	0.893082	1.82	0.989943
0.48	0.502750	1.16	0.899096	1.84	0.990736
0.50	0.520500	1.18	0.904837	1.86	0.991473
0.52	0.537898	1.20	0.910314	1.88	0.992156
0.54	0.554939	1.22	0.915534	1.90	0.992791
0.56	0.571616	1.24	0.920505	1.92	0.993378
0.58	0.587923	1.26	0.925236	1.94	0.993923
0.60	0.603856	1.28	0.928734	1.96	0.994427
0.62	0.619411	1.30	0.934008	1.98	0.994892
0.64	0.634586	1.32	0.938065	2.00	0.995323
0.66	0.649376	1.34	0.941913	2.02	0.995720

Series expansion for erf x:

$$\text{erf } x = \frac{2}{\sqrt{\pi}}\left(x - \frac{x^3}{3} + \frac{x^5}{5\cdot 2!} - \frac{x^7}{7\cdot 3!} + \cdots\right)$$

TABLE A4 Bessel Functions[a]

x	$J_0(x)$	$Y_0(x)$	$J_1(x)$	$Y_1(x)$
0.0	1.00000	$-\infty$	0.00000	$-\infty$
0.2	0.99002	–1.08114	0.09950	–3.32382
0.4	0.96040	–0.60602	0.19603	–1.78087
0.6	0.91200	–0.30851	0.28670	–1.26039
0.8	0.84629	–0.08680	0.36884	–0.97814
1.0	0.76520	0.08825	0.44005	–0.78121
1.2	0.67113	0.22808	0.49829	–0.62114
1.4	0.56686	0.33790	0.54195	–0.47915
1.6	0.45540	0.42043	0.56990	–0.34758
1.8	0.33999	0.47743	0.58152	–0.22366
2.0	0.22389	0.51038	0.57672	–0.10703
2.2	0.11036	0.52078	0.55596	+0.00149
2.4	+0.00251	0.51041	0.52019	0.10049
2.6	–0.09680	0.48133	0.47082	0.18836
2.8	–0.18504	0.43592	0.40971	0.26355
3.0	–0.26005	0.37685	0.33906	0.32467
3.2	–0.32019	0.30705	0.26134	0.37071
3.4	–0.36430	0.22962	0.17923	0.40102
3.6	–0.39177	0.14771	0.09547	0.41539
3.8	–0.40256	0.06450	+0.01282	0.41411
4.0	–0.39715	–0.01694	–0.06604	0.39793
4.2	–0.37656	–0.09375	–0.13865	0.36801
4.4	–0.34226	–0.16334	–0.20278	0.32597
4.6	–0.29614	–0.22346	–0.25655	0.27375
4.8	–0.24043	–0.27230	–0.29850	0.21357
5.0	–0.17760	–0.30852	–0.32758	0.14786
5.2	–0.11029	–0.33125	–0.34322	0.07919
5.4	–0.04121	–0.34017	–0.34534	+0.01013
5.6	+0.02697	–0.33544	–0.33433	–0.05681
5.8	0.09170	–0.31775	–0.31103	–0.11923
6.0	0.15065	–0.28819	–0.27668	–0.17501
6.2	0.20175	–0.24831	–0.23292	–0.22228
6.4	0.24331	–0.19995	–0.18164	–0.25956
6.6	0.27404	–0.14523	–0.12498	–0.28575
6.8	0.29310	–0.08643	–0.06522	–0.30019
7.0	0.30008	–0.02595	–0.00468	–0.30267
7.2	0.29507	0.03385	0.05433	–0.29342

TABLE A4 Bessel Functions[a] (*Cont.*)

x	$J_0(x)$	$Y_0(x)$	$J_1(x)$	$Y_1(x)$
7.4	0.27860	0.09068	0.10963	−0.27311
7.6	0.25160	0.14243	0.15921	−0.24280
7.8	0.21541	0.18723	0.20136	−0.20389
8.0	0.17165	0.22352	0.23464	−0.15806
8.2	0.12222	0.25012	0.25800	−0.10724
8.4	0.06916	0.26622	0.27079	−0.05348
8.6	+0.01462	0.27146	0.27275	+0.00108
8.8	−0.03923	0.26587	0.26407	0.05436
9.0	−0.09033	0.24994	0.24531	0.10431
9.2	−0.13675	0.22449	0.21741	0.14911
9.4	−0.17677	0.19074	0.18163	0.18714
9.6	−0.20895	0.15018	0.13952	0.21706
9.8	−0.23228	0.10453	0.09284	0.23789
10.0	−0.24594	0.05567	0.04347	0.24902
10.2	−0.24962	+0.00559	−0.00662	0.25019
10.4	−0.24337	−0.04375	−0.05547	0.24155
10.6	−0.22764	−0.09042	−0.10123	0.22363
10.8	−0.20320	−0.13264	−0.14217	0.19729
11.0	−0.17119	−0.16885	0.17679	0.16371
11.2	−0.13299	−0.19773	−0.20385	0.12431
11.4	−0.09021	−0.21829	−0.22245	0.08074
11.6	−0.04462	−0.22987	−0.23200	0.03477
11.8	+0.00197	−0.23216	−0.23228	−0.01179
12.0	0.04769	−0.22524	−0.22345	−0.05710
12.2	0.09077	−0.20952	−0.20598	−0.09942
12.4	0.12956	−0.18578	−0.18071	−0.13714
12.6	0.16261	−0.15506	−0.14874	−0.16888
12.8	0.18870	−0.11870	−0.11143	−0.19347
13.0	0.20693	−0.07821	−0.07032	−0.21008
13.2	0.21669	−0.03524	−0.02707	−0.21817
13.4	0.21773	+0.00848	+0.01660	−0.21756
13.6	0.21013	0.05122	0.05896	−0.20839
13.8	0.19434	0.09130	0.09839	−0.19116
14.0	0.17107	0.12719	0.13338	−0.16664
14.2	0.14137	0.15754	0.16261	−0.13592
14.4	0.10648	0.18123	0.18503	−0.10026
14.6	0.06786	0.19742	0.19985	−0.06115

TABLE A4 Bessel Functions[a] (*Cont.*)

x	$J_0(x)$	$Y_0(x)$	$J_1(x)$	$Y_1(x)$
14.8	0.02708	0.20557	0.20660	−0.02016
15.0	−0.01422	0.20546	0.20510	0.02107

Polynomial Approximations[b]

$$3 \le x \le 3$$

$$J_0(x) = 1 - 2.2499997(x/3)^2 + 1.2656208(x/3)^4 - 0.3163866(x/3)^6 + 0.0444479(x/3)^8 - 0.0039444(x/3)^{10} + 0.0002100(x/3)^{12} + \epsilon$$

$$|\epsilon| < 5 \times 10^{-9}$$

$$0 < x \le 3$$

$$Y_0(x) = (2/\pi)\ln\left(\tfrac{1}{2}x\right)J_0(x) + 0.36746691 + 0.60559366(x/3)^2 - 0.74350384(x/3)^4 + 0.25300117(x/3)^6 - 0.04261214(x/3)^8 + 0.00427916(x/3)^{10} - 0.00024846(x/3)^{12} + \epsilon$$

$$|\epsilon| < 1.4 \times 10^{-8}$$

$$-3 \le x \le 3$$

$$x^{-1}J_1(x) = \tfrac{1}{2} - 0.56249985(x/3)^2 + 0.21093573(x/3)^4 - 0.03954289(x/3)^6 + 0.00443319(x/3)^8 - 0.00031761(x/3)^{10} + 0.00001109(x/3)^{12} + \epsilon$$

$$|\epsilon| < 1.3 \times 10^{-8}$$

$$0 < x \le 3$$

$$xY_1(x) = (2/\pi)x\ln\left(\tfrac{1}{2}x\right)J_1(x) - 0.6366198 + 0.2212091(x/3)^2 + 2.1682709(x/3)^4 - 1.3164827(x/3)^6 + 0.3123951(x/3)^8 - 0.0400976(x/3)^{10} + 0.0027873(x/3)^{12} + \epsilon$$

$$|\epsilon| \le 1.1 \times 10^{-7}$$

$$3 < x < \infty$$

$$J_0(x) = x^{-1/2} f_0 \cos\theta_0 \qquad Y_0(x) = x^{-1/2} f_0 \sin\theta_0$$

$$f_0 = 0.79788456 - 0.00000077(3/x) - 0.00552740(3/x)^2 - 0.00009512(3/x)^3 + 0.00137237(3/x)^4 - 0.00072805(3/x)^5 + 0.00014476(3/x)^6 + \epsilon$$

$$|\epsilon| < 1.6 \times 10^{-8}$$

$$\theta_0 = x - 0.78539816 - 0.04166397(3/x) - 0.00003954(3/x)^2 + 0.00262573(3/x)^3 - 0.00054125(3/x)^4 - 0.00029333(3/x)^5 + 0.00013558(3/x)^6 + \epsilon$$

$$|\epsilon| < 7 \times 10^{-8}$$

$$3 \le x < \infty$$

TABLE A4 Bessel Functions[a] (*Cont.*)

$$J_1(x) = x^{-1/2} f_1 \cos \theta_1, \qquad Y_1(x) = x^{-1/2} f_1 \sin \theta_1$$

$$f_1 = 0.79788456 + 0.00000156(3/x) + 0.01659667(3/x)^2 + 0.00017105(3/x)^3 - 0.00249511(3/x)^4 + 0.00113653(3/x)^5 - 0.00020033(3/x)^6 + \epsilon$$

$$|\epsilon| < 4 \times 10^{-8}$$

$$\theta_1 = x - 2.35619449 + 0.12499612(3/x) + 0.00005650(3/x)^2 - 0.00637879(3/x)^3 + 0.00074348(3/x)^4 + 0.00079824(3/x)^5 - 0.00029166(3/x)^6 + \epsilon$$

$$|\epsilon| < 9 \times 10^{-8}$$

[a] $J_0(x) = 0$ at $x = 2.4048, 5.5201, 8.6537, 11.7915, 14.9309, 18.0711, \ldots$
$Y_0(x) = 0$ at $x = 0.8936, 3.9577, 7.0861, 10.2223, 13.3611, 16.5009, \ldots$
$J_1(x) = 0$ at $x = 3.8317, 7.0156, 10.1735, 13.3237, 16.4706, 19.6159, \ldots$
$Y_1(x) = 0$ at $x = 2.1971, 5.4297, 8.5960, 11.7492, 14.8974, 18.0434, \ldots$

[b] From *Handbook of Mathematical Functions*, National Bureau of Standards, p. 369, 1964.

Appendix B
Introduction to MATLAB

Outline

B.1 Introduction
B.2 Real and Complex Numbers
B.3 Vectors and Matrices
B.4 Format and Scientific Notation
B.5 Programming Loops
B.6 Plotting
B.7 String Arrays
B.8 MATLAB Files, Input, and Output
B.8.1 Setting the Path
B.8.2 Script Files, or m-Files
B.8.3 Input Files and Output Files
B.8.4 Interactive Input and Output
B.9 Functions
B.10 The Workspace Browser
B.11 Final Remarks

B.1 INTRODUCTION

MATLAB is one of the most widely used computational packages. Its potential depth of computation and the sophistication of graphical display can be extensive. This appendix provides a very brief introduction, just meant to get the reader started. This introduction might best be reviewed while MATLAB is running on the reader's computer, so that the reader can actively follow along, as probably the best way to learn the commands and coding is actually do them. There are examples in the book that use MATLAB, and these will introduce more details as needed. As the reader faces further needs, the abundant MATLAB documentation can be accessed using an internet search engine.

Once MATLAB is invoked on the computer, several MATLAB windows can be active. We will discuss the command window (MATLAB window) for interactive use, and the file editor window for batch runs.

We will start interactively in the command window. The command window reads "MATLAB" at the top, and has a command prompt, indicated by the symbol

```
>>
```

on the left margin. Commands can be input at the prompt symbol, and followed by a return (or enter), after which the output, or result of the command input, is displayed on the next line. Figure B.1 shows a command window. The command prompts are on the left margin of a large open field in the image.

M. C. Potter and B. F. Feeny, *Mathematical Methods for Engineering and Science*,
https://doi.org/10.1007/978-3-031-26151-0

For example, the addition command and its output are shown below, where it is assumed that return is entered at the end of the command:

```
>> 3+2
ans =
   5
```

Likewise, multiplication, division, and power operations are executed by typing 3*2, 6/4, and 3^2, respectively.

Each output is proceeded by "ans =" (this is a default setting that can indeed be suppressed). The variable ans takes the value of the answer from the previous command. Thus if ans is commanded, and the output is the value of ans, which is the answer from the previous command, as in

```
>> ans
ans =
   5
```

The variable ans can be quite useful as a "last chance" at retaining a value, if foresight was not made to assign the last result to a variable name. The command window in Fig. B.1 shows a simple command input and the resulting output. Following the output, a new prompt is ready for the next command input.

In the remainder of this appendix, "ans = " will sometimes be omitted from the output for brevity.

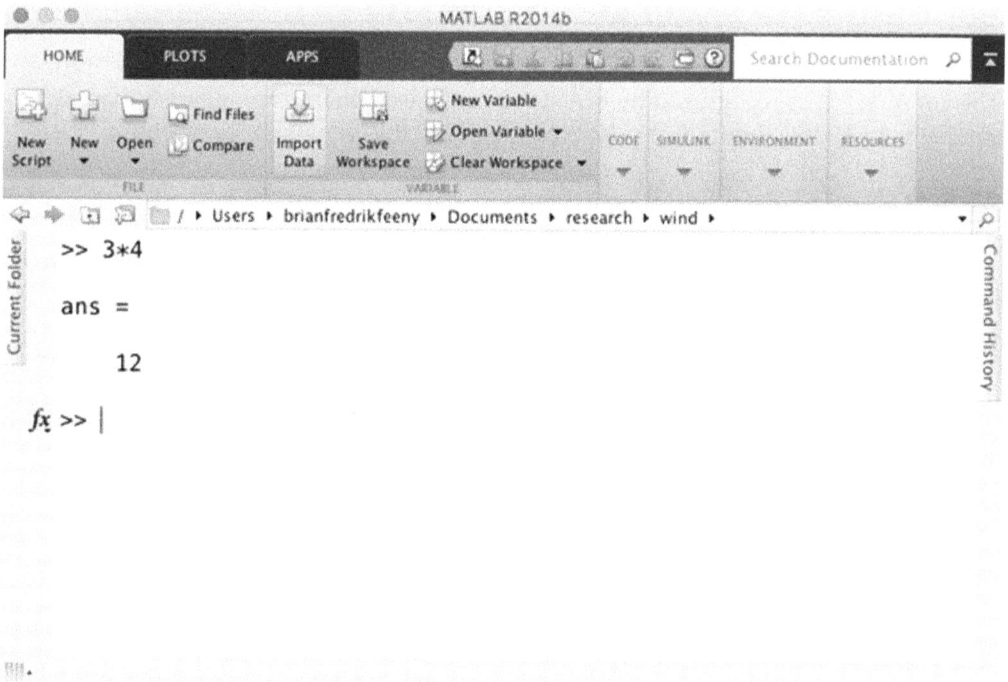

FIGURE B.1 An image of the command window.

The display of the output of the answer can also be suppressed using a semicolon. As an example of suppressed output,

```
>> x = 2;
>> y = 5;  % x and y now have these assigned values until otherwise specified
>> x*y
   10
```

In the above example commands, the symbol "%" was used to indicate a "comment". Everything typed on the right of "%", up until the line return (hard return), will not be processed by MATLAB, and can be included to provide programmer's information and documentation.

The examples above are simulating an interactive entry of commands, with immediate assignment of values and output. On the other hand, an "m-file", or script file, can be used to run a program or a series of commands. The m-file has a file type .m, such as `myfile.m`. The m-file can be programmed in the MATLAB editor, which we will discuss more later. A main program can be in an m-file, and it might call a subprogram, like a function, which is housed in another m-file.

In most of our discussion of MATLAB commands, functions, and formats, we will present them in the context of a command line with the prompt >>. We do this so that the input command is clear, and the output of the result is clear. But all of the command inputs can be, and are more typically, written as lines in an m-file.

We discuss m-files and other MATLAB files in more detail in a later section.

B.2 REAL AND COMPLEX NUMBERS

It is often necessary to work with quantities represented by real and complex numbers. The variable "`i`" is recognized by MATLAB as the imaginary unit quantity, namely the square root of –1. Caution: It is possible to reassign this value. For example,

```
>> i = 2;
```

will now make the variable `i` have the value of 2. If it is desired to reestablish the imaginary number, it can be done by typing

```
>> i = sqrt(-1);
```

where `sqrt(y)` is a defined MATLAB function that takes the square root of a variable `y`. A complex number can be entered into the work space, for instance,

```
>> z = 3 + 4*i;
```

Then

```
>> real(z)
   3
>> imag(z)
   4
```

```
>> abs(z)
   5
>> angle(z)  % angle in radians
   0.9273
>> conj(z)
   3 - 4i
```

are commands that produce the associated properties of the complex number. Computations of complex numbers can be made, such as

```
>> (1 - i)*z        % Calculation: (1 - i)*(3 + 4 i) = 3 - 3 i + 4 i + 4 = 7 + i
   7 + i
>> (1 - i) + ans
   8
>> z^3              % (3+4i)*(3+4i)*(3+4i) = (3+4i)*(9 + 24i - 16)
   -117 + 44i
```

Many functions can be evaluated with complex entries. For example,

```
>> sin(3*i)
   0.0000 + 10.0179i
>> sinh(3)
   10.0179
>> log(i)
   0.0000 + 1.5708i
```

Note that `log(x)` produces the natural log. The commands `log10(x)` and `log2(x)` give logarithms with base 10 and base 2.

It is important to know that a user can clear the assigned variables using the command `clear`. For instance,

```
>> i = 1;    % i is no longer the unit imaginary number.
>> exp(i)    % exp(x) is the exponential function.
   2.7183
>> clear i   % i is now reset to the default unit imaginary number.
>> exp(i)
   0.5403 + 0.8415i
```

To clear several variables at once,

```
>> clear i, x, y
```

To clear all of the variables,

```
>> clear
```

B.3 VECTORS AND MATRICES

Arrays in the form of vectors and matrices can be assigned in a variety of ways. A row vector can be manually written as

```
>> vrow = [1 2 3 4];
```

Commas can also be used between the elements. A column vector can be assigned by using semicolons to indicate each new row. For example,

```
>> vcol = [-1; 1; 3; 5];
```

In this context a vector is just a matrix with a single row or column. Matrices can be assigned by using these ideas for rows and columns. For instance,

```
>> A = [1 2 3 4; 5 6 7 8];
```

produces a 2x4 matrix.

MATLAB distinguishes elements by the space or comma between them. However, regarding the space rule, caution is needed when filling in a matrix using computational expressions. While the expressions

```
>> 4 + 5
ans = 9
>> 4+5
ans = 9
```

are robust to the existence of spaces, having spaces in expressions of matrix elements could be confusing for MATLAB. Thus,

```
>> [4+9 5 10-1];
```

is safe, but entries like

```
>> [4+9 5 10 - 1];
```

might not produce the intended matrix. As an example of using commas instead of just spaces, consider

```
>> vrow = [1, 2, 3, 4];
```

Once assigned, the elements of a vector or matrix can be recalled:

```
>> A(2,3)
ans = 7
```

Specific elements of a matrix can also be reassigned:

```
>> A(2,4) = 0;
```

The colon symbol ":" can represent all of the indices of a column or row. Thus, `A(2,:)` represents all elements in row 2, which is a row vector that is the second row of A. Likewise, `A(:,3)` grabs the third column of matrix A, and has the form of a column vector. Value assignments can therefore be made for a complete row or column, if the dimensions are compatible. For instance,

```
>> A(2,:) = [5 6 7 8];
```

assigns the whole second row of matrix A.

The transpose of a matrix can be obtained by using `transpose(A)`. The transpose of A is therefore

```
>> Atrans = transpose(A);
```

An apostrophe provides the conjugate-transpose of a matrix, which is sometimes known as a Hermitian operation. If the matrix is real, then the apostrophe reduces to a simple transpose. Since matrix A as defined above is real, then

```
>> Atrans = A';
```

Applying a transpose to a column vector, as with

```
>> vcol';
```

produces a row vector.

If A is complex, the apostrophe also involves a complex conjugation. Thus the same result is obtained from the following two commands:

```
>> A';
>> conj(transpose(A));
```

We can also make matrices by concatenating existing, dimensionally compatible matrices. For example,

```
>> B = [vrow; vcol'];
```

Since `vcol'` (the transpose of a column vector) is a row vector, the result is a matrix B with dimensions of 2 rows by 4 columns, and is the same result as

```
>> [1 2 3 4; -1 1 3 5];
```

Larger matrices can be concatenated, too, such as

```
>> C = [A; B];
```

which produces a 4x4 matrix, the top half of which is the matrix A, and the lower half consisting of matrix B.

Vectors with values that fit convenient regular patterns can sometimes be assigned as in this example:

```
>> t = 0:1:9
t = [0 1 2 3 4 5 6 7 8 9]
```

If we type `t = 0:9` the increments are 1 by default, but we can choose other increments, e.g.,

```
>> x = 0.2:0.2:8;
```

which produces a 1x40 vector of evenly spaced values at increments of 0.2. The last value in the command line need not be precisely indicated. For instance,

```
>> x = 0.2:0.2:8.1;
```

produces the same result.

The command `diag` can be used to either list the diagonal elements of a diagonal matrix in a vector, or to produce a diagonal matrix whose elements come from a vector. For example,

```
>> D = [1 0 0 0; 0 2 0 0; 0 0 4 0; 0 0 0 -1];
>> diag(D);
```

produces the same matrix as

```
>> d = [1; 2; 4; -1];
```

Note that the result is a column vector. Conversely,

```
>> diag(d);
```

will regenerate a matrix equal to `D`.

The identity matrix of desired dimension can be generated using the command `eye(n)`, as in

```
>> eye(4);
```

which produces the 4x4 identity matrix. The command

```
>> zeros(4,6);
```

creates a 4x6 matrix whose elements are all zeros. Similarly,

```
>> ones(3,2)
ans =
       1     1
       1     1
       1     1
```

Matrix multiplication is simple. The matrices must be dimensionally compatible. For example,

```
>> D*d;
```

results in a column matrix whose elements are 1, 4, 16, and 1.

There is one case where the matrices need not be dimensionally compatible, and that is when one of them is a scalar. MATLAB will recognize this as the scalar multiplication of a matrix, and treat it accordingly. For instance,

```
>> 4*[1 -1 2; 1 0 -1]
ans =
       4    -4     8
       4     0    -4
```

Another very handy tool is the element-by-element product, which is specified with the ".*" symbol combination (that inside the quotes). An illustration of this is

```
>> E = [1 2 3; -2 -4 -6];
>> F = [0 1 2; 1 0 -1];
>> EF = E.*F
EF =
       0     2     6
      -2     0     6
```

In this way, it is also possible to take element by element powers, using the ".^" symbols. Matrix addition always takes place element by element. As such,

```
>> E + F
ans =
       1     3     5
      -1    -4    -7
```

An important matrix computation is the eigenvalue problem. Given a square matrix

```
>> A = [1 -3 1; 3 0 0; 0 1 1];
```

we can compute the eigenvalues using

```
>> eig(A)
ans =
          0.3408 + 2.9976i
          0.3408 - 2.9976i
          1.3185 + 0.0000i
```

To capture both the eigenvalues and eigenvectors and put them in square matrices `lam` and `v`, we type

```
>> [v lam] = eig(A)
v =
    0.6911 + 0.0000i      0.6911 + 0.0000i      0.1322 + 0.0000i
    0.0776 - 0.6828i      0.0776 + 0.6828i      0.3008 + 0.0000i
   -0.2227 + 0.0231i     -0.2227 - 0.0231i      0.9445 + 0.0000i

lam =
    0.3408 + 2.9976i      0.0000 + 0.0000i      0.0000 + 0.0000i
    0.0000 + 0.0000i      0.3408 - 2.9976i      0.0000 + 0.0000i
    0.0000 + 0.0000i      0.0000 + 0.0000i      1.3185 + 0.0000i
```

The eigenvectors are the columns of `v`, and the eigenvalues are the diagonal elements of `lam`. The eigenvector `v(:,n)` corresponds to the eigenvalue `lam(n,n)`. As such,

```
>> A*v(:,2) - lam(2,2)*v(:,2)
ans =

   1.0e-14 *

   -0.0361 - 0.0888i
    0.1776 + 0.0111i
    0.0111 + 0.0222i
```

which is essentially the expected zero vector, with roundoff error.

Below are some examples of other commonly useful commands for matrix calculations.

```
>> abs(A);  % a matrix whose elements are absolute values of elements of A
>> Ainv = inv(A)

Ainv =

          0      0.3333      0
    -0.2500      0.0833      0.2500
     0.2500     -0.0833      0.7500
```

Then `Ainv*A` produces the identity matrix.

It can be necessary to know the dimensions of a matrix or vector, for example in examining compatibility for multiplication. There are two ways to check the dimensions. One is the command `size`. If the matrix `E` is a 2x3 matrix, then

```
>> size(E)
ans =
     2     3
```

If you need the number of rows and columns for subsequent calculations, you can retain them with

```
>> [rows, cols] = size(E);
```

Similarly, the length of a vector, for example, `vrow`, can be obtained by `length(vrow)`.

Another way to check the dimensions, and also to inspect values, is to look at the variable in the "Workspace." This topic goes beyond matrices and vectors, and is therefore given its own section a bit later.

Some other useful commands for working with vectors and matrices include

```
>> vrow = [1 2 3 4];
>> vcol = [-1; 1; 3; 5];
>> sum(vrow)
ans = 10
>> sum(vcol)
ans = 8
>> max(vrow)
ans = 4
>> min(vrow)
ans = 1
>> norm(vrow)
ans = 5.4772
```

Some of these commands applied to a matrix are performed column by column, unless it is a row vector:

```
>> max(A)
ans =
     3     1     1
>> min(A)
     0    -3     0
>> max(max(A))
ans = 3
>> min(max(A))
ans = 1
>> max(min(A))
ans = 0
>> sum(A)
ans =
     4    -2     2
>> rank(A)
ans = 3
```

There are many more things to do with arrays and matrices, and the reader can search the internet for additional operations as the need arises.

B.4 FORMAT AND SCIENTIFIC NOTATION

Numbers and values can have a variety of formats. There can be real and complex numbers, integers, and strings, among other things. Different formats are sometimes convenient for expressing real numbers.

The `format` *style* command specifies how real numbers are to be displayed. The default is the same as what is specified by `format short`. With the `short` style, four decimal places are given, and very large and small values are expressed in scientific notation. For instance,

```
>> format short
>> pi
   3.1416
>> y = 27.90129456          % y has the value 27.90123456
   27.9013                  % but is displayed to four decimals with the
                            % short format.
>> exp(10)                  % e^10
   2.2026e+04
>> exp(-10)
   4.5400e-05
>> e^10                     % MATLAB does not know the value of e
                            Undefined function or variable 'e'.
```

If more digits are needed, an option is

```
>> format long
>> pi                       % MATLAB does know the value of pi
   3.141592653589793
>> exp(10)
   2.202646579480672e+04
```

The default format can be re-established by simply typing

```
>> format
```

Then

```
>> pi
   3.1416
```

The same format is achieved by setting

```
>> format short
```

In this format, a matrix with extreme values is displayed in scientific notation with a common exponent. To apply scientific notation individually in matrix elements use

```
>> format shortG
```

A table from Mathworks displays the format options in Table B.1. (The contents of the table are quoted from https://www.mathworks.com/help/MATLAB/ref/format.html.)

TABLE B.1 Format options (quoted from Mathworks)

Style	Result	Example
`short (default)`	Short, fixed-decimal format with 4 digits after the decimal point.	`3.1416`
`long`	Long, fixed-decimal format with 15 digits after the decimal point for `double` values, and 7 digits after the decimal point for `single` values.	`3.141592653589793`
`shortE`	Short scientific notation with 4 digits after the decimal point.	`3.1416e+00`
`longE`	Long scientific notation with 15 digits after the decimal point for `double` values, and 7 digits after the decimal point for `single` values.	`3.141592653589793e+00`
`shortG`	Short, fixed-decimal format or scientific notation, whichever is more compact, with a total of 5 digits.	`3.1416`
`longG`	Long, fixed-decimal format or scientific notation, whichever is more compact, with a total of 15 digits for `double` values, and 7 digits for `single` values.	`3.14159265358979`
`shortEng`	Short engineering notation (exponent is a multiple of 3) with 4 digits after the decimal point.	`3.1416e+000`
`longEng`	Long engineering notation (exponent is a multiple of 3) with 15 significant digits.	`3.14159265358979e+000`
`+`	Positive/Negative format with +, −, and blank characters displayed for positive, negative, and zero elements.	`+`
`bank`	Currency format with 2 digits after the decimal point.	`3.14`
`hex`	Hexadecimal representation of a binary double-precision number.	`400921fb54442d18`
`rational`	Ratio of small integers.	`355/113`

B.5 PROGRAMMING LOOPS

Fundamental programming involves programming loops. The most common loops used in MATLAB are the for, while, if and if-else loops. Programming loops are most conveniently written in an m-file, so that the programming sequence will not be executed until the programmer runs the program. However, for illustration, we present basic examples of these loops as typed directly into the command window.

For example, consider the series of commands

```
>> for j = 1:4
   num = 2*j-1;
   den = 2*j;
   victor(j) = num/den;
end
```

Let's explain each ingredient here. After the prompt, the for command induces a loop for a predefined set of values of j, specifically here for values of j = 1, 2, 3, and 4. Starting with the for command, MATLAB will not prepare an output until after the end command. Therefore, the next lines after the prompt can be typed without expecting an output. MATLAB does not produce a new prompt until the loop is complete. (This is not an issue when we write loops in an "m-file" rather than in the command window.) The lines num = 2*j-1 and den = 2*j will get evaluated for each value of j. The line victor(j) = num/den will assign values to the vector victor, element by element, for elements j = 1, 2, 3, and 4. Thus, victor(1) = 1/2, victor(2) = 3/4, victor(3) = 5/6, and victor(4) = 7/8. When the end line is encountered, the program will go back to the beginning, increment the value of j as instructed, and repeat the calculations, until it has worked through each of the specified values of j. The result of this for loop script is thus the vector victor. Since each executable line in the loop ends with a semicolon, none of the expressions will be displayed during the execution of this loop.

If it is desired to execute a for loop with, say, even integer values, then the leading line could read, for j = 0:2:10. In this case, the for loop will make use of values of j = 0, 2, 4, 6, 8, and 10.

A while loop can be constructed to carry out calculations while a variable is within specified limits. This variable needs to change for each execution of the loop, so that at some point the conditions for ending the loop can be met. An example is

```
>> j = 1;
>> while j < 5
   num = 2*j - 1;
   den = 2*j;
   hector(j) = num/den;
   j = j + 1;
end
```

Each time the end line is reached, the while condition, j < 5, is checked. If the condition is met, the calculation sequence is performed. If the condition is not met, the action is ended. In this example, the vector hector is produced. Here, the vector hector equals the vector victor from the previous for loop example.

Another important command loop is the `if` loop. The `if` loop can be coordinated with an `else` logical structure. The "simple if" loop does not use `else`. For example,

```
>> r1 = rand;
>> r2 = rand;
>> if r1 < r2
   disp('Now we are in trouble');
end
```

In the above commands, the built-in function `rand` produces a random number between 0 and 1. Thus, `r1` and `r2` are random. If `r1 < r2`, the `disp` command is activated, and the program will tell us, `Now we are in trouble`. The `disp` command will display what is written in the quotes when the line is encountered. However, if the random numbers are such that `r1 >= r2`, we are safe from such a proclamation.

An example of an "`if-else`" construct is

```
>> you = rand;
>> me = rand;
>> if you < me
   disp('I win!');
else
   disp('You win.');
end
```

Notice that if it's a tie (`you = me`), then I win. If I really want to win, I can increase my chances with a nested loop:

```
>> you = rand;
>> me = rand;
>> if you < me
   disp('I win!');
else
   you = rand;
   me = rand;
   if you < me
      disp('I win!');
   else
      disp('You win.');
end
```

Or, if I really want to cheat,

```
>> you = rand;
>> me = rand;
>> if you < me
   disp('I win!');
else
   disp('I win.');
end
```

In an `if` or `if-else` loop, you can include as many commands or calculations as needed, or nest other loops.

The first line of the `if` command includes a condition. This is a logical statement that internally has a value of TRUE or FALSE. Our example used `r1 < r2`. If `r1` is indeed less than `r2`, the condition is true, while otherwise it is false, and this determines which part of the loop will get executed. Other common logical conditions include <=, >=, == and ~= (not equal to). Caution: If you want to check equality, and use, for example, `if i = j`, an error is developed. For a logical condition of equality, it is necessary to type `if i == j`.

If two conditions need to be checked, && indicates "and", namely that both conditions are must be true, and || (two vertical bars) indicates "or", namely that only either of the conditions can be true.

B.6 PLOTTING

There are many graphics functions in MATLAB. The primary plotting command is `plot`, which can be used to plot values from one array against values in another array. In the commercial software Excel, this corresponds to the scatter plot.

The plot command in its simplest form plots the values of an array against the index values of the array. For instance, consider

```
>> t = 0.1*(1:1:20*pi); % this is a row vector of 62 elements.
>> victor = cos(t);
>> plot(victor);         % plot the vector "victor"
```

Here, `plot(victor)` will plot the elements of vector `victor` against its indices. The vertical axis will be the values of `victor`, and the horizontal axis will cover the values from 1 to the greatest integer of `20*pi`. By default, the values will be connected by a thin solid blue line.

If instead we type

```
>> x = sin(t);
>> plot(t,x);
```

the horizontal axis will cover the time values from `t = 0.1` to nearly `2*pi`. The result is shown on the left in Fig. B.2. We can choose to plot symbols or lines of various types. An example is

```
>> plot(t,x,'.');
```

which defines each plotted value a dot, and the values are unconnected. Want another symbol or color? Options are, for example, '`r.`' for a red dot, '`k*`' for a black asterisk, '`bo`' for a blue circle, or '`g--`' for a green dashed line. The symbol and color codes must be presented between single quotes. By default, '`r`' produces a red line, and '`o`' produces a blue circle (although when multiple plots are overlaid, MATLAB may choose another color).

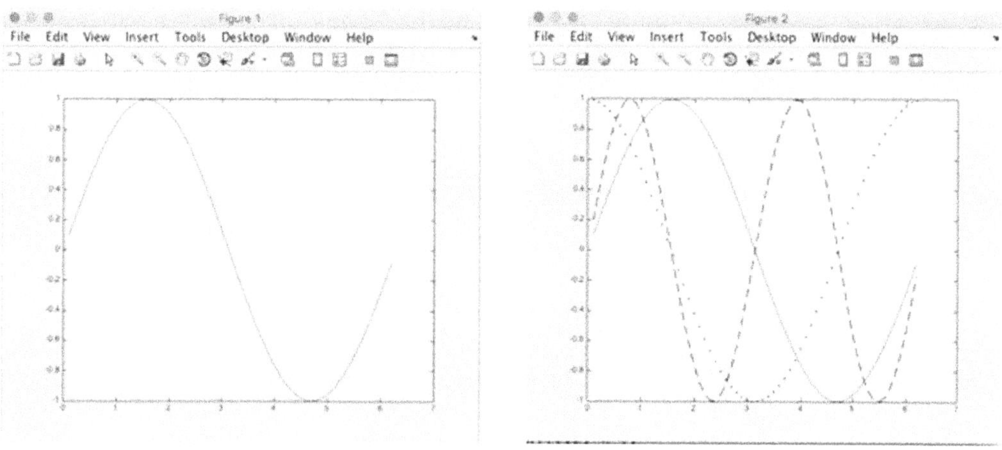

FIGURE B.2 Images of figure windows. Left, after `plot(t,x)` where `x = sin(t)` has sampled sinusoidal values. Right, after using the `hold on` command, to overlay three sinusoids.

We can also manipulate the axes, as with

```
>> plot(t,x); axis([0 pi -0.5 1.5]);
```

which defines the plot range such that the horizonal axis goes from `0` to `pi`, and the vertical axis ranges from -0.5 to 1.5. The general command is `axis([xmin xmax ymin ymax])`, for which `x` and `y` refer to the horizontal and vertical axes.

Zooming in and out of the plot can also be done in the plot's figure window. See the + magnifier icon under the menu item "insert" in Fig. B.2. Zooming can also be done by clicking and dragging a range box directly on the figure. Double click to restore the original axes.

Labeling the axes is of utmost importance when presenting your results. It is the pet peeve of many bosses and professors to see a plot without labeled axes. Example labeling commands are

```
>> plot(t,x);
>> xlabel('time (seconds)'); ylabel('x (cm)'); title('sin(t) vs. t');
```

Suppose we have

```
>> y = cos(t); z = sin(2*t);
```

We may want to control where we plot these sampled functions. To overlay the plots, try

```
>> plot(t,x);
>> hold on;
>> plot(t,y,'b.'); plot(t,z,'k--');
>> hold off;
```

The result is shown on the right in Fig. B.2. Instead, we can plot on separate figure windows:

```
>> figure(1); plot(t,x);
>> figure(2); plot(t,x);
>> hold on; plot(t,y); plot(t,z,'k'); hold off;
```

Or we may want to use sub figures:

```
>> subplot(2,2,1); plot(t,x); xlabel('t'); ylabel('x');
>> subplot(2,2,2); plot(t,y); xlabel('t'); ylabel('x');
>> subplot(2,2,3); plot(t,z); xlabel('t'); ylabel('x');
```

LaTeX commands for Greek fonts can be accepted, as in `ylabel('\omega')`. Many of the formatting and labeling functions can be done in the graphics editor available from the menu items in the figure window, which is a preferred approach for many MATLAB users.

The command `plot` can be used to simultaneously overlay graphs if the data is in a matrix, such as

```
>> data = [x; y; z];     % x, y, and z are in row vectors, so data has three
rows.
>> plot(data);
>> plot(data');
```

One of these will overlay three plots of the arrays `x`, `y`, and `z`, and give them different colors according to a default. The other will plot `x(1)`, `y(1)`, and `z(1)`, then overlay a plot of `x(2)`, `y(2)`, and `z(2)`, then overlay a plot of `x(3)`, `y(3)`, and `z(3)`, and so on up until `x(62)`, `y(62)`, and `z(62)`, if `x`, `y`, and `z` have length 62. Each set of three points will be connected by a line with a default color.

B.7 STRING ARRAYS

Sometimes it is handy for the graph `title`, `xlabel`, or `ylabel`, or output text, to depend on a changing value. This can be done with "strings". A string is an array of characters, in contrast to numerical values. In this section, we will look at the string assignment, size, concatenation, output, and some example uses.

For example, `'Hello world'` is an array of 11 characters. The single quotes define it as a "string", which has no numerical value, and therefore cannot be used in calculations. The command

```
>> num2str(1234);
```

creates the string `'1234'`, i.e., an array of four characters.

The commands `title`, `xlabel`, and `ylabel` used above actually handle strings. The example,

```
>> title('Frequency plot')
```

can be achieved by typing

```
>> my_title = 'Frequency plot';
>> title(my_title);
```

This can be useful if the user wants the title to specify a parameter value. As an example,

```
>> f = 50;
>> title(['Frequency plot, case of f = ',num2str(f)])
```

The string handled by the title command is `'Frequency plot, case of f = 50'`. If the script commands are executed again, but this time with `f = 100`, the title can reflect this information. In the above example, we performed concatenation. Much like with the concatenation of matrices, we combined two strings, `'Frequency plot, case of f = '`, and the result of '`num2str(f)`', as if they were dimensionally compatible arrays. In fact, they are dimensionally compatible row vectors, whose elements are string characters. The concatenated string is contained within array brackets.

B.8 MATLAB FILES, INPUT, AND OUTPUT

In contrast to interactive calculations, programs can be coded in files and then run as a batch. The files are edited in the Editor window. They can then be run by using the run arrow under the EDITOR tab, or by invoking the file at the prompt in the command window. The default fonts in the command window and the editor window will be different. We try to reflect the font difference in this text.

B.8.1 Setting the Path

Files are kept in directories, or folders. To have proper access between MATLAB and the file directory, it is important to "set the path". This is done in the horizontal folder bar in the command window. Referring to Fig. B.1, to set the path, click on the folder icon at the start of the horizontal folder bar (right above the command space in the figure), and browse to the desired folder. If your MATLAB files are in the selected folder, you can access them with commands in the command window or in an executed script file. The down arrow on the right end of the browser bar provides a dropdown menu of past path selections for quick access to previously used paths.

B.8.2 Script Files, or m-Files

Computations with even the slightest level of sophistication will be most conveniently programmed into a script file, often called an "m-file". The lines in the m-file get executed in sequence when the programmer runs the program. In the m-file, there is no visible prompt ">>". Each line after a hard return is to be executed as a command, as if typed after the >> symbol.

One way to open a new m-file is to use the menu in the MATLAB window. In the command window (with the command space that has the >> prompt), under the "home" tab (which is open in the sample image in Fig. B.1), there is a button on the left called "New Script". Also, next to this button, is a button labeled "New" with an arrow below it. If you press "New", a drop-down menu appears in which "Script" is an option. Slide down and click "Script". A tab will be generated in the Editor window. The Editor window also has menu tabs, and under the "Editor" tab, there is also a button, "New", from which a new script can be selected. The "Open" button allows the user to browse to an existing script file.

The new or opened script file shows up under a tab in the Editor window. An example is shown in Fig. B.3. In this example window, we see that the file "LScensusUS.m" is open for editing. Other file tabs are depicted for quick access to the files indicated in the tabs. Notice that the "new" and "open" buttons are visible when the EDITOR menu tab is open. These are the same as those discussed in the command window. The "Save" button allows the user to save the file. The dropdown menu from the "Save" button also includes "save as".

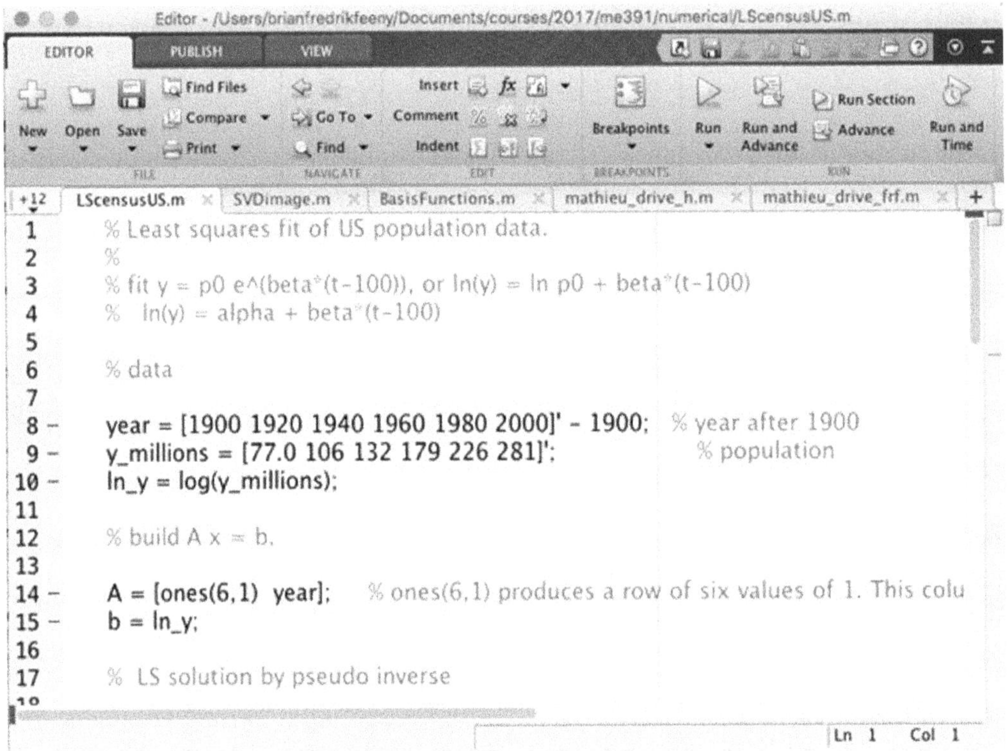

FIGURE B.3 Editor window.

The m-file has a file name in the format `filename.m`. That is, the extension `.m` indicates that the file is a set of commands to be executed by MATLAB. As a simple example of contents in an m-file, we apply a previous example, but this time as line entries to the m-file, rather than as command-prompt entries:

```
you = rand;
me = rand;
if you < me
    disp('I win!');
else
    disp('You win.');
end
```

Suppose these lines are written in the file "`game.m`". The file "`game.m`" will be kept in the folder to which the path is set, as described above. In the command window, you can then type the filename and observe the output:

```
>> game
   I win!
```

You can also run a file by clicking the run button, visible in Fig. B.3 as the green triangle. The user can choose to run a selection of text by highlighting the desired commands and clicking "run selection" in the dropdown menu revealed by the down arrow under the run button. A secret trick for Mac users: fn-shift-F7 is a convenient key-command shortcut for "run selection". If there is a problem and the code is running indefinitely, use ctrl-c (Windows or macOS) or command-. (macOS) to abort. Also, ctrl-c and ctrl-v (Windows) or command-c and command-v (macOS) can be used to copy and paste from selected text in the editor window to the command window.

B.8.3 Input Files and Output Files

It is often necessary to work on data that is in another file, or with parameters that are to be chosen at the time of running MATLAB commands. First, make sure the path is set to the directory housing the desired file needs to be specified. Alternatively, relative to this specified path, variations in the path can be specified in in-line commands. (That is, use folder navigation prefixes such as "../foldername/filename" to arrive at the correct file.) The command "`load`" will bring the contents of a file into your MATLAB workspace. When loading a desired file, it is probable that the file will be formatted to put the data into a usable form, for example with n rows and m columns. The script-file command

```
x = load('datafile.dat');
```

pulls the data from the file "`datafile.dat`" and puts them into the variable `x`. E.g., if `datafile.dat` is formatted to include elements of an $n \times m$, matrix, then the variable `x` will be $n \times m$.

There are several ways to store the data in a file. To store a matrix, Mathworks recommends the command `writematrix`. For instance,

```
A = rand(5,6);
writematrix(A)
```

will give the filename "`A.txt`", and it will be delimited with commas. To take control of filenames and formats, some examples are

```
writematrix(A, 'matrixdata.txt')
writematrix(A, 'matrixdata.dat')
writematrix(A, 'matrixdata.csv')
```

which produce delimited files. And

```
writematrix(A, 'matrixdata.xls')
writematrix(A, 'matrixdata.xlsm')
writematrix(A, 'matrixdata.xlsx')
```

produce Excel spreadsheet files. MATLAB identifies the file extension, and formats the file appropriately.

B.8.4 Interactive Input and Output

When running code in a script file, it may be useful to prompt the user input, for example to interactively define parameter values. This can be done with the "input" command. For instance, the lines

```
h = input('Enter the value of h ');
```

will display the line 'Enter the value of h ' to the user, and await a response followed by the return key. If this line is in the file my_program.m, the result in the command window may look like this:

```
>> my_program
Enter the value of h 2
>>
```

In this example, the user responded to the rest of Enter the value of h by typing 2-return. Since the code says, "h = input( );", the variable h took on the input value of 2. That is, now h = 2. Then the program executed the rest of its code, and after completion, presented the new prompt symbol.

By using the command, "fprintf", the program can display output text to the command line without waiting for input. For example,

```
fprintf('Have a good day \n');
```

will display the statement without anticipating input. If this is the only output command in the m-file my_program.m, the result in the command window may look like this:

```
>> my_program
Have a good day
>>
```

Here, the program displayed the sentence and continued to complete its commands. The symbols '\n' instruct a line return.

An example where this might be useful is in printing a status update while executing a program that may take a long time. This way, the user knows that the program is making progress, and is not stuck somehow. To this end, consider these lines:

```
nsteps = 1000;
for j = 1:nsteps
    if (rem(j,200) == 0), fprintf('Progress is %d out of %d steps\n', j, nsteps); end
end
fprintf('Done! \n')
```

In this case, fprintf displays a sentence which uses the current values of j and nsteps, and puts them in the placeholders indicated by %d. According to the if statement, this display will be made when j has values that are multiples of 200. The "\n", again, means line return, and makes it so that each incident of fprintf in the for loop happens on a new line. Here's how it looks:

```
>> my_program
Progress is 200 out of 1000
Progress is 400 out of 1000
Progress is 600 out of 1000
Progress is 800 out of 1000
Progress is 1000 out of 1000
Done!
>>
```

A pause command can be used to pause the program, for example, if the user wants to observe a graph before the next plot command overwrites the graph. This can be done by inserting

```
pause('Press the spacebar to continue\n')
```

The program will pause until any key is depressed. Here, the statement asks for the spacebar, but it actually doesn't matter which key the user strikes. The command

```
pause(10)
```

will cause the program to pause for 10 seconds. This can be useful if the user needs to see sequential output, and the visualization benefits from slowing things down between quick calculations, e.g., in animating plots of dynamically changing states.

B.9 FUNCTIONS

Functions are a big part of any programming language. In MATLAB, functions are set up in separate files. The "main" file calls the function. The function takes variables from the main program, does a task, and sends back values to the main file, as defined by the programmer. For example, we can write a script file volume.m to include the lines

```
r = input('Enter the radius ');
thickness = input('Enter the thickness ');
v = disk_volume(r,thickness);
fprintf('The volume of the disk is %d \n',v);
```

In order for this program to work, we need a function called disk_volume. If it is not a predefined MATLAB function (such as "sin"), we need to define it in a script file with the same name. Specifically, disk_volume.m has to be the file name. This file might have the code

```
function y = disk_volume(s,t)
y = t*pi*s^2;
return
```

When function disk_volume(r,thickness) is called in the main program, the values of r and thickness are recognized in the function to be in the positions of s and t. So now in the function, and only in the function, s = r and t = thickness. The volume of the disk is computed and assigned to the value y in the function. The calling program, with v = disk_volume(r,thickness), is anticipating the returned value and assigns it to v. The value is then printed by the main program.

The variables y, s , and t are only known inside the function, and the function only knows the values of these three variables. Sometimes it is desired to have the function know values of variables that are not passed in the calling and returning of the function. In such case, a global variable is created. Global variables are recognized, without being passed, to all functions that have a global variable statement that matches such a statement in the calling program. For example, suppose the script file volume.m is rewritten to call a function in the following manner:

```
global thickness
thickness = input('Enter the thickness ');
r = input('Enter the radius ');
v = disk_volume(r);
fprintf('The volume of the disk is %d \n',v);
```

The file called disk_volume.m has the lines

```
function y = disk_volume(s)
global thickness
y = pi*s^2*thickness;
return
```

Running the program in volume.m, with interactive user input, could result in

```
>> volume
Enter the thickness 3
Enter the radius 5
The volume of the disk is 235.6194
>>
```

Typically a programmer will not pass variables with a global statement, unless there is something difficult about passing the desired variables. Difficulty might occur if using a pre-defined MATLAB function as an intermediary (a function that calls another function specified by the programmer), and the programmer would like to assign the values in the main program and not in the defined function. An example is in using MATLAB functions to numerically solve differential equations with parameters.

MATLAB has many, many functions already defined. Common built-in functions are trigonometric functions like sin, cos, tan, and tanh. The inverse trig functions produce principal values. For instance, theta = atan(y/x) will yield a value in the range $-\pi/2 \leq$ theta $\leq \pi/2$. A very handy variation is theta = atan2(y,x), which identifies which quadrant the pair (y,x) is in, and produces a value of theta in the range $-\pi <$ theta $\leq \pi$.

Many of the MATLAB commands that we have seen can be thought of as built-in functions, for example, `det`, `inv`, and `eig`. There are so many other mathematical and computational functions in the MATLAB library: `besselj`, `bessely`, `gamma`, `erf`, `step`, `cumtrapz`, `ellipke` (elliptic integrals K and E), `airy`, and `polyfit`, just to name a very few. To visualize a built-in mathematical function, such as a Bessel function, a command sequence like

```
>> x = 0:0.01:50;
>> b1 = besselj(0,x);
>> b2 = besselj(1,x);
>> plot(x,b1,x,b2)
```

could be used to produce a figure similar to one drawn in this text. The user can learn the functions as needed.

B.10 THE WORKSPACE BROWSER

The Workspace was mentioned in Section B.3, and gets more description here. The Workspace provides a quick tool for assessing variables. To use it, it is necessary to open the "Workspace browser". The Workspace browser shows a list of active variables, with information about them. It can be displayed in the MATLAB command window or in its own window. If docked in the MATLAB command window, it will be in a subsection titled "Workspace". If it is not already displayed, use the "Home" tab, then the "Environment" tab, and select "Workspace". If the workspace browser shows up as a separate window, and the user would like to embed it into a sub window in the MATLAB command window, it can be done by "docking" the workspace. Do this by clicking the small triangle in the top right of the workspace window, and selecting "Dock". It will be docked on the right or left of the command window, with a side tab. To move it to the other side, drag the side tab to the right or left edge of the MATLAB command window. To undock, and return it to its own window, click the triangle in the top right of the Workspace sub window, and select "Undock".

Another way to display the Workspace browser is to type workspace at the command prompt, and it will appear. Whether in its own window or in a sub window of the command window, the user can specify what information is displayed in the Workspace browser. At the top of the browser are headers, which may include items such as "Name", "Size", and "Value". The header bar can be right-clicked to show options of what information should be displayed. Then information can be selected or deselected. There are lots of options, including "Mean", "Max", and others that may be of momentary interest.

Many users of MATLAB simply thrive on the Workspace. In addition to the Workspace, there are many tools and features that can be browsed and accessed through the GUI MATLAB windows. This includes the MATLAB command window, the Editor window, and also figure windows. In Figs. B.1, B.2, and B.3, it is apparent that the windows have many menu items, buttons, and tabs. We will leave it to the user to explore the many opportunities.

B.11 FINAL REMARKS

We have introduced some very basic aspects of MATLAB, and hope that the reader can further pursue the capabilities of MATLAB as needed, whether it involves advanced graphics, computational packages, data analysis, or symbolic computation. Probably the best way to learn MATLAB is first by example, and next by applying it to a new problem. In the text, there are examples of problems solved by MATLAB, either with a few simple commands, or with short programs. The programs introduce additional details, such as plot fonts and settings, or new computational commands or manipulations.

MATLAB is periodically updated. But the command structure does not change much over long periods of time, and even when the format of a command changes, MATLAB can still understand input using an old format. That is, if commands or built-in-function formats change, the old formats carry on. The form of window menus may also change, but the main ideas persist. If there is some difference between this introduction and the user's MATLAB version, a little bit of intuitive trial and error, or investigation of online documentation, might be needed.

Answers to Selected Problems

CHAPTER 1

1.2 250 m

1.4 1.277 s

1.6 a) $\left(\ln\sqrt{\dfrac{x+2}{Cx}}\right)^{-1}$

c) $u(u+4x)^3 = C$

1.7 a) $C/(x^2+2)$

c) $C/\sin 2x$

1.8 a) $x(1-x^2)/2$

c) $(\sin x - \cos x + e^x)/2$

1.9 0.2175 amp

1.10 $0.01\,(e^{2t} - e^{-2\times 10^4 t})$

1.12 1.386×10^{-4} s

1.14 $0.03456 - 0.03056e^{-1.111\times 10^{-4}t}$

1.16 $(9.84 \times 10^{-4}t + 0.707)^{-2}$, 298 s

1.18 354.3 m/s

1.20 805 s

1.22 a) $\dfrac{1}{\alpha}\tanh(\alpha g t)$, where $\alpha = \sqrt{c/mg}$

b) 0.40 seconds.

1.23 7.45 m/s.

1.24 a) $\dfrac{du}{d\theta}u + \dfrac{g}{l}\sin\theta = 0$, with $u(A) = 0$.

b) $\pm\sqrt{\dfrac{2g}{l}(\cos\theta - \cos A)}$, valid if $\cos\theta - \cos A \ge 0$.

1.26 $\pm\sqrt{\dfrac{2ga\sin\gamma x + 2k}{(1 + a^2\gamma^2\cos^2\gamma x)}}$

1.29 a) $c_1e^{3x} + c_2e^{-2x}$

c) $A\cos 3x + B\sin 3x$

e) $c_1e^{2x} + c_2xe^{2x}$

g) $c_1e^{4.828x} + c_2e^{-0.8284x}$

i) $c_1 + c_2e^{4x}$

k) $e^{-x}(A\cos 2x + B\sin 2x)$

1.30 $A\cos\sqrt{\dfrac{g}{L}}t + B\sin\sqrt{\dfrac{g}{L}}t$

1.32 $A\cos 5t + B\sin 5t$, 0.7958 hz

1.34 a) $\dfrac{1}{2}\cos 4t$

1.36 $\dfrac{25}{3}(e^{-8t} - e^{-2t})$

1.38 $e^{-5t/2}\,[\cos 3.12t + 0.801\sin 3.12t]$

1.40 $\dfrac{2M}{\sqrt{C^2 - 4KM}} \times \tanh^{-1}\dfrac{\sqrt{C^2 - 4KM}}{C}$

1.42 0.9734 s

1.44 $(0.5 + 0.452i)[e^{(-3+3.32i)t} + e^{-(3+3.32i)t}]$,
$e^{-3t}(\cos 3.32t + 0.904\sin 3.32t)$,
$1.35e^{-3t}\cos(3.32t - 0.737)$

1.46 2.19 ohms

1.48 $M \to I$, $C \to c$, $K \to k$

1.49 a) x

c) $\dfrac{1}{2}e^{-x}$

e) $\dfrac{5}{9}\sin x$

g) $\dfrac{x^2}{2}e^{-2x}$

1.50 a) $A\cos x + B\sin x + \dfrac{1}{5}e^{2x}$

c) $A\cos 3x + B\sin 3x + \dfrac{x^2}{9} - \dfrac{2}{81} + \dfrac{1}{5}\sin 2x$

e) $c_1e^{4x} + c_2e^{-4x} + \dfrac{x}{8}e^{4x}$

1.51 a) $\left(-\dfrac{3}{8} + \dfrac{1}{4}x\right)e^{-2x} + \dfrac{1}{8}(2x^2 - 4x + 3)$

c) $e^{-2x}(-1.176\cos x - 2.032\sin x) + \dfrac{x^2}{5}$
$- 0.32x + 1.176$

e) $-\dfrac{1}{30}[e^{-3x}(\cos x + 7\sin x) - 2\sin 2x - \cos 2x]$

1.52 $c_1 + c_2e^{-(C/M)t} - \dfrac{Mg}{C}t$

1.54 5.56 s, 8.8 s

M. C. Potter and B. F. Feeny, *Mathematical Methods for Engineering and Science*,
https://doi.org/10.1007/978-3-031-26151-0

1.56 0.00533 hz, 2.1 m

1.61 a) $-\cos 2t$
c) $-2\cos t + \sin t$
e) $\frac{1}{5}\sin t - \frac{1}{10}\cos t - \frac{3}{13}\sin 3t + \frac{2}{13}\cos 3t$

1.62 a) $c_1e^{-4t} + c_2e^{-t} - \frac{1}{5}\cos 2t$
c) $(c_1 + c_2t)e^{-2t} + \frac{12}{25}\sin t - \frac{16}{25}\cos t$

1.63 a) $12e^{-3t} - 13e^{-2t} + \cos 2t + 5\sin 2t$
c) $e^{-t}(3\cos 3t - \sin 3t) + 3\sin 2t - 2\cos 2t$
e) $5e^{-t} - 2e^{-2t} + \sin t - 3\cos t$

1.64 b) $\pi/2$, 2.00 m

1.66 2.5 kg/s

1.68 $2.59\times10^{-4}\sin 120\pi t + 11.4\times10^{-4}\cos 120\pi t$

1.70 $-0.04\sin 200t + 10^{-5}\cos 200t$

1.72 a) $\frac{\pi}{2} + \sum_{n=1}^{\infty}\frac{2}{n\pi^2}[(-1)^n - 1]\times\cos nt$
b) $\frac{\pi^2}{3} + \sum_{n=1}^{\infty}\frac{4\pi(-1)^n}{n^2}\cos nt$
c) $\frac{2}{\pi} + \sum_{n=1}^{\infty}\frac{(-1)^n}{\pi\left(\frac{1}{4} - n^2\right)}\cos nt$
d) $2\pi - \sum_{n=1}^{\infty}\frac{4(-1)^n}{n}\sin\frac{nt}{2}$

1.74 a) $-\sum_{n=1}^{\infty}\frac{(-1)^n 8}{n}\sin nt,$
$2\pi + \sum_{n=1}^{\infty}\frac{8}{\pi n^2}[(-1)^n - 1]\times\cos nt$
b) $-\sum_{n=1}^{\infty}\frac{20}{n\pi}\left(\cos\frac{n\pi}{2} - 1\right)\sin\frac{nt}{2},$
$\sum_{n=1}^{\infty}\frac{20}{n\pi}\sin\frac{n\pi}{2}\cos\frac{nt}{2}$
c) $\sin t,\ \frac{2}{\pi} + \frac{2}{\pi}\sum_{n=1}^{\infty}\frac{1}{1-n^2}\times[1-(-1)^{n+1}]\cos nt$

1.76 $\frac{1}{6} - \sum_{n=1}^{\infty}\frac{2}{n^2\pi^2}(\cos n\pi - 1)\times\cos n\pi t$
$-\sum_{n=1}^{\infty}\frac{4}{n^3\pi^3}\times(\cos n\pi - 1)\sin n\pi t$

1.78 $\sum_{n=1}^{\infty}\frac{20(-1)^{n+1}}{(10-n^2)n\pi}\sin nt$

1.80 $\frac{3}{5000} + \sum_{n=1}^{\infty}[A_n\sin(1000n\pi t0 - B_n\cos(1000n\pi t)]$

where $A_n = \frac{120}{n\pi}\left[\frac{10^5 - 1000\pi^2n^2}{(10^5 - 1000\pi^2n^2)^2 - (20000\pi n)^2}\right]$

$B_n = \frac{120}{n\pi}\left[\frac{20000\pi n}{(10^5 - 1000\pi^2n^2)^2 - (20000\pi n)^2}\right]$

1.81 a) $c_1x^{-4} + c_2x^{-2}$
c) $c_1x^4 + c_2x^{-3} - 2x$

1.82 $\ln x$

1.83 a) $\frac{\sin 2t}{8}(\sin t + 2t\cos t)$
$+\frac{\cos 2t}{8}(\cos t - 2t\sin t) - \frac{t^2}{4}\cos t$
c) $\frac{t^3}{6}e^{-2t}$

1.84 $\partial^2u/\partial\xi\,\partial\eta = 0$

1.86 $\Delta\alpha\cos\alpha$

CHAPTER 2

2.1 a) $1 + x + x^2 + x^3 + \cdots$
c) $x - \frac{x^3}{3!} + \frac{x^5}{5!} + \cdots$
e) no expansion

2.2 a) $1 - x + x^2 - x^3 + \cdots$
c) $\frac{1}{2}\left(1 - \frac{3}{2}x + \frac{7}{4}x^2 - \frac{15}{8}x^3 + \cdots\right)$
e) $e\left(1 + 2x + 2x^2 + \frac{4}{3}x^3 + \cdots\right)$
g) $x^2 - \frac{x^6}{3!} + \frac{x^{10}}{5!} + \cdots$
i) $-0.69315 + x - \frac{x^2}{2} + \frac{x^3}{3} + \cdots$
k) $\frac{1}{4}\left(1 + \frac{3}{4}x + \frac{5}{16}x^2 + \frac{17}{192}x^3 + \cdots\right)$

2.3 a) $x - \frac{x^2}{2} + \frac{x^3}{3} - \frac{x^4}{4} + \cdots$
c) $\frac{x^2}{2}\left(1 - \frac{x^2}{2} + \frac{x^4}{3} - \frac{x^6}{4} + \cdots\right) = \frac{1}{2}\ln(1 + x^2)$
e) $\frac{x^2}{2}\left(1 + \frac{x^2}{6} + \frac{2x^4}{45} + \cdots\right) + C$

2.4 a) 1
c) 2

2.5 a) no singular pts., $R=\infty$
c) (0, 0), (0, 2i), (0, −2i), $R=0$
e) (−1, 0), $R=1$

2.6 a) 1
c) 2
e) ∞

2.7 a) $b_0\left(1-x+\dfrac{x^2}{2!}-\dfrac{x^3}{3!}+\cdots\right),\ R=\infty$

c) $b_0(1-x)+\dfrac{x^2}{2}\left(1+\dfrac{x}{3}+\dfrac{x^2}{6}+\cdots\right),\ R=1$

e) $b_0\left(1+2x^2+\dfrac{2}{3}x^4+\cdots\right)$
$+b_1\left(x+\dfrac{2x^3}{3}+\dfrac{2x^5}{15}+\cdots\right),\ R=\infty$

g) $b_0\left(1-\dfrac{x^2}{2}+\dfrac{x^3}{3}+\cdots\right)+b_1\left(x-x^2+\dfrac{x^3}{2}+\cdots\right)$
$+\dfrac{x^4}{12}-\dfrac{x^5}{30}+\dfrac{x^6}{120}+\cdots,\ R=\infty$

2.8 a) $1-x+\dfrac{x^2}{2}-\dfrac{x^3}{9}+\cdots$, all x

c) $1-\dfrac{x^2}{2}+\dfrac{5}{6}x^3+\dfrac{x^4}{24}+\cdots$, all x

2.10 $1-5(x-2)+4(x-2)^2-4(x-2)^3+\cdots$,
1.544, 1.5444

2.11 a) $b_0\left[1+\dfrac{(x-1)^2}{2}+\dfrac{(x-1)^3}{6}+\cdots\right]$
$+b_1\left[x-1+\dfrac{(x-1)^3}{6}+\dfrac{(x-1)^4}{12}+\cdots\right]$

c) $b_0\left[1-(x-2)^2-\dfrac{(x-3)^3}{6}+\cdots\right]$
$+b_1\left[x-2-\dfrac{(x-2)^3}{3}-\dfrac{(x-2)^4}{12}+\cdots\right]$
$+2(x-2)^2+\dfrac{2}{3}(x-2)^3-\dfrac{1}{4}(x-2)^4+\cdots$

2.12 $4-2x+\dfrac{x^3}{3}+\dfrac{x^4}{3}-\dfrac{x^5}{10}+\dfrac{x^7}{126}+\cdots$, 6.95

2.14 8.115, 7.302

2.16 $\dfrac{1}{128}(6435x^8-12012x^6+6930x^4$
$-1260x^2+35)$

2.20 a) $\dfrac{1}{2}(5x^3-3x)\left(c_1+c_2\ln\dfrac{1+x}{1-x}\right)$
$-c_2\left(\dfrac{5x^2}{2}-\dfrac{2}{3}\right)$

c) $c_1\left(1-\dfrac{3x^2}{8}-\dfrac{21x^4}{128}+\cdots\right)$
$+c_2\left(x+\dfrac{5x^3}{24}+\dfrac{15x^5}{128}+\cdots\right)$

2.22 a) $a_0(1-x)+d_0\left(1-\dfrac{x}{6}\right.$
$\left.-\dfrac{x^2}{120}-\dfrac{x^3}{1680}+\cdots\right)x^{1/2}$

c) $a_0\left(1+x+\dfrac{x^2}{6}+\cdots\right)+$
$d_0\left(1+\dfrac{x}{3}+\dfrac{x^2}{6}+\cdots\right)x^{1/2}$

e) $a_0(1+x/7+9x^2/161+\cdots)x$
$+d_0(1+x/4+9x^2/160+\cdots)x^{1/4}$

g) $a_0\left(1-\dfrac{x}{3}+\dfrac{x^2}{5}+\cdots\right)x$
$+d_0\left(1-\dfrac{x}{2}+\dfrac{x^2}{8}+\cdots\right)x^{1/2}$

2.23 a) $\left(1-x+\dfrac{x^2}{4}+\cdots\right)\times(B+C\ln x)$
$+C\left(2x-\dfrac{3x^2}{4}+\dfrac{11x^3}{108}+\cdots\right)$

c) $\left(x^2+x^3+\dfrac{x^4}{4}+\cdots\right)\times(B+C\ln x)$
$+C\left(2x^3+\dfrac{3x^4}{4}+\dfrac{11x^5}{108}+\cdots\right)$

2.24 a) $\left(x+\dfrac{x^2}{2}+\dfrac{x^3}{12}+\cdots\right)$
$\times\left[A+B\left(\ln x+\dfrac{1}{x}-x+\dfrac{59}{144}x^2+\cdots\right)\right]$

c) $\left(1-\dfrac{3x}{4}+\dfrac{5x^2}{16}+\cdots\right)x^{3/2}$
$\times\left[A+B\left(\ln x-\dfrac{1}{2x}+\dfrac{x}{8}+\cdots\right)\right]$

2.25 a) $u_1(x) = a_0\left[1 + \frac{3(x-1)}{8} - \frac{15(x-1)^2}{256} + \cdots\right]$

$u_2(x) = u_1(x)\ln(x-1)$

$-a_0\left[\frac{5}{4}(x-1) + \frac{29}{256}(x-1)^2 + \cdots\right]$

c) $(x-1)^2\left[1 - \frac{x-1}{3} + \frac{5(x-1)^2}{24} + \cdots\right]$

$\times\left\{A + B\left[\frac{1}{2(x-1)^2} - \frac{1}{6(x-1)} + \frac{1}{72}\ln(x-1) + \cdots\right]\right\}$

2.26 a) 0.886
c) 6
e) −2.258
g) 5.437

2.27 a) $1 - \frac{x^2}{4} + \frac{x^4}{64} - \frac{x^6}{2304} + \cdots$

2.28 0.2222, 0.5764
Compare with 0.2239, 0.5767

2.30 a) $AJ_1(x) + BY_1(x)$
c) $AJ_{1/2}(x) + BJ_{-1/2}(x)$
e) $AJ_0(2\sqrt{x}) + BY_0(2\sqrt{x})$

2.31 a) 0.1289
c) −0.5767
e) 0.3252

2.32 a) $(8x - x^3)J_1(x) - 4x^2J_0(x) + C$
c) $\left(\frac{1}{x} + \frac{12}{x^3}\right)J_1(x) + \frac{6}{x^2}J_0(x) + C$
e) $3x^2J_1(x) - (x^3 + 3x)J_0(x) + 3\int J_0(x)dx$

CHAPTER 3

3.1 a) $2/s^2$
c) $1/(s-3)$
e) $s/(s^2+16)$
g) $2.659\, s^{-5/2}$
i) $2/(s^2-4)$
k) $s/(s^2-16)$
m) $\frac{2}{s}(e^{-5s} - e^{-10s})$
o) $\left(\frac{2}{s} - \frac{1}{s^2}\right)e^{-4s} + \frac{1}{s^2}e^{-6s}$

3.2 a) $s/(s-3)^2$
c) $(s+2)/(s^2+4s+20)$
e) $6/(s^2+2s+5)$
g) $(s-7)/(s^2+2s+17)$
i) $(5s^2+24s+30)/(s+2)^3$

3.3 a) e^{-2s}/s
c) $2\left(\frac{1}{s^2} - \frac{2}{s}\right)e^{-2s} - 2\left(\frac{1}{s^2} + \frac{4}{s}\right)e^{-4s}$
e) $-e^{-4s}\left(\frac{1}{s^2} - \frac{2}{s}\right) + \frac{1}{s^2}e^{-6s}$
g) $\frac{2}{s^2+4}(1 - e^{-2\pi s})$

3.4 a) $2/(s^2-4s+8) + 2/(s^2+4s+8)$
c) $6/(s^2-4s-5) + 6/(s^2+4s-5)$
e) $8/(s^2-4s-12) - 8/(s^2+4s-12)$

3.5 a) $t^2 + t - 2$
c) $2(1-3t)e^{-3t}$
e) $1 - e^{-t}$
g) $e^{2t}/3 - e^{-t}/3$
i) $\left(\frac{1}{2} + t\right)e^t - \frac{1}{2}e^{-t}$
k) $u_2(t)[1 - te^{-(t-2)} + e^{-(t-2)}]$
m) $e^{-2t}\left(4\cos 3t - \frac{5}{3}\sin 3t\right)$
o) $e^{2t}\left(3\cosh 3t + \frac{7}{3}\sinh 3t\right)$
q) $\sinh t \sin t$

3.6 $s/(s^2+16)$

3.9 a) $s^4\mathcal{L}(f) - s^3f(0) - s^2f'(0) - sf''(0) - f'''(0)$

3.11 a) $\omega/(s^2+\omega^2)$
c) $a/(s^2-a^2)$
e) $1/(s-2)$

3.12 $\frac{1}{s^2}(1-e^{-s}), \frac{1}{s^2}(1-e^{-s})$

3.14 a) $1/(s-1)^2$
c) $(s^2-1)/(s^2+1)^2$
e) $\frac{2(s^2-2s+1)}{(s-1)(s^2-2s+2)^2}$
g) $(s^2+4)/(s^2-4)^2$
i) $4s/(s^2-4)^2$

3.15 a) $e^{-t}\sinh t$
c) $1 - \cos 2t$
e) $-\frac{2}{3}(1 - \cosh 3t)$
g) $t - \frac{1}{\sqrt{2}}\sin\sqrt{2}t$
i) $2 - t - 2e^{-t}$

3.16 a) $12s/(s^2+9)^2$
c) $4(3s^2-4)/(s^2+4)^3$
e) $(s^2-2s+3)/(s^2-2s+5)^2$
g) $3(2s+1)/[(s-1)^2(s+2)^2]$
i) $2(3s^2+6s+2)/(s^2+2s+2)^3$
k) $\ln[(s^2+4)/s^2]$
m) $\ln[(s+2)/(s-2)]$

3.18 a) te^{-2t}
c) $\frac{t}{4}\sinh 2t$
e) $\frac{1}{t}(e^{-3t}-e^{-2t})$
g) $\frac{2}{t}(\cos 2t-\cos t)$
i) $\frac{2}{t}(e^{-t}\cos 2t-e^{-2t}\cos t)$

3.19 a) $\frac{1}{s^2+1}\frac{1+e^{-\pi s}}{1-e^{-\pi s}}$
c) $\frac{1}{s^2(1-e^{-2s})}(e^{-2s}+2s-1)$
e) $\frac{-1}{s^3(1-e^{-\pi s})}\times[e^{-\pi s}(\pi^2 s^2+2\pi s+2)-2]$
g) $\frac{1}{s^2(1-e^{-4s})}\times[e^{-2s}(2s+1)-e^{-4s}(4s+1)]$
i) $\frac{2}{s(1-e^{-4s})}\times(1-e^{-s}+e^{-3s}-e^{-2s})$

3.20 a) $-10e^{t}+30e^{-2t}+9e^{3t}-15e^{-t}$
c) $-3e^{-2t}+\frac{1}{2}e^{-t}+\frac{33}{10}e^{-3t}+\frac{1}{5}e^{2t}$
e) $-\frac{3}{2}-t+\left(\frac{3}{2}-2t+t^2\right)e^{2t}$
g) $\frac{5}{6}\sin 2t-\frac{2}{3}\sin t$
i) $-\frac{2}{5}e^{-t}+\frac{2}{5}\cos 2t+\frac{3}{10}\sin 2t$
j) $-\frac{6}{25}e^{-t}+\frac{2}{5}te^{-t}-\frac{6}{25}\cos 2t-\frac{9}{50}\sin 2t$
l) $\frac{10}{144}(\sin 2t-2t\cos 2t)+\frac{40}{27}\cos 2t$
$-\frac{40}{27}\cos t+\frac{5}{9}(\sin t-t\cos t)$

3.21 a) $5\sin 2t$
c) $-2\cos t+2\sin t+2$
e) $t\sin 2t$
g) $-20e^{-3t}+20e^{-2t}$
i) $\frac{1}{3}e^{4t}+\frac{2}{3}e^{-2t}$
k) $-4+2t+(2+2t)e^{-t}$
m) $-\frac{1}{10}e^{-2t}\sin 10t$
$-\frac{5}{990}\cos 10t+\frac{5}{99}\sin 10t$

3.22 a-1) $\frac{1}{36}(1-\cos 6t)$
a-3) $\frac{5}{72}(\sin 6t-6t\cos 6t)$
a-5) $\frac{5}{36}\left(e^{-0.2t}-\cos 6t-\frac{1}{30}\sin 6t\right)$
b-1) $\frac{1}{36}-\frac{1}{36}e^{-t/2}\left(\cos 5.98t+\frac{1}{12}\sin 5.98t\right)$
b-3) $-\frac{5}{6}\cos 6t+\frac{5}{6}e^{-t/2}$
$\times\left(\cos 5.98t+\frac{1}{12}\sin 5.98t\right)$
c-2) $-\frac{3}{40}\cos 2t+\frac{1}{10}\sin 2t-\frac{3}{40}e^{-6t}+\frac{1}{4}te^{-6t}$
c-6) $50te^{-6t}$
d-3) $-\frac{1}{4}\cos 6t-\frac{1}{192}e^{-18t}+\frac{3}{64}e^{-2t}$
d-4) $\frac{5}{36}+\frac{5}{288}e^{-18t}-\frac{5}{32}e^{-2t}$
$-\left[\frac{5}{36}+\frac{5}{288}e^{-18(t-4\pi)}-\frac{5}{32}e^{-2(t-4\pi)}\right]$
$u_{4\pi}(t)$

3.23 a-1) $\sin 10t$
a-3) $\frac{t}{4}\sin 10t$
a-5) $10\cos 10t$
b-1) $\frac{1}{15}e^{-8t}(25\sin 6t-24\cos 6t)$
b-3) $-\frac{4}{5}\sin 10t-\frac{1}{2}\cos 10t$
$+\frac{1}{48}e^{-8t}(24\cos 6t-7\sin 6t)$
b-5) $\frac{10}{3}e^{-8t}(3\cos 6t-4\sin 6t)$
c-2) $\frac{1}{25}(8\sin 20t-6\cos 20t$
$+6e^{-10t}-100te^{-10t})$
c-4) $10te^{-10t}-10(t-2\pi)\times e^{-10(t-2\pi)}u_{2\pi}(t)$
d-5) $\frac{10}{3}(4e^{-20t}-e^{-5t})$
d-6) $-\frac{5}{19}e^{-t}-\frac{80}{27}e^{-20t}+\frac{5}{3}e^{-5t}$

3.24 a-1) $\frac{1}{25}(1-\cos 5t)[u_{2\pi}(t)-u_{4\pi}(t)]$

a-3) $\frac{t}{10\pi}-\frac{t-2\pi}{10\pi}u_{2\pi}(t)$
$-\frac{1}{50\pi}\sin 5t[1-u_{2\pi}(t)]$
$-\frac{1}{5}[1-\cos 5t]u_{2\pi}(t)$

a-5) $\frac{1}{25}(1-\cos 5t)[1+u_{4\pi}(t)]$

a-7) $\sin 5t+\frac{1}{5}(1-\cos 5t)\times u_{2\pi}(t)$

b-1) $\frac{1}{10}\sin 10t[u_{2\pi}(t)-u_{4\pi}(t)]$

b-3) $\frac{1}{40\pi}(1-\cos 10t)\times[1-u_{2\pi}(t)]-\frac{1}{2}u_{2\pi}(t)\sin 10t$

b-5) $\frac{1}{10}\sin 10t[1+u_{4\pi}(t)]$

b-7) $5\cos 10t+\left(\frac{1}{2}\sin 10t\right)u_{2\pi}(t)$

3.25 a-1) $\frac{1}{50}u_{2\pi}(t)[1+e^{-25(t-2\pi)}]$
$-\frac{1}{50}u_{4\pi}(t)[1-e^{-25(t-4\pi)}]$

a-3) $\frac{1}{500\pi}\{25t+e^{-25t}-1-[25(t-2\pi)$
$+e^{-25(t-2\pi)}-1]u_{2\pi}(t)\}$

a-5) $\frac{1}{50}[1+u_{4\pi}(t)-e^{-25t}-e^{-25(t-4\pi)}u_{4\pi}(t)]$

a-7) $\frac{5}{2}e^{-25t}+\frac{1}{10}[1-e^{-25(t-2\pi)}]\times u_{2\pi}(t)$

b-1) $\frac{1}{50}u_{2\pi}(t)[1+e^{-25(t-2\pi)}]$
$-\frac{1}{50}u_{4\pi}(t)[1-e^{-25(t-4\pi)}]$

b-3) $\frac{1}{500\pi}\{25t+e^{-25t}-1-[25(t-2\pi)$
$+e^{-25(t-2\pi)}-1]u_{2\pi}(t)\}$

b-5) $\frac{1}{50}[1+u_{4\pi}(t)-e^{-25t}-e^{-25(t-4\pi)}u_{4\pi}(t)]$

b-7) $\frac{5}{2}e^{-25t}+\frac{1}{10}[1-e^{-25(t-2\pi)}]\times u_{2\pi}(t)$

3.26 $c_1x+c_2\frac{x^3}{6}+\frac{P}{6EI}\left(x-\frac{L}{2}\right)^3\times u_{L/2}(x)$
$+\frac{w}{24EI}\times\left[x^4-\left(x-\frac{L}{2}\right)^4u_{L/2}(x)\right]$

$c_1=\frac{PL^2}{16EI}+\frac{3wL^3}{128EI},$

$c_2=-\frac{P}{2EI}-\frac{3wL}{8EI}$

CHAPTER 4

4.1 a) yes
c) no
e) no
g) yes
i) yes
k) yes

4.2 a) *a, b, h, i*
c) *a, h*

4.3 a) $\begin{bmatrix}-2&7&0\\0&6&8\\4&0&9\end{bmatrix}$

4.4 a) –2
c) 7
e) –3

4.5 a) $\begin{bmatrix}5\\4\\1\\-2\end{bmatrix},\begin{bmatrix}2\\-2\\0\\4\end{bmatrix},\begin{bmatrix}0\\7\\6\\0\end{bmatrix},\begin{bmatrix}-3\\0\\8\\9\end{bmatrix}$

4.6 a) $\begin{bmatrix}3&2&1\\1&-1&-2\\6&3&-3\end{bmatrix},$
$\begin{bmatrix}3&2&1\\1&-1&-2\\6&3&-3\end{bmatrix}$

c) $\begin{bmatrix}1&-1&0\\1&-3&-2\\7&1&-2\end{bmatrix},$
$\begin{bmatrix}1&-1&0\\1&-3&-2\\7&1&-2\end{bmatrix}$

e) $\begin{bmatrix}-4&-10&-4\\2&-10&-4\\12&-4&-4\end{bmatrix},$
$\begin{bmatrix}-4&-10&-4\\2&-10&-4\\12&-4&-4\end{bmatrix}$

4.10 a) $\begin{bmatrix} 2 & 2 & 1 \\ 2 & 4 & 0 \\ 1 & 0 & 2 \end{bmatrix}, \begin{bmatrix} 0 & 2 & 5 \\ -2 & 0 & -2 \\ -5 & 2 & 0 \end{bmatrix}$

c) $\begin{bmatrix} 2 & -3 & 5 \\ -3 & 4 & -1 \\ 5 & -1 & 6 \end{bmatrix},$

$\begin{bmatrix} 2 & 7 & -3 \\ 7 & 4 & 1 \\ -3 & 1 & -2 \end{bmatrix}$

4.12 a) $\begin{bmatrix} 2 & 4 & -1 \\ -2 & 4 & 1 \\ 4 & 8 & -2 \end{bmatrix}$

c) $\begin{bmatrix} -3 & 4 & -3 \\ -13 & -4 & -5 \\ -6 & 8 & -6 \end{bmatrix}$

e) $\begin{bmatrix} -3 & 4 & -3 \\ 3 & -4 & 3 \\ -6 & 8 & -6 \end{bmatrix}$

g) $\begin{bmatrix} 1 & 3 & 7 \\ 0 & 1 & -3 \\ 1 & -1 & 1 \end{bmatrix}$

i) $\begin{bmatrix} 3 \\ 1 \\ 1 \end{bmatrix}$

4.14
$$\begin{aligned} 6y_1 - y_2 - 2y_3 &= r_1 \\ 2y_1 + 2y_2 - y_3 &= r_2 \\ -\ y_2 \qquad &= r_3 \end{aligned}$$

4.15 a) $\begin{bmatrix} 1 \\ -5 \\ 5 \end{bmatrix}$

c) $\begin{bmatrix} -3 \\ -8 \\ 6 \end{bmatrix}$

e) $\begin{bmatrix} 0 & -3 & 3 \\ 7 & 4 & 1 \\ 3 & 2 & 0 \end{bmatrix}$

g) $[-3, -8, 6]$

i) not defined

k) not defined

m) $\begin{bmatrix} -2 & 6 & 1 \\ 0 & 0 & -3 \\ -6 & -2 & 4 \end{bmatrix}$

4.16 $\begin{bmatrix} 6 & -3 & 9 \\ 0 & 3 & 6 \\ -6 & 0 & 0 \end{bmatrix}$

4.18 $\begin{bmatrix} 4 & 2 & 6 \\ -1 & 1 & -2 \\ 3 & 9 & 6 \end{bmatrix}, \begin{bmatrix} 4 \\ -1 \\ -3 \end{bmatrix}$

4.19 a) 6
c) 0
e) 36

4.20 a) −36
c) −36

4.21 a) 3, −3
c) 1.87, 0.348
e) $-2, -\frac{1}{3}, -1$

4.22 a) −276
c) −276

4.24 a) −12
c) −5
e) −24
g) −244

4.26 $k_1\, k_2\, k_3 \ldots k_n$

4.27 a) $\begin{bmatrix} 1/2 & 1/2 \\ -1/2 & 1/2 \end{bmatrix}$

c) $\begin{bmatrix} 1/2 & 0 \\ 0 & 1 \end{bmatrix}$

e) $\begin{bmatrix} -3/4 & 5/8 \\ -1/4 & 3/8 \end{bmatrix}$

g) A^{-1} does not exist

i) $\begin{bmatrix} 0 & -1 & 1 & 0 \\ 0 & 0 & -1 & 1 \\ 1 & 0 & 1 & -1 \\ -1 & 1 & -1 & 1 \end{bmatrix}$

4.28 a) $\begin{bmatrix} 1/2 & 0 \\ 0 & -1 \end{bmatrix}$

c) $\begin{bmatrix} 1/2 & 0 & 0 \\ 0 & 1/2 & 0 \\ 0 & 0 & 1 \end{bmatrix}$

4.29 a) $\begin{bmatrix} 1 & -1 \\ -1 & 2 \end{bmatrix}$

c) $\begin{bmatrix} -1/6 & 1/4 & 1/6 \\ 1/4 & -3/8 & 1/4 \\ 1/6 & 1/4 & -1/6 \end{bmatrix}$

4.31 a) $\begin{bmatrix} 1 \\ -1 \end{bmatrix}$

c) $\begin{bmatrix} 6 \\ -3 \\ -2 \end{bmatrix}$

4.32 a) -3

b) 1

c) 1

d) $4/3$

4.33 a) $\begin{bmatrix} 2 \\ -3 \end{bmatrix}$

c) $\begin{bmatrix} -1 \\ 1 \\ -2 \end{bmatrix}$

4.34 a) $\begin{bmatrix} 2 \\ -3 \end{bmatrix}$

c) $\begin{bmatrix} -1 \\ 1 \\ -2 \end{bmatrix}$

4.36 a) $\alpha \begin{bmatrix} 0.4082 \\ -0.8165 \\ 0.4082 \end{bmatrix}$

b) $\beta \begin{bmatrix} 1 \\ -2 \\ 1 \end{bmatrix}$

4.38 $85.8015\dfrac{C}{A} + 4.1250\dfrac{Y}{A} + 341.6720\dfrac{T}{A} - 399.9026\dfrac{I}{A}$

4.40 a) $2, \ -2, \begin{bmatrix} 2/\sqrt{5} \\ 1/\sqrt{5} \end{bmatrix}, \begin{bmatrix} -2/\sqrt{5} \\ 1/\sqrt{5} \end{bmatrix}$

c) $9, \ -1, \begin{bmatrix} 1/\sqrt{10} \\ 3/\sqrt{10} \end{bmatrix}, \begin{bmatrix} -3/\sqrt{10} \\ 1/\sqrt{10} \end{bmatrix}$

e) $7, \ -3, \begin{bmatrix} 2/\sqrt{5} \\ 1/\sqrt{5} \end{bmatrix}, \begin{bmatrix} 1/\sqrt{5} \\ -2/\sqrt{5} \end{bmatrix}$

g) $2, 2, -1, \begin{bmatrix} x_1 \\ 0 \\ x_3 \end{bmatrix}, \begin{bmatrix} x_1 \\ 0 \\ x_3 \end{bmatrix}, \begin{bmatrix} 0 \\ 1 \\ 0 \end{bmatrix}$

4.41 a) $4, -2$

d) $64, -8$

4.42 a) $2 + 2i, 2 - 2i$

b) $2 + 2i, 2 - 2i$

4.43 a) $8, 6, -2$

d) $36, 64, 4$

4.44 a) $4, 2, -2$

b) $4, 2, -2$

4.45 a) $i_1'' = \dfrac{C_1 - C_2}{L_1 C_1 C_2} i_1 + \dfrac{1}{L_1 C_2} i_2$

$i_2'' = \dfrac{1}{L_2 C_2} i_1 + \dfrac{1}{L_2 C_2} i_2$

d) $120.7, -20.7,$

$\begin{bmatrix} 0.816 \\ 0.577 \end{bmatrix}, \begin{bmatrix} 0.816 \\ -0.577 \end{bmatrix}$

f) $\dfrac{1}{4}(e^{11t} + e^{-11t}) + \dfrac{1}{2}\cos 4.55t,$

$\dfrac{1}{4}(e^{11t} + e^{-11t}) - \dfrac{1}{2}\cos 4.55t$

4.46 a) $-12, -2, \begin{bmatrix} -2/\sqrt{5} \\ 1/\sqrt{5} \end{bmatrix}, \begin{bmatrix} 1/\sqrt{5} \\ 2/\sqrt{5} \end{bmatrix}$

c) $y_1(t) = -\dfrac{2}{\sqrt{3}} \sin \sqrt{12}\, t + 2\sqrt{2} \sin \sqrt{2}\, t$

$y_2(t) = \dfrac{1}{\sqrt{3}} \sin \sqrt{12}\, t + 4\sqrt{2} \sin \sqrt{2}\, t$

4.47 a) $0.912(e^{2\sqrt{30}\,t} - e^{-2\sqrt{30}\,t}) + 5.48 \sin \sqrt{30}\, t,$

$0.912(e^{2\sqrt{30}\,t} - e^{-2\sqrt{30}\,t}) - 3.65 \sin \sqrt{30}\, t$

4.48 $0, \dfrac{mL^2}{3}, \dfrac{\sqrt{2}mL^2}{3}, \begin{bmatrix} \sqrt{2} \\ \sqrt{2} \\ 0 \end{bmatrix}, \begin{bmatrix} 0 \\ 0 \\ 1 \end{bmatrix}, \begin{bmatrix} -\sqrt{2} \\ \sqrt{2} \\ 0 \end{bmatrix}$

4.50 $2.7312mL^2, 21.6012mL^2, \dfrac{71mL^2}{3},$

$\begin{bmatrix} -0.8063 \\ 0 \\ -0.5915 \end{bmatrix}, \begin{bmatrix} -0.5915 \\ 0 \\ 0.8063 \end{bmatrix}, \begin{bmatrix} 0 \\ 1 \\ 0 \end{bmatrix}$

4.52 $-30, 0, 40, \dfrac{1}{\sqrt{3}} \begin{bmatrix} 1 \\ 1 \\ 1 \end{bmatrix}, \dfrac{1}{\sqrt{6}} \begin{bmatrix} -2 \\ 1 \\ 1 \end{bmatrix}, \dfrac{1}{\sqrt{2}} \begin{bmatrix} 0 \\ -1 \\ 1 \end{bmatrix}$

4.54 The system is unstable for all σ

4.55 The system is stable for $\sigma \geq 5/3$

CHAPTER 5

5.1 a) No
c) vector
e) vector
g) No
i) vector
k) No

5.2 a) 23.17, 17.76°
10.62, 318.3°
c) 11.18, 333.4°
11.18, 26.57°

5.3 a) $0.5\hat{i} + 0.707\hat{j} + 0.5\hat{k}$
c) $0.6427\hat{i} - 0.3419\hat{j} + 0.6852\hat{k}$

5.4 a) $6\hat{i} + 3\hat{j} - 8\hat{k}$
c) $3\hat{i} + 3\hat{j} - 4\hat{k}$
e) −4
g) 53.85
i) 0
k) $100\hat{i} + 154\hat{j} - 104\hat{k}$

5.5 a) 8
c) 107.3

5.8 a) 5/9
c) −2.941

5.10 $-0.3914\hat{i} - 0.5571\hat{j} - 0.7428\hat{k}$

5.12 −11.93

5.14 26 N · m

5.16 $80\hat{i} - 5\hat{j} + 145\hat{k}$ m/s

5.18 a) 100 N · m

5.20 a) $2\hat{i} + 4\hat{k}$
c) −30.80
e) 74.79

5.22 $-210\hat{i} + 600\hat{j}$

5.24 0.00925 m/s², 2.39×10^{-5} m/s²

5.25 $16\hat{i} + 8\hat{j} + 4\hat{k}$

5.27 a) $2x\hat{i} + 2y\hat{j}$
c) $2x\hat{i} + 2y\hat{j} + 2z\hat{k}$
e) $(2x + 2y)\hat{i} + 2x\hat{j} - 2z\hat{k}$
g) $-\vec{r}/r^3$
i) $nr^{n-2}\vec{r}$

5.28 a) $\frac{3}{5}\hat{i} + \frac{4}{5}\hat{j}$
c) $0.97\hat{i} - 0.243\hat{j}$
e) $0.174\hat{i} + 0.696\hat{j} - 0.696\hat{k}$

5.29 a) $3x + 4y = 25$
c) $y = 2$

5.30 a) 5
c) $4\sqrt{3}$

5.31 a) 3
c) 3
e) $\sqrt{6}/3$

5.34 $x^2\hat{i} - 2xy\hat{j}$

5.35 a) 0
c) $-\hat{j} + 2\hat{k}$
e) $-0.1353\hat{j}$

5.36 a) 7
c) $\hat{k}$
e) 4
g) $-5\hat{i} + 10\hat{j} - 38\hat{k}$
i) $14\hat{i} + 4\hat{j} + 6\hat{k}$
k) $14\hat{i} - 9\hat{j} + 8\hat{k}$

5.37 a) irrotational
c) irrotational, solenoidal
e) irrotational
g) neither
i) irrotational, solenoidal

5.39 a) $\frac{x^2}{2} + \frac{y^2}{2} + \frac{z^2}{2} + C$
c) $xy^2 + \frac{z^2}{2} + C$
e) $x^2 \sin y + \frac{z^3}{3} + C$

5.42 a) $\cos\phi$
c) $\sin\phi\sin\theta$
e) 0
g) $\cos\phi$
i) $\cos\phi$

5.44 a) two planes and a cylinder

5.45 a) $(2x - y + xz)\hat{i} + (2y + x + yz)\hat{j} + (2z - x^2 - y^2)\hat{k}$
c) $r\sin\phi[2\cos\phi\cos\theta + \cos\theta\sin\theta\sin\phi + \cos\phi\sin\theta]\hat{i}_r + r[\sin\phi\cos^2\theta - 2\cos\phi\sin\theta]\hat{i}_\theta + r[2\cos^2\phi\cos\theta + \sin\phi\cos\phi\cos\theta\sin\theta - \sin^2\phi\sin\theta]\hat{i}_\phi$

5.46 a) $ds^2 = dx^2 + dy^2 + dz^2$
c) $ds^2 = dr^2 + r^2\sin^2\phi\, d\theta^2 + r^2 d\phi^2$

5.48 a) $\frac{r^2}{2} + z + C$

c) $\left(Ar + \frac{B}{2r^2}\right)\cos\phi + C$

5.49 a) 108π

c) 288π

5.50 a) 32π

c) 128π

5.52 $\oiint_S \hat{n} \times \vec{v}\, dS = \iiint_V \vec{\nabla} \times \vec{v}\, dV$

5.54 $\frac{\partial \rho}{\partial t} + \vec{\nabla} \cdot (\rho \vec{v}) = 0$

5.56 $\oint_C \phi\, dx + \psi\, dy = \iint_S \left(\frac{\partial \psi}{\partial x} - \frac{\partial \phi}{\partial y}\right) dS$

5.57 a) -813

5.58 a) -2π

5.59 a) 2

c) 0

CHAPTER 6

6.4 $\frac{\partial^2 u}{\partial t^2} = \frac{P}{m}\frac{\partial^2 u}{\partial x^2} - g - \frac{c}{m}\frac{\partial u}{\partial t}$

6.9 $\frac{1}{2}\left[\phi(x - at) + \phi(x + at) + \frac{1}{a}\int_{x-at}^{x+at} \theta ds\right]$

6.11 $f(x + at) = \sin\left[\frac{n\pi}{L}(x + at)\right]$

$g(x - at) = \sin\left[-\frac{n\pi}{L}(x - at)\right]$

6.15 $f = \frac{n}{2L}\sqrt{\frac{P}{m}}$

6.16 a) $0.1\cos\frac{\pi at}{2}\sin\frac{\pi x}{2}$

c) $0.1\left[\cos\frac{\pi at}{2}\sin\frac{\pi x}{2} - \cos\frac{3\pi at}{2}\sin\frac{3\pi x}{2}\right]$

6.17 $0.2\pi a$, $3/2$ m or $1/2$ m

6.19 $(0.424 \sin 15\pi t + 0.2 \cos 15\pi t) \times \sin\frac{\pi x}{4}$

6.20 a) $0.2 \cos 5\pi t \sin\frac{\pi x}{4}$

c) 0.334 m at 2 m

6.21 $\frac{4k}{\pi}\left(\cos\frac{\pi at}{L}\sin\frac{\pi x}{L} + \frac{1}{3}\cos\frac{3\pi at}{L}\right.$
$\left.\sin\frac{3\pi x}{L} + \frac{1}{5}\cos\frac{5\pi at}{L}\sin\frac{5\pi x}{L} + \cdots\right)$

6.23 0.526 m

6.25 $\sum_{n=1}^{\infty} \frac{8\alpha}{n^2\pi^2}\sin\frac{n\pi}{2}\cos nat \sin nx$

6.27 $C\rho\frac{\partial T}{\partial t} = K\frac{\partial^2 T}{\partial x^2} + \frac{4h}{D}(T - T_f)$

6.29 $C\rho\frac{\partial T}{\partial t} = K\frac{\partial^2 T}{\partial x^2} + \frac{\partial K}{\partial x}\frac{\partial T}{\partial x}$

6.31 a) $T = 100°C$ everywhere

6.33 -247 W, -126 W

6.35 $25 + \sum_{n=1}^{\infty} B_n e^{-9.6\times 10^{-6} n^2 t} \cos\frac{n\pi x}{2}$

where $B_n = \frac{400}{n^2\pi^2}\cos\frac{n\pi}{2} - \frac{200}{n^2\pi^2}$
$[1 + (-1)^n]$, 25°C

6.37 $\sum_{n=1}^{\infty} e^{-\left[(2n-1)^2/4\right]kt} A_n \sin\frac{2n-1}{2}x$,

$A_n = \frac{2}{\pi}\int_0^{\pi} f(x)\sin\frac{2n-1}{2}x\, dx$

6.39 a) $0.43x^4 - 3.44x^3 + 27.5x$

6.41 $\sum_{n=1}^{\infty} \frac{200[1 - (-1)^n]}{n\pi(e^{n\pi/2} - e^{-n\pi/2})}$
$\times \sin\frac{n\pi x}{2}(e^{n\pi y/2} - e^{-n\pi y/2})$

6.43 Inside: 250°C
Outside: $50/r$

6.45 $200(1 - 1/r)$

6.47 $\sum_{n=1}^{\infty} A_n e^{-k\mu_n{}^2 t} J_0(\mu_n r)$,

$A_n = \frac{2}{r_0^2 J_1(\mu_n r_0)} \times \int_0^{r_0} r f(r) J_0(\mu_n r)\, dr$

6.49 $\sum_{n=1}^{\infty} A_n e^{-\mu_n^2 kt} J_0(\mu_n r)$,

$A_1 = 200/3$,

$A_2 = 1230\int_0^1 r^2 J_0(3.83r)dr$,

$A_3 = 2220\int_0^1 r^2 J_0(7.02r)dr$

CHAPTER 7

7.1 a) 36.87°, 0.6435 rad
c) 323.1°, 5.640 rad

7.2 a) $-7-24i$
c) $-\frac{7}{25}-\frac{24}{25}i$
e) $-8-32i$
g) $1.68+0.26i$
i) $-2+i, 2-i$
k) $-2.66-1.217i, 0.2754+2.912i,$
$2.383-1.695i$

7.3 a) $1, 0.309+0.9511i, -0.809+0.5878i,$
$-0.89-0.5878i, 0.309-0.9511i$
c) $0.866+0.5i, -0.866+0.5i, -i$

7.4 $15(x^2+y^2)+34x+15=0$
$3(x^2+y^2)-6y+3=0$

7.6 Inside and on $y^2-4x+4\le 0$

7.8 a) $2e^{i\pi}$
c) $2e^{(3\pi/2)i}$
e) $13e^{5.107i}$
g) $13e^{1.966i}$

7.10 a) i
c) $0.707-0.707i$
e) -7.39

7.11 a) $1, 0.309+0.951i,$
$-0.809+0.588i,$
$-0.809-0.588i,$
$0.309-0.951i$
c) $0.5+0.866i, -1,$
$0.5-0.866i$
e) $-527-336i$

7.12 a) $2.193i$
c) $1.737i$
e) $1.774-1.627i$
g) 0.7622

7.13 a) $(\pi/2)i$
c) $1.609+0.9273i$
e) $1+(\pi/2)i$
g) i

7.14 a) 0.2079
c) $0.0841+0.0282i$
e) $6.704-2.421i$

7.15 a) $\pi/2+1.317i$ and $\pi/2-1.317i$
c) $1.099+\pi i$
e) $\pi+1.317i$ and $\pi-1.317i$

7.19 b) $\frac{\partial^2 v}{\partial r^2}+\frac{1}{r}\frac{\partial v}{\partial r}+\frac{1}{r^2}\frac{\partial^2 v}{\partial \theta^2}$

7.20 a) $xy+\left(\frac{y^2}{2}-\frac{x^2}{2}\right)i$
c) $e^y\sin x+i\,e^y\cos x$

7.22 a) -2
c) $16/3$
e) 0

7.23 a) 4

7.24 a) $4i$

7.25 a) 0
c) 0
e) 0

7.26 a) $2\pi i$
c) $2\pi i$
e) 0

7.27 a) 0
c) $-\pi i$
e) $2\pi i$

7.28 a) $\pi(i-1)$
c) 0
e) $2\pi i$

7.29 a) $2\pi i$
c) $2\pi i$
e) $3.395+5.287i$

7.30 a) $1-\frac{z^2}{2!}+\frac{z^4}{4!}-\frac{z^6}{6!}+\cdots, R=\infty$
c) $z-\frac{z^2}{2}+\frac{z^3}{3}-\frac{z^4}{4}+\cdots, R=1$
e) $1+\frac{z^2}{2!}+\frac{z^4}{4!}+\cdots, R=\infty$

7.31 a) $-\frac{1}{2}\left(1+\frac{z}{2}+\frac{z^2}{4}+\frac{z^3}{8}+\cdots\right), R=2$
c) $-\frac{2+i}{5}\left[1+\frac{2+i}{5}(z-1)+\frac{3+4i}{25}(z-1)^2\right.$
$\left.+\frac{2+11i}{125}(z-1)^3+\cdots\right], R=\sqrt{5}$
e) $4-z+(3-z)^2+(3-z)^3+\cdots, R=1$

7.32 a) $1+z^2+z^4+z^6+\cdots$
c) $\frac{3}{2}+\frac{3}{4}z+\frac{7}{8}z^2+\frac{7}{16}z^3+\frac{7}{32}z^4+\cdots$
e) $1-z^2+\frac{1}{21}z^4-\frac{1}{31}z^6+\cdots$

g) $\pi z - \frac{1}{31}\pi^3 z^3 + \frac{1}{51}\pi^5 z^5 + \cdots$

i) $z + z^2 + \frac{5}{6}z^3 + \frac{5}{6}z^4 + \cdots$

k) $z + \frac{1}{3}z^3 + \frac{11}{120}z^5 + \cdots$

7.33 a) $z - \frac{z^3}{3} + \frac{z^5}{10} - \frac{z^7}{42} + \cdots$

c) $z - \frac{z^3}{18} + \frac{z^5}{600} + \cdots$

7.34 $z - \frac{z^3}{3} + \frac{z^5}{5} - \frac{z^7}{7} + \cdots$

7.35 a) $e\left[z + \frac{(z-1)^2}{2!} + \frac{(z-1)^3}{3!} + \cdots\right]$

c) $1 - \frac{(z-\pi/2)^2}{2} + \frac{(z-\pi/2)^4}{12} + \cdots$

e) $-\frac{1}{2}\left(1 - \frac{1}{2}z + \frac{3}{4}z^2 - \frac{5}{8}z^3 + \cdots\right)$

7.36 a) $\frac{1}{z} - \frac{z}{6} + \frac{z^3}{120} + \cdots, R = \infty$

c) $\frac{1}{2z} - \frac{3}{4} + \frac{7}{8}z - \frac{15}{16}z^2 + \cdots, R = 1$

7.37 a) $1 + (1-z) + (1-z)^2 + \cdots, 0 \le |z-1| < 1$

c) $\frac{1}{z-1} - \frac{1}{(z-1)^2} + \frac{1}{(z-1)^3} + \cdots, |z-1| > 1$

e) $\frac{1}{2-z} + \frac{1}{(2-z)^2} + \frac{1}{(2-z)^3} + \cdots, 1 < |z-2|$

g) $-1 + (z-2) - (z-2)^2 + (z-2)^3 + \cdots,$
$0 \le |z-2| < 1$

i) $-\frac{1}{2} + \frac{z}{4} - \frac{3}{8}z^2 + \frac{5}{16}z^3 + \cdots,$
$0 \le |z| < 1$

$\cdots - \frac{1}{3z^3} + \frac{1}{3z^2} - \frac{1}{3z} - \frac{1}{6}$
$-\frac{z}{12} - \frac{z^2}{24} + \cdots,$
$1 < |z| < 2$

$\frac{1}{z^2} + \frac{1}{z^3} + \frac{1}{z^4} + \cdots,$
$2 < |z|$

7.38 a) $i/4$ at $(0, -2i)$, $-i/4$ at $(0, -2i)$
c) 2 at (0, 0)
e) 0.841 at (−1, 0)

7.39 a) $(\pi/3)i$
c) $-2\pi i$
e) $-\pi i$
g) $2\pi i$
i) $-28.9i$
k) 0

7.40 a) $2\pi i$
c) $2\pi/3$
e) $3.93i$

7.41 a) 2.22
c) $2.3(1-i)$
e) $2.617 + 0.1511i$

7.42 a) π/e^2
c) π/e
e) 0

CHAPTER 8

8.1 $\nabla^4 f_i = f_i - 4f_{i-1} + 6f_{i-2} - 4f_{i-3} + f_{i-4}$

8.6 $\delta + \frac{\delta^2}{2} + \frac{\delta^3}{8} - \frac{\delta^5}{128} + \cdots$

8.10 $h^3 D^3 + \frac{h^5 D^5}{4} + \frac{h^7 D^7}{40} + \cdots$

8.12 $\frac{1}{6h}(2f_{i+3} - 9f_{i+2} + 18f_{i+1} - 11f_i)$

8.14 a) 0.08728

8.16 $\frac{h}{3}(f_i + 4f_{i-1} + f_{i-2})$

8.17 a) 73

8.18 a) −81.4

8.19 a) 1.537

8.20 $\frac{2h}{45}(7f_{i+2} + 32f_{i+1} + 12f_i + 32f_{i-1} + 7f_{i-2})$

8.21 a) 0.29233
c) 0.29267

8.22 a) 0.53534

8.23 a) −5.952
c) 1.226

8.24 0.327

8.26 1.513

8.27 a) 2, 2.048, 2.235, 2.622, 3.20
c) 2.01, 2.12, 2.393, 2.845, 3.464

8.28 a) 6.2, 5.44, 4.03, 2.58, 1.64

8.29 a) 1.99, 3.89, 5.65, 7.17, 8.41

8.30 $y_i + \frac{h}{12}(23\dot{y}_i - 16\dot{y}_{i-1} + 5\dot{y}_{i-2})$

8.32 2.012, 2.118, 2.390, 2.846, 3.470

8.34 a) 2, 3.756, 4.980, 5.383, 4.893
c) 1.947, 3.587, 4.660, 4.998, 4.546

8.36 a) Assume $\dot{y}_0 = -0.001$:
−0.004, −0.0064, −0.0072,
−0.0064, −0.004
Assume $\dot{y}_0 = -0.0008$:
−0.0032, −0.0048,
−0.0048, −0.0032, −0.0.

8.38 Assume $T_0 = 10, \dot{T}_0 = 0$:
10, 10, 9.968, 9.904, 9.808, 9.68
Assume $T_0 = 0, \dot{T}_0 = 10$:
0, 4, 7.952, 11.862, 15.74, 19. 59.
Combine:
0, 30.4, 60.6, 90.5, 120.3, 150

8.40 200, 0, 0, 0, 0, 0
200, 100, 0, 0, 0, 0
200, 100, 50, 0, 0, 0
200, 125, 50, 25, 0, 0
200, 125, 75, 25, 12.5, 12.5

8.42 0, 0.02, 0.04, 0.04, 0.02, 0
0, 0.02, 0.02, 0.02, 0.02, 0
0, 0, 0, 0, 0, 0

8.44 Starting at the lower left:
66, 78, 109, 127, 139, 161, 184,
189, 195, 241, 241, 234

Index

A

Absolute
  acceleration, 243–244, 246
  value of complex number, 338
  velocity, 243, 245–246
Acceleration, 228, 242, 248, 266
  absolute, 243–244, 246
  Coriolis, 244
  definition, 247
  of gravity, 13, 26
  in Newton's second law, 285
  rectilinear, 209
Adams' method initial-value problem, 414
Addition of matrix, 161–163
  commutative and associative laws, 162
  sum of *k* matrices A, 162
Addition of vectors, 230
Adjoint matrix, 177–178
Aerodynamics, 272, 283
Analytic functions, 349–353
Angular frequency, 27
Angular velocity, 18, 235, 242–243, 245–246, 253
Approximation of function, 66
Argument of complex number, 338
Augmented matrix, 183
Auxiliary equation, 62
Averaging operator, 390

B

Backward difference operator, 388
Bernoulli's equation, 13
Bessel functions, 106–110, 322, 408, 450–453
  of the first kind, 106, 108
  of higher negative order, 110
  of higher order, 110
  of lower order, 110
  of the second kind, 108–109
Bessel's differential equation, 103–112
Binomial theorem, 395
Boundary conditions, 23, 66, 283, 297–299, 312–313, 316, 324–325
Boundary-value problems, 3, 283, 428–432
  iterative method, 428
  simultaneous equations, 429–430
  superposition principle, 429

C

Cartesian coordinate systems, 231, 259
Cartesian reference frame, 229
Cauchy equation, 76, 324
  of order 2, 61–63
Cauchy–Riemann equations, 350–353, 356
Cauchy's integral theorem, 356–360, 367, 373
  formula, 360–362, 365–366
  integration variables for, 362
Cauchy's residue theorem, 376
Central difference operator, 388
Characteristic equation, 21–25, 27, 29, 37–38, 42, 201–204, 409
Characteristic-value problem, 201
Circulation of vector, 272
Cofactor, 174, 178
Column vector, 160, 164, 167, 178, 181–182
Complex conjugates, 216
Complex integration, 353–365
Complex numbers, 337–343
  absolute value of, 338
  addition, subtraction, multiplication, or division, 339–340
  conjugate of, 339
  equal, 341
  imaginary part, 338
  *n*th root of, 340
  pure imaginary, 339
  real part, 338
Complex plane, 78, 338, 353
Complex variables, 337
Conjugate of complex numbers, 339
Conservation of energy, 309
Conversion factors, 447
Coordinate systems
  Cartesian, 259
  cylindrical, 259–263, 265, 310–311, 327
  rectangular coordinates, 251
  relationships, 260, 265
  spherical, 259, 263–266, 310–311, 324
Coriolis acceleration, 244
Cramer's rule, 182, 186
Critical damping, 31
Cross product, 233–234, 262
  by determinant, 234
  in rectangular coordinates, 251
Curl, 251–252, 262, 264
  vector function, 251–254, 257

M. C. Potter and B. F. Feeny, *Mathematical Methods for Engineering and Science*,
https://doi.org/10.1007/978-3-031-26151-0

Cylindrical coordinate systems, 259–263, 311
acceleration in, 266
differential changes in, 261

D

D'Alembert solution of wave equation, 292–296
Damped motion, 29–34
Damping
coefficient, 26, 29, 32, 41, 42
criteria, 35
critical, 31
forced oscillations with, 45–48
magnitude of
critically damped, 29–30
overdamped, 29–30
underdamped, 29, 31
viscous, 25–26
Density, 229
Dependent variable, 2
Derivative of a function, 349
Determinants, 171–177
calculation using MATLAB, 175–176
cross product of, 234
definition, 171
expansion of, 174
*n*th-order, 173–174
second-order, 171–172
third-order, 172–173
value of, 174–175
Diagonal matrix, 165
Difference operators, 390
backward, 388
central, 388
finite, 388–392
forward, 388
Differential equations, 1
boundary conditions, 23, 66
of first order, 3–9
higher-order, 421–427
homogeneous, 2
*n*th-order, 3
order of, 2
ordinary, 2
partial, 2
solution of, 2, 144–151
system of, 216–218
Differential operator, 392–397
backward difference operator to, 393
central difference operator to, 393, 396–397
relationship between operators, 394
Diffusion, 308–311
Diffusion equation, 310, 433–434
of long, totally insulated rod, 315–319
of long insulated rod, 311–313
solution of, 311–323
of two-dimensional heat conduction in a long, rectangular bar, 319–323
Direction cosines of vector, 231
Direct method initial-value problem, 418, 420–421
Discontinuity, 283
Displacement of string, 302–308
Divergence, 251, 253, 257, 261–262, 264
Divergence theorem, 267–269, 273
Dot product, 232–233, 238–239, 249
Dynamics, 14–19
Dynamic systems, linear, 216–218

E

Eigenvalue problem, 201, 206–208
in displacements of two masses, 206–207
in engineering, 209–218
mass moment of inertia, 209–213
stress of an element, 213–214
Eigenvalues, 201–208, 316
of principal stresses, 214
properties, 202
of symmetric matrix, 203–205
using MATLAB, 205, 461–462
Eigenvectors, 201–208, 217–218
of symmetric matrix, 203–205
using MATLAB, 205, 462
Electrical circuits, 9–11
Electrical circuit analog, 34–36
damped circuit, 35
damping criteria, 35
undamped circuit, 35
Elementary functions, 343–348
Energy balance, 309
Engineering units, 447
E operator, 389
Error
bounds on, 403
function, 449
order of magnitude of, 397
round-off, 387–388
truncation, 387, 397–400
Euler's constant, 108
Euler's formula, 22
Euler's method for initial-value problem, 413, 418–419
Even function, 53–57
Exact differential, 6
Exact equations, 6–8
Expansion
of determinants, 174
Expansions
half-range, 57–59

F

Finite-difference operators, 388–392
  derivatives, 398
First differences, 389
First-order differential equations
  applications
    in dynamics, 14–19
    of electrical circuits, 9–11
    fluid flow, 13–14
    in mechanics, 16–17
    Newton's second law, 14–16
    rate equation, 11–13
  exact equations, 6–8
  integrating factors, 8–9
  linear, 8–9
  nonlinear, 2
  separable equations, 3–6
Fixed-point system, 387
Floating-point system, 387
Fluid flow, 13–14
Force, 228
Forced motion of spring-mass systems, 39–48
  without damping, 42–45
Forced oscillations
  with damping, 45–48
    amplitude of oscillation, 46
    general solution, 45
  on spring–mass system, 59–61
Forward difference operator, 388
Fourier series, 49–61
  cosine, 55–57
  of even functions, 53–57
  forced oscillations, 59–61
  half-range expansions, 57–59
  of odd functions, 53–57
  sine, 55–57, 301, 321–322
Fourier theorem, 49
Free-body diagram, 26
Free motion of spring–mass system, 25–26
  in circuit parameters, 34–36
  damped motion, 29–34
  undamped motion, 26–28
Frobenius's method, 93–103

G

Gamma function, 105, 448
  of negative integer, 105
  of positive integer, 105
Gaussian elimination, 182
  homogeneous solution, 184
Gauss's theorem, 267–269
General solution, 20, 22–25, 37–38, 62–63, 298
  of an *n*th-order differential equation, 3–5, 8
  of Bessel's equation, 107
  for forced oscillations with damping, 45
  for linear second-order differential
      equation, 19
  for resonance, 42
Gradient, 249–258
  divergence of, 253, 257
Gradient of a scalar function, 263
Gravitational attraction, 331
Gravitational potential, 330–332
Green's theorem, 273–274, 353–356
  in the plane, 275

H

Half-range expansions, 57–59
Harmonic motion, 59, 144–145, 300
Harmonic oscillation, 28
Harmonic oscillator/harmonic
      oscillation, 27–28
Heat equation, one-dimensional, 310, 312,
      315–316, 318, 322–323
Heat flow, 308
  by conduction, 308, 319–323
  by convection, 308
  by radiation, 308
Heat flux, 308, 311, 316
Heat transfer, 311
  in cylindrical body, 326–330
Higher-order differential equations, 421–427
  using MATLAB, 425–427
Homogeneous linear differential equations,
      2, 19–20
  second-order differential equations, 21–25
Hyperbolic equation, 283, 286
Hyperbolic sine and cosine, 344

I

Identity matrix, 165–166
Imaginary axis, 338
Independent variable, 2
Indicial equation, 94
  roots of, 94–103
Initial condition, 3
Initial-value problem, 3, 283
Integral theorem
  Cauchy's, 356–360, 367, 373
  divergence, 267–276
Integrating factors, 8–9
Integration
  complex, 353–365
  Liebnitz' rule of, 403
  numerical, 400–406
  process, 3
  Simpson's one-third rule of, 402, 404
  Simpson's three-eights rule of, 402
  trapezoidal rule of, 401, 405
  variables for Cauchy's integral theorem, 362

Inverse Laplace transform, 118, 135
Inverse matrix, 177–180
definition, 177
inverse of, 179
of nonsingular diagonal matrix, 179
of product of two square matrices, 179–180
properties, 179
transpose of, 179
using MATLAB, 179, 186–188, 462
Inverse transforms
partial fractions, 139–143
repeated linear factor, 139–140
repeated quadratic factor, 140
unrepeated linear factor, 139
unrepeated quadratic factor, 140
Irrotational (or conservative) vector field, 254
Iterative method, 428

K

Kernel, 185
Kirchhoff's law, 9–10, 35–36
Kronecker delta, 166

L

Laplace's equation, 253, 310, 324, 326, 332, 435–436
Laplace transform, 117–128
definition, 118
of derivatives and integrals, 128–135
first shifting property, 119
of half-wave rectified sine wave, 137–138
inverse, 118, 135
linear operator of, 119
of periodic functions, 135–138
second shifting property, 119
of square-wave function, 126, 136–137
Laplacian of ϕ, 253
Laplacian of a scalar function, 264
Laurent series, 366–368, 371–374
Least-squares fit, 192–200
Left-hand limit, 118, 129
Legendre polynomials, 322
Legendre's equation, 89–93, 324
Liebnitz' rule of integration, 403
Liebnitz's rule of differentiating an integral, 132
Limit
left-hand, 118, 129
right-hand, 118, 129
Linear algebraic equations, 159
homogeneous sets, 184–186
nonhomogeneous sets, 182–184
nonzero solutions, 184
using MATLAB, 186–192
Linear differential equations, 19–21
definition, 2
first-order, 20
homogeneous, 19–20
ordinary, 216–218
second-order, 19
system of, 216–218
with variable coefficients, 80–89
Linear dynamic systems, 216–218
Linear interpolation, 408, 428
Logarithmic decrement, 31
Lower triangular matrix, 165

M

Maclaurin series, 366
Magnitude of vector, 229, 231, 235
Main diagonal, 160
Mathematical shorthand, 160
MATLAB, 2, 454–478
determinants, calculation of, 175–176
files, input, and output, 471–475
format and scientific notation, 464–465
functions, 475–477
higher-order differential equations using, 425–427
inverse matrix, 179, 186–188
numerical integration, 404–405
numerical solution using, 439–442
plotting, 468–470
programming loops, 466–468
real and complex numbers, 456–457
simultaneous linear algebraic equations, solution of, 186–192
string arrays, 470–471
vectors and matrices, 458–463
Workspace browser, 477
Matrix
addition of, 161–163
adjoint, 177–178
augmented, 183
definition, 160
diagonal, 165
elements, 160
equal, 161
of finite-difference form, 429–430
identity, 165–166
inverse, 177–179
least-squares fit, 192–200
lower triangular, 165
main diagonal, 160
m by *n*, 160–162
moment of inertia, 209–211
multiplication of scalar with, 162
multiplication, 167–171
null, 161
nullity, 185
partitioned, 161
pseudo inverse, 192–200
rank, 185
scalar, 165
singular, 178

skew-symmetric, 164–165
square, 160, 170
submatrix, 161
symmetric, 164–165
transpose of, 163–166
unit, 165–166, 179
upper triangular, 165
vector, 160
zero, 161
Maximum directional derivative, 250
m by *n* matrix, 160–162
Method of undetermined coefficients, 37
Minor of element, 174
Moment of inertia, 209–213
Multiplication of matrices, 167–171
postmultiplier, 168
premultiplier, 168
properties of, 169–171
Multiplication of vector, 232–241

N

Natural frequency, 42–43
of undamped system, 47, 61
of underdamped system, 59
Near-resonance solution, 43–45
Newton–Raphson method, 409
Newton's method, 409–410
Newton's second law, 14, 39
Nonhomogeneous second-order linear equations, 36–39
Nonlinear differential equation
definition, 2
second-order equation, 20
Nonsingular matrix, 178–179
Nonzero normal stresses, 214
Normalized equation, 66
*n*th-order determinants, 173–174
*n*th-order differential equation, 3
general solution of, 3
Nullity of matrix, 185
Null matrix, 161
Null space, 185–186
Null vector, 185
Numerical integration, 400–406
Liebnitz' rule of integration, 403
Simpson's one-third rule of integration, 402, 404
Simpson's three-eights rule of integration, 402
trapezoidal rule of integration, 401, 405
Numerical interpolation, 407–408
Numerical stability, 432

O

Odd function, 53–57
One-dimensional heat equation, 310
Operator of Laplace transform, 119
Order
Bessel functions, 110
Cauchy equation of order 2, 61–63
determinants
*n*th-order, 173–174
second-order, 171–172
third-order, 172–173
differential equations, 2
first order, 3–9
higher-order, 421–427
*n*th-order, 3
higher negative, 110
lower, 110
of magnitude of error, 397
Ordinary differential equations, 297
boundary-value problems, 428–432
initial-value problems, 412–421
Adams' method, 414
direct method, 418, 420–421
Euler's method, 413, 418–419
Runge–Kutta method, 414–417, 419–420
Taylor's method, 413
linear, 216–218
of second order, 2, 298
Ordinary point, 78
Overdetermined equations, 193

P

Parabolic equation, 283
Parallelogram area, 233, 235–237
Partial differential equations, 282, 314
of fourth order, 2
general form, 283
numerical solution of, 432–442
second-order, 283
separation of variables, 296–308
Partitioning, 161
Pascals (Pa), 213
Periodic input functions, 49–61
Phase angle, 46
Phase lag, 46
Plane, complex, 78, 338, 353
Point
ordinary, 78
singular, 78
Polynomial function, 343
Position vector, 235, 249, 260, 270–271
Postmultiplier, 168
Power series, 76–80, 343
convergence of, 77
*m*th partial sum, 77
remainder of, 77
Premultiplier, 168
Principal axes, 210, 214
Principal coordinate system, 210
Principal moments of inertia, 210–211
Principal stresses, 214

Principal value, 338
Pseudo inverse, 192–200
Pure imaginary, 339

R

Radius of convergence, 78, 366
Rank of matrix, 185
Rate equation, 11–13
Real axis, 338
Rectangular coordinates, 232, 242, 265
  cross product of, 251
  dot product, 233
  scalar triple product, 236
  curl, 251
  divergence, 251
  Laplacian, 253
Rectangular rule, 404
Recurrence relations, 82, 111
Relaxation method, 436
Residuals, 193, 197–200
Residues, 373–381
Resonance, 42–43, 45–46, 147
  in forced oscillations with damping, 45–46
  near, 43–44
Riemann. *see* Cauchy–Riemann equations
Right-hand limit, 118, 129
Roots of equation, 409–411
Rotated coordinate system, 210, 213–214
Round-off error, 387–388
Row vector, 160–161, 164
Runge–Kutta method initial-value problem, 414–417, 419–420

S

Scalar
  definition, 229
  matrix, 165
  potential function, 254, 257–258
  product, 169, 232
  triple product, 235
    rectangular coordinates of, 236
Second backward difference, 389
Second central difference, 389
Second forward difference, 389
Second moment of mass, 209
Second moments of area, 210
Second-order determinants, 171–172
Second-order differential equation, 3
Second shifting property, 119, 121, 125, 128, 149
Sectionally continuous function, 118, 128–129
Secular equation, 201–204
Self-inductance, 290–291
Separable equations, 3–6
Separation constant, 297
Separation of variables, 296–308, 313–314
Series
  Fourier, 49–61
    cosine, 55–57
    of even functions, 53–57
    forced oscillations, 59–61
    half-range expansions, 57–59
    of odd functions, 53–57
    sine, 55–57, 301, 321–322
  Laurent, 366–368, 371–374
  Maclaurin, 366
  power, 343
    convergence of, 77
    *m*th partial sum, 77
    remainder of, 77
  Taylor, 365–366, 368–370
Shear stress, 213
Simpson's one-third rule of integration, 402, 404
Simpson's three-eights rule of integration, 402
Simultaneous linear algebraic equations, solution of, 182–192
  homogeneous sets, 184–186
  nonhomogeneous sets, 182–184
  using MATLAB, 186–192
Singular matrix, 178
Singular point, 78
SI Units, 447
Skew-symmetric matrix, 164–165
Solenoidal vector field, 254
Solution, 282. *see also* General solution
  Cramer's rule, 182, 186
  D'Alembert, 292–296
  of differential equations, 2, 144–151
  of diffusion equation, 311–323
  near-resonance, 43–45
  nonzero, 184
  numerical, 387–388, 432–442
  of simultaneous linear algebraic equations, 182–192
  specific, 3, 44
Span, 186
Specific heat, 308
Specific solution, 3
  for near resonance, 44
Spherical coordinate systems, 259, 263–265, 311
  differential changes in, 263
  of spherical surface, 324–326
Spring–mass system
  damped harmonic motion of, 144–145
  with a forcing function, 39–48
  undamped motion, 26–36
Square matrix, 160
Stability, 216
  numerical, 432
Stokes's theorem, 270–272

Stresses, 213–214, 229
  nonzero normal, 214
  principal, 214
  shear stress, 213
  two-dimensional, 213
Submatrix, 161
Subsidiary equation, 144
Subtraction of vectors, 230
Sum
  of complex numbers, 339–340
  of *k* matrices A, 162
  of vectors, 230
Superposition principle, 429
Surface integrals, 267
Symmetric matrix, 164–165

T

Tacoma Narrows bridge, 42, 45
Taylor series, 66, 77, 79, 365–366, 368–370, 392
Taylor's method initial-value problem, 413
Temperature, 229
  diffusion equation, 433–434
  distribution in laterally insulated rod, 436–437
  distribution in long, insulated rods, 311–319
  as a function, 314–315
  steady-state temperature distribution in plate, 322–323
Theorems
  binomial theorem, 395
  Cauchy's integral theorem, 356–360
  Cauchy's residue theorem, 376
  Fourier theorem, 49
  Gauss's theorem, 267–269
  Green's theorem, 273–275, 353–356
  Stokes's theorem, 270–272
Thermal conductivity, 308
Thermal diffusivity, 310
Thermodynamics, 308
Third-order determinants, 172–173
Transformation of variables, 66
Transmission-line equations, 289–292
Transpose, 163
  of a column vector, 164
  of a row vector, 164
Trapezoidal rule of integration, 401, 405
Triangular matrix
  lower, 165
  upper, 165, 183
Trigonometric identities, 301, 317
Triple product
  scalar, 235–236
  vector, 240–241
Truncation error, 387, 397–400
Two-dimensional stresses, 213

U

Undamped motion, 26–28
Undamped spring–mass system, 42
Undetermined coefficients, 37
Unit(s)
  engineering, 447
  impulse function, 124–125
  matrix, 165–166
  SI, 447
  vector, 229, 232, 234–235, 238–240, 254–255
  vectors in cylindrical coordinates, 259–260, 262
  vectors in spherical coordinates, 259, 263
Unknown, 3
Upper triangular matrix, 165, 183

V

Value
  of determinants, 174–175
  principal, 338
Variables
  complex, 337
  dependent, 2
  independent, 2
  integration, for Cauchy's integral theorem, 362
  separation of, 296–308, 313–314
  transformation of, 66
Variation of parameters, 64–65
Vector, 160, 228–229
  addition, 230
  column, 160, 164, 167, 178, 181–182
  components of, 230–231
  in cylindrical and spherical coordinates, 259
  definitions, 228–229
  differential operator, 250
  differentiation, 241–248
  direction cosines of, 231
  direction of, 229
  divergence of, 251, 257, 262, 264–265
  identities, 254
  magnitude of, 229, 231, 235
  multiplication, 232–241
  position, 235, 249, 260, 270–271
  product, 233
  resultant, 230
  row, 160–161, 164
  subtraction, 230
  triple product, 240–241
  unit, 229, 234–235, 238–240, 255

Velocity, 300
  angular, 18, 235, 242–243, 245–246, 253
Vibration
  membrane, 286–288
  of an elastic bar, 288–289
  of a stretched, flexible string, 284–286

W

Wave equation, 434–435
  one-dimensional, 286
Wave motion
  in elastic bar, 288–289
  transmission-line equations, 289–292
  vibration membrane, 286–288
  vibration of stretched, flexible strings, 284–286
Well-posed problem, 283
Work, 232

Z

Zero determinant, 178
Zero matrix, 161

GPSR Compliance
The European Union's (EU) General Product Safety Regulation (GPSR) is a set of rules that requires consumer products to be safe and our obligations to ensure this.

If you have any concerns about our products, you can contact us on

ProductSafety@springernature.com

In case Publisher is established outside the EU, the EU authorized representative is:

Springer Nature Customer Service Center GmbH
Europaplatz 3
69115 Heidelberg, Germany

www.ingramcontent.com/pod-product-compliance
Ingram Content Group UK Ltd.
Pitfield, Milton Keynes, MK11 3LW, UK
UKHW051646031026
15588UKWH00007B/1362
* 9 7 8 3 0 3 1 2 6 1 5 2 7 *